Reliability Assessment of Cyclically Loaded Engineering Structures

NATO ASI Series

Advanced Science Institutes Series

A Series presenting the results of activities sponsored by the NATO Science Committee, which aims at the dissemination of advanced scientific and technological knowledge, with a view to strengthening links between scientific communities.

The Series is published by an international board of publishers in conjunction with the NATO Scientific Affairs Division

A Life Sciences **B Physics**	Plenum Publishing Corporation London and New York
C Mathematical and Physical Sciences **D Behavioural and Social Sciences** **E Applied Sciences**	Kluwer Academic Publishers Dordrecht, Boston and London
F Computer and Systems Sciences **G Ecological Sciences** **H Cell Biology** **I Global Environmental Change**	Springer-Verlag Berlin, Heidelberg, New York, London, Paris and Tokyo

PARTNERSHIP SUB-SERIES

1. Disarmament Technologies	Kluwer Academic Publishers
2. Environment	Springer-Verlag / Kluwer Academic Publishers
3. High Technology	Kluwer Academic Publishers
4. Science and Technology Policy	Kluwer Academic Publishers
5. Computer Networking	Kluwer Academic Publishers

The Partnership Sub-Series incorporates activities undertaken in collaboration with NATO's Cooperation Partners, the countries of the CIS and Central and Eastern Europe, in Priority Areas of concern to those countries.

NATO-PCO-DATA BASE

The electronic index to the NATO ASI Series provides full bibliographical references (with keywords and/or abstracts) to more than 50000 contributions from international scientists published in all sections of the NATO ASI Series.
Access to the NATO-PCO-DATA BASE is possible in two ways:

– via online FILE 128 (NATO-PCO-DATA BASE) hosted by ESRIN, Via Galileo Galilei, I-00044 Frascati, Italy.

– via CD-ROM "NATO-PCO-DATA BASE" with user-friendly retrieval software in English, French and German (© WTV GmbH and DATAWARE Technologies Inc. 1989).

The CD-ROM can be ordered through any member of the Board of Publishers or through NATO-PCO, Overijse, Belgium.

3. High Technology – Vol. 39

Reliability Assessment of Cyclically Loaded Engineering Structures

edited by

Roderick A. Smith
Department of Mechanical Engineering,
University of Sheffield, U.K.

Springer Science+Business Media, B.V.

Proceedings of the NATO Advanced Research Workshop on
Reliability Assessment of Cyclically Loaded Engineering Structures
Varna, Bulgaria
June 6–8, 1996

A C.I.P. Catalogue record for this book is available from the Library of Congress.

ISBN 978-0-7923-4695-1 ISBN 978-94-011-5556-4 (eBook)
DOI 10.1007/978-94-011-5556-4

Printed on acid-free paper

TABLE OF CONTENTS

PREFACE

As Directors of this NATO Workshop we welcome this opportunity to record formally our thanks to the NATO Scientific Affairs Division for making our meeting possible through generous financial support and encouragement.

This meeting had two purposes: the first obvious one because we have collected key scientists from East and West together to discuss the latest developments in the design against fatigue for structures and components. The second is less obvious but perhaps in the longer term more important; that is the building of bridges between East and West Europe, bridges cemented in the first place by personal friendship between scientists.

Fatigue is the process by which structures subjected to cyclical loads deteriorate. The advent of the industrial revolution and in particular the spread of railways caused this phenomena to be recognised and studied some 150 years ago. Despite intensive efforts over the years and despite a huge increase in our theoretical and practical understanding of fatigue, failures still occur causing economic disruption and even loss of life. Some of the reasons lying behind this apparent failure to apply advanced knowledge to everyday engineering were explored during this Workshop. Economic pressures to extend the lives of existing plant are becoming more intense. Therefore of vital importance is the generation of knowledge used to assess the performance of existing structures and machines and to produce guidance on their continued safe operation.

The papers contained in these proceedings represent a wide range of views and approaches to the topic of fatigue. Readers will readily distinguish the, in general, more theoretical and mathematical approaches developed in the East and the practical, pragmatic methods of the West. The balance will, we hope, offer useful insights and topics of interest to the wide range of engineers and scientists concerned with fatigue and associated problems.

Although physical bridges are subject to fatigue, we hope the bridges built during our Workshop will endure to have infinite life. The last decade has brought exciting changes to the political structures of Eastern Europe and in their wake much easier contact between scientists from East and West. The continuing contacts which have occurred in the period that has elapsed between our meeting and the appearance of these proceedings encourage us to think that our hopes have been met.

Finally our grateful thanks are recorded to N. R. Harris and Christine Young of the Mechanical Engineering Department at the University of Sheffield for the considerable and cheerful assistance they have given in the production of these proceedings.

ARW Co-Directors: Prof. R. A. Smith, Sheffield, UK.
Dr. St. B. Vodenicharov, Sofia, Bulgaria.

ADDRESS LIST OF AUTHORS

PRINCIPAL CONTACTS

Professor J H Beynon
Department of Mechanical Engineering
The University of Sheffield
Mappin Street
Sheffield
S1 3JD

Professor M Bily
Institute of Materials and Machine Mechanics
Slovak Academy of Sciences
Racianska 75, 83008
Bratislava 38
SLOVAKIA

Professor J G Blauel
Fraunhofer Institut für Werkstoffmechanik
Wohlerstrasse 11
D-79108 Freiburg
GERMANY

Professor A Carpinteri
Universita Degli Studi di Parma
Dipartimento di Ingegneria Civile
Viale delle Scienze
43100 Parma
ITALY

Professor G Glinka
Department of Mechanical Engineering
University of Waterloo
Waterloo
Ontario
CANADA N2L 3GI

Professor J Knott
School of Metallurgy and Materials
Faculty of Engineering
The University of Birmingham
Edgbaston
BIRMINGHAM
B15 2TT

Professor A J Krasovsky
Institute of Problems of Strength
National Academy of Science
2 Timiryazevskaya Str.
252014 Kiev
UKRAINE

Professor E Macha
Technical University of Opole
ul. St. Mikolajczyka 5
45 - 233 Opole
POLAND

Professor N Makhutov
International Institute of Engineering Safety
4 Griboedov Str.
101 830 Moscow
RUSSIA

Professor J Petit
Laboratoire de Mechanique et de Physique des Materiaux
ENSMA, Sita du. Futuroscope
BP 109, 86960 Cedex
FRANCE

Professor A Pieracci
Department of Aerospace Engineering
Via Diotisalvi 2
56126 Pisa
ITALY

Professor G Pluvinage
Laboratoire de Fiabilité Mécanique
Universite de Metz
Ile du Sauley
57 045 Metz Cedex 01
FRANCE

Professor R Ritchie
Lawrence Berkeley Laboratory
University of California
282 Hearst Mining Building
Berkeley, CA 94720
USA

Professor R A Smith
Department of Mechanical Engineering
The University of Sheffield
Mappin Street
SHEFFIELD
S1 3JD

Dr L Toth
Department of Mechanical Engineering
University of Miskolc
Egyetemvaros
H-3515 Miskolc
HUNGARY

Dr S Vodenicharov
Deputy Director of Institute of Metal Science
Bulgarian Academy of Sciences
67 Shipchensky Prohod Str.
1574 Sofia
BULGARIA

CO-AUTHORS

Dr A Kapoor
Department of Mechanical Engineering
The University of Sheffield
Mappin Street
Sheffield
S1 3JD

Dr R Brighenti
Universita Degli Studi di Parma
Dipartimento di Ingegneria Civile
Viale delle Scienze
43100 Parma
ITALY

Dr A Sanguanini
Universita Degli Studi di Parma
Dipartimento di Ingegneria Civile
Viale delle Scienze
43100 Parma
ITALY

Dr A Spagnoli
Universita Degli Studi di Parma
Dipartimento di Ingegneria Civile
Viale delle Scienze
43100 Parma
ITALY

Dr T D Liebster
Department of Mechanical Engineering
University of Waterloo
Waterloo
Ontario
CANADA N2L 3GI

Dr M M Gadenin
International Institute of Engineering Safety
4 Griboedov Str.
101 830 Moscow
RUSSIA

Dr G Henaff
Laboratoire de Mechanique et de Physique des Materiaux
ENSMA, Sita du. Futuroscope
BP 109, 86960 Cedex
FRANCE

Dr S Lesterlin
Laboratoire de Mechanique et de Physique des Materiaux
ENSMA, Sita du. Futuroscope
BP 109, 86960 Cedex
FRANCE

Dr C Sarrazin-Baudoux
Laboratoire de Mechanique et de Physique des Materiaux
ENSMA, Sita du. Futuroscope
BP 109, 86960 Cedex
FRANCE

Dr T Boukharouba
Université des Sciences et de Technologie
H. Boumediene (IGMUSTHB)
B.P. 32 El-Alia Alger
ALGERIA

Dr J Gilgert
École Nationale d'Ingénieurs de Metz (ENIM)
Ile du Saulcy 57045
FRANCE

Dr K T Venkateswara Rao
Lawrence Berkeley Laboratory
University of California
282 Hearst Mining Building
Berkeley, CA 94720
USA

Dr T Lagoda
Technical University of Opole
ul. St. Mikolajczyka 5
45 - 233 Opole
POLAND

THE INTERACTION OF WEAR AND ROLLING CONTACT FATIGUE

JOHN H. BEYNON and AJAY KAPOOR
Department of Mechanical Engineering
The University of Sheffield
Mappin Street
Sheffield S1 3JD
U.K

1. Abstract

The wear of material from a surface occurs, other than in the most severe of operations normally classified as machining, after repeated cycles of loading. This has lead many authors to describe wear in terms of low cycle fatigue. Closer inspection of the cyclic nature of the surface deformation reveals the strain cycles to be open, leading to accumulated plastic strain by ratchetting. Using the mechanism of ratchetting failure to describe the wear data provides much better correlation with theory than does low cycle fatigue. A crucial result obtained by reversing the direction of a series of tests is demonstrated to support this mechanism.

The presence of a fluid can give rise to cracks propagating from the surface. This mechanism is described and supported by experiments involving a reversal in the test direction. Finally, the role of prior damage accumulated during dry rolling/sliding contact in reducing the subsequent "wet" fatigue life is discussed. Ratchetting theory is shown to provide a quantified description of the life. Although the basis for the paper is the railway wheel - rail contact, the conclusions are applicable to other circumstances.

2. Introduction

The wear of material from a surface occurs, other than in the most severe of operations normally classified as machining, after repeated cycles of loading. This has lead many authors to describe wear in terms of fatigue. Since wear usually involves considerable plastic flow in the surface, low cycle fatigue is the most popular model. Attempting to apply the classical Coffin-Manson analysis to wear data leads to discrepancies which are difficult to argue away. Closer inspection of the cyclic nature of the surface deformation reveals the strain cycles to be open, leading to accumulated plastic strain by ratchetting. This paper considers several proposed mechanisms for the wear of surfaces which undergo plastic deformation.

Using the mechanism of ratchetting failure can provide a much better description of wear data than low cycle fatigue. This mechanism is further used to describe how

R.A. Smith (ed.), Reliability Assessment of Cyclically Loaded Engineering Structures, 1–26.

surfaces begin to wear and later achieve steady state wear rates. The crucial result of a decreasing wear rate with reversal of the loading direction can be explained only by ratchetting theory, and not by low cycle fatigue.

The presence of a fluid can give rise to cracks propagating from the surface. An example of such a circumstance is described, together with powerful evidence for the mechanism of fluid pressurisation propagating the crack. Finally, the effect of prior wear damage on the subsequent fatigue life in the presence of a liquid is discussed, together with a quantitative description.

The first part of this paper reviews some existing wear models. Experiments to probe the validity of these models will be considered next and then a new wear model based on "ratchetting" will be suggested. Rolling Contact Fatigue will be reviewed next and the role of ratchetting investigated. Since ratchetting is being suggested as the key factor the paper opens with a description of ratchetting.

3. Ratchetting

Consider a rough, hard surface (1) which slides repeatedly on a softer plane (2) as shown in Figure 1a. The contact is made at higher points of the surface, known as asperities. One such interaction is shown in Figure 1b. In the elastic state the contact pressure distribution is Hertzian and the stress components acting on a typical material element beneath the surface are also shown. For an asperity whose top can be approximated to a cylindrical surface (appropriate for surfaces produced by turning or grinding) these comprise a shear stress τ_{zx} , and compressive direct stresses σ_{xx} and σ_{zz}. As sliding proceeds, the variation of the magnitudes of these stresses will be as shown in Figure 1b. Provided the normal load or contact pressure on the junction is less than some critical value, no material element will undergo plastic flow and the half-space will return to its original geometry once the load has passed. This particular severity of loading therefore represents the "elastic limit" of the contact. If the contact pressure is greater than this, so that the elastic limit is exceeded, then some material elements will undergo plastic flow with two significant consequences. First, residual stresses will develop in the upper region of the half-space and second, the material may strain harden, so increasing its effective yield strength.

In a situation of repeated contact, further passes of the asperity will subject the half-space to a combination of the stress field shown in Figure 1b and the residual stresses developed in the earlier passes. These residual stresses are essentially protective in the sense that they make subsequent yielding less likely, and this, together with any effects of strain hardening, may be sufficient to ensure that after many cycles of loading the steady state deformation of the material is entirely elastic. This is the process of "shakedown" and the contact pressure limit below which it occurs is known as the "elastic shakedown limit". This type of behaviour for a component which is subjected to repeated loading may be conveniently illustrated in schematic plots of applied load versus resultant deflection, as shown in Figure 2. Up to the elastic limit, regime (a), the response of the structure is entirely elastic. Above the elastic limit, plastic flow is encountered - at least in early applications of the load. However, the process of shakedown, referred to above, can lead to the steady state being elastic. The upper limit for this behaviour is shown as the elastic shakedown limit in regime (b). If the cyclic load is increased still further, then plastic flow is encountered even in the steady state condition. This deformation can take two forms: If the load is below the "plastic shakedown limit", as in regime (c), then the steady state cyclic response will be represented by a closed cycle of plastic deformation. Plastic flow then occurs at two instants in each cycle, although the strains at these points are equal and opposite, so that in one cycle there is no net accumulation of deformation. Finally, if the load lies above the plastic shakedown limit, then the steady state response is represented by an open cycle of plastic deformation and the material accumulates small increments of plastic deformation with each application of the load. This is the behaviour known as "ratchetting".

4. Wear Models

4.1. WEAR BY PLASTIC FLOW

Wear of ductile materials under mild conditions (low friction and moderate loads) is found to generate lamellar wear debris in the form of thin platelets [1-4]. In particular, Akagaki and Kato [5] revealed how such debris can be formed by progressive plastic extrusion from the edges of the irregularities on the softer of the two sliding surfaces, Figure 3. A hard steel ball with a ground flat as shown in Figure 3a was slid over a nominally flat rough surface of softer material. When the asperities of the softer surface were perpendicular to the sliding motion, wear particles were extruded at the trailing edge of the asperity contact (Figure 3b). On the other hand, when the asperities were aligned longitudinally, i.e. parallel with the direction of sliding, wear particles with a similar morphology were extruded laterally, that is at the sides of the flattened ridges, as illustrated in Figure 3c. These slivers were formed progressively over thousands of cycles.

This behaviour was reproduced by Kapoor and Johnson in their large scale model asperity experiment [6]. They suggested that the slivers are formed by plastic ratchetting of the material (regime (d) in Figure 2) in a thin surface layer which is progressively extruded out. A noteworthy feature of this wear mechanism is that the material is extruded out of the contact by plastic flow and, even though the ultimate separation of debris is caused by fracture, *wear itself is not caused by fracture*.

Figure 4 shows the contact of a rough hard surface with a smoother softer surface. Note that even if the softer surface were rough to start with it would quickly smoothen by "running-in" during sliding and contact would take place between a rough hard surface and a smoother softer plane [7,8]. Because of the roughness of the hard slider, contact takes place at the asperity summits, where the contact pressure is much higher than the nominal contact pressure. The contact stresses are severe immediately below the asperity contact but they decay rapidly as the square of the distance [9]. In conditions of mild loading, there are just a few asperity contacts at any cross-section of the softer surface and the contact stresses (averaged over the full cross-section) are not high enough to cause plastic extrusion.

As sliding starts, these asperity contacts, with high contact pressures and severe contact stresses, start to traverse the softer surface. If the asperity were perfectly aligned with the direction of sliding, then the softer surface would be loaded continuously along the same tracks. It would develop permanent grooves and the stresses would shakedown to an entirely elastic state [10]. But a real rough surface is random and the roughness is not correlated in the direction of sliding. In these circumstances contacts are made randomly across the width of the softer surface, as illustrated in Figure 4. When sliding has proceeded for sufficient time, every point on the soft surface has been subjected to at least one pass of the asperity contact and the associated high contact stresses. It is this action of pummelling by hard asperities which subjects a thin layer of the softer surface to severe contact stresses, even at low nominal contact pressures, and causes extrusion of slivers. Kapoor and Johnson have analysed pummelling by a hard surface with asperities whose tops could be modelled as hemispheres. Machining operations such as turning and grinding produce surfaces where asperity tops are cylindrical rather than hemispherical and this model was analysed by Kapoor and Cocks [11]. The load at which wear by this process would start has also been considered [8].

4.2. WEAR BY FRACTURE

Another mechanism relies on fracture to generate debris and cause wear. Suh and his co-workers [12] observed fine cracks close and parallel to the surface which produced wear debris in the form of thin platelets and described this as "delamination wear". Application of Mode II Linear Elastic Fracture Mechanics (LEFM) to model the crack propagation has not been successful, primarily because the crack is closed by very high compressive stresses. Also the deformation in the sub-surface layer is plastic and not elastic as assumed by LEFM.

The mechanism which would be expected to cause failure in the wearing surface may be identified in the load-deformation diagram in Figure 2. Below the elastic limit the material behaves elastically, regime (a), and the material would fail by (high cycle) fatigue. Between the elastic and elastic shakedown limits some plastic flow is obtained in the early cycles but the steady state behaviour is elastic. Provided the early plastic flow is small, then the failure would again be expected to be by high cycle fatigue. Above the elastic shakedown limit but below the plastic shakedown limit, plastic flow is observed even in the steady state and the failure would be expected to be by (low cycle) fatigue. The view that wear is a fatigue process is long-standing; one of its chief exponents being Kragelski et al [13]. Cyclic stressing occurs both through repeated sliding and repeated asperity encounters, and the material fails by fatigue. However, not many models exist which correlate the wear rate to a material's fatigue properties in a fundamental way.

4.3. THE OXLEY MODEL

In their model of sliding wear, Oxley and co-workers [14] modelled a hard asperity by a rigid wedge which pushes a wave across the mating, softer (rigid-perfectly plastic) surface as illustrated in Figure 5a. By using the slip line field theory the load was related to the depth of plastically deforming material. Typical plastic strain cycles of material in this layer, shown in Figure 5b, comprise cyclic strain $\Delta\varepsilon^{p}_{xx}$ and a unidirectional shear strain $\Delta\varepsilon^{p}_{zx}$. After some cycles the material fails producing wear particles. Challen, Oxley and Hockenhull [15]related the wear rate to the plastic deformation involved by utilising the Coffin-Manson law of low cycle fatigue (subsequently referred to as LCF), that is by using the following relation:

$$\left(\frac{\Delta\varepsilon_f}{2}\right).N_f^n = C \qquad (1)$$

where $\Delta\varepsilon_f$ is the range of alternating plastic strain occurring in the wave and equals $\Delta\varepsilon^{p}_{xx}$ in Figure 5b. N_f is the number of cycles to failure, the exponent n is approximately 0.5 and C is a strain related to the failure strain in static loading. Similar models have been used by Lacey and Torrance [16].

However, the strain cycle shown in Figure 5b is not closed as is normally associated with fatigue. The shear strain $\Delta\varepsilon^{p}_{zx}$ is unidirectional and will accumulate with each cycle. A similar open strain cycle is obtained when a cylinder rolls and slides on an elastic-plastic surface under a load that exceeds the elastic limit with a coefficient of sliding friction m > 0.25, Bower and Johnson [17]. Under these rather severe conditions a thin near-surface layer becomes plastic. Again the ratchetting strain is the accumulative shear which is observed in the surface layer. Its magnitude was confirmed by experiments in a rolling contact disc machine.

4.4. FAILURE OF MATERIALS SUBJECTED TO RATCHETTING

A question now arises: what is the mechanism of rupture when a metal is subjected to open plastic strains of the sort shown in Figure 5b? Two possibilities immediately suggest themselves: (i) fatigue failure driven by the cyclic component of plastic strain, or (ii) ductile fracture, similar to that in monotonic loading, driven by the accumulated unidirectional plastic strain. This question was considered in detail by Kapoor [18], who found that during repeated loading, involving plastic straining of the material, failure occurs by one of the two processes. If the cycle of plastic strain is closed, i.e. there is no accumulation of strain, the material fails by conventional low cycle fatigue and the life may be estimated by using the Coffin-Manson relationship, Eq.(1). On the other hand, if the cycle of plastic strain is open such that the material accumulates plastic deformation with each cycle, a different type of failure termed "ratchetting failure" is possible. It occurs when the accumulated strain reaches a critical value. For a ratchetting strain of $\Delta\varepsilon_r$ the number of cycles to failure N_f would be given by the Eq(2) below,

$$N_r = \varepsilon_c / \Delta\varepsilon_r \qquad (2)$$

Here ε_c is the critical strain to failure and was found to be comparable with the strain to failure in a monotonic test. It is also suggested that LCF and RF (ratchetting failure) are competitive mechanisms and whichever corresponds to earlier failure governs the life of the specimen, i.e.

$$N = \mathrm{Min}(N_f, N_r) \tag{3}$$

here N is the actual number of cycles to failure and N_f and N_r are given by Eqs.(1) and (2), respectively. A wide range of tests in the literature is revealed to follow the above hypothesis and the results have been reproduced in Figure 6. These tests include some of Coffin's early experiments [18].

Kapoor evaluated the theoretical strain cycles experienced by material elements subjected to rolling and/or sliding contact, presented by Bower and Johnson [17] and Challen et al [15]. On the basis of Eq. (3) it was predicted that the material would fail by ratchetting. Experiments by Oxley and co-workers on fracture of an aluminium alloy due to repeated sliding of hard wedges were found to be in complete accord with the predictions based on ratchetting [14,15].

4.5. A WEAR MODEL BASED ON RATCHETTING FAILURE

Kapoor proposed a hypothesis to model wear based on ratchetting failure of the material [19]. Figure 7 shows a surface being slid by a line contact with a coefficient of sliding friction, m, greater than one third. The surface has been divided into discrete layers which fail, one by one, causing wear. As a result of repeated sliding these layers undergo the following processes:

a) Development of residual stresses and shakedown will ensure that the direct stresses become hydrostatic.

b) Based on the theoretical work of Ponter and Cocks [20], and Johnson and co-workers [21,22] the ratchet strain per cycle, $\Delta\gamma_{zx}$ can be given as

$$\Delta\gamma_{zx} = f(\tau_{zx} - k) \tag{4}$$

where τ_{zx} is the shear stress and k the shear strength of the material. Note:

- Due to plastic flow the material work hardens, i.e. the value of k increases causing a drop in the ratchet rate.
- Due to wear the material element moves closer to the surface and the intensity of shear stress τ_{zx} increases and ratchetting continues.

c) Continuing rolling/sliding causes the accumulated shear strain to increase and when it reaches the critical value the material fails and wear particles are formed.

After the top layer has "delaminated", newer layers are exposed and the process repeats.

4.6. THE STEADY STATE WEAR RATE

Let us apply the above hypothesis to a wearing surface. Consider the first layer at the start of rolling/sliding. It has no accumulated strain. As a result of rolling/sliding the material is subjected to a cycle of plastic strain and let the number of cycles to failure, as estimated by using Eq.(3), be N. Depending on the thickness of the layer (which would be a function of the semi-contact width) this corresponds to a certain wear rate. During failure of this layer (the first N cycles) other layers (2, 3 etc. in Figure 7) also accumulate strain. When layer 2 reaches the surface it has accumulated some shear strain and so the number of cycles to failure would be expected to be less than N and the wear rate would be correspondingly higher. Likewise, layer 3 would have an even higher wear rate. After some time all the layers reaching the surface will have the same accumulated strain as their predecessor and so the wear rate would become constant; i.e., steady state would be achieved.

Tyfour, Beynon and Kapoor [23] conducted wear experiments on a twin disc machine in which a pearlitic rail steel (BS11) was run dry against a railway wheel steel (W8A) under 1500 MPa maximum contact pressure (line contact condition) and 1% slip. Wear was caused by both plastic flow and fracture. Plastic flow caused the material to accumulate in the lips to the side of contact (as illustrated in Figure 11 in the reference). Wear caused by fracture was separated from the total wear by weighing the disc at regular intervals. The results of the test have been reproduced in Figure 8. The material can be seen to strain harden with the progress of rolling/sliding - the surface hardness nearly doubles after about 17500 cycles. The accumulated strain can also be seen to increase from zero to a constant value over the same number of cycles. Also, the wear rate can be seen to increase from a low value to a steady state value in about 17500 cycles.

4.7. EFFECT OF REVERSAL ON WEAR

Consider the surface in the previous section which, after some cycles running, is then subjected to a reverse in the direction of rolling/sliding. Figure 9 illustrates the process and the effect of reversal on the accumulated strain. It is clear that the material would start accumulating strain in the other direction and would require a larger number of cycles to reach the critical limit at which ratchetting failure takes place. This would manifest itself in a reduced wear rate.

Tyfour and Beynon [25] conducted experiments on a twin disc machine in which pearlitic rail steel (BS11) was run dry against wheel material (W8A) under 1500 MPa maximum contact pressure (line contact condition) and 1% slip. The direction of rolling/sliding was reversed at different stages and the results have been reproduced in Figure 10. As expected the wear rate drops at the point of reversal of the rolling/sliding.

If the rolling/sliding direction is reversed only once, the wear rate drops and then creeps up as the material with accumulated strain in the opposite direction is worn out.

However, if the direction is continuously reversed then the wear rate would be expected to remain low. Just this was observed in the above experiments.

An important point to note is that even under the reversed direction the amplitude of alternating plastic strain remains the same and so the Coffin-Manson Eq.(1) is not able to distinguish between the two cases. *Any model based on the LCF approach, therefore, would not predict the wear rate accurately in this situation.*

5. Rolling Contact Fatigue

Rolling Contact Fatigue (RCF) is a generic name for cracking and damage in a sub-surface layer, as a result of prolonged rolling/sliding. Such cracks in the heavily plastically deformed sub-surface layer generally cause spalling of the material, but occasionally a crack turns into the body and could cause serious damage, such as a complete fracture of a rail. RCF is a significant cause of preventative rail replacement on most railway systems (for example, M. Ishida [25] and G.M. Galvin [26]).

There are a few stages in the development of RCF. First, the presence of a crack is essential. It may be in the form of a pre-existing flaw in the material or a micro-crack which developed as a result of damage accumulation in the sub-surface material. Next the crack grows in the short crack regime and finally as a long crack. The early part of the crack propagation takes place in the thin sub-surface layer, which is heavily plastically deformed. As mentioned earlier, the application of LEFM to crack propagation in this near-surface layer has not been successful because the crack is closed by very high compressive stresses (in the range 1000 - 1500 MPa in case of railways). Also the material around the crack, in its early life, is deforming plastically and not elastically as is assumed in LEFM.

It is now generally accepted that the crack propagates by the fluid pressure mechanism, which was first proposed by Way [27] and is reviewed next.

5.1. THE FLUID PRESSURE MECHANISM

Figure 11d-f shows a cracked surface being rolled by a driving wheel. As the contact approaches the crack its mouth is opened by the traction forces and the fluid enters the crack. Next the contact moves over the crack mouth, effectively closing it by the large normal pressure (approximately 1000 - 1500 MPa in the case of railways). The fluid transmits the contact pressure to the crack tip and causes the crack to grow, even in the presence of high compressive stresses perpendicular to the crack faces. Note that this mechanism will not work if the wheel was braking; in such a case the contact traverses from the tip to the mouth of the crack, thereby draining the fluid. This expectation is observed in experiments [28].

Way's proposition [27] that the dominant cause of crack growth is the hydraulic pressure generated in the lubricant was extended by Bower [29] to include entrapment of low viscosity fluid within the crack and friction between the crack faces. He also estimated the stress intensity factor at the crack tip as the contact passes over the crack. Further studies, carried out by Hanson and Keer [30], and by Kaneta and Murakami [31], demonstrated that fluid pressure is indeed necessary for crack propagation.

Recently Tyfour and Beynon have introduced a similar mechanism with friction varying along the crack face [32].

5.2. EFFECT OF ROLLING DIRECTION REVERSAL ON RCF LIFE

As with the dry wear tests which were conducted with a reversal [24], tests run with water present throughout were also interrupted to reverse the rail disc [32]. Since those cracks which had begun to grow in the first direction were now oriented in an unfavourable direction, their propagation had to continue in a different direction, or new cracks were required to grow. Either event leads to an increase in the fatigue life of the disc (measured as a given maximum crack size in these tests). Figure 12 shows a schematic representation of the types of crack morphology which were observed for different reversal regimes. The effect of these different regimes on the fatigue life is shown in Figure 13. A reversal factor of about 0.3, i.e., reversing the rail disc after 30% of the fatigue life expected in a uni-directional test, produced the largest increase in fatigue life: nearly double the uni-directional value; ("life factor" is the ratio of the number of cycles to failure in a test to the life of the uni-directional test).

An important observation in this work was the absence of a measurable difference in the sub-surface work hardening with or without a reversal in the direction of the test [33]. Thus simply using work hardening as a measure of the damage in the material is inadequate to describe the fatigue behaviour. It is vital to include the direction of loading in the analysis. Taking strain path effects into account is a novel aspect which cannot be incorporated into an approach based on low cycle fatigue.

5.3. EFFECT OF PRIOR DRY ROLLING/SLIDING ON RCF LIFE

A crack has to be present for the fluid pressure mechanism to work. It may be a pre-existing flaw in the material or a micro-crack which initiated as a result of repeated loading of the material. The use of clean steels and improved manufacturing procedures have greatly reduced the occurrences of pre-existing flaws in materials, leaving ratchetting failure as probably the dominant route to crack initiation. Rolling/sliding under a high coefficient of friction leads to quicker accumulation of plastic deformation and therefore rapid initiation of cracks. In the presence of a fluid these cracks propagate and lead to RCF failure. A combination of dry and wet conditions may be met by rail / wheel contact many a times in a day.

Tyfour, Beynon and Kapoor have investigated this problem [33] in a twin-disc machine. The RCF life of pearlitic rail steel was studied under rolling/sliding conditions at 1% slip, and either 1200 or 1500 MPa normal pressure. The discs were first run dry and then drip fed with water. An eddy current probe was used, in situ, to check the cracking of the disc and failure was taken to be when the crack reached a certain depth (about 200 μm). Results of the RCF life have been reproduced in Figure 14, where the abscissa represents the number of dry cycles and the ordinate the RCF life. It can be noticed that the effect of initial dry cycling on the subsequent RCF life is characterised by three stages:

1. No deterioration: if the number of dry cycles is below a critical level. The early increase in RCF life has been confirmed (by experiments in an inert atmosphere) to be due to oxidation effects.

2. A transition stage: which is characterised by a sharp and sudden drop in RCF life if the critical level of dry cycles is exceeded.

3. A severe effect stage: if the critical level of dry cycles is significantly exceeded.

A physical model to describe the effect of dry cycling on the subsequent RCF life is schematically illustrated in Figure 11. A *driving* wheel is rolling on a rail in the direction shown. The shear traction on the rail is in a direction opposite to that of the motion of the wheel. Note that if the wheel were braking, then the direction of shear traction would be in the same direction as the motion of the wheel and the fluid pressure mechanism would not be able to propagate the crack.

Dry cycling under a coefficient of friction higher than 0.25 will cause ratchetting of the surface layer, provided the ratchetting threshold load is exceeded, Figure 11a. When the accumulated plastic strain reaches a critical value surface cracks will initiate, Figure 11b. Further cycling will propagate these cracks to a critical length, Figure 11c.

Upon application of the fluid lubricant, at the start of the wet phase, liquid enters the crack cavity, Figure 11d. The approaching loaded contact closes the crack mouth, raising the pressure of trapped fluid and causing mode I crack propagation, Figure 11e. When the maximum load passes the crack mouth the pressurised fluid escapes from the crack cavity to lubricate the crack faces, causing mode II crack propagation, Figure 11f.

Tyfour, Beynon and Kapoor estimated the accumulated plastic strain as a result of dry rolling/sliding: it occurs for the part of dry rolling/sliding which has a coefficient of friction higher than 0.25. The plot of accumulated strain versus the deterioration in RCF life provided a convenient means of estimating the effect of dry rolling/sliding on the RCF life, as shown in Figure 15.

6. Conclusions

Various wear mechanisms have been reviewed, in particular the use of low cycle fatigue to explain the wear of surfaces subject to plastic deformation. Accumulation of plastic deformation by ratchetting has been demonstrated to provide a quantitatively accurate model for many wear circumstances. Is also provides a sound explanation of how a test achieves a steady state wear rate.

The critical test involving a reversal in the direction of traction provides convincing support for the model based on ratchetting.

The fluid pressure mechanism for rolling contact fatigue has been described and then supported by further critical tests involving the reversal in the direction of the test.

The effect of prior dry rolling/sliding on subsequent "wet" rolling contact fatigue life has been explained and quantitatively described using ratchetting theory.

Although the basis for the paper is the railway wheel - rail contact, the conclusions are applicable to other circumstances.

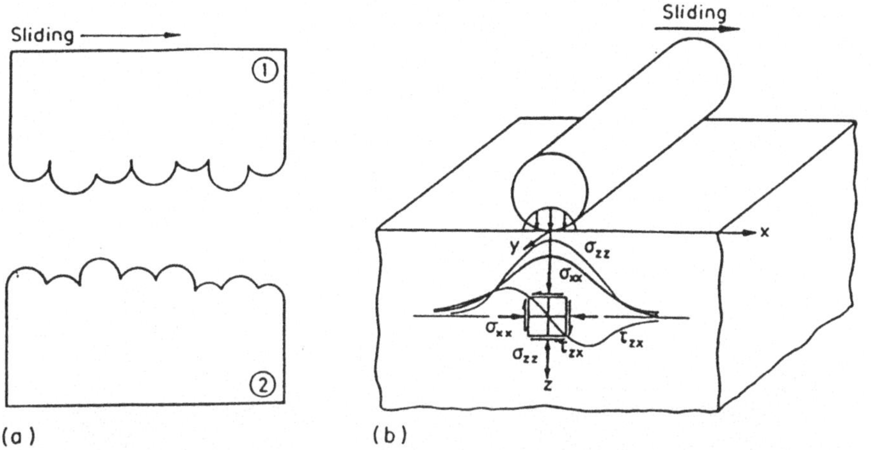

Figure 1. (*a*) Repeated sliding of two rough surfaces. The roughness is two-dimensional, extending infinitely perpendicular to the page. (*b*) An individual asperity from (*a*) has been isolated and replaced by its equivalent geometry, *i.e.* a cylinder sliding on a half-space. In the elastic state the contact pressure distribution is Hertzian. A material element of the half-space is subjected to the shear stress τ_{zx} and direct compressive stresses σ_{xx} and σ_{zz}. For normal pressure, their variation with x is shown by the curves.

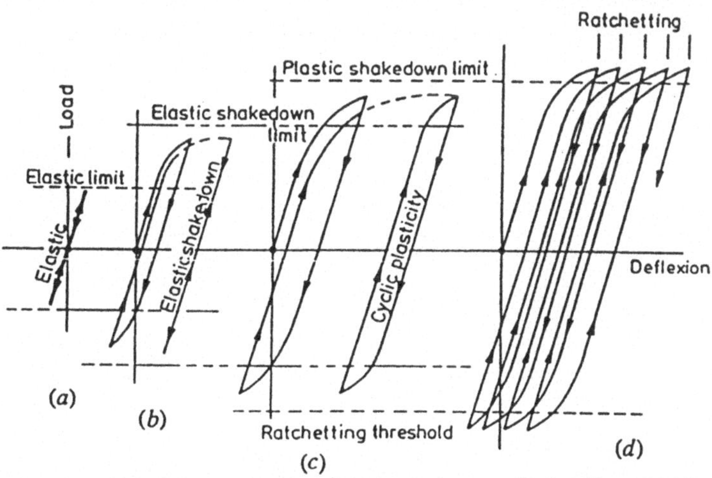

Figure 2. Response of a structure or material element to cyclic loading: (*a*) Up to elastic limit the deformations are elastic. (*b*) On increasing the load the structure undergoes some plastic flow and develops protective residual stresses. The steady-state deformation, if the load is within the elastic shakedown limit, is elastic. (*c*) If the load is increased still further the steady-state deformation comprises a closed cycle of plastic deformation. (*d*) If the load is increased above the plastic shakedown limit or the ratchetting threshold, then the structure accumulates a net plastic deformation in each cycle.

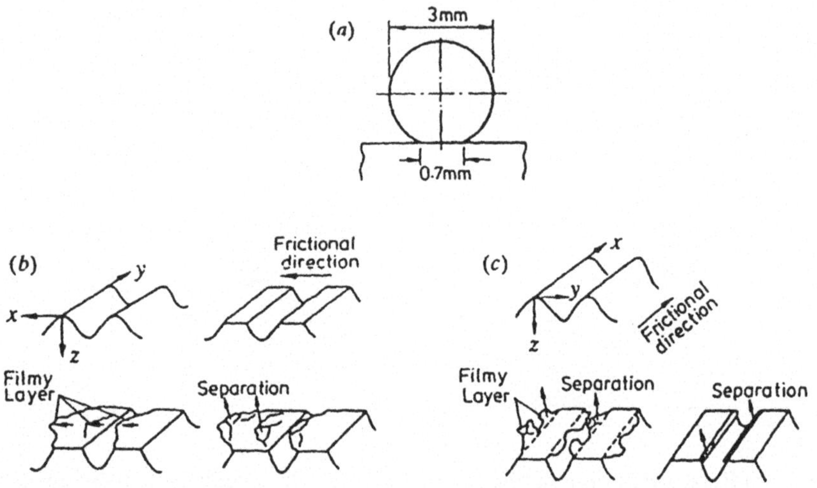

Figure 3. Formation of 'filmy' wear particles in Akagaki & Kato's experiments [5]: (*a*) The experimental set-up consisted of a hard ball with a flat ground on it sliding on a nominally flat softer counterface; (*b*) when asperities on the softer surface were perpendicular to the direction of sliding, thin slivers of material extruded out at the downstream side of the asperities; and (*c*) when asperities on the softer surface were parallel to the direction of sliding, slivers extruded laterally to the sides.

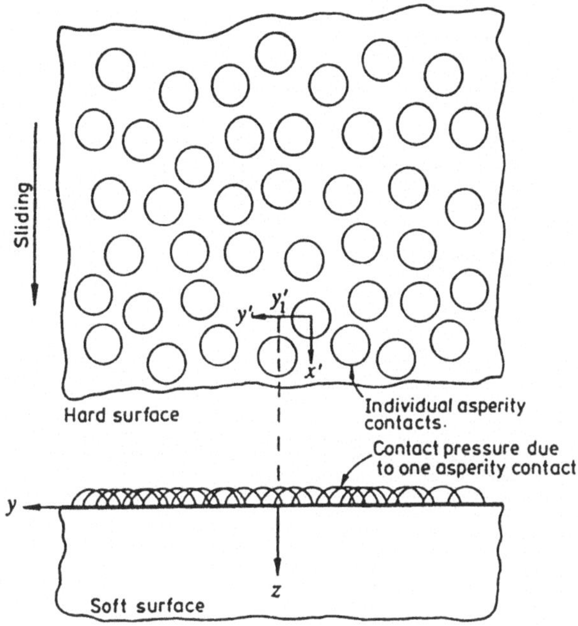

Figure 4. Pummelling action by hard asperities. The plan view, top, shows randomly distributed asperity contacts on the softer track. As sliding proceeds in the x-direction, a cross-section of the softer surface will see asperity contacts passing over randomly spaced points on its width. If sliding is continued for a sufficient length of time, each point on the cross-section would be passed by at least one micro-contact.

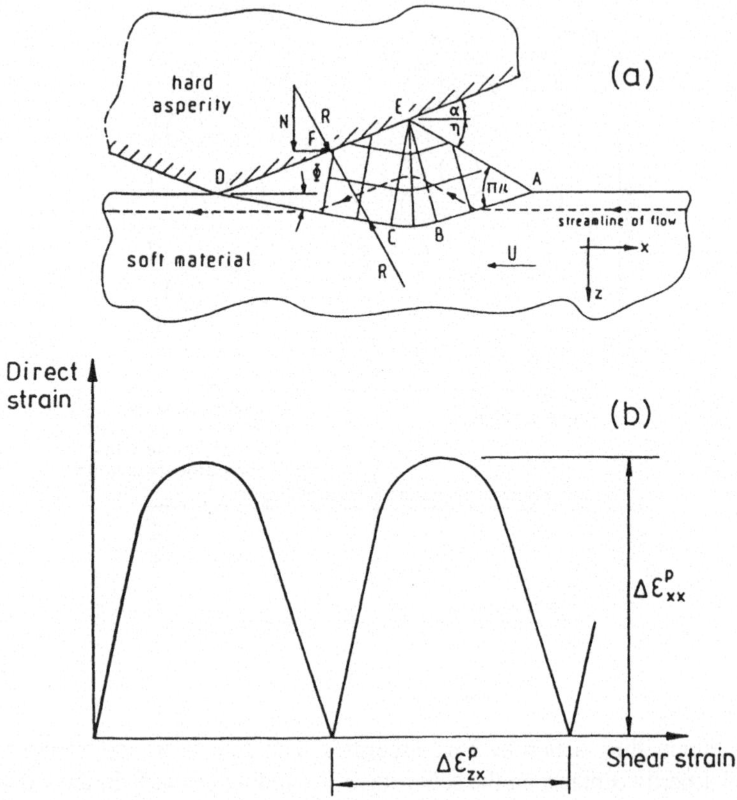

Figure 5. Plastic cyclic strain of a surface element subjected to sliding contact. (*a*) Slip line field in the Oxley model of a rigid wedge sliding on a rigid perfectly-plastic half-space [14]. (*b*) Plastic strain cycle for the slip line field of *Fig.5(a)*.

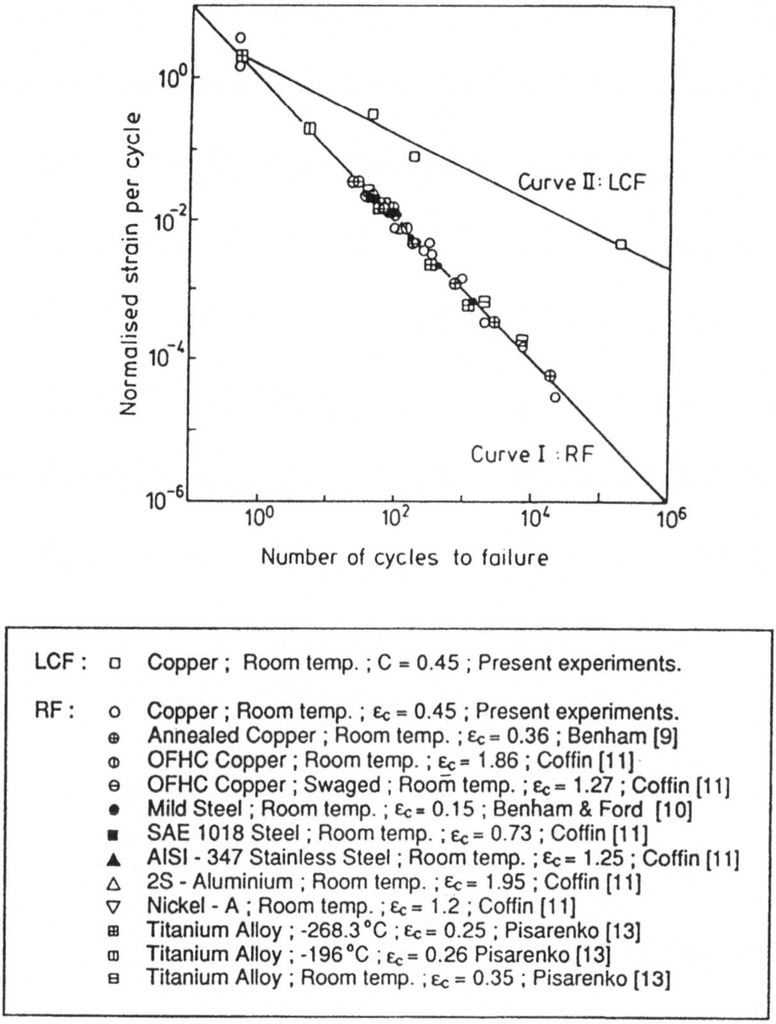

Figure 6. Cyclic loading data evaluated according to the hypothesis of competing modes of low cycle fatigue failure and ratchetting failure (references in the key can be found in [17]).

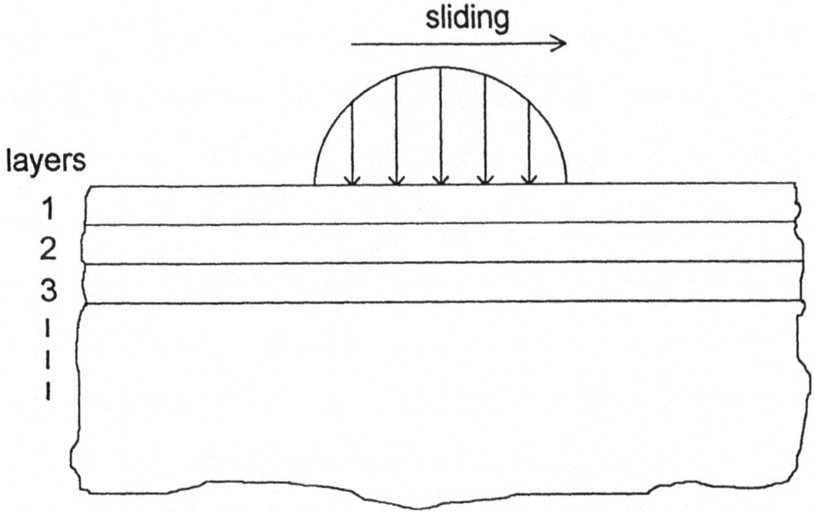

Figure 7. Failure of a surface due to ratchetting.

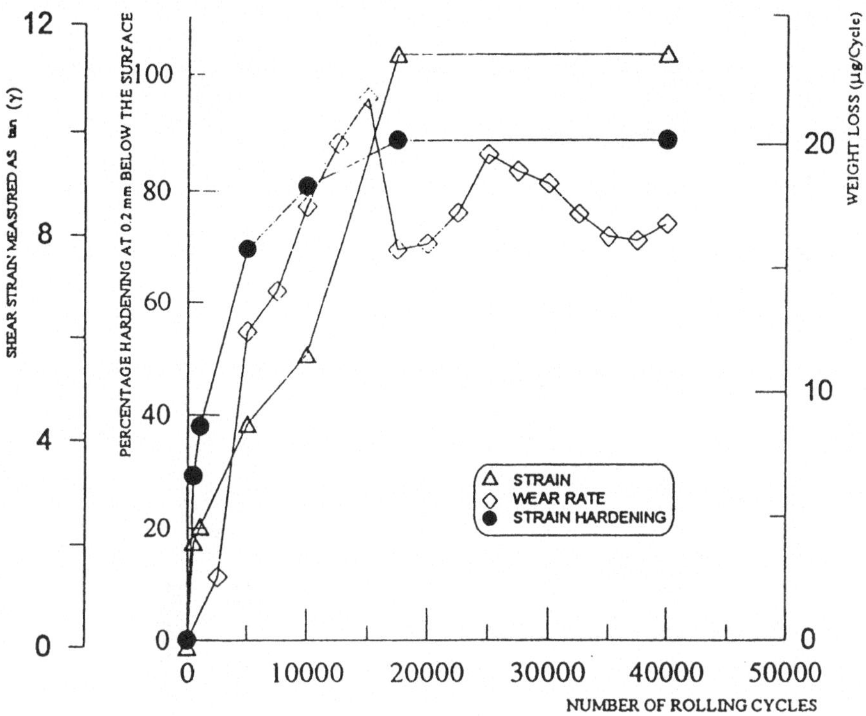

Figure 8. Effect of rolling distance on the accumulated shear strain, percentage hardening and wear rate of rail steel.

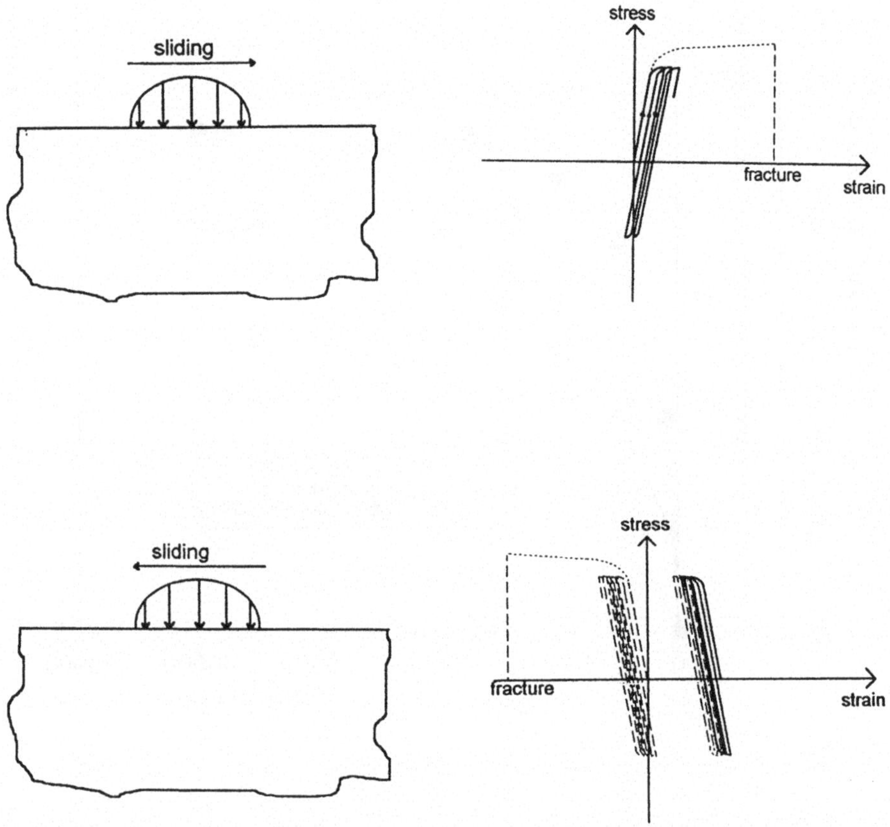

Figure 9. Effect of reversal on accumulation of plastic deformation.

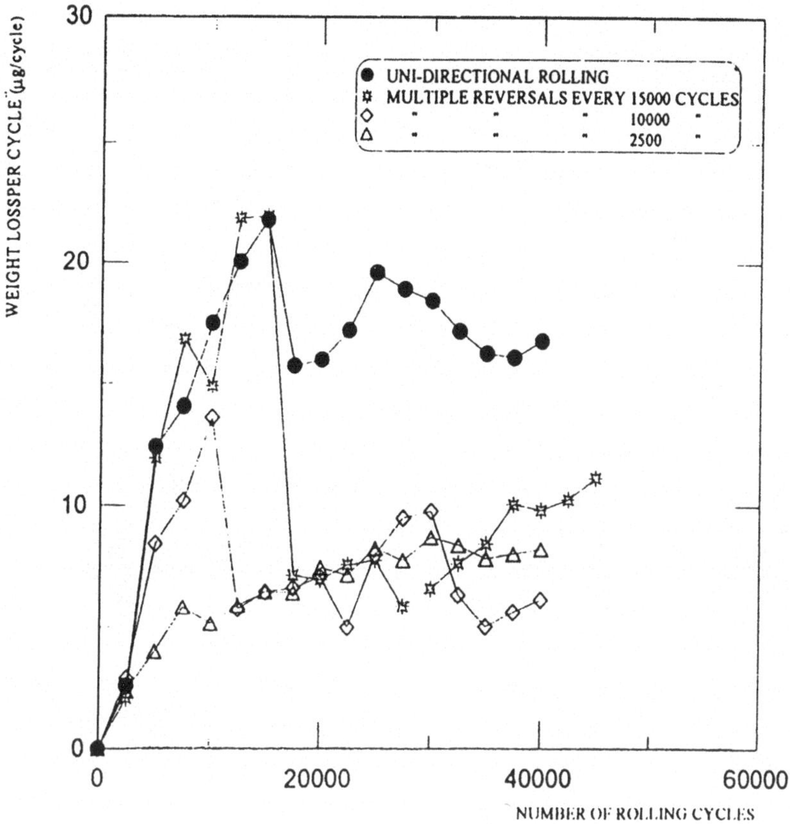

Figure 10. Effect of multiple rolling direction reversal on the rail disc weight loss per rolling cycle.

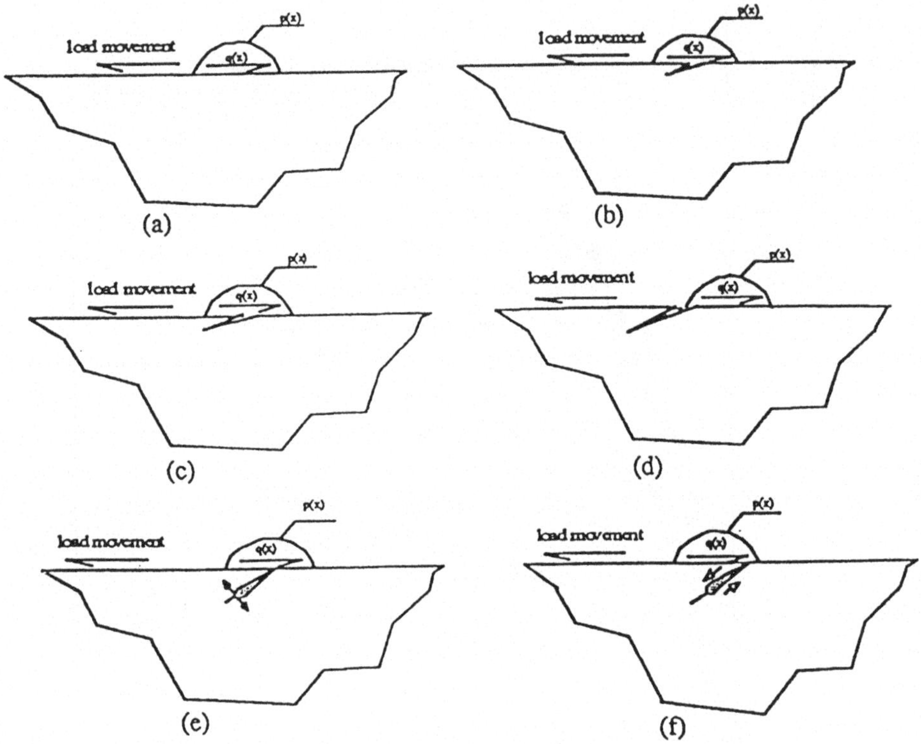

Figure 11. The physical model of the effect of ratchetting on RCF life. (d), (e) and (f) after Bower [29] ($p(x)$ is the spatial distribution of the contact pressure and $q(x)$ the consequent traction distribution).

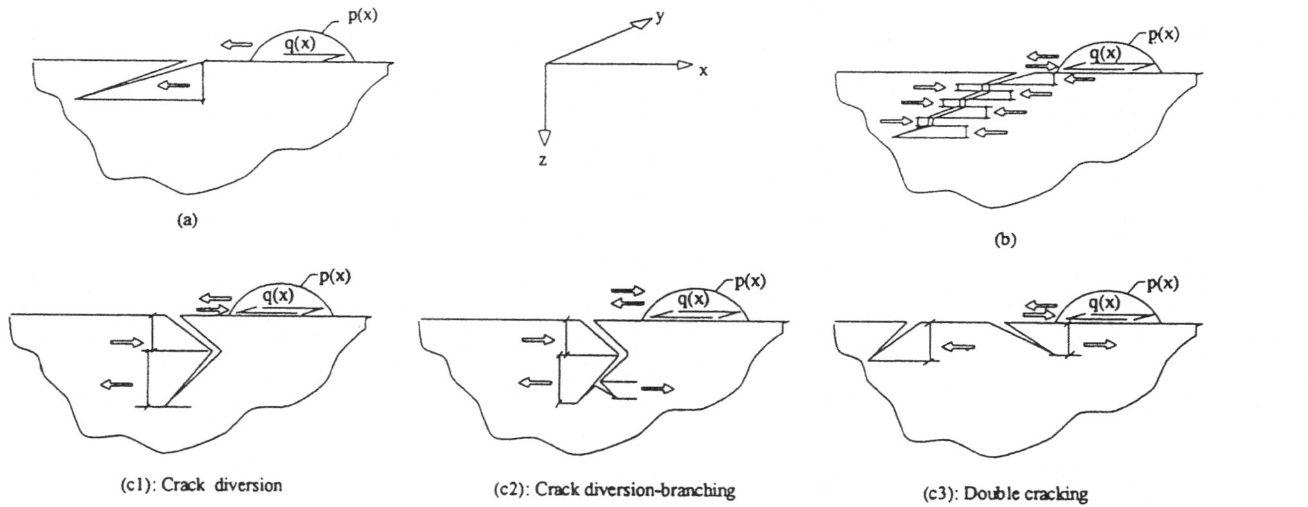

Fig. No.	Reversal factor	Life factor	Micrograph
(a)	1*	1.00	fig. (7)
(b)	0.053	1.36	fig. (11)
(c)	0.267	1.87	fig. (8&9)
(d)	0.5	1.36	fig. (10)

* Uni-directional

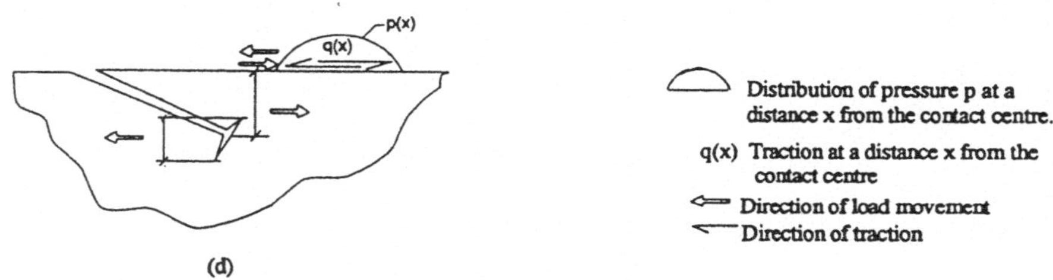

Figure 12. Schematic representation of cracking morphology under different reversal factors.

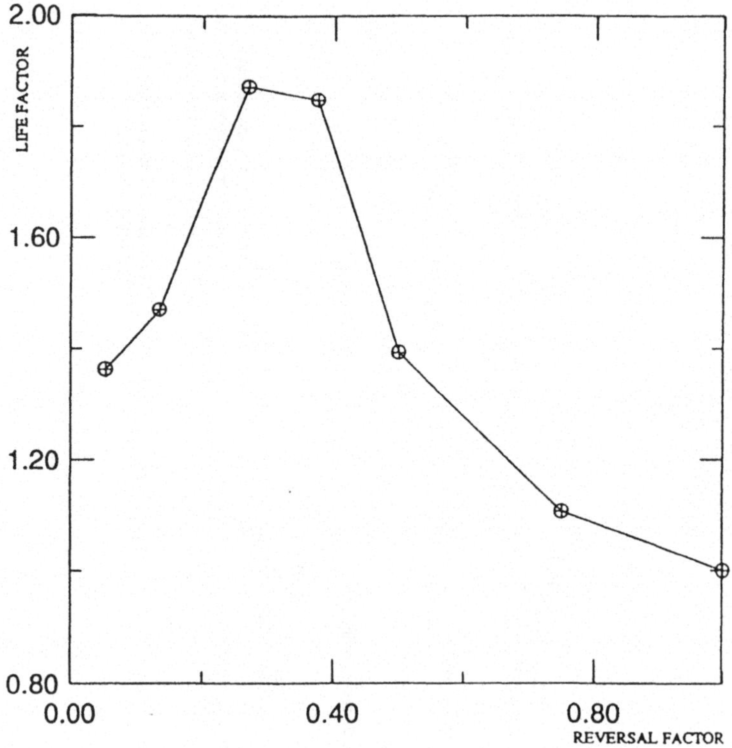

Figure 13. Effect of rolling direction reversal on RCF life; unity life factor represents the uni-directional fatigue life.

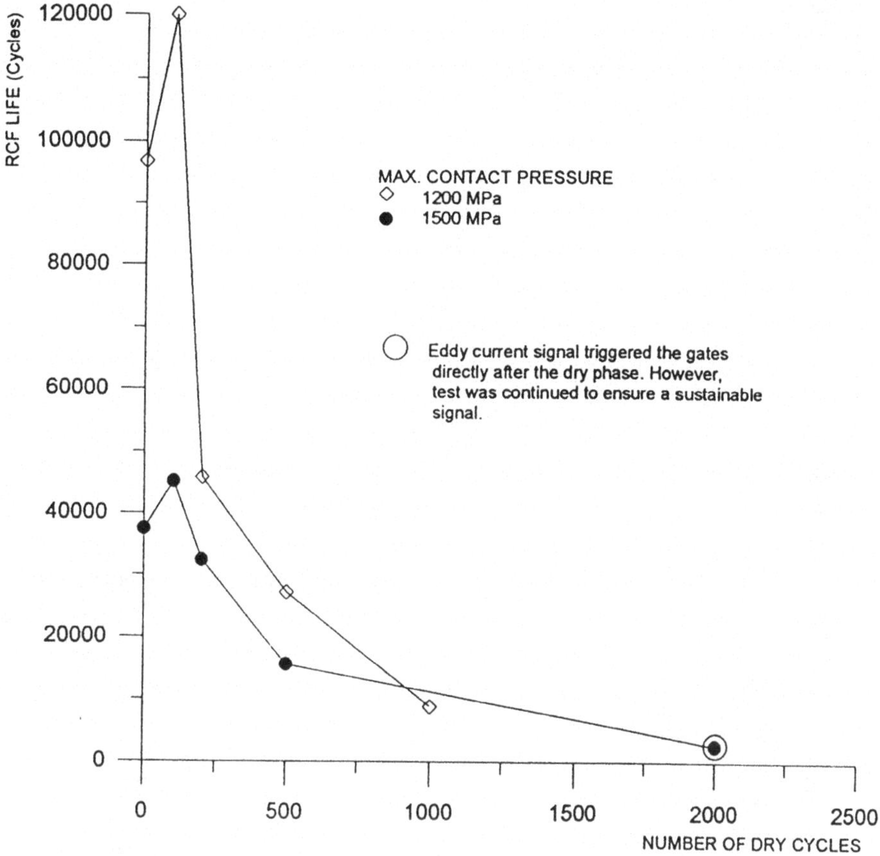

Figure 14. Effect of initial dry cycling on the subsequent RCF life.

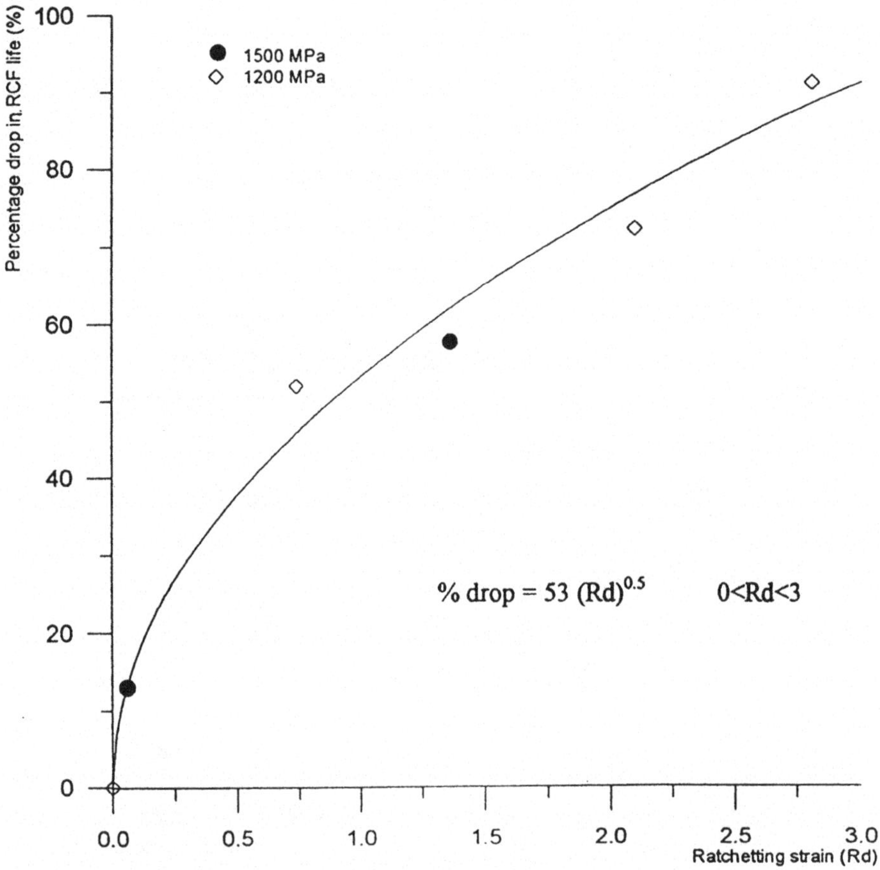

Figure 15. Deterioration in RCF life as a function of ratchetting (*Rd* is the ratchetting strain accumulated during the initial dry phase).

7. References

1. Reda, A., Bowen, R. and Westcott, V.C. (1975) Characteristics of particles generated at the interface between sliding steel surfaces, *Wear* **34**, 261-273.
2. Shetty, H.R., Kosel, T.H. and Fiore, N.F. (1982) A study of abrasive wear mechanisms using diamond and alumina scratch tests, *Wear* **80**, 347-376.
3. Martin, J.M., Mansot, J.L., Bebezier, I. and Dexpert, H. (1984) The nature and origin of wear particles from boundary lubrication with a zinc dialkyldithiophosphate, *Wear* **93**, 117-126.
4. Kuo, S.M. and Rigney, D.A. (1992) Sliding behaviour of aluminium, *Mater. Sci. Engng. A* **157**, 131-143.
5. Akagaki, T. and Kato, K. (1987) Plastic flow process of surface layers in flow wear under boundary lubricated conditions, *Wear* **117**, 179.
6. Kapoor, A. and Johnson, K.L. (1994) Plastic ratchetting as a mechanism of metallic wear, *Proc. Royal Soc. Lond. A* **445**, 367 - 381.
7. Kapoor, A. and Johnson, K.L., in Ed. D. Dowson, C.M. Taylor, T.H.C. Childs & M. Godet, G. Dalmaz, *Thin Films in Tribology* - Proc. of 19th Leeds-Lyon Symp. held at the Inst. of Tribology, Leeds, 8-11 Sept. 1992, pp. 81-90, Elsevier.
8. Kapoor, A., Williams, J.A. and Johnson, K.L. (1994) The steady state sliding of rough surfaces, *Wear* **175** 81-92.
9. Johnson, K.L. (1987) *Contact Mechanics*, Cambridge University Press.
10. Kapoor, A. and Johnson, K.L. (1992) Effect of changes in contact geometry on shakedown of surfaces in rolling/sliding contact, *Int. J. Mech. Sci.* **34**, 223-239.
11. Kapoor, A. and Cocks, A.C.F. (1994) Wear through the plastic interaction of cylindrical asperities in sliding, *Int. J. Mech. Sci.* **36**, 1045-1059.
12. Suh, N.P. (1977) An overview of the delamination theory of wear, *Wear* **44**, 1-16.
13. Kragelski, I.V., Dobychin, M.N. and Kombalov, V.S. (1982) *Friction and Wear: Calculation Methods*, Pergamon Press.
14. Hockenhull, B.S., Kopalinsky, E.M. and Oxley, P.L.B. (1991) An investigation of the role of low cycle fatigue in producing surface damage in sliding metallic friction, *Wear* **148**, 135-146.
15. Challen, J.M., Oxley, P.L.B. and Hockenhull, B.S. (1986) Prediction of Archard's wear coefficient assuming a low-cycle fatigue mechanism, *Wear* **111**, 275-288.
16. Lacey, P. and Torrance, A.A. (1991) Calculation of wear coefficients of plastic contacts, *Wear* **145**, 367-383.
17. Bower, A.F. and Johnson, K.L. (1989) The influence of strain hardening on cumulative plastic deformation in rolling and sling line contact, *J.Mech.Phys.Solids.* **37**, 471-493.
18. Kapoor, A., (1994) A re-evaluation of the life to rupture of ductile metals by cyclic plastic strain, *Fatigue Fract. Engng. Mater. Struct.* **17**, 201 - 219.
19. Kapoor, A., 'Mechanics of failure in a hardening and wearing surface', The Mechanics and Materials Science of Contact: Issues and Opportunities, workshop held at Vanderbilt University, Nashville, US, 18-20 July 1994, Proc. (as Report No. 94-17) by Institute of Mechanics and Materials, University of California, San Diego, La Jolla, CA 92093-0404, USA.

20. Ponter, A.R.S. and Cocks, A.C.F. (1984) An incremental strain growth of an elastic-plastic body loaded in excess of the shakedown limit, *Trans. ASME - J. of Appl. Mech.* **51**, 465-469.
21. Hearle, A.D. and Johnson, K.L. (1987) Cumulative plastic flow in rolling and sliding line contact, *Trans. ASME - J. of Appl. Mech.* **54**, 1-7.
22. Tyfour, W.R., Beynon, J.H. and Kapoor, A. (1995) The steady state wear behaviour of pearlitic rail steel under dry rolling/sliding contact conditions, *Wear* **180**, 79-89.
23. Tyfour, W. R. and Beynon, J.H. (1994) The effect of rolling direction reversal on the wear rate and wear mechanism of pearlitic rail steel, *Tribology International* **27**, 401-412.
24. Ishida, M. (1990) Relationship between rail shelling and track surrounds, Quarterly Report of the Railway Technical Research Institute Japan **31**, 22-28.
25. Glavin, W.M. (1989) Rail grinding - the BN experience, *American Railway Engineering Association, Bulletin 772,* **90**, 237-255.
26. Way, S. (1935) Pitting due to rolling contact, *Trans. ASME - J. Appl. Mech.* **2**, A49-A58.
27. Beynon, J.H., Garnham, J.E. and Sawley, K.J. (1996) Rolling contact fatigue of three pearlitic rail steels, *Wear* **192**, 94-111.
28. Bower, A.F. (1988) The influence of crack face friction and trapped fluid on surface initiated rolling contact fatigue cracks, *Trans. ASME - J. of Tribology* **110**, 704-711.
29. Hanson, M.T. and Keer, L.M. (1992) An analytical life prediction model for the crack propagation occurring in contact fatigue failure, *Tribology Transactions* **35**, 451-461.
30. Kaneta, M. and Murakami, Y. (1991) Propagation of semi-elliptical surface cracks in lubricated rolling/sliding elliptical contacts, *Trans. ASME - J. of Tribology* **113**, 270-275.
31. Tyfour, W.R. and Beynon, J.H. (1994) The effect of rolling direction reversal on fatigue crack morphology and propagation, *Tribology International* **27**, 273-282.
32. Tyfour, W.R., Beynon, J.H. and Kapoor, A. (1996) Deterioration of rolling contact fatigue life of pearlitic rail steel due to dry/wet rolling/sliding line contact, *Wear*, to appear.

RELIABILITY ASSESSMENT OF RANDOMLY LOADED CRITICAL COMPONENTS

M. BILY
Institute of Materials and Machine Mechanics of the Slovak Academy of Sciences
Racianska 75, P.O. Box 95, 830 08 BRATISLAVA, Slovakia

1. Introduction

The estimation of reliability of a complex mechanical system, i.e. its ability to perform a required function under given conditions for a given time interval, is more than a complicated problem and has undoubtedly taken the first place in the contemporary evolution of technical scientific disciplines. Despite the evident progress in many directions, it is not easy to guarantee the usually required level of reliability because the true damage processes of complex structures do not coincide exactly with our simplified physical and mathematical models.

Furthermore, the reliability estimation (especially experimental) is a time consuming and as usually costly process and so a suitable compromise should be accepted, balancing the initial acquisition costs against the costs of subsequent ownership, as schematically depicted in Fig. 1. It is obvious that the more the producer spends to develop and produce a product, the higher the resulting reliability level and at the same time the lower the expenses of operating, maintaining and repairing it. Thus a certain optimizing strategy should be adopted suggesting which components are to be assessed, what kind of information is required and how this assessment is to be carried out having in mind the crucial fact: whereas a change in the product design at the drawing board may cost pounds, in the development stage hundreds of pounds, but when the product has been released onto the market, millions of pounds [1].

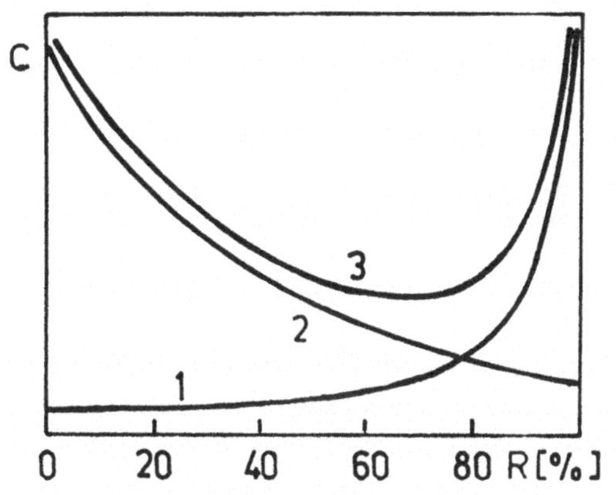

Figure 1. Relations between costs C and reliability level R: (a) - costs of concept, design and development; 2 - costs of use; 3 - resulting costs.

The intention of this paper is,

R.A. Smith (ed.), Reliability Assessment of Cyclically Loaded Engineering Structures, 27–45.

therefore, to bring together the problems that are encountered when defining and creating a new mechanical system with a prescribed reliability. Considering further, that practically all mechanical structures are randomly loaded and consequently the overwhelming amount of service item failures exhibit the fatigue damage character, we shall concentrate our attention on the fatigue life estimation understanding that decreasing the fatigue failure occurrences will be a substantial contribution to the system reliability increase.

Thus, from this point of view various areas of problems are to be tackled, schematically shown in Fig. 2.

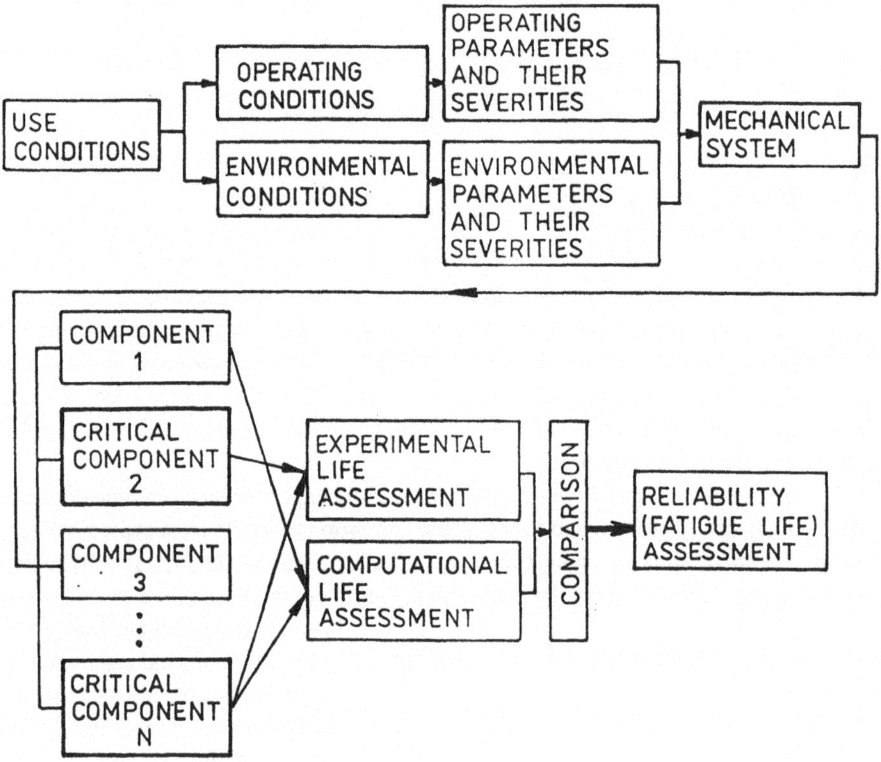

Figure 2. Chain of problems encountered when estimating reliability (fatigue endurance) of mechanical systems.

2. Assessment of Use Conditions

The primary input to designing procedures is the external (use, service) loading due to structure operation; this can be conceived as the transformation of use conditions by the transfer properties of the structure design. In this sense the use conditions represent the primary input not only for the reliability estimation but also for other activities in design and development phases or during operation. Our effort should be naturally aimed at such a set of use conditions which would guarantee a good correspondence between calculated (test) and use reliability.

It is useful to divide the use conditions into operating and environmental conditions, appearing simultaneously or sequentially.

The **operating conditions** are related to functioning of the structure and comprise a combination of single operating parameters and their severity.

Operating parameters are, for example, functional modes, input signals, loads, manipulation by operator, various supplies, etc.

The **environmental conditions** represent physical and chemical conditions external to the structure and comprise a combination of single environmental parameters and their severity.

Environmental parameters may be divided into a few groups such as climatic parameters (indoor and outdoor conditions), mechanical parameters (earthquake, shock, free fall) and other parameters (radiation, biological agents, wear, atmospheric corrosion).

The **severity** of the operating and environmental parameters is the value of each quantity characterizing them (say, stress level).

In many practical cases the environmental conditions are supposed to be constant with no or limited influence; this may be profoundly incorrect, however, in cases when possible synergetic effects may occur (a car moving on salty roads, in severe frost in Siberia-type conditions, etc.).

Considering that the use conditions are the "source" of the use loads, their ascertainment and definition must precede the stage of determining the load acting on a structure and its components.

The first step in determining the use (operating) conditions is to analyze the use conditions of similar existing structures. It may be done starting from the working day (week, hour, flight) division into activities (states of conditions and exposure to environment and usage), covering both useful up times and free runs. These activities are then composed of the relevant operating and environmental parameters (as usually appearing simultaneously), each of them possessing a few levels of severity. All activities should be characterized by their relations to the total life, i.e. by their percentage or probabilities of appearance. The combination of all possible activities, their parameters and severities, forming a set of use vectors, then determines the true condensed working day of the structure, or in the experimental terminology the test cycle. This is further split into up times of assemblies and components which allow to express the probability of structure functioning (or functioning of its items).

If possible, the resulting test cycle should be presented in a diagram where all activities, parameters and their severities are shown as simultaneous time functions. Fig. 3 presents a simple example of this, describing two activities of a lorry delivering goods and running unloaded. Its use conditions (actually operating conditions) are characterized by four parameters, viz. by power performance (2 levels of severity), road conditions (3 levels of severity), load size (3 levels of severity) and travelling speed (3 levels of severity). As a result 17 vectors were created with various percentage (probability) of existence in one test cycle.

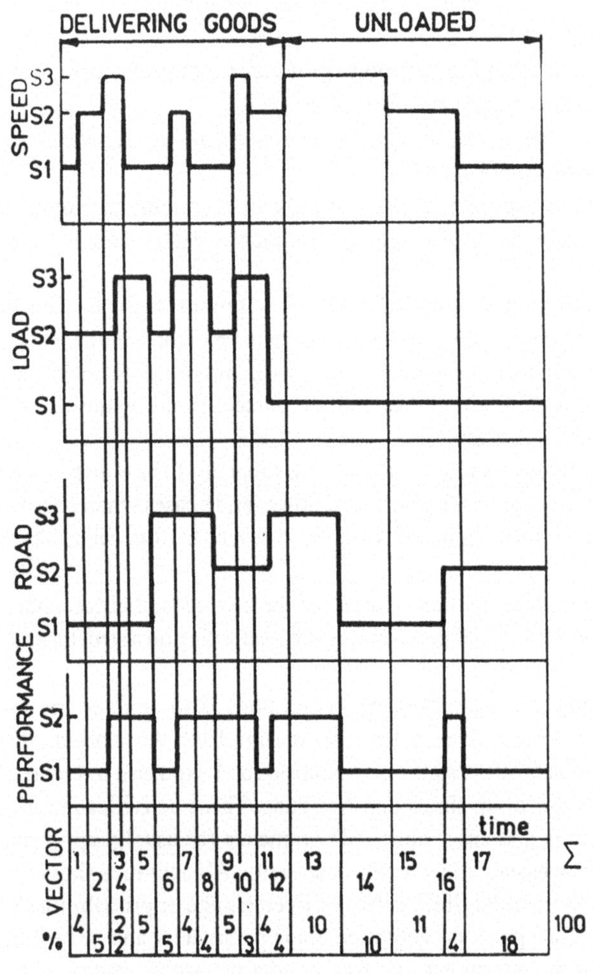

Figure 3. Example of working day of lorry composed of 17 vectors.

3. Assessment of criticality level

Fig. 1 illustrates that the implementation of a specified value of reliability into the complex structure is a game in which the costs of acquisition and ownership are to be balanced. Even if the optimal value of reliability of the whole structure is successfully substantiated, it still remains at the designer's shoulders to decide how to split it into thousands of assemblies, components and their locations. Not all of them have the same importance for the integral reliability. Some failures may belong to the category of "nuisance" and their elimination may be postponed to a planned maintenance, but failures of the critical components may lead to catastrophic consequences.

3.1. FAULT TREE ANALYSIS

In order to quantify these verbally described facts one can adopt a few approaches. The most significant and effective of them is the Failure (Event) Tree Analysis (FTA).

3.1.1. Graphical Construction of Failure Tree (FT)

The FT construction starts from definition of the Top Event (main, total failure, accident) and then proceeds by inductive determination of sub-item failures which it condition. This procedure goes deeper along the tree branches until primary (elementary) failures, requiring no further dissection, are reached. The resulting arrangement is a tree-like structure with information flow from the branch tips to the top event; the construction uses some standardized symbols defined, e.g., by the Standard IEC 1025 [2].

An example of the FT for the Top Event 4-CYLINDER ENGINE DOES NOT START is shown in Fig. 4.

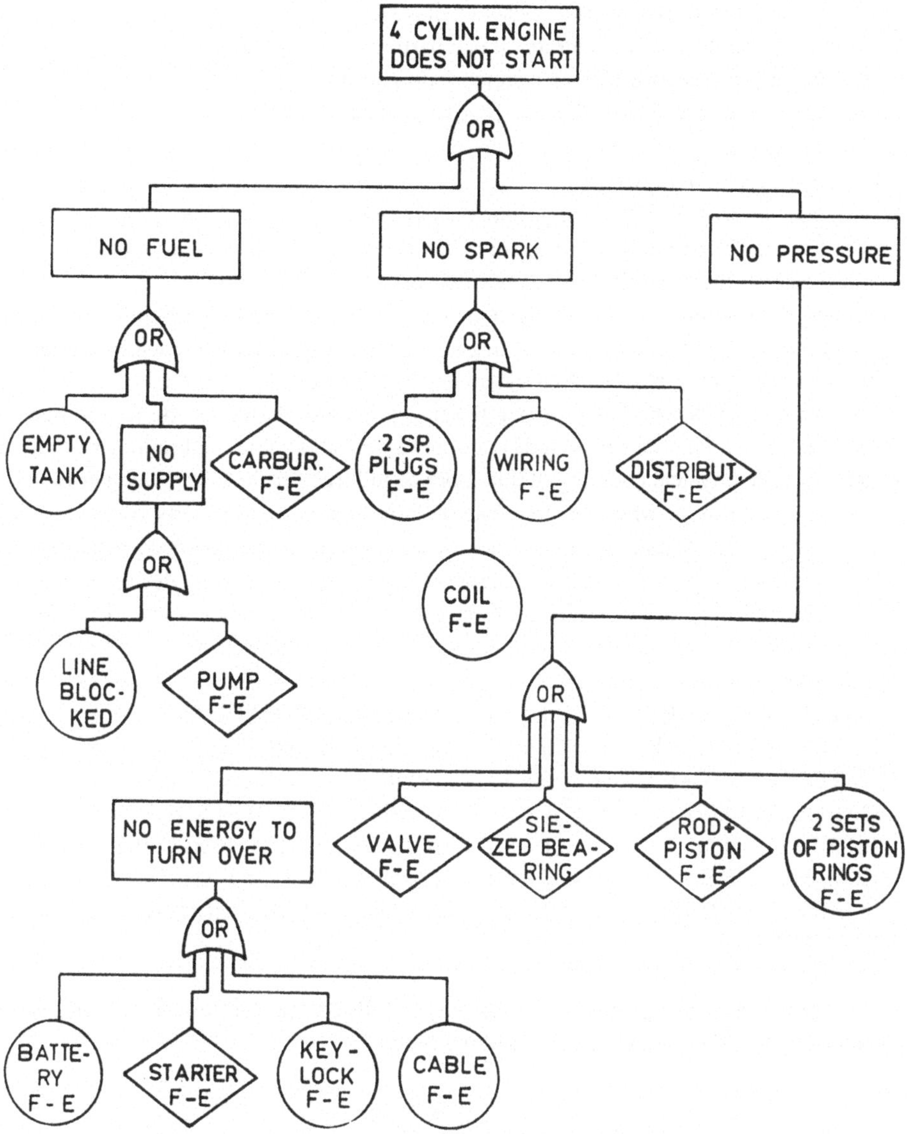

Figure 4. FT for Top Event 4-CYLINDER ENGINE DOES NOT START

Because the actual tree construction is rather tedious, though critical part of FT analysis, an automatic approach has been worked out which makes this a routine procedure, being at the same time extremaly fast.

The benefits of FT analysis include:

- forcing the analyst to actively seek out failure events deductively;
- providing a graphical aid which displays how the system can mulfunction, offering a clear picture of the system to people other than the designer;

- poiting out the critical aspects of system behaviour;
- allowing the analyst to concentrate on one particular system failure at a time;
- providing a reference for further system modifications; and
- providing a systematic basis for quantitative analysis.

3.1.2. Quantitative analysis

Having constructed the FT of the system considered it is now possible either

- to compute the Top Event probability or failure rate based on knowledge of the elementary failure probabilities or failure rates, or vice versa;
- knowing the required value of the system (Top Event) reliability (probability of failure) or its failure rate, to distribute it among sub-branches until the elementary failures of the "simplest" items, into which they are implemented by the designer.

In what follows we shall illustrate the first procedure based on the failure rates of elementary items. This approach starts from the failure rates values of the elementary items which have the physical meaning of a number of failures per time unit (say, per hour).

In the design stage when justified data on failure rates are as usually very scarce it is reasonable to suppose that the failure rates are constant [this assumption correspods to the central and longest part of a bathtub failure rate function of a properly designed structure and correspondingly to the exponential probability law $R(t) = \exp(- \lambda t)$]. Thus for the system in Fig. 4 the failure rates of the elementary events per hour (in circles and diamonds) could make

λ(BATTERY FAILURE) $= 4 \times 10^{-5}$ λ(VALVE MECHANISM) $= 5 \times 10^{-5}$
λ(STARTER FAILURE) $= 15 \times 10^{-5}$ λ(SEIZED BEARING) $= 1 \times 10^{-5}$
λ(SWITCH BOX FAILURE) $= 6 \times 10^{-5}$ λ(ROD + PISTON FAILURE) $= 1 \times 10^{-5}$
λ(CABLE FAILURE) $= 3 \times 10^{-5}$ λ(2 SETS OF PISTON RINGS) $= 2 \times 10^{-5}$
λ(PIPE BLOCKED) $= 11 \times 10^{-5}$ λ(CARBURATTOR FAILURE) $= 10 \times 10^{-5}$
λ(COIL FAILURE) $= 2 \times 10^{-5}$ λ(2 SPARK PLUGS FAILURE) $= 10 \times 10^{-5}$
λ(WIRING FAILURE) $= 3 \times 10^{-5}$ λ(DISTRIBUTOR FAILURE) $= 5 \times 10^{-5}$
λ(EMPTY TANK) $= 20 \times 10^{-5}$

From the elementary theory of reliability it follows that the failure rate of a series structure (gate OR) l_s with n independent units equals

$$\lambda_s = \sum_{i=1}^{n} \lambda_i$$

and so

λ(NO ENERGY TO TURN OVER) $= 4 \times 10^{-5} + 15 \times 10^{-5} + 6 \times 10^{-5} + 3 \times 10^{-5} = 28 \times 10^{-5}$/h.

Similarly

λ(NO SUPPLY) $= 16 \times 10^{-5}$/h, λ(NO FUEL) $= 46 \times 10^{-5}$/h,
λ(NO SPARK) $= 20 \times 10^{-5}$/h, λ(NO PRESSURE) $= 37 \times 10^{-5}$/h,

and consequently

λ(4-CYLINDER ENGINE DOES NOT START) $= 103 \times 10^{-5}$/h.

In other words it means that every 1.5 month we can expect problems when starting the analysed engine.

It is now the designer's problem to assess whether this result is adequate or whether the elementary failure rates or even a system structure are to be changed in order to get a more appropriate result.

This kind of failure rate (probability of failure) assessment provides quantified information about the importance of every failure and so about the importance (criticality) of components considered and about their influence on the resulting top failure rate. Consequently, these components should then attract attention of the team of their creators and evoke application of more "accurate" computational and/or experimental methods.

4. Transformation of External Loads to Critical Components

Practically all operating mechanical systems are exposed to operating loads of a random nature, representing in general random processes of acceleration, velocity, pressure, deformation, strain, temperature, etc. Even if we were able to assess their "source", i.e. their typical use (operating) conditions, we still face the problem of how to transform the external load conditions of the whole structure (say, road uneveness causing car wheel movements) to the critical locations.

Depending on the life cycle phase of a structure and many other circumstances (means available, time, qualification of personnel, certain habits, etc.) this can be done in two separate or supplementing ways: computationally or experimentally.

4.1. TRANSFORMATION EXPLOITING FREQUENCY RESPONSE FUNCTIONS

A mechanical system is represented by a theoretical dynamic model *multiple input x(t) - one output y(t)* (Fig. 5) and is characterized by E constant parameter linear dynamic systems described by the frequency response functions $H_m(if)$. All processes $x_i(t)$ $(i = 1,2,..., E)$ and y(t) are considered to be stationary random and may represent any physical quantity of interest. In reliability problems the input processes are, as a rule, chosen according to their physical meaning and the possibility of measuring them. Thus they represent the actual external excitation (e.g., car wheel displacement) that may be relatively easily measured and changed, if needed (riding on a smooth or rough surface road, at high or low speed along a curved or straight line, etc.). The output processes usually represent strains/stresses at critical locations.

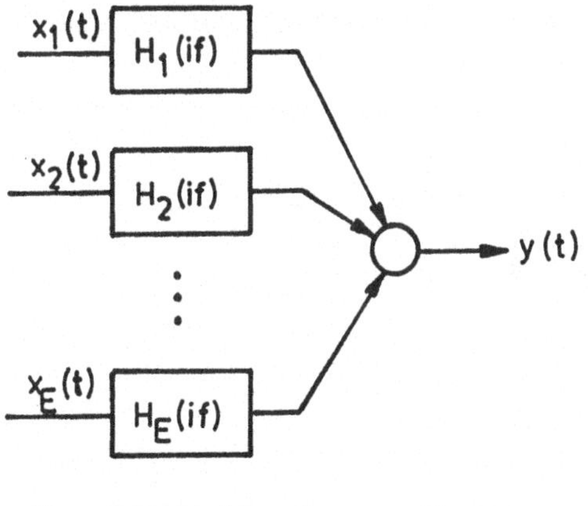

Figure 5. Model of dynamic system with multiple inputs $x_i(t)$ and one output y(t).

Knowing the input processes $x_i(t)$ and the frequency response functions $H_i(if)$($i = \sqrt{-1}$) we can determine the power spectral density $S_{yy}(f)$ of the output process in a critical location [3]. Moreover, providing all input processes are Gaussian then the output process is also Gaussian (this is, fortunately, a very frequent case because for non-Gaussian input processes no possibility to determine the probability density function of the output process exists).

Having known the power spectral density $S_{yy}(f)$ of the output process y(t) and its Gaussian probability density function, one can easily reconstruct the corresponding random output process using an appropriate algorithm [4].

Although at first glance this approach seems to be straightforward and easily realizable at personal computers, it contains so many sources of possible errors (e.g., a questionable assumption of the complex mechanical system linearity) that its application is restricted to theoretical studies of behaviour of dynamic systems. Nevertheless, in combination with experimental measurements it may yield useful information especially during innovations of an existing structure.

4.2. EXPERIMENTAL DETERMINATION OF CRITICAL LOADS

Similarly as for the input processes the output processes can also have any physical meaning. If the operating measurements are intended to be an initial step in reliability estimation and specifically in dimensioning against fatigue, then we are concerned with strain and deformation.

Strain measurements are especially easy thanks to the low price of resistance strain gauges, their ease of maintenance during long- time operations in dusty and humid environments. Moreover, in the life assessment applications they are the only means capable of measuring the cyclic material properties and cyclic plastic deformation indispensable for the fatigue life estimation.

In some rather simple cases when there is no doubt about the principal strain direction a single strain gauge can be used. But as usually this is not the case and so a strain gauge rosette is to be used composed of three strain gauges inclined to each other at, say, 60°. The mutually perpendicular strains ε_x and ε_y are then obtained from the well known equations

$$\varepsilon = \frac{\varepsilon_x + \varepsilon_y}{2} + \frac{\varepsilon_x - \varepsilon_y}{2}\cos 2\alpha + \frac{\gamma}{2}\sin 2\alpha$$

into which for ε the measured ε_{60} (for $\alpha = 60°$), ε_{-60} ($\alpha = -60°$) and ε_0 ($\alpha = 0°$) values are inserted.

After obvious arrangements we get

$$\varepsilon_x = \varepsilon_0 \ , \ \varepsilon_y = \frac{2}{3}\left(\varepsilon_{60} + \varepsilon_{-60} - 0.5\varepsilon_0\right) \ , \ \gamma = \frac{1}{\sqrt{3}}\left(\varepsilon_{60} - \varepsilon_{-60}\right)$$

and so the principal strains ε_1 and ε_2 are obtained from equation

$$\varepsilon_{1,2} = \frac{\varepsilon_0 + \varepsilon_{60} + \varepsilon_{-60}}{3} \pm \frac{1}{3}\sqrt{\left(2\varepsilon_0 - \varepsilon_{60} - \varepsilon_{-60}\right)^2 + 3\left(\varepsilon_{60} - \varepsilon_{-60}\right)^2}.$$

The maximum principal strain ε_1 direction determined by the angle φ is obtained from equations

$$\sin 2\varphi = \frac{\sqrt{3}\left(\varepsilon_{60} - \varepsilon_{-60}\right)}{\sqrt{\left(2\varepsilon_0 - \varepsilon_{60} - \varepsilon_{-60}\right)^2 + 3\left(\varepsilon_{60} - \varepsilon_{-60}\right)^2}}$$

and

$$\cos 2\varphi = \frac{2\varepsilon_0 - \varepsilon_{60} - \varepsilon_{-60}}{\sqrt{\left(2\varepsilon_0 - \varepsilon_{60} - \varepsilon_{-60}\right)^2 + 3\left(\varepsilon_{60} - \varepsilon_{-60}\right)^2}}.$$

4.2.1. Manson - Coffin Approach

When the fatigue life estimation is based on the Manson - Coffin curve (see further) then this information should be sufficient to characterize the operating loads with a word of causion, however:

- from the previous equations it is easily seen that the principal strains randomly change not only in their magnitudes but also in their directions. Consequently, the fatigue damage which is clearly correlated with the maximum stress also changes its slip plane. Unfortunately, information about the fatigue damage accumulation under this condition is not sufficient for the time being;
- the situation is relatively simpler when only one strain is dominant and even more, when it does not substantially change its direction;
- according to the results of the surface strain measurements it is often impossible to decide whether a given component is loaded in tension - compression, bending or torsion (in practice as usually all modes of loading are present, anyway). Relating the operating stresses to the tensile - compressive fatigue life curve means in such cases that the endurance estimation is on the safe side;
- strain gauge rosettes can be located on a smooth surface only, as usually in the close vicinity of a critical spot (notch, weld) into which the strain (stress) processes must be transformed (Fig. 6). It can be done in various ways (see further).

4.2.2. Wöhler Approach

This approach to the fatigue life estimation relies on stresses and the S/N (Wöhler) curve and so the measured strain processes are to be transformed to the stress processes $\sigma_{1,2}$. This can be done, for example, using the equations of elasticity

$$\sigma_{1,2} = \frac{m^2 E}{m^2 - 1}\left(\varepsilon_{1,2} + \frac{\varepsilon_{2,1}}{m}\right),$$

where E is the modulus of elasticity and m is the Poisson ratio. Their directions coincide with the principal strain directions.

For random loading it may be more appropriate to base this transformation on the cyclic stress - strain curve expressed as

$$\varepsilon = \frac{\sigma}{E} + \left(\frac{\sigma}{k}\right)^{\frac{1}{n}} \tag{1}$$

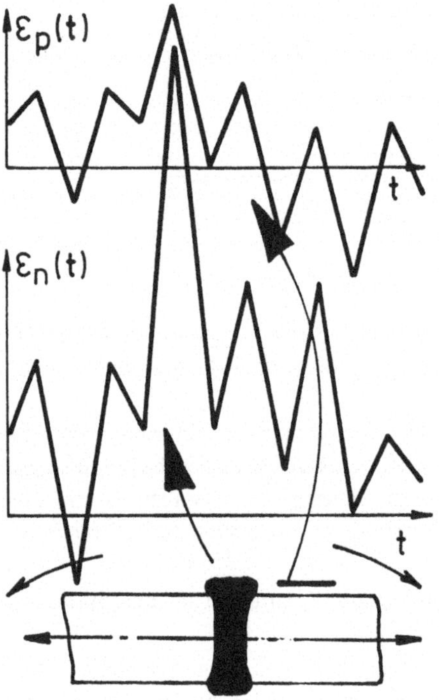

Figure 6. Relation between measured random process on smooth surface and its pair in weld.

which results in equation

$$\varepsilon_{1,2} = \frac{1}{E}\left(\sigma_{1,2} + \frac{\sigma_{2,1}}{m}\right) + k^{-\frac{1}{n}}\left(\sigma_1^{1/n} - \frac{\sigma_1^{1/n}}{m}\right)$$

from which the principal stresses $\sigma_{1,2}$ can be computed. Here k stands for the cyclic strength coefficient and n is the cyclic strain hardening exponent.

Some investigators suggest that the problem of random changes of the principal stress direction could be overcome by computing a reduced stress, say, according to the HMH hypothesis. According to our view this has no justification, however.

5. Fatigue Life Assessment of Critical Locations

Suppose we have measured or in some way estimated strains (stress) at critical components and want to assess the corresponding fatigue endurance (life-time). This can be carried out either experimentally or computationally.

5.1. EXPERIMENTAL FATIGUE LIFE ASSESSMENT

Experimental fatigue life assessment of critical components or their locations presupposes that a measured operating process can be conveniently transformed to become the driving signal of a loading system.

Modern electrohydraulic loading simulators (such as MTS or Schenck) allow to simulate practically any operating situation. In practice it results in two approaches: *full scale testing* when the whole structure is loaded in various locations, and *component (specimen) testing* when the component/material specimen operating conditions are reproduced.

Whereas the full scale tests seem to be *à priori* "more accurate" because the whole structure behaviour is reproduced including all non-linearities, clearences, dry friction, assembling drawbacks, dependent failures, etc., it requires for the structure to be completely finished before testing, it is usually more expensive than the component testing, it requires a complicated testing system, etc. Nevertheless, in many cases it justifies all exerted effort and may serve as a final non-destructive commissioning proof of the attained reliability.

The component/specimen tests are usually simpler and their philosophy is based on modelling the critical spot behaviour (in our cases processes of strain or stress).

Both these procedures contain one common problem: the choice of simulated process characteristics as the criterion of *quality of simulation*. The ideal universal method should guarantee the identity of all available characteristics which within the *correlation theory of random processes* would mean the identity of power spectral density and probability density function of amplitudes. Within the *counting method language*, familiar to the fatigue community, the occurrences of amplitudes, peaks, frequencies, ranges and means should also be respected. Other views may strenghten the Markovian chain properties. So in practice we have to choose a certain criterion of equivalence which is supposed to satisfy, in a given specific case, all the requirements imposed by equivalence of the operating and simulated processes. The whole situation would involve no problems if a completely rigorous theory of fatigue were available. Unfortunately, this is not the case and so different opinions on the relative importance of various characteristics of the simulated process must be taken into account.

Random Process Simulation. Basic possibilities of operating process simulations are illustrated in Fig. 7. Generally we can simulate

- a block of harmonic cycles produced, as a rule, by the Rain Flow Method (RFM),
- a probability density function and power spectral density in the frame of the correlation theory, and
- a transition probability density matrix between two or three successive process ordinates.

In each category there are varieties of the simulating algorithms [4] and their choice is controlled by various views: phase of existence of a structure/component (compare design phase with prototype testing), means available, testing equipment capability, fatigue

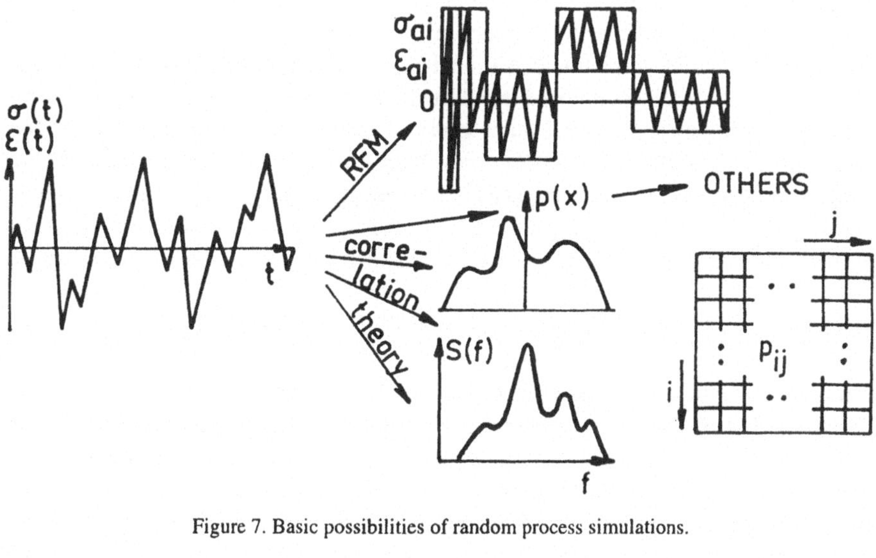

Figure 7. Basic possibilities of random process simulations.

damage accumulation hypothesis preferred, technical level of personnel, testing time available, etc.

A wide verification of the quality of simulation algorithms performed in the author's Institute revealed [9] that the best simulation method **from the point of view of the resulting fatigue life** is the simulation based on the second order Markovian matrices (containing transitions between three successive process ordinates). This approach can be easily applied also to strongly non-stationary processes (with suddenly changing means, for example) that are difficult to be analysed otherwise.

Uniaxial and Multiaxial Loading. Another problem which during a few last years has attracted attention of the research community is the multiaxial loading. Because the uniaxial operating stress is rather rare, it is natural that this kind of tests is motivated by the need to perform the reliability assessment of the most critical components.

Whereas for static loading the two- or three-axial stress states are sufficiently well described by some strength hypothesis (for example, Huber-Mises-Hencky for metallic materials), in case of random multiaxial loading the situation is far from being clear. The tests are not easy to be realized and the results may above all depend on correlation between the principal stresses. Nevertheless, up to now one qualitative conclusion could be possibly formulated: in static cases the acceptance of a multiaxial stress state and application of a strength hypothesis yield results which are on the safe side compared with the uniaxial maximum stress assessment. This does not seem to be true for repeated loading, however, where the multiaxial stress state probably accelerates the corresponding fatigue damage.

Control Mode. Last but not least point worth mentioning here is the control mode of testing machines. In the full scale reliability (fatigue) tests this problem is more or less clear as the external excitation is always either force or stroke (deformation). In the component/ material tests the control mode can be either stroke (for large deformations) or force or strain. Technically this makes no difference but the results obtained for each mode usually differ and may be hardly comparable. Although in the fatigue applications the strain control mode is for various reasons preferred (resulting in the Manson - Coffin curve, see further), components with welds, holes and other kinds of notches are usually tested under force controlled mode (resulting in the S/N curve). Moreover, one must be aware that the load control mode may generate phenomena which cannot principally occur under strain control (for example, cyclic creep, a discontinuity in the S/N curve, etc.). So before realizing any test its aim and physical nature of the problem investigated should be always consulted.

5.2. COMPUTATIONAL ESTIMATION OF FATIGUE RESISTANCE

The theoretical estimation of fatigue endurance has its sense not only in the stage of preliminary dimensioning of a structural component, but also during the following more advanced stages when the input information about operating loads are more complete, and so it may become indispensable to check and eventually re-work the previous proposal with respect to the new input conditions. Moreover, economic, time and exploitation restrictions may not permit the verification of every component experimetally under laboratory conditions, although the endurance determined in this way is obviously the closest approximation to the true operating endurance.

Among various requirements of the computational methods, the following ones should be emphasized:

- inclusion of those characteristics which best describe the material response to random loading;
- consideration of operating load parameters; and
- a simple, non-problematic practical application.

Further we shall suppose that all necessary material properties are available. This means the statistically evaluated Manson - Coffin curve in the form

$$\varepsilon_a = \frac{\sigma_f' - \sigma_m}{E}(2N_f)^b + \varepsilon_f'(2N_f)^c, \tag{2}$$

S/N (Wöhler) curve in the form

$$(\sigma_a - \sigma_c)^m N_f = A \qquad \text{and} \tag{3}$$

cyclic stress-strain curve in the form of equation (1), where ε_a and σ_a are the strain and stress amplitudes, resp., E is the modulus of elasticity, σ_f' and ε_f' stand for the fatigue strength and fatigue ductility coefficients, resp., b and c are the fatigue strength and fatigue ductility exponents, resp., σ_m is a mean stress, m and A are material constants, and σ_c is the stress fatigue limit.

According to fatigue damage accumulation conditions as well as mathematical formulation, various fatigue damage accumulation hypotheses have been proposed bringing to mutual relations the operating random process characteristics with the relevant material characteristics, and yielding the operating fatigue life.

5.2.1. Deterministic Approach

Macroblock of Harmonic Cycles. Suppose that the operating strain/stress random process has been analyzed by the RFM, for example, and so transformed to the macroblock of harmonic cycles with various amplitudes ε_a or σ_a and possibly their corresponding mean levels (as schematically shown on the top of Fig. 7 or in Fig. 8).

The oldest, simplest and so far commonest hypothesis is:

(a) *Palmgren-Miner linear damage rule* (PM) according to which the damage caused by N_i stress cycles equals

$$D_{N_i} = \frac{N_i}{N_{fi}}. \tag{4}$$

The limit condition for the fatigue fracture is then expressed by the relation

$$\sum_{(i)} D_{N_i} = 1, \tag{5}$$

where the limit number of cycles N_{fi} at the ith stress level σ_{ai} is taken from the median S/N curve (3) for the probability of failure 50%. Experiments prove, however, that more accurate results are obtained if the load history is represented by the strain process and consequently N_{fi} then corresponds to the Manson - Coffin curve (2). This practice is also recommended by the SAE Fatigue Design and Evaluation Committee [6].

There are many other hypotheses which fairly well consider various other material and load parameters. Let us mention at least three of them which are often applied in practice because in some cases they are advantageous compared with the PM rule:

(b) *Corten - Dolan hypothesis given as*

$$D_{N_i} = \frac{N_i}{N_{f\,min}} \left(\frac{\sigma_{ai}}{\sigma_{a\,max}} \right)^d \tag{6}$$

with the same limit condition (5) as for the PM rule. Here $\sigma_{a\,max}$ is the maximum stress amplitude in the macroblock for which the number of cycles to fracture is $N_{f\,min}$, and d is a material parameter.

(c) *Serensen - Kogaev hypothesis* [5] has the same form as the PM rule (4) but with the limit condition

$$\sum_{(i)} D_{N_i} = a\,,$$

where $a = (\sigma_{a\,max}\xi - K\sigma_c)/(\sigma_{a\,max} - K\sigma_c)$ for $a \geq 0.2$, otherwise $a = 0.2$; $K = 0.5$ to 0.7 is a correction factor taking into account the influence of stresses below the fatigue limit σ_c, ξ is a shadowed area below the macroblock contour in Fig. 8 plotted for the relative coordinate $(\sigma_{ai}/\sigma_{a\,max})$ versus $N_i/\sum N_i = N_i/N_m$, where $N_m = \sum N_i$ is the total number of cycles in the macroblock.

Naturally, this hypothesis is more accurate than the previous one because it contains more experimental parameters.

(d) *Kliman hypothesis* [7] for
- *stress controlled loading* (macroblock of stress amplitudes with r levels) in the form

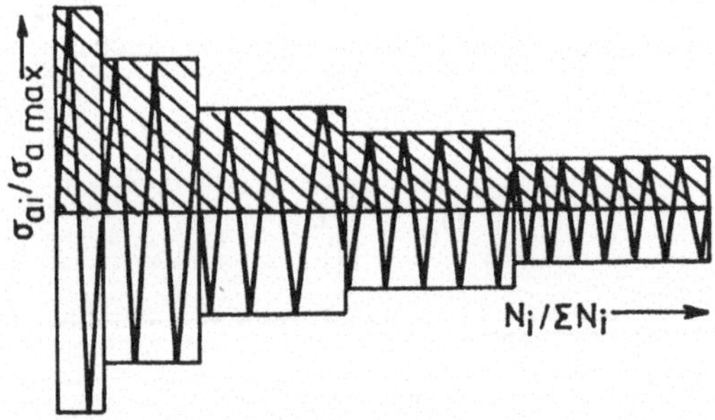

Figure 8. Organized macroblock of harmonic cycles in relative coordinates.

$$D=\frac{1}{N_{f\min}}\sum_{i=1}^{r}N_i\left(\frac{\sigma_{ai}}{\sigma_{a\max}}\right)^{(1+n)/n}$$

or for

- *strain controlled loading* (macroblock of strain amplitudes with r levels) in the form

$$D=\frac{1}{N_{f\min}}\sum_{i=1}^{r}N_i\left(\frac{\varepsilon_{api}}{\varepsilon_{ap\max}}\right)^{1+n}$$

with the same limit condition (number of macroblock repetitions) as for the PM rule (5).

It is worth mentioning that the Kliman's approach based on continuous counting of the hysteresis energy reflects both the influence of the arrangement of amplitudes within the macroblock as well as the amplitudes below the fatigue limit on the resulting fatigue life and so it is clearly more descriptive than, for example, the PM rule.

Continuous Process. If the random process analysis is carried out in the framework of the correlation theory, a different type of hypotheses is used. Unfortunately, most of them are so sophisticated that they are difficult to be used in design practice. One directly applicable is the Raikher hypothesis according to which the fatigue life T in seconds is computed from the relation

$$T=\frac{A}{L(m)\left[2\pi\int_{(f)}S(f)f^{2/m}df\right]^{m/2}}$$

where m and A are parameters of the S/N curve (3), S(f) is the power spectral density, $L(m)=2^{m/2}\Gamma\left(\frac{m+2}{m}\right)$ and $\Gamma\left(\frac{m+2}{m}\right)$ is a gamma function. The integral in the denominator is taken over the whole frequency range.

There is probably no good reason for recommending any of these or other hypotheses for fatigue life estimation. Intuitively one perhaps feels that the more experimentally determined parameters a hypothesis contains, the better the approximation of experimental results should be, but then this usually turns out to be the most difficult, expensive and time-consuming method. In other words, the more complicated hypotheses require more data to be analysed before their parameters can be reliably determined. If these data are not available or are doubtful then the expected "accuracy" of a multiparametric model is problematic. This is why a chosen hypothesis should not be more complicated than is necessary for a given task, taking into account, for example, the design phase and corresponding information about operating loads and functional properties. During the preliminary concept, design and development phase, when neither operating loads nor final design form or material are definitely known, it is sufficient to use the PM rule and the material S/N curve. Prototype tests and/or its reliability verification offer more information about measured operating loads and verified material properties and so the fatigue life can be estimated according to more descriptive relations. Our long experience suggests that in any case it is useful to apply all the fatigue damage accumulation hypotheses mentioned above (and may be even some others); this yields a certain "scatter" and "mean value" of the fatigue life, and simple calculator programmes make it an easy job.

5.2.2. Probabilistic Approach

Random character of use conditions, operating loads and material properties suggest that the reliability (fatigue life) assessments should be also expressed in a random way, i.e. in the probabilistic sense. Thus having known the required probability of occurrence of the elementary failures (or reliability or failure rate, resp.) our task is to design a given component satisfying this condition. Two possible ways are feasible here.

Interference Theory of Reliability Approach. Estimation of the component reliability emanates from equation [3,10]

$$R = \int_{-\infty}^{\infty} f_A(a)\left[1 - F_S(s)\right] da \tag{7}$$

illustrated in Fig. 9.

Its use is straightforward, it only requires the knowledge of a probability density function of operating loads $f_A(a)$ or its distribution function $F_A(a)$, and a probability density function of strength resistance $f_S(s)$ (for example, the fatigue limit) or its distribution function $F_S(s)$, resp.

Application of equation (7) to fatigue problems looks to be questionable, however. The main reason is that it is obviously not sufficient to describe the material and operating load

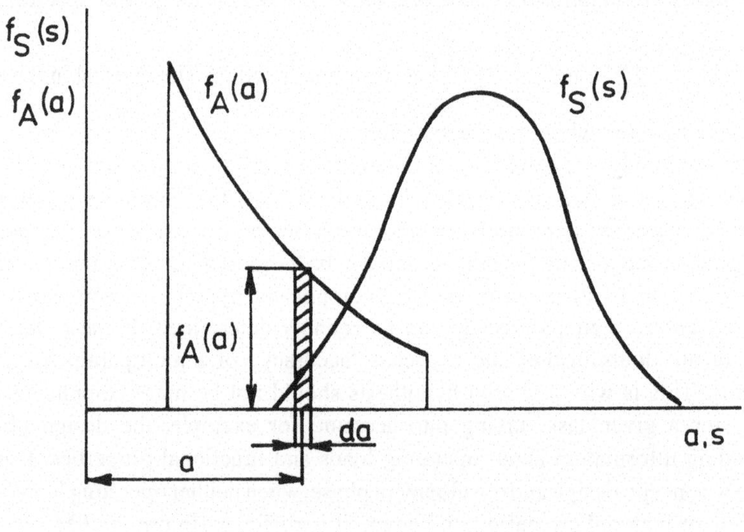

Figure 9. Reliability estimation using ITR approach.

properties by their probability density functions only. Although in theory even more variables may be involved [10], it requires adoption of further assumptions and the computation becomes problematic. Nevertheless, this area of research seems to be promising.

Fatigue Life Distribution Function. Estimation of the fatigue life distribution function looks to be more appropriate as it considers a random character of operating loads, random character of the fatigue curve with its constant probability intervals and also certain randomness of results obtained from the fatigue damage accumulation hypotheses exploited. This approach has been elaborated by Kliman [11] for the median 50% and extreme 2.5 and 97.5% values of the input parameters, resp., stationary and non-stationary processes and various fatigue damage accumulation hypotheses. The random process is for this purpose analysed by the RFM and the result is in the form of the fatigue life distribution functions, also with their probability intervals.

5.2.3. Stress/strain Concentration Effects

As mentioned before, practicaly all operating processes are measured at plain locations, as usually in the close vicinity of notches (holes, welds, grooves, etc.) whose roots are the most critical spots for the fatigue crack initiation. Taking into account that the Manson - Coffin and S/N curves also usually apply to plain material specimens it is indispensable either

- to transfer the operating random process to the notch root using some stress/strain concentration factor and then to estimate the fatigue life for this process and material S/N or Manson - Coffin curve; or
- to use the S/N curve for a component with a given notch understanding that it does not actually characterize the material but component properties (application of the Manson - Coffin curve is also possible but technically rather problematic because it is not easy to measure peak strains in notch roots); or
- to use the Manson - Coffin curve for a notched component as proposed by Lukáš [5] in the form

$$\left(\sigma_{an}\varepsilon_{an}E\right)^{1/2} = \frac{1}{\alpha_f}\left[\sigma_f^{'}\left(2N_f\right)^{2b} + E\sigma_f^{'}\varepsilon_f^{'}\left(2N_f\right)^{b+c}\right]$$

where σ_{an} and ε_{an} are plain stress and strain amplitudes, resp., and α_f is the fatigue notch factor.

Transformation of the plain strain process to its “notch” pair can be done in various ways:

(a) the plain strain process is multiplied by the theoretical strain concentration factor α and related to the Manson - Coffin material curve. Although the notch root is strained in at least two directions, one strain process is usually dominant and so the assumption of unidirectional loading is justified here; the same argument applies to stresses. This approach works in static cases but it probably overestimates the random load fatigue effect and so the fatigue lives obtained are on the safe side.

(b) the plain strain (stress) random process is multiplied by the "cyclic" stress (strain) concentration factor α_σ (α_ε) derived by Polák [5] using three approximate procedures, viz. the Hardrath - Ohman concept, Neuber concept and equivalent energy concept, resp. Comparison with the measurement of the notch root strain and with the finite element calculations shows that the best agreement is given by the last approach in which the relative stress (strain) concentration factors $x_\sigma = \alpha_\sigma/\alpha$ ($x_\varepsilon = \alpha_\varepsilon/\alpha$) are computed from equations

$$A_2 \alpha^{\frac{1}{n}-1} x_\sigma^{\frac{1}{n}+1} + x_\sigma^2 - (1 + A_2) = 0,$$

$$x_\varepsilon = \frac{(1-n)x_\sigma + (1+n)(1+A_2)\frac{1}{x_\sigma}}{2 + (1+n)A_2},$$

$$A_2 = \frac{2}{1+n} A_1 \, , \; A_1 = \frac{E \sigma_n^{\frac{1}{n}-1}}{K^{1/n}},$$

where σ_n is the nominal stress in the smallest cross section, α is the theoretical stress concentration factor and n, k are parameters of the cyclic stress-strain curve (1).

A rather wide verification of all three procedures for random processes with various statistical properties (including non-stationarities) revealed, however [8], that based on the fatigue life results all procedures mentioned hitherto are in the range of the experimental scatter equivalent irrespective whether the a_s or a_e are applied to

- the plain random process ordinates and the resulting process is further analysed by the RFM, or whether
- the transformation is based on the cyclic stress - strain curve (1) or on the static stress - strain curve.

One fact is worth mentioning here, however: the non-linear transformation of plain random processes to their notched pairs may cause that the originally stationary process may become non-stationary, say, with a step-wise changing mean level (see Fig. 6). Although, generally speaking, in a long run the influence of non-stationarities on the fatigue life may not be so pronounced [9] (using the RFM) it is wise to examine their local effects (for example, sudden jumps into the elastic - plastic range).

ACKNOWLEDGEMENT

This research has been partially supported by the Slovak Academy of Sciences under grant 2/1136/94 which is gratefully acknowledged.

6. References

[1] Farnworth, N. (1989) How to make safer products by the management of design *Product Liability* International, **November**,166-170.

[2] IEC 1025 (1990) *Fault Tree Analysis.*

[3] Bílý, M. (1989) *Dependability of Mechanical Systems*, Elsevier, Amsterdam.

[4] Čačko, J., Bílý, M. and Bukoveczky, J. (1988) *Random Processes, Measurement, Analysis and Simulation*, Elsevier, Amsterdam.

[5] *Cyclic Deformation and Fatigue of Metals (1993)*, M. Bílý (ed.), Elsevier, Amsterdam.

[6] *SAE Information Report J 1099 (1979)* SAE Handbook.

[7] Kliman, V. (1985) Fatigue life estimation under random loading using the energy criterion, *Int. J. Fatigue* 7,39-44.

[8] Prohácka, J. and Bílý M. (1992) Notch behaviour under random loading, in M.H.Aliabadi, D.J.Cartwright and H.Nisitani (eds.), *Localized Damage, Vol. I: Fatigue and Fracture Mechanics*, Elsevier,London,pp.141-155.

[9] Bílý, M., Prohácka, J. and Illés, Š. (1992) Fatigue damage under non-stationary stochastic strain, in S. Sedmak, A. Sedmak and D. Ružič (eds.), *Reliability and Structural Integrity of Advanced Materials, ECF9*, EMAS, Warley,pp.345-350.

[10] Thoft-Christensen,P. and Baker,J.M. (1982) *Structural Reliability Theory and Its Application*, Springer-Verlag, Berlin.

[11] Kliman, V. (1993) Prediction of the random load fatigue life distribution, in J. Solin et al. (eds.), *Fatigue Design, ESIS*, Mech. Engng.Publications, London,pp.241-255.

FITNESS FOR PURPOSE ASSESSMENT OF STRUCTURAL INTEGRITY

J.G. BLAUEL
Fraunhofer-Institut für Werkstoffmechanik
Wöhlerstrasse 11, D-79108 Freiburg, Germany

1. Introduction and overview

Technical components very rarely can be fabricated without deviations from their ideal geometry as laid down in the design. Especially in weldments additional imperfections may occur. They could develop into cracks due to critical service conditions and would then possibly influence in a negative way the usage properties (e.g. life time, load carrying capacity) of the component.

Fitness-for-purpose (FFP) is an integral concept for structural assessment which allows to demonstrate in a quantitative way that a (welded) structure will safely fulfill its usage properties even if imperfections have been detected.

Within this concept (Fig. 1) a loading parameter is calculated for the imperfection in the component under load which is then compared with an experimentally determined value

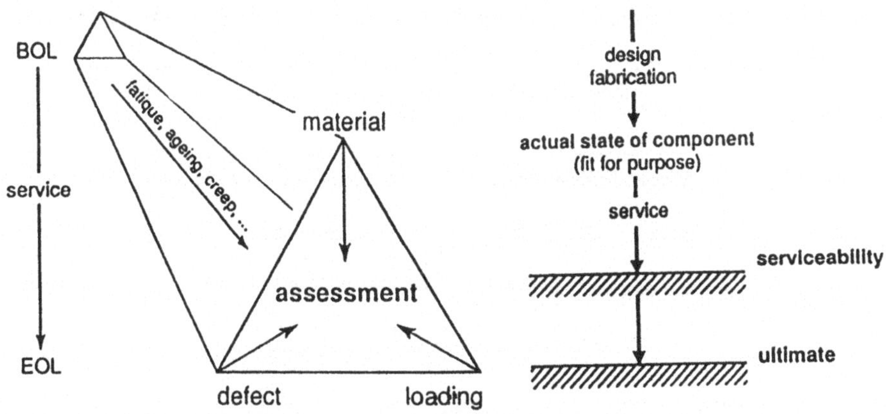

Figure 1. The Fitness-for-Purpose concept.

of material resistance. From that comparison conclusions are drawn whether failure will occur or not. Depending on the begin of life (BOL) conditions concerning fabrication defects, design stresses, material properties and possible changes during service before end of life (EOL) different margins are possible between the actual and the limit states of the structure. Serviceability limit states depending on the kind and purpose of the structure and ultimate limit states resulting in fracture or plastic collapse with different consequences for the structure can be considered.

R.A. Smith (ed.), Reliability Assessment of Cyclically Loaded Engineering Structures, 47–64.

FFP-analyses deliver for specified single cases quantitative results on critical defect sizes, limiting loads, and minimum toughness. On the basis of them decision can be made on design and choice of materials, quality control during fabrication, non-destructive examination (NDE-) strategies during service, safety and availability assessment, repair and exchange actions as well as extension of life.

The main elements of a FFP-analysis, for which detailed and quantitative information is required and which determine the quality and reliability of the results are:

- description of defect status
- calculation of stresses and strains
- material characterization
- structural assessment.

The requirements for the FFP-elements are more detailed and more stringent than an experience based NDE defect categorization, or a standard calculation of permissible stresses, or just normal material testing on tension bars or Charpy specimens. The status of knowledge and examples of ongoing developments for improved structural assessment are described in the following.

2. FFP-element: Status of defects

Structures and components cannot be produced without imperfections. Examples from the fabrication phase are casting-, rolling-, forging-defects, segregations; especially in weldments material inhomogeneities, slag inclusions or lack of fusion can occur, in addition. The catalogues of IIW [1] and DVS [2] differentiate between imperfections of shape and metallurgical inhomogeneities (Fig. 2). Imperfections of a more global kind are misalignement and angular out of shape; local imperfections are called discontinuities and can be of

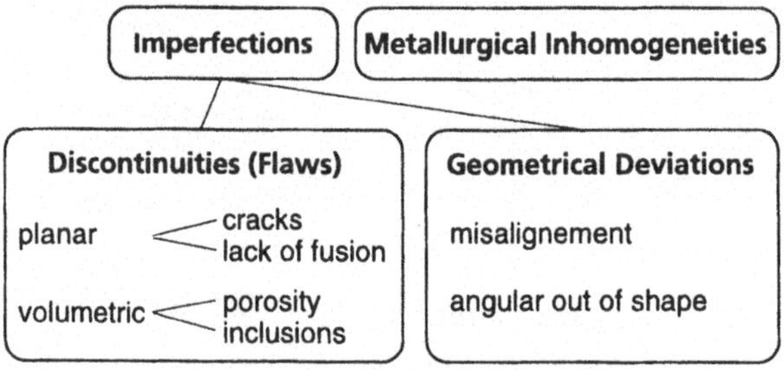

Figure 2. Terminology for weld joint imperfections.

volumetric shape (like pores and inclusions) or can be planar (cracks, lack of fusion). Fabrication discontinuities may grow during service and new defects can be generated, for instance creep cracks, stress corrosion cracks, thermal shock cracks.

Figure 3 shows a practical case: One single fatigue crack has been extended into the wall of a large cast steel tube from a machined notch by alternating internal pressure; using only the information from the final pre-test NDE the burst test load would have been seriously overestimated - unsafely predicted. As in this case, it is not possible without extra information and increased effort to distinguish between different types of imperfections and to follow the possible growth under service loads. In only a few cases fractography and other means allow to reconstruct the different phases of failure behaviour as schematically shown in Fig. 4.

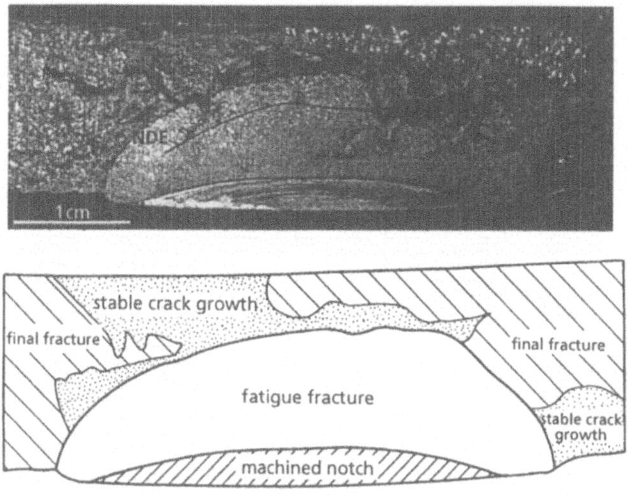

Figure 3. Crack extension in GS-25 tube experiment.

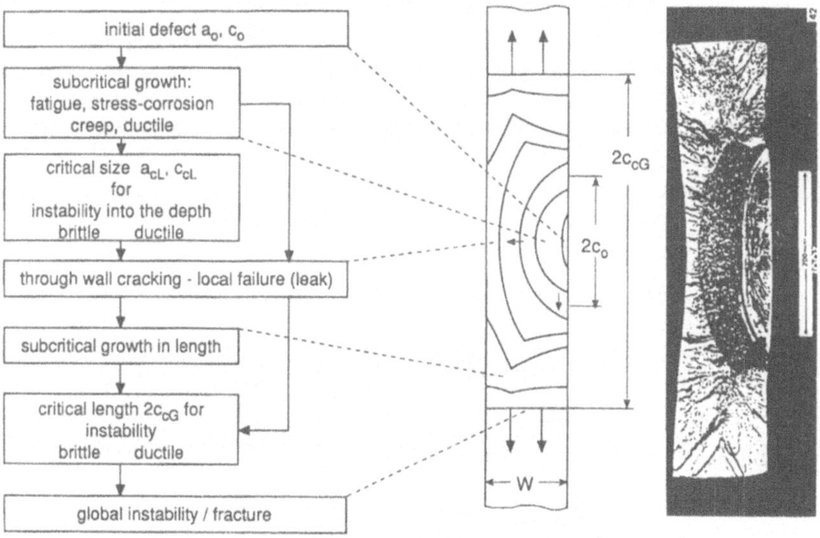

Figure 4. Possible phases of failure of a pressurized tube with axial defect.

To avoid such situations that imperfections degradate the load carrying capacity and/or lifetime of a component present technical rules require that the imperfections are limited in number and size to a level which can be realized in normal fabrication and makes sense in terms of economy. NDE is used to demonstrate that these requirements are fulfilled, and this results in qualitative fabrication quality classes. An example is shown in Fig. 5 from the German DIN 8563 [3] for external imperfections like joint overfill and joint undercut.

Lfd. Nr	Ordnungs-nummer *)	Merkmal ¹)	Bemerkung	Werkstückdicke t	Bewertungsgruppe AS	BS	CS	DS
Äußerer Befund der Nahtoberfläche von ein- oder beidseitig geschweißten Stumpfnähten								
1	2.502	Nahtüberhöhung Δa_1 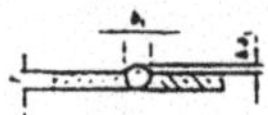	Aus Gründen der Prüfbarkeit kann eine geringere Nahtüberhöhung, notwendig sein.	0.5 bis 3.6	$\Delta a_1 \leq 1 + 0{,}10\,b_1$	$\Delta a_1 \leq 1 + 0{,}15\,b_1$, örtlich geringe Überschreitungen zulässig	$\Delta a_1 \leq 1 + 0{,}20\,b_1$, örtlich Überschreitungen zulässig	$\Delta a_1 \leq 1 + 0{,}25\,b_1$, örtlich Überschreitungen zulässig
				> 3,6	$\Delta a_1 \leq 1 + 0{,}05\,b_1$	$\Delta a_1 \leq 1 + 0{,}10\,b_1$, örtlich geringe Überschreitungen zulässig	$\Delta a_1 \leq 1 + 0{,}15\,b_1$, örtlich Überschreitungen zulässig	
2	2.511	Decklagenunterwölbung Δa_2 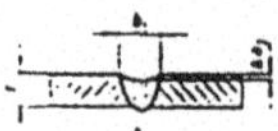 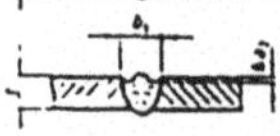	beide Fugen-Längskanten angeschmolzen, Nahtquerschnitt in der Mitte oder an den Rändern nicht erreicht	0,5 bis 3,6	nicht zulässig	flach örtlich zulässig, jedoch $\Delta a_2 \leq 0{,}1 + 0{,}07\,t$	örtlich zulässig, jedoch $\Delta a_2 \leq 0{,}2 + 0{,}05\,t$	flach durchlaufend zulässig, örtlich auch tiefer, jedoch $\Delta a_2 \leq 0{,}3 + 0{,}07\,t$
				> 3,6	nicht zulässig	flach örtlich zulässig, jedoch $\Delta a_2 \leq 0{,}2 + 0{,}02\,t$ $\leq 0{,}5$	örtlich zulässig, jedoch $\Delta a_2 \leq 0{,}2 + 0{,}04\,t$ $\leq 1{,}0$	flach durchlaufend zulässig, örtlich auch tiefer, jedoch $\Delta a_2 \leq 0{,}4 + 0{,}08\,t$ $\leq 1{,}5$

Figure 5. Quality assessment groups for external weld imperfections.

For butt and for fillet joints there are some quantitative requirements if a joint is to be classified better than just the lowest level D, but there is no load parameter in the assessment. When indications of internal imperfections like pores, inclusions, lack of fusion defects have to be assessed again only qualitative requirements are given - only few, isolated, not crack like - which obviously depend on the method and applied efforts in NDE. The German Merkblatt (Published Document) AD HP 5/3 [4] therefore discusses NDE test classes and ends up with criteria for permissible US indications, which are maximum permitted registration lengths for any given sensitivity threshold in db. But also these rules do not include a criterion for crack depth and there is no correlation with the applied loads.

The situation concerning detectability of discontinuities, quality assurance levels and

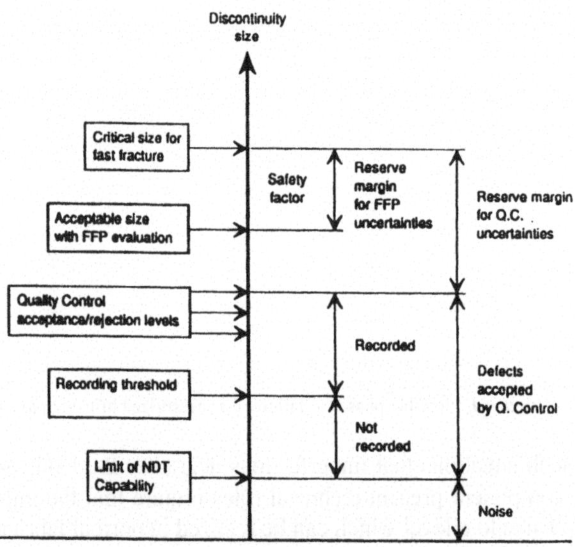

Figure 6. Significance of different sizes of discontinuities in relation to NDE capability, weld quality and fitness-for-purpose.

criticality under load is summarized in Fig. 6 from [5]. The quality assurance criteria based on experience are meant to safely exclude critical situations. Depending on the specific application and safety implications such acceptance/rejection levels can be different for different industries. The scheme indicates that beyond fabrication related limits there can be quite a margin of safety, which can now be quantified by fracture mechanics and FFP-methods if enough is known about the defect status and - see the following chapters - the stresses as well as material properties.

Figure 7 describes as an example the results of a study which has been performed at

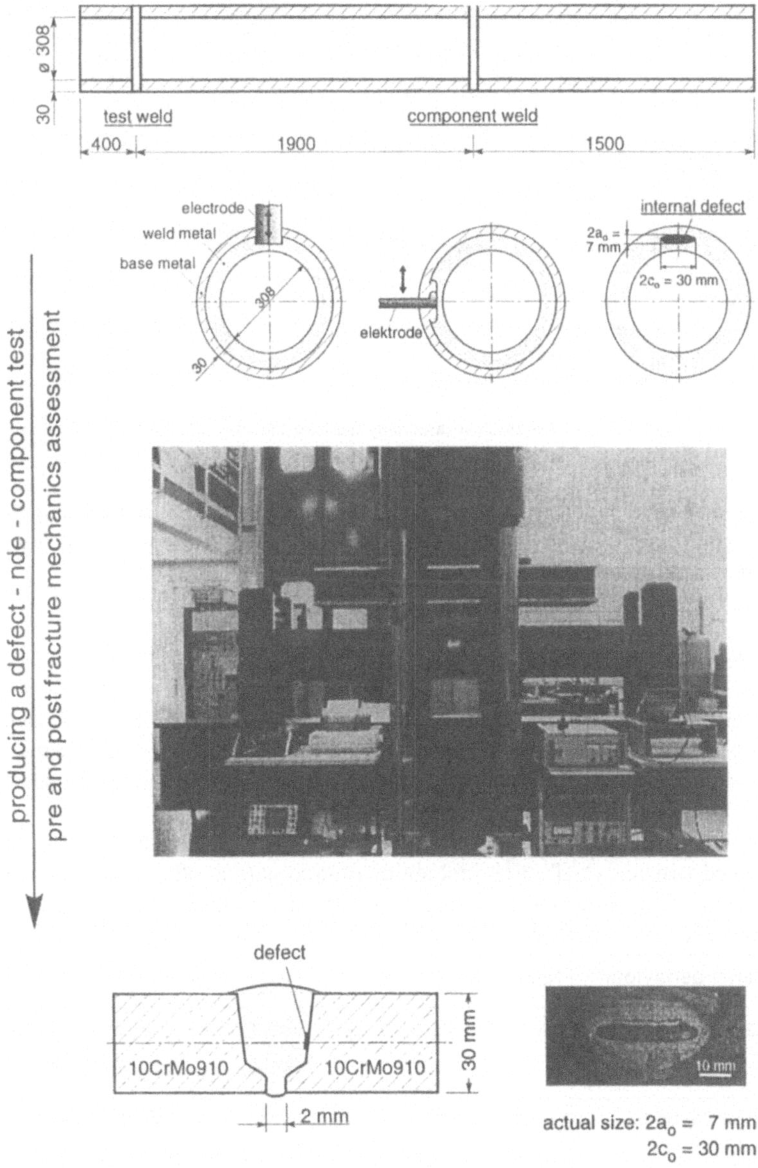

Figure 7. Study into the interrelation of NDE and fracture mechanics assessment.

IWM [6]: Artificial defects have been introduced in circumferential pipe welds by a special spark erosion process; the defects were analysed using standard and advanced US techniques before the pipes were fatigue loaded and burst tested; fracture mechanics assessments were done on the basis of the pre-test NDE and of the actual defect size measured on the opened up fracture surfaces. As the results on defect lengths show in Fig. 8 the standard technique in nearly all cases has underestimated the actual length.

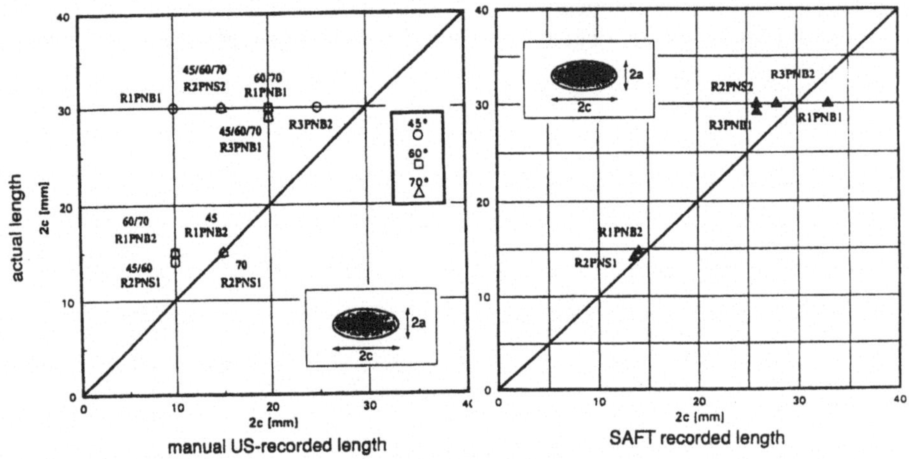

Figure 8. Defect lengths 2c measured by standard and advanced US-techniques compared with actual lengths.

Using an advanced reconstructive technique - synthetic aperture focussing technique (SAFT) - could improve the situation very much and gave very convincing results of the fracture mechanics predictions.

3. FFP-element: Stress-strain analysis

Within the standard design process the stresses are calculated in the component for the relevant geometry and the applied loads. Whether these stresses are permissible or not is decided acc. to the rules by comparing with a material yield strength. In such strength oriented concepts no defects are permitted.

For specific cases with high safety requirements (e.g. in nuclear technology) or in large scale welded structures with high probability of defects (e.g. offshore industry) the traditional design is supplemented by fracture mechanics considerations to analyze the behaviour especially in the neighbourhood of defects (real or hypothetical ones). Models have

material behaviour	failure mode	loading parameter
linear-elastic	brittle fracture, fatigue	stress concentration factor α stress intensity factor K_I
elastic-plastic	plastic collapse ductile fracture	von Mises-stress, -strain crack tip opening displacement

Figure 9. Characteristic features of a FFP-assessment.

been developed to describe the defect as a limited stress concentrator or a crack like discontinuity and the corresponding loading parameters to be selected (see Fig. 9) depend on the deformation behaviour of the material and the type of failure to be conservatively assumed. For standard situations of component geometry (e.g. beam, plate, tube, ...) of defect (through the wall-, semielliptical surface-, corner crack, ...), and of loading (uniaxial tension, bending, internal pressure, ...) solutions are available in handbooks (e.g. [7]) and are stored in PC-programs. They allow first estimates even of more complex cases and to

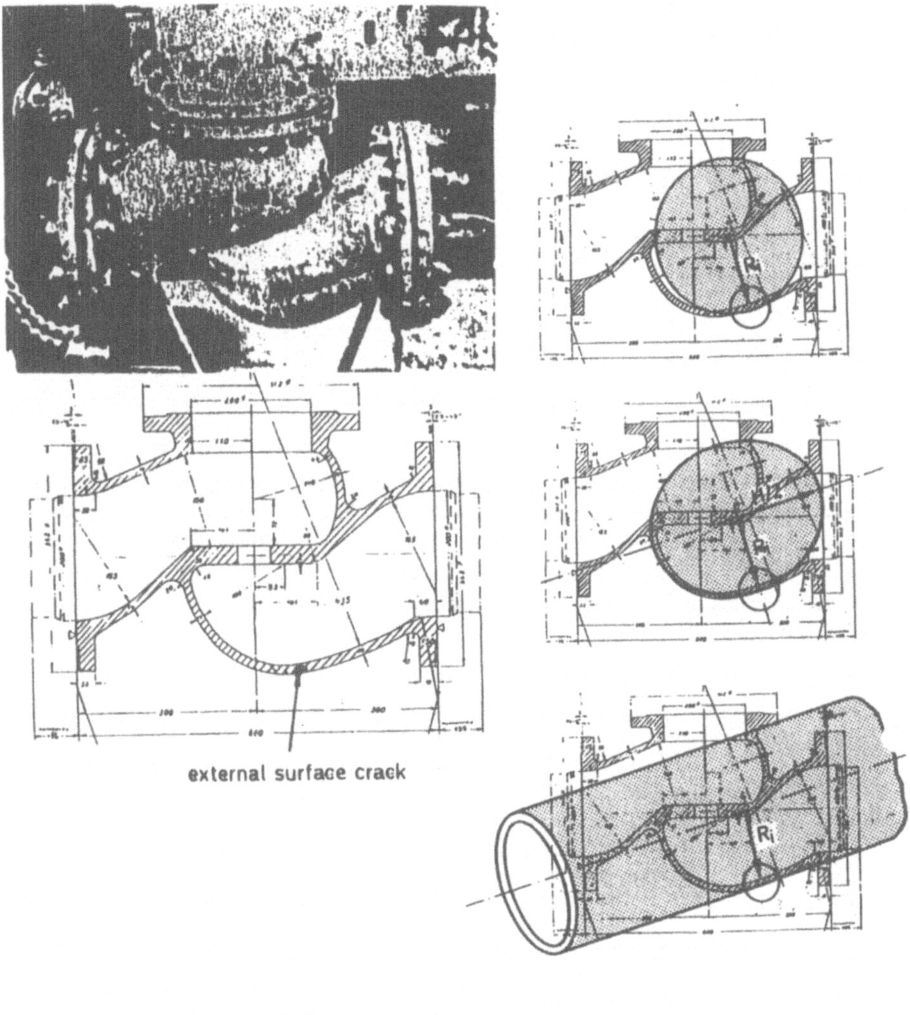

Figure 10. Defect assessment in a cast steel control valve casing.

demonstrate the influence of certain parameters. This is shown in Fig. 10 and 11 for the assessment of possible defects in the wall of a large control valve casing made from cast steel. Depending on the defect size - an elliptical crack being assumed to circumscribe a field of detected small casting defects - the permissible internal pressure is overestimated by the model cylinder and underestimated by the model sphere as compared to the true ellipsoidal shape of the defected area of the vessel.

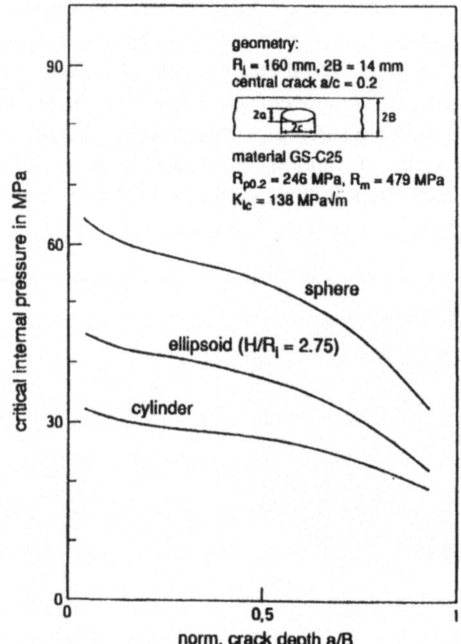

Figure 11. Influence of choice of model for the defected component on failure load.

This parameter study has been performed using the PC defect assessment program IWM-VERB [8]; as in other such programs IWM-VERB is based on either closed form analytical solutions or solutions using the weight functions approach [9] or solutions interpolating between a limited number of verification calculations using FEM [10] or BEM-methods [11]. The examples in Fig. 12 and 13 demonstrate what is possible today with such advanced numerical methods:

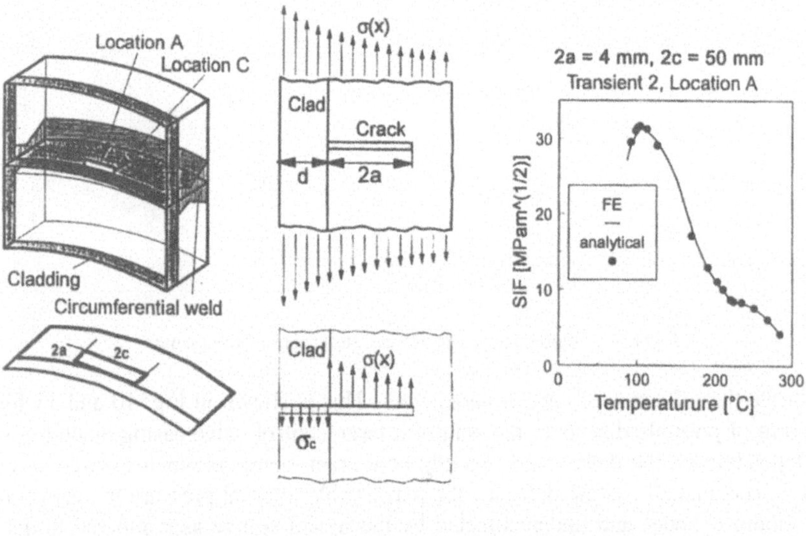

Figure 12. Defect assessment for subclad cracks in RPV-wall.

Figure 12 describes the analysis of a hypothetical near surface crack in a reactor pressure vessel (RPV) at the interface of an austenitic cladding on the ferritic base material under thermal shock and internal pressure loading as occurs in so called loss of coolant accidents. To avoid expensive 3D elastic plastic FE calculations for many different load cases and crack geometries a simplified engineering method has been developed which uses the stresses in the defect free cylinder as an input to calculate stress intensity factors for the crack apex and at the interface through a weight functions approach; the influence of the cladding is taken into account by a "crack-closure force" correction, the parameters of which are fitted to some complete FE-calculations. A very good agreement can be seen between the analytical approximation and the numerical solution [12].

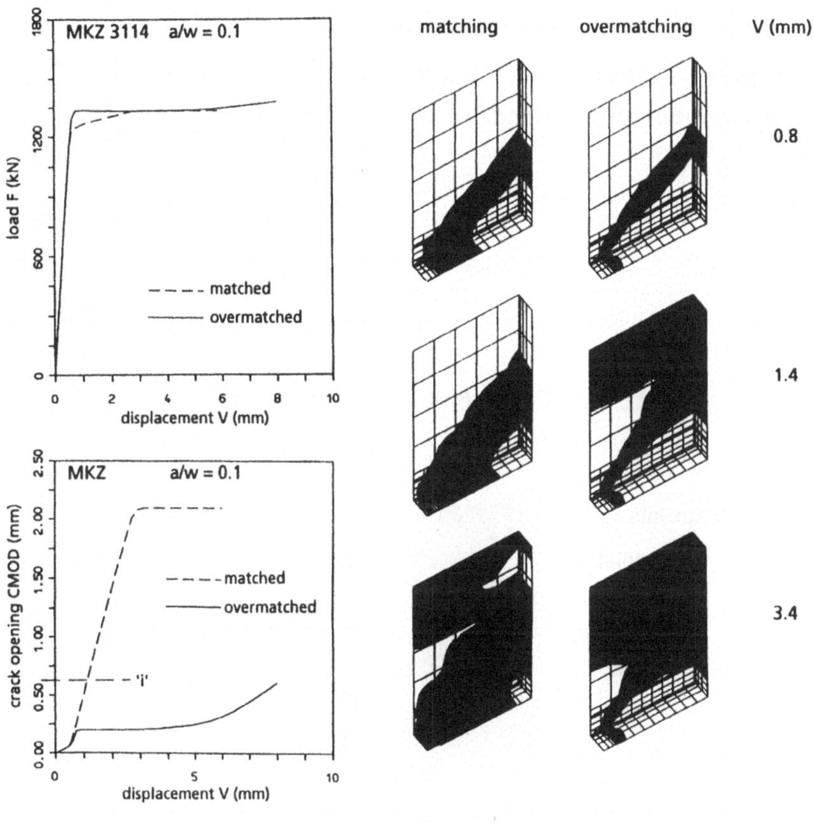

Figure 13. Influence of mismatch effects on defect assessment in weldments.

The second example in Fig. 13 gives results from a FE-analysis for a flat plate under tension with a small (a/W = 0.1) central crack in a transversal weld joint with equal (= matching) or higher (= overmatching) strength properties than the base material [13]. The development of the plastic deformation in the plate with increasing load (opening displacement) clearly shows that in the overmatching case the cracked ligament is protected from loading and - other than in the matching case - all deformation is taken by the lower strength base material. This beneficial effect of overmatching has been found in many experiments; numerical analyses of structural models are able to explain the mismatch effect in a quantitative way.

4. FFP-element: Material characterization

When characterizing the material resistance the different modes of failure and the different limit states as in Fig. 14 have to be taken into account and material specific values have to

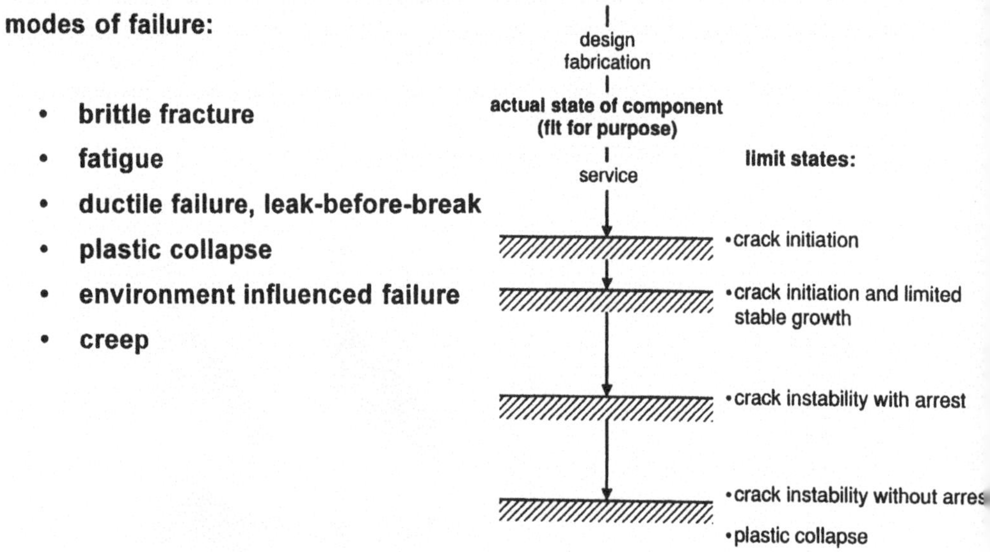

Figure 14. Failure modes and limit states.

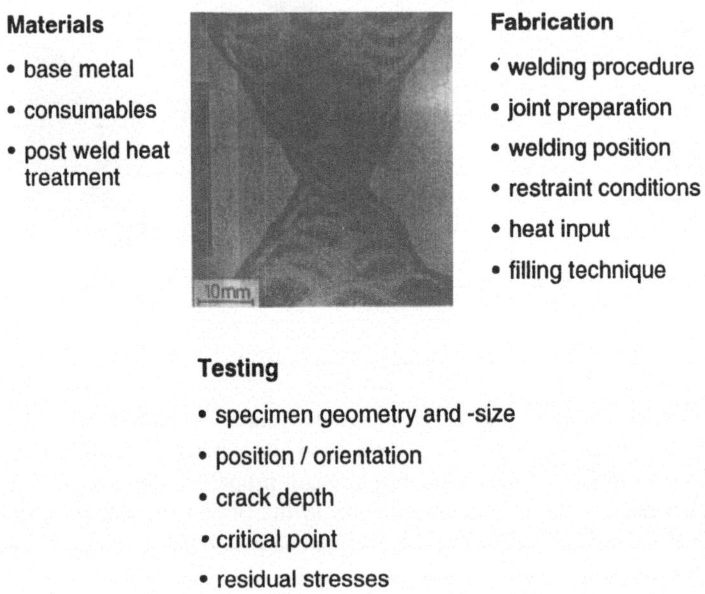

Figure 15. Parameters influencing weld joint toughness CTOD.

be determined from laboratory tests as described in ASTM or other national standards. In tables, e.g. [14], a number of specific and/or minimum values are available; but for quantitative answers testing of the original material is required. The material to be tested must have seen the same steps of fabrication as the structure to be assessed, the relevant dimensions, temperatures, atmospheric conditions and loading rate have to be representative for the service conditions. To simulate a real component situation new application oriented tests had to be developed.

For a complete assessment of the material behaviour in a welded joint, see Fig. 15, the properties of the base metal and the consumables play a role but also many details of the fabrication (what procedure is used with which parameters; how is the joint prepared and so on) and even the conditions of testing (only a few of the parameters are mentioned ...) are important.

The fracture mechanics toughness test according to the CTOD-concept then requires high additional efforts in specimen extraction, careful prefatigueing, and possibly post test metallographic evaluation. This is especially true when the HAZ properties have to be characterized because of possible coarse grain formation. Some details of the test procedure and a typical result are depicted in Fig. 16: The value of the critical HAZ-CTOD at

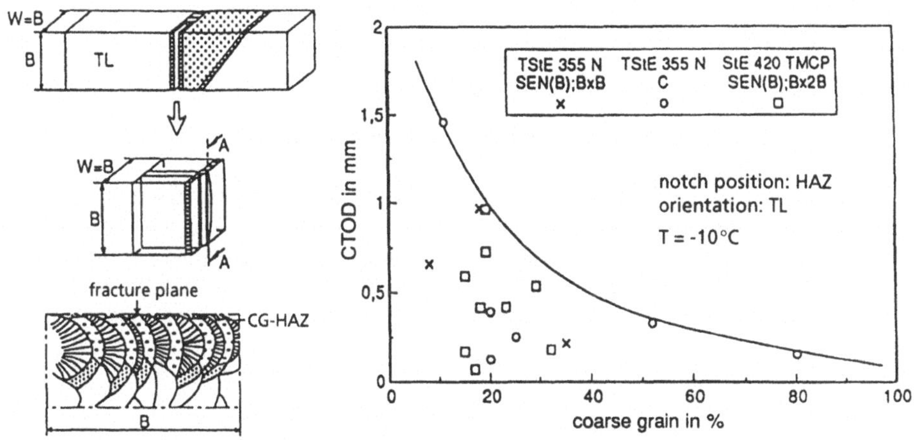

Figure 16. Weld joint characterization by CTOD tests.

brittle fracture initiation for all three steels depends on the amount of coarse grained microstructure collected by the primary fatigue starter crack [15]. Weld thermal simulation on bigger specimens and testing them without the influence of neighbouring zones can deliver a lower bound for HAZ properties. But when transferring to components the local stress state as influenced by the possibly higher strength weld metal has to be taken into account. For a multipass weld metal the measured toughness in terms of critical CTOD depends on the amount of reheated microstructure and the angle that the dendrites of the as deposited material form with the crack plane.

Many of such details like material gradients and local variations in microstructure can be taken into account by a new hybrid experimental/numerical approach called damage mechanics/local approach, the principle of which is illustrated in Fig. 17. Simple small scale tension or Charpy tests are evaluated to deliver a set of material specific basic data which on the one hand side are coupled to the parameters of the microstructure and which on the other side can be used in FE structural models to predict the deformation and failure behaviour of specimens and complex components.

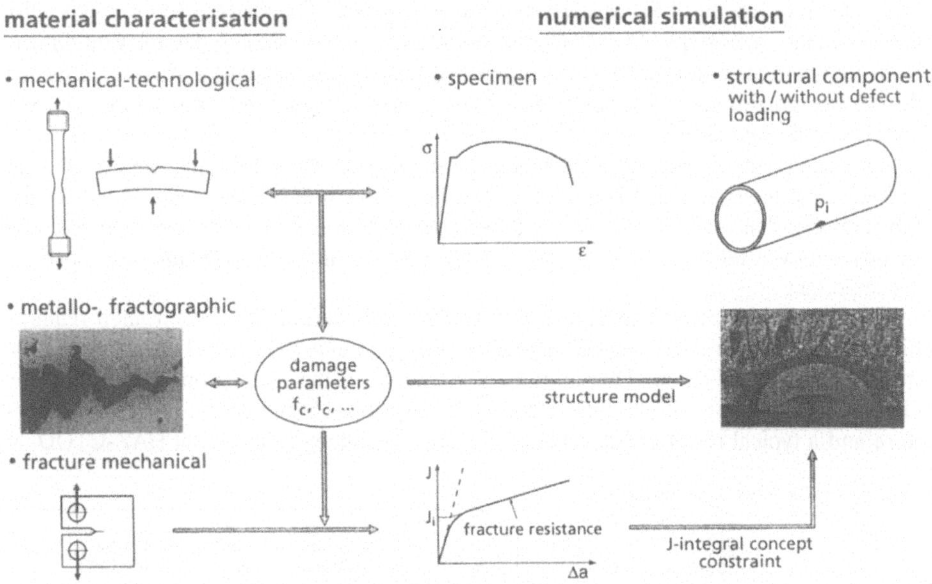

Figure 17. **Principle of damage mechanics concepts.**

For ductile material behaviour, for instance steels, this concept is based on the experimental observation, that damage occurs in loaded specimens with or without notches by void formation and void growth. And these processes can be described in a material law acc. to Gurson, Needleman and Tvergaard [16] the free parameters of which can be derived from smooth round tension bars or the Charpy test. The model from the beginning

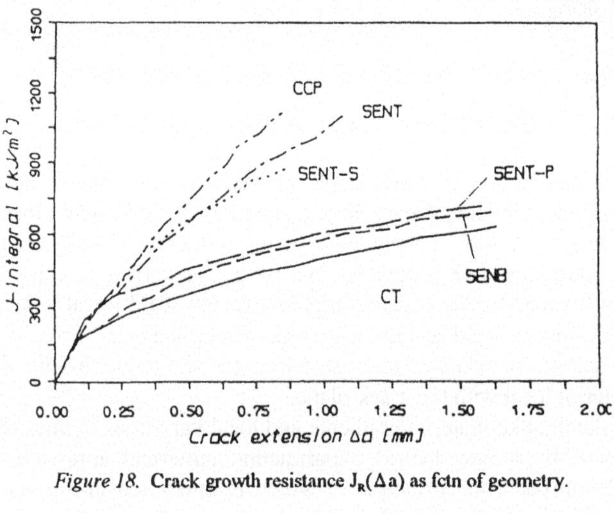

Figure 18. **Crack growth resistance $J_R(\Delta a)$ as fctn of geometry.**

includes the multiaxiality of the stress/strain state and therefore can deliver the experimentally well known geometry dependency of J_R-curves for elastic plastic material behaviour

[17]. Therefore damage mechanics is a tool to allow transferability from the specimen to the component.

5. FFP-element: Component assessment

The final comparison of the calculated global or local driving force with the material resistance, which has been determined before in a laboratory test, allows to state whether

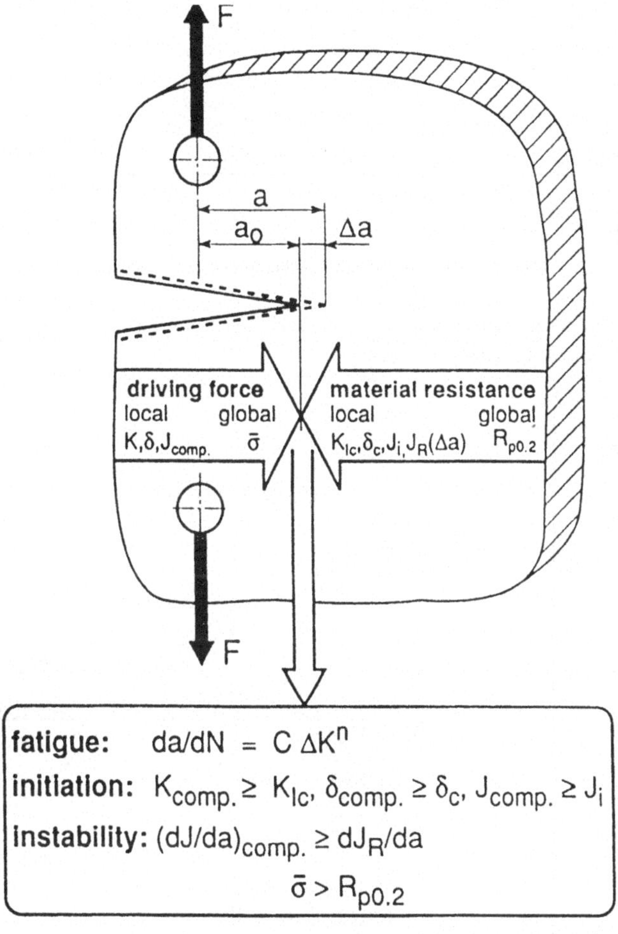

Figure 19. Principle of fracture mechanics component assessment.

failure of the component will occur or not and what kind of failure event will occur. The major events - i.e. fatigue, initiation of growth and instability - are indicated in Fig. 19 and the governing fracture mechanics equations are given. For the failure case "fracture" many different methods are available which have developed from different national experiences and traditions. Different loading and toughness parameters are used. There are direct and indirect methods available: Direct methods in principle allow a failure prediction, the necessary efforts can be rather high; indirect methods rely on design curves which include

Fracture Mechanics Based Assessment Methods

method	country	parameters and concepts			failure	
		loading	toughness	assessment	exclusion	prediction
ASME III	USA	K_I	K_{Ic}	LEFM	X	
ASME XI	USA			(LL)	X	
KTA- Regeln	BRD	K_I	K_{Ic}	LEFM	X	
PD 6493	UK	K_I, δ_I	K_{Ic}, δ_c	LEFM+EPFM+LL	X	
WES 2805	J	δ_I, ε	δ_c	EPFM	X	
CEGB R6	UK	K_r, L_r	K_{Ic}, J_R	LEFM+EPFM+LL	X	X
R Curve	USA	J_I, δ_R, T	J_R, δ_R, T_{mat}	EPFM		X
EPRI-EA	USA	J_I	J_R	EPFM		X
DVS 2401	BRD	K_I, δ_I, J_I	K_{Ic}, δ_c, J_R	LEFM+EPFM+LL	X	X
RCC-M	F	K_I, J_I	K_{Ic}, J_R	LEFM+EPFM	X	
DNV	N	K_I, δ_I, J	K_{Ic}, δ_c, J_R	LEFM+EPFM+LL	X	
NIL 3805	NL	K_I, J_I	K_{Ic}, J_R	LEFM+EPFM+LL	X	
Crack arrest		K_I	K_{Ia}	LEFM dyn.	X	X

LEFM Linear elastic fracture mechanics
EPFM Elastic plastic fracture mechanics
LL Limit load

Figure 20. **Fracture mechanics based assessment methods.**

simplification and conservative assumptions and aim at an exclusion of failure rather than trying to quantify existing safety margins.

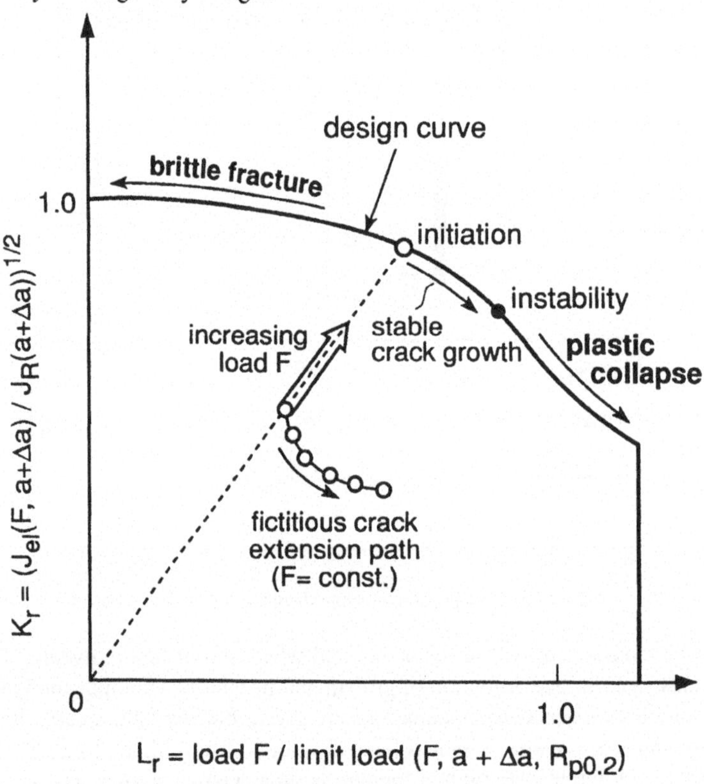

Figure 21. **Two criteria failure assessment diagram in R6.**

The most complete defect assessment concept is the so-called R6-routine [18]. In its lowest assessment level it is based on a simple interpolation between the limiting cases brittle fracture and plastic collapse. The higher assessment options require more information about the material behaviour and finally complete Finite Element analyses in order to allow quantitative prediction of ductile crack initiation or instability.

The regime of validity and the accuracy of the results derived from different assessment methods can only be determined by comparison with experiments and this means evaluation of service experience, analyses of observed failures or controlled component tests. Such results are still very limited. Instead of quantitative statements the IIW-Guidance [1] in § 7.6 "Validation and Verification Requirements" gives a check list concerning input data and procedures which are meant to help the user to avoid mistakes and to judge completeness and reliability. In addition a collection of case studies is under development [19].

But for a final assessment of the fitness-for-purpose i.e. a decision whether a weld imperfection in a loaded component is tolerable or not, safety factors have to be defined and quantified for the specific application, which take into account the consequences of a possible failure, the uncertainties of the modelling and the variability of the input data. First steps into that direction have been done within the development of Eurocode EC3 [20], but work still remains to be done.

The following examples describe some of the chances and the problems in connection with fracture mechanics based defect assessments:

In Fig. 22 the influence of input data concerning the size of the starter crack and the material fatigue crack growth rate on the life time prediction for welded tubes (see Fig. 7).

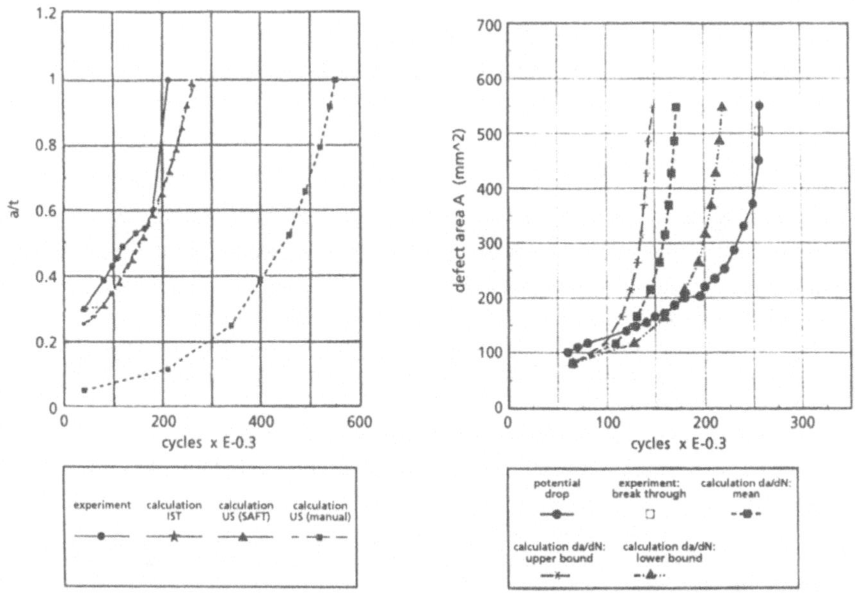

Figure 22. **Tube weld with internal defect under bending fatigue - experiment and calculation.**

The parameter studies conducted with the PC-program [8] deliver a high over estimate of the real lifetime when using the manual US measurements: conservative predictions as compared to the experiments are received when lower bounds of the observed fatigue da/dN-data scatter band are used [6].

In Fig. 23 the results of an analysis are summarized showing that an unexpected loss of

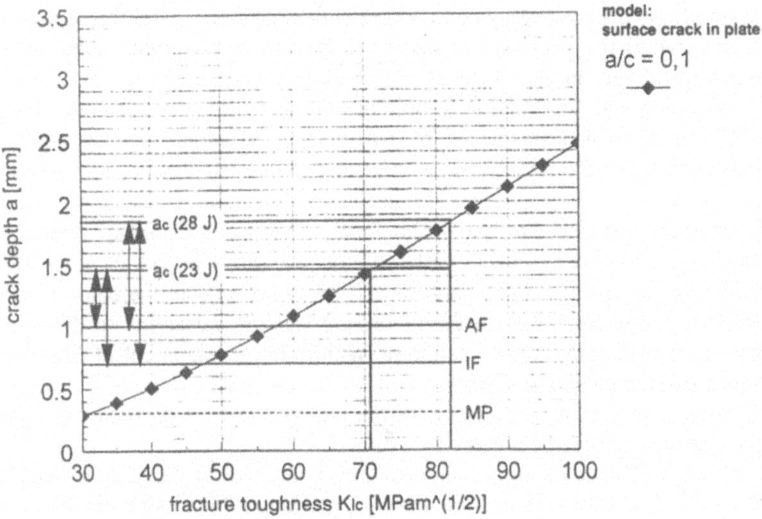

Figure 23. Change of safety margin in friction weld.

toughness in the range of 5 J CVN energy in a friction weld gave a tolerable decrease in safety margin between safely detectable and critical crack size.

6. Summary and conclusions

Fitness for Purpose is an integral and quantitative concept to demonstrate structural integrity of a specific welded component even if imperfections have been detected. Conditions for an application in praxis are

- detailed knowledge and quantitative description of the status of defects
- global and local stress/strain analysis for all relevant loads in the specific component
- advanced characterization including fracture toughness testing of the "true" material for all service conditions
- use of verified fracture mechanics based assessment concepts.

Improvements have been gained but need further elaboration in the following areas:

- defect status
  recording of length and depth of internal defects
  develop process/fabrication quality control into load and toughness oriented fitness for purpose quality control
- stress/strain analysis
  use PC based defect assessment for sensitivity studies and prediction
  analytical solutions should be FE verified
  constraint-modified crack growth analysis

in welds mismatch effects have to be considered through FE structural models.

- material characterization
  details of HAZ testing have to be defined, thermal weld simulation should be used
  damage mechanics facilitates transferability
  dynamic material testing required
- component assessment
  stepwise assessment relative to availability of input data
  verification experiments
  leak before break analyses
  cooperation of experts.

7. References

1. SST-1157 (1990): IIW Guidance on Assessment of the Fitness-for-Purpose of Welded Structures, *Draft for Development*, Publ. P 90.09, Svejsecentralen, DK-2605 Broendby
2. DVS-Merkblätter 2401 (1989): Bruchmechanische Bewertung von Fehlern in Schweißverbindungen, Teil 1: Grundlagen und Vorgehensweise, Teil 2: Praktische Anwendung, *Fachbuchreihe Schweißtechnik* Bd. 101, DVS-Verlag, Düsseldorf
3. DVS Merkblatt 0704/DIN 8563, Teil 3 (1990): Sicherung der Güte von Schweißarbeiten
4. AD HP 5/3 (1989): Herstellung und Prüfung der Verbindungen, zerstörungsfreie Prüfung der Schweißnähte, VDTÜV Essen
5. see [1]
6. Burget, W. et al. (1988): Nachweis der Übertragbarkeit von an Laborproben ermittelten bruchmechanischen Kennwerten für eine Lebensdauerberechnung rißbehafteter Bauteile des Maschinenbaus, *FKM-Forschungshefte*, Heft 136, Forschungskuratorium Maschinenbau, Frankfurt/M.
7. Tada, H., Paris, P.C., Irvin, G.R. (1973): *The Stress Analysis of Cracks Handbook*, Del. Research Cooperation, Hellertown
8. Hodulak, L. (1993): Mit dem PC-Bauteilfehler bewerten, *Materialprüfung* 35, **3**, 73-75
   Fehlerbewertungsprogramm "IWM-VERB", Version 4.8, FhG-Software 19411043, Fraunhofer-Institut für Werkstoffmechanik, Wöhlerstrasse 11, D-79108 Freiburg
9. Bueckner, H.F. (1973): Field singularities and related integral representations, *Methods of Analysis and Solutions of Crack Problems, Mechanics of Fracture*, Vol. **1** (Ed. G.C. Sih), Noordhoff Int. Publ., Leyden, 239-314
10. Hodulak, L. and Siegele, D. (1993): Calculation of Stress Intensity Factors for Cracks Under Thermal Shock Transients, ASTM Nat. Symp. Fracture Mechanics
11. Busch, M., Maschke, H.-G. and Kuna, M. (1990): A novel BEM-approach to weight functions based on Bueckner's fundamental fields, In: *Numerical Methods in Fracture Mechanics*, Vol. **5**, 5-16
12. Hodulak, L. and Siegele, D. (1995): Fracture Behaviour of Subclad Cracks, *ASTM STP 1256*, Eds. Reuter, Underwood and Newman, ASTM, Philadelphia, USA
13. Burget, W. and Memhard, D. (1991): Experimental and Numerical Investigation on Weld Metal Strength Mismatch Effects in Fracture Toughness Testing of Butt Welded Joints, *IIW-Doc. X-1228-91*
14. Schwalbe, K.H. (1980): Bruchmechanik metallischer Werkstoffe, C. Hanser Verlag

15. Burget, W. and Blauel, J.G. (1990): Weld Metal and HAZ Toughness in SA-welded Joints of New TMCP Steels, *Proc. IIW Intern. Conf. on Advances in Joining Newer Structural Materials*, Montreal Ca. July 1990
16. Gurson, L. (1977): Continuum Theory of Ductile Rupture by Void Nucleation and Growth: Part I - Yield Criteria and Flow Rules for Porous Ductile Media, *J. Engng. Mat. Tech.* **99**, 2-15
Needleman, A. and Tvergaard, V. (1984): An Analysis of Ductile Rupture in Notched Bars, *J. Mech. Phys. Solids* **32**, 461-490
17. Sun, D.-Z., Kienzler, R., Voss, B. and Schmitt, W. (1992): Application of Micro-mechanical Models to the Prediction of Ductile Fracture, Fracture Mechanics: 22. Symp. (Vol. II), *ASTM STP 1131*, Eds.: S.N. Atluri et al., American Soc. for Testing and Materials, Philadelphia, 1992, 368-378
18. Milne, I. et al. (1988): Assessment of the Integrity of Structures Containing Defects, *International Journal of Pressure Vessels and Piping* **32**, 3-104
19. Blauel, J.G. (1993): IIW-Case Study Collection on the Assessment of the Significance of Weld Imperfections, *IIW-Doc. X-1280-93*
20. Eurocode III Annex C-Part 2 (1995): Design Against Brittle Fracture, Rev. Sept. 1995, G. Sedlacek: Institut für Stahlbau der RWTH Aachen, *IIW-Doc. X-1329-95*

PART-THROUGH CRACKED STRUCTURES UNDER CYCLIC LOADING

Andrea CARPINTERI, Roberto BRIGHENTI,
Alberto SANGUANINI and Andrea SPAGNOLI
Dipartimento di Ingegneria Civile, Università di Parma
Viale delle Scienze, 43100 Parma, Italy

Abstract

The purpose of the present paper is to review some recent advances in the field of *fatigue fracture mechanics* of engineering metallic structures. In particular, the failure of cyclically loaded structures often develops from part-through cracks. The analysis of this type of defects is complex, since the stress-intensity factor varies along the crack front. Moreover the shape of these flaws changes during the fatigue propagation, and this significantly affects the structure life prediction. The behaviour of surface cracks in metallic plates, round bars and pipes under constant amplitude cyclic loading is examined.

1. Introduction

As is well-known, a cyclically time-varying loading reduces the material strength and can provoke the fatigue failure of engineering metallic structures. This phenomenon consists of three stages: (a) crack initiation (from microcrack initiation to engineering-size flaw formation), (b) crack propagation, and (c) catastrophic failure. A better understanding of the fatigue fracture phenomenon can lead to an improvement of the structural reliability [1-3].

The conventional (or classical) fatigue design approach, which involves the use of the "stress - fatigue life" (S-N) curves developed from endurance tests on laboratory specimens, does not distinguish between crack initiation and crack propagation. More precisely, this approach is based on smooth specimen tests where the major portion of the life is spent in the formation

R.A. Smith (ed.), Reliability Assessment of Cyclically Loaded Engineering Structures, 65–99.

of engineering-size flaws, and therefore it can lead to gross overestimates of life in the case of real structures where the crack initiation stage is often small due to pre-existing defects.

The approach currently used is the defect-tolerant approach (based on *fatigue fracture mechanics*), the aim of which is to understand the fatigue crack initiation and propagation phenomena and to define when a flaw can be judged tolerable and, on the contrary, when it has to be repaired to prevent a catastrophic failure.

2. Part-through Cracks in Plates

2.1. STRESS CONCENTRATION IN PART-THROUGH CRACKED PLATES

Consider a plate with a through-the-thickness crack. The stress field (σ, τ) near the crack tip presents a square-root singularity (Fig.1):

$$(\sigma, \tau) = \frac{K}{(2\pi r)^{1/2}} f(\theta) \tag{1}$$

where

r, θ = polar coordinates (of a generic point of the body) in the coordinate system with the origin in the tip and the $\theta = 0°$ axis collinear to the crack line;

$f(\theta)$ = function of the polar coordinate θ;

K = stress-intensity factor, depending on the stucture and crack geometry and on the acting loading.

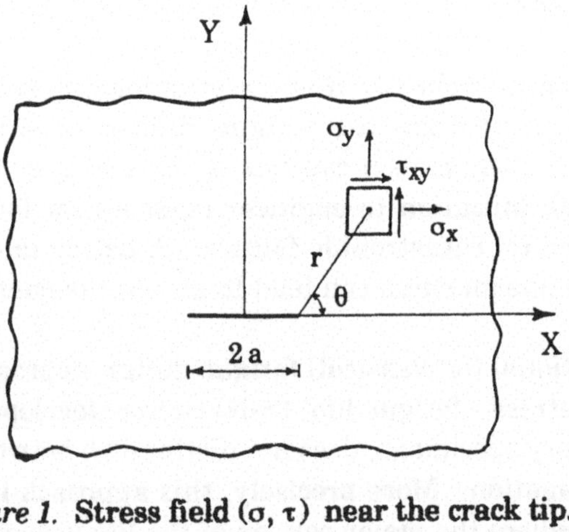

Figure 1. Stress field (σ, τ) near the crack tip.

An external loading symmetrical to the crack line gives rise to the "opening" Mode, or Mode I of loading, which tends to open the two surfaces of the crack. In this case, the stress-intensity factor can be written as follows :

$$K_I = Y_I \ \sigma \ (\pi a)^{1/2} \tag{2}$$

where the subscript I indicates the mode of loading, σ is the opening stress related to the external loading, a represents the crack half-length (or the crack length in the case of an edge crack), and Y_I is a dimensionless correction function depending on the geometry of the crack and the cracked component and on the loading condition. Several handbooks supply a tabulation of Y_I for all the crack and loading configurations most significant from an engineering point of view [4,5]. If the external loading belongs to Mode II ("sliding") or to Mode III ("tearing"), the stress-intensity factor is called K_{II} or K_{III}, respectively, and can be expressed by equations similar to eqn (2).

The crack propagation failure of structural components often develops from part-through flaws (Fig.2). The stress square-root singularity along the crack front was experimentally determined for these defects by photoelastic stress freezing [6], but practical difficulties made it impossible to measure the singularity power at the crack border point R on the free surface. It has theoretically been shown [7-10] that, when the front of the flaw intersects the free surface perpendicularly, the power of the stress singularity at the intersection point R is equal to 0.5 only if the Poisson ratio ν of the material is equal to 0.0 (Fig.3).

From fracture energy considerations, it was also deduced that the crack growth process requires a square-root singularity [9]. Therefore, the angle β at the intersection point for $\nu = 0.3$ is equal to about 100° for Mode I and to about 67° for Modes II and III (Fig.3).

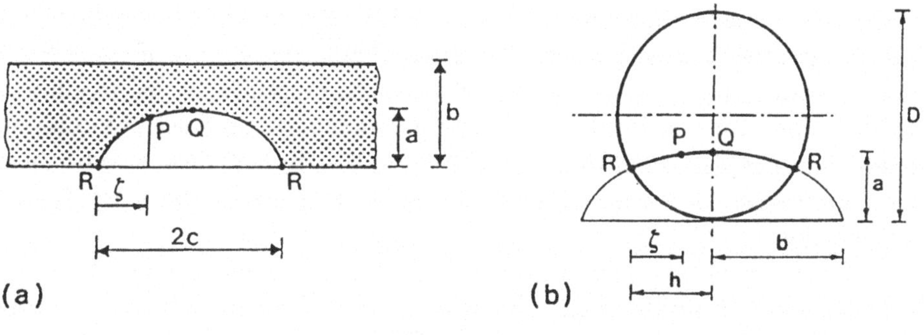

Figure 2. Elliptical-arc surface flaws : geometrical parameters.

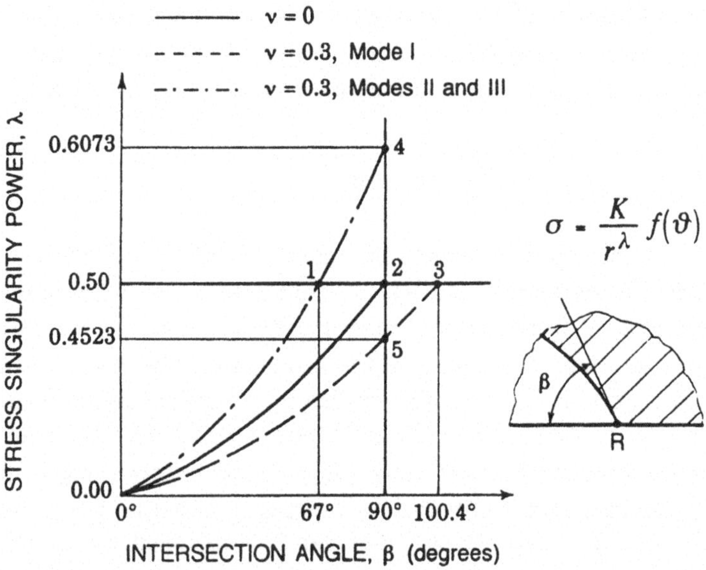

Figure 3. Qualitative diagram of stress singularity power λ against crack front intersection angle β for different values of Poisson ratio ν (results taken from [7-10]).

In conclusion, the stress analysis of the surface defects is complex for two main reasons. Firstly, these flaws are three-dimensional in nature, that is, the stress-intensity factor varies along the crack front. Secondly, the singularity power at the intersection point R depends on the Poisson ratio ν of the material, the mode of loading and the intersection angle β.

Stress-intensity factor calculations for surface cracks in metallic plates have been carried out by many authors, and the results of these analyses are very useful for reliable prediction of structure fatigue life. A large amount of research is related to semi-elliptical flaws (Fig.2a), since the real surface defects can often be assumed to have this shape. This hypothesis is supported by many experimental analyses [3]. Moreover, it was shown [11] that fatigue crack growth predictions carried out on the basis of crack front models with either 2 or 21 degrees of freedom, respectively, led to similar results. Therefore, several recommendations [12,13] suggest the rules of actual defect replacement by an equivalent semi-elliptical flaw, also because such approximation remarkably simplifies the stress-intensity factor computation.

A review of the stress-intensity factor solutions proposed in the literature for a single semi-elliptical surface flaw in a metallic plate subjected to tension and/or bending loading can be found in chapters 18 and 19 of Ref. [3]. Some results obtained through the finite element method are presented in the following sub-section.

2.1.1. *3D Finite Element Analysis for Surface Cracks in Plates*

A three-dimensional finite element analysis has been carried out to examine the semi-elliptical surface flaw in Fig.2a, subjected to tension or bending loading [14]. Because of symmetry, only one-fourth of the part-through cracked structural component is modelled by 20-node isoparametric solid elements (Fig.4). The aspect ratio a/c of the flaw ranges from 0.2 to 1.0, while the relative depth $\xi = a/b$ of the deepest point Q on the crack front is made to vary from 0.1 to 0.4.

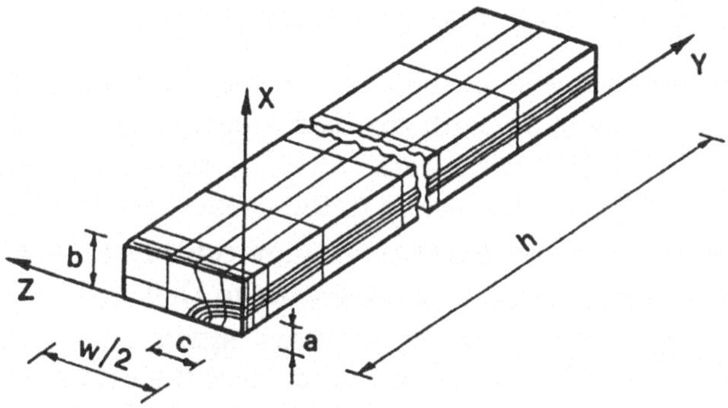

Figure 4. Finite element mesh for a plate with a semi-elliptical surface flaw.

The square-root singularity of the stress field near the crack is modelled by shifting the finite element midside nodes near the crack front to quarter-point positions [15-17]. As has been discussed in the previous section, when the crack front intersects the free surface perpendicularly, the stress singularity power at the intersection point R (Fig.3) is different from 0.5, in the case of Poisson ratio $\nu \neq 0.0$. Therefore, the use of quarter-point finite elements does not produce reliable results in a boundary layer near the above intersection point, but this effect is confined to only a small region because the singularity power does not deviate remarkably from the value 0.5 , i.e. the stress distribution in the interior zone of the crack front is almost insensitive to whatever occurs in the boundary layer.

After obtaining the stress field for both tension and bending loading by means of the finite element analysis, the dimensionless stress-intensity factors $\tilde{K}_{I,F}$ and $\tilde{K}_{I,M}$ are obtained as follows:

$$\tilde{K}_{I,F} = \frac{K_{I,F}}{\sigma_F (\pi a)^{1/2}} \tag{3}$$

$$\tilde{K}_{I,M} = \frac{K_{I,M}}{\sigma_M (\pi a)^{1/2}} \tag{4}$$

where

$K_{I,F}$ = stress-intensity factor for tension = $\sigma_{Y,F}\,(2\pi r)^{1/2}$, with $r \longrightarrow 0$,

$K_{I,M}$ = stress-intensity factor for bending = $\sigma_{Y,M}\,(2\pi r)^{1/2}$, with $r \longrightarrow 0$,

$\sigma_{Y,F}$, $\sigma_{Y,M}$ = stress σ_Y (for tension and bending, respectively) at the distance r evaluated perpendicular to the crack front,

$\sigma_F = F/(b\,w)$ = applied uniform tensile stress,

$\sigma_M = 6\,M/(b^2\,w)$ = maximum bending stress.

The dimensionless factor $\widetilde{K}_{I,F}$ along the crack front under tension is plotted in Figs 5a to 5e, with the flaw aspect ratio a/c made to vary from 0.2 for case (a) to 1.0 for case (e). The maximum value of each curve displayed in Fig.5a corresponds to the deepest point Q ($\zeta/c = 1.00$) on the crack front. If greater aspect ratios are considered, e.g. cases (b) and (c), the curves show a behaviour analogous to that in Fig. 5a , although they tend to flatten. When a/c is increased again, e.g. cases (d) and (e), the greatest values of $\widetilde{K}_{I,F}$ are attained near the external surface, i.e. for $\zeta/c = 0.10$.

Figure 6 shows the dimensionless stress-intensity factor $\widetilde{K}_{I,M}$ along the crack front under bending. For the case $a/c = 0.2$, the results in Fig.6a are qualitatively similar to those in Fig.5a, even if it can be noticed that the curves for bending are flatter than those for tension, particularly for the highest values of ξ being examined. If the relative crack depth ξ is equal to 0.4, it can be pointed out that the curves for $a/c \geq 0.4$ (Fig.6b to 6e) show the greatest values near the free surface and decrease monotonically by increasing the normalized coordinate, with the slope getting higher and higher as the aspect ratio is increased. On the other hand, the greatest stress-intensity factors are attained in the vicinity of the external surface for $a/c \geq 0.6$ in the case of $\xi = 0.3$, for $a/c \geq 0.8$ in the case of $\xi = 0.2$, and for $a/c > 0.8$ in the case of $\xi = 0.1$.

The trends of the diagrams in Figs 5 and 6 are very similar to those described by other authors [5,18-20].

2.2. FATIGUE PROPAGATION OF PART-THROUGH CRACKS IN PLATES

When a cracked structural joint or component is subjected to cyclic loading [3], the crack propagation rate, da/dN, is related to the stress-intensity factor range ΔK, as was first pointed out in Ref. [21] for a standard through-thickness cracked specimen under constant amplitude fatigue loading :

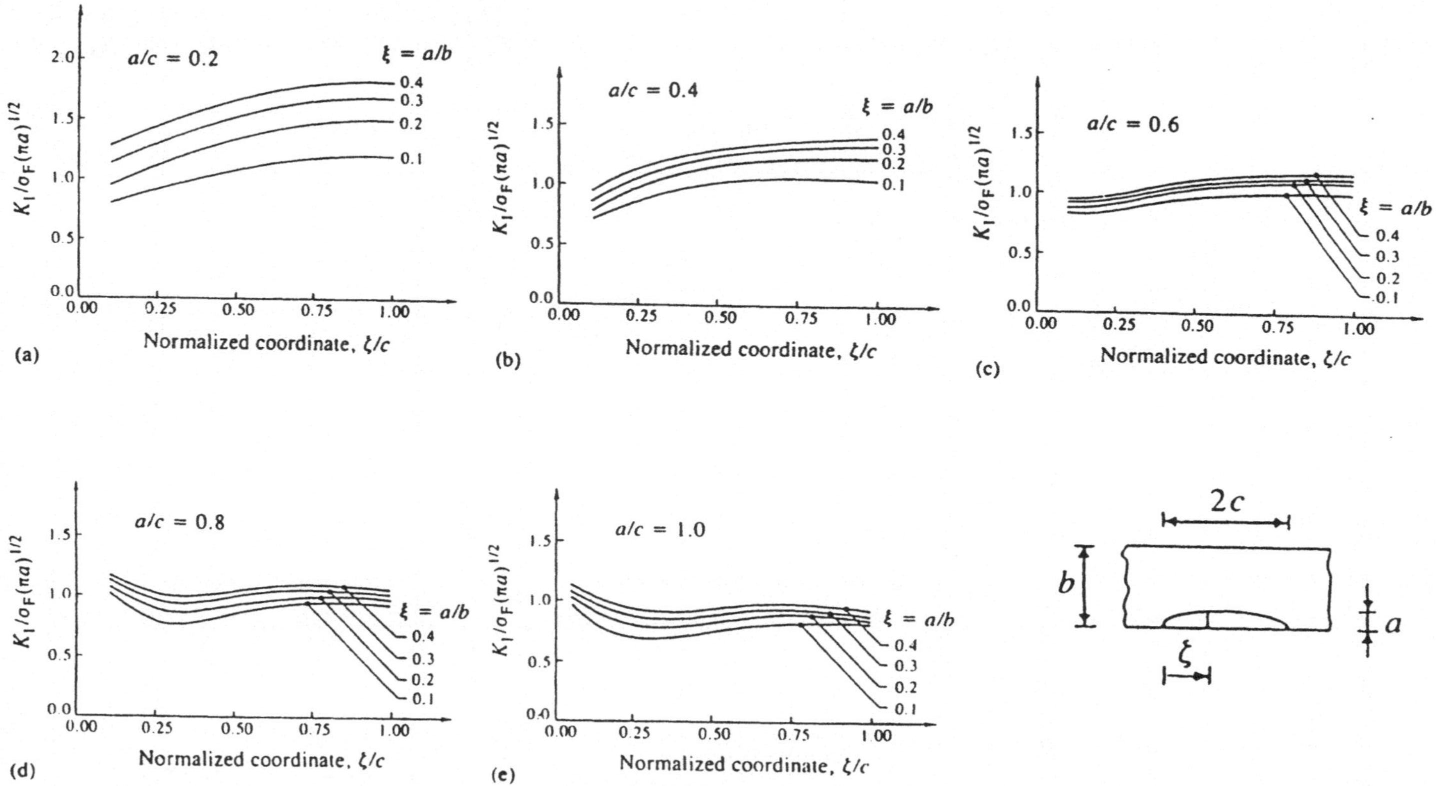

Figure 5. **Dimensionless stress-intensity factor along the crack front under tension, with aspect ratio a/c made to vary from 0.2 (case a) to 1.0 (case e) [14].**

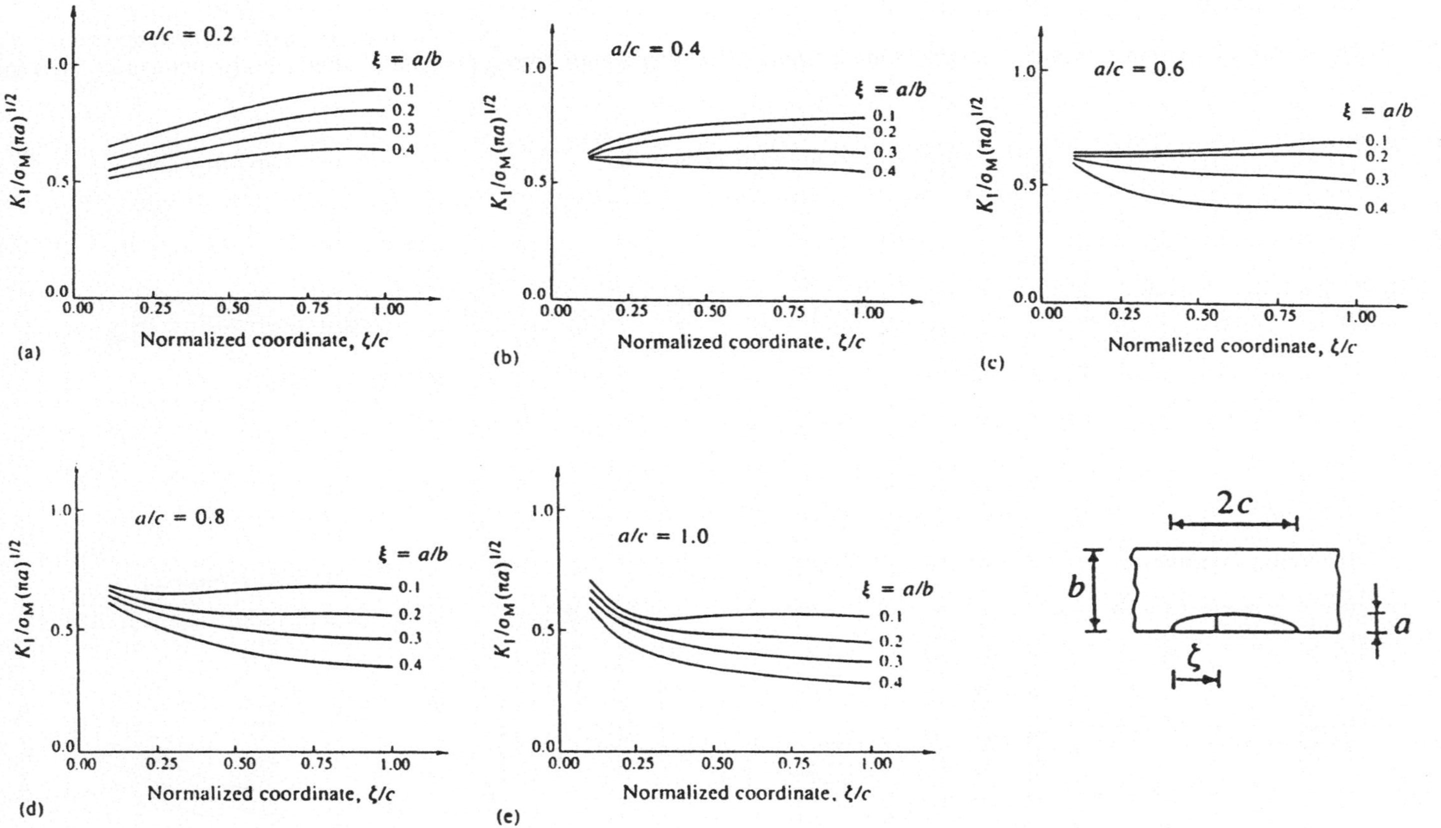

Figure 6. **Dimensionless stress-intensity factor along the crack front under bending, with aspect ratio a/c made to vary from 0.2 (case a) to 1.0 (case e) [14].**

$$da / dN = A (\Delta K)^m \tag{5}$$

where N is the number of loading cycles, while A and m are constants dependent on the specimen material; ΔK is equal to $Y \Delta\sigma (\pi a)^{1/2}$ according to eqn (2), with the stress range $\Delta\sigma$ equal to $\sigma_{max} - \sigma_{min}$. Note that eqn (5) represents a straight line with slope m in a bi-logarithmic diagram of da/dN against ΔK, while the experimental plot is sigmoidal. Other crack propagation laws were proposed to better describe the actual fatigue behaviour of cracked structures [3], but they require some more material constants seldom provided in the literature, and are more complex to be used for fatigue life evaluations.

For cyclic loading with variable amplitude, remarkable interaction effects occur between cycles of different amplitude. In particular, the application of an overload decreases the crack growth rate during the next loading cycles, since the overload introduces a large plastic zone at the crack tip with residual compressive stresses after unloading (crack closure phenomenon) and a consequent delay in the crack propagation [22]. Negative loads (underloads) can partially decrease the compression due to previous overloads, therefore reducing the positive effect of the overloads on the crack growth delay. The behaviour of structures under complex loading conditions (variable amplitude loading, mixed mode loading, impact loading) is analysed in chapters 26 to 30 of Ref.[3].

Fatigue phenomena can remarkably be influenced by the surrounding environment. This is particularly true for structures which often work at low or high temperatures and under corrosive conditions. A complete description of the environmental effects on the fatigue behaviour is out of the present paper purpose. This topic is extensively treated in chapters 31 to 37 of Ref.[3].

The fatigue analysis of the part-through flaws is complex since the shape of these defects changes during the propagation, and this significantly affects the structure life prediction (chapters 18 and 19 in Ref.[3]). The crack front of a surface flaw can approximately be characterized by means of a finite number of degrees of freedom. For example, a lot of papers on plates deal with semi-elliptical defects by assuming a two-parameter crack front model, with the fatigue crack growth rates at points Q and R (Fig. 2a) given by

$$da / dN = f_1 (\Delta K_Q{}^*, const_1) \tag{6}$$

$$dc / dN = f_2 (\Delta K_R{}^*, const_2) \tag{7}$$

where the crack propagation functions f_1 and f_2 depend on the empirical material constants $const_1$ and $const_2$ and on a linear-elastic fracture mechanics parameter ΔK^*. Such parameter can be calculated in different ways. For example, it can be assumed that the crack growth rate at the generic point P of the front depends on the stress-intensity factor range at that point, that is, ΔK_P^* is equal to ΔK_P. On the other hand, some authors proposed to use an effective value $\Delta K_{e,P}$ related to the crack closure phenomenon. More precisely, in the last case the stress-intensity factor range at the generic point P is given by

$$\Delta K_P^* = \Delta K_{e,P} = \gamma_P \, \Delta K_P \tag{8}$$

where the coefficient γ_P is experimentally determined.

By employing the crack propagation law expressed in eqn (5) and the effective value ΔK_e, eqns (6) and (7) can be written as follows:

$$\mathrm{d}a \, / \, \mathrm{d}N = A \, (\Delta K_{e,Q})^m = A \, (\gamma_Q \, \Delta K_Q)^m \tag{9}$$

$$\mathrm{d}c \, / \, \mathrm{d}N = A \, (\Delta K_{e,R})^m = A \, (\gamma_R \, \Delta K_R)^m \tag{10}$$

where the coefficients γ characterize the difference of crack growth rates in depth and length directions, and the exponent m is assumed to be independent of the crack front point position and equal to the value determined from standard through-the-thickness cracked specimens, as is confirmed from the data reported in the literature. Some authors have experimentally obtained the following results for semi-elliptical surface flaws in metallic plates : $\gamma_Q = 1.1$ and $\gamma_R = 1.0$.

The above crack growth laws can be applied by developing fatigue flaw propagation models, to verify the reliability of structures and to predict their residual life. For example, an automated procedure to deduce the growth history of a generic semi-elliptical flaw under cyclic loading was proposed in Refs [23,24]. Several studies have been carried out to examine the fatigue propagation of surface defects [25].

Some recommendations suggest to analyse the part-through cracked components under cyclic loading by assuming semi-elliptical flaws with either a constant aspect ratio [12] or with the ellipse semi-major axis being kept constant until the crack depth equals the semi-major axis value, and then a semi-circular shape is used during the following growth [13], but these hypotheses can lead to predictions remarkably different from the real behaviour. As a matter of fact, many experimental and theoretical analyses of surface cracks under both cyclic axial and bending loading [3] have shown that the semi-elliptical shape is preserved during the whole growth, but the aspect ratio changes according to preferred paths in the diagram of the crack aspect ratio a/c against the relative crack depth ξ (Fig.7). These propagation paths tend to converge to an inclined asymptote, independent of the initial crack configuration (a_0/c_0 and $\xi_0 = a_0/b$, where a_0 and c_0 are the initial flaw semi-axes), with the asymptote for bending lower than that for axial loading. Analogous conclusions will be drawn for surface flaws in round bars and pipes in the following sections.

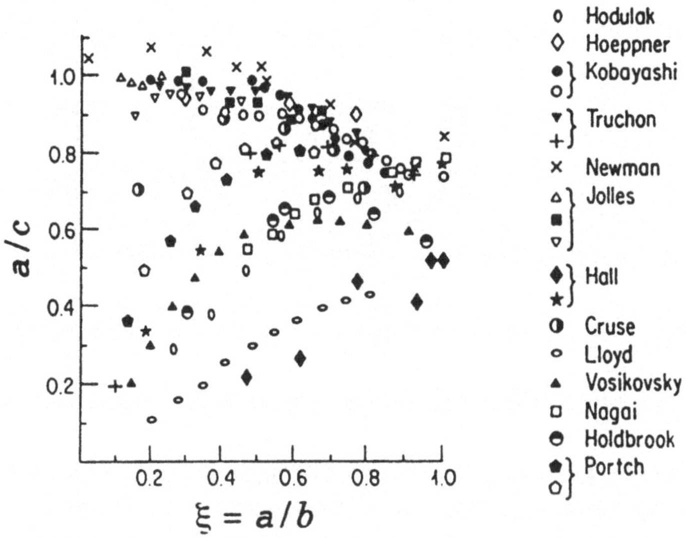

Figure 7. Propagation paths for surface flaws in plates under cyclic axial loading : experimental results (see references in chapter 18 of Ref.[3]).

3. Part-through Cracks in Round Bars

3.1. STRESS CONCENTRATION IN PART-THROUGH CRACKED ROUND BARS

Many authors have examined the stress-intensity factor variation along the front of surface flaws in cylindrical structural components with circular cross-section (see Refs[5,26-41] and chapters 5, 18 and 38 in Ref.[3]).

3.1.1. *Two-parameter model for stress analysis*

The behaviour of an elliptical-arc surface crack in a round bar subjected to tension or bending loading (Fig.8) has been analysed in Refs[34,35,37,39] by pointing out the influence of the flaw aspect ratio $\alpha = a/b$ on the stress-intensity factor. The relative crack depth ξ, equal to the ratio between the maximum crack depth a and the bar diameter D, ranges from 0.1 to 0.6, while the flaw aspect ratio is made to vary from 0.0 (straight front) to 1.0 (circular-arc front). The generic point P along the crack front is identified by the normalized coordinate ζ/h. The axial force F acts perpendicular to the crack plane, while the bending moment M acts about an axis parallel to the semi-major axis b of the elliptical front.

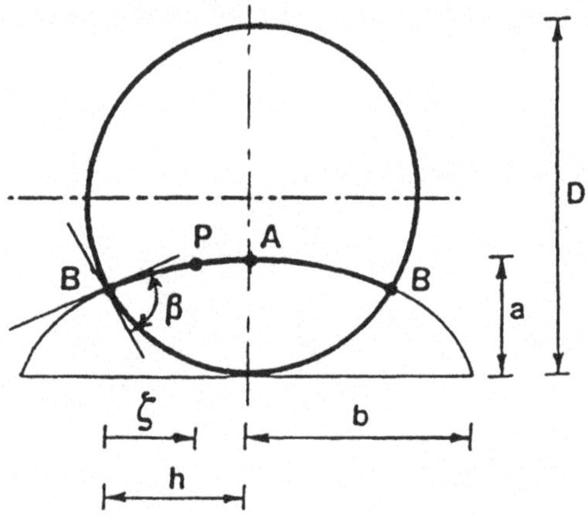

Figure 8. Elliptical-arc surface crack in a round bar.

The stress field is determined through a finite element analysis carried out by employing 20-node isoparametric solid elements. Quarter-point finite elements are used near the crack front to model the stress square-root singularity. Then the dimensionless stress-intensity factors $\widetilde{K}_{I,F}$ and $\widetilde{K}_{I,M}$ are calculated from eqns (3) and (4), where

$\sigma_F = F/(\pi D^2/4)$ = applied uniform tensile stress,

$\sigma_M = M/(\pi D^3/32)$ = maximum bending stress.

The influence of the parameter α in the case of $\xi = 0.4$ is shown in Fig.9 [40]. For tension loading, the slope of the curve $\widetilde{K}_{I,F}$ becomes negative for $\alpha = a/b$ greater than or equal to about 0.6, that is to say, the maximum stress-intensity factor is attained at point A for $\alpha \leq 0.6$ and near point B for $\alpha \geq 0.6$ (see points A and B in Fig.8). An analogous transition is observed for $\alpha = 0.5$ in the case of bending loading (Fig.9). For other values of ξ, the change of slope sign occurs in correspondence to different values of aspect ratio, and the transition value of α for tension is greater than that for bending at the same ξ.

Some of the results deduced through the above finite element analysis are compared with those determined by other authors [5,27,30]. For $\alpha = a/b = 0.6$ (Fig.10), the agreement between the results plotted is quite satisfactory for both tension and bending. More precisely, the values

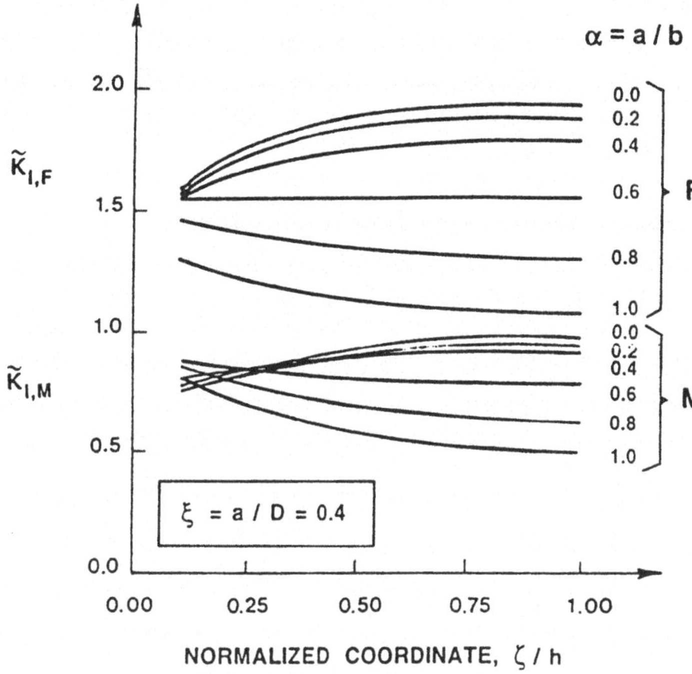

Figure 9. **Transition from straight-fronted crack (α = *a*/*b* =0.0) to circular arc crack (α = *a*/*b* = 1.0), for both tension and bending, in the case of ξ = 0.4 [40].**

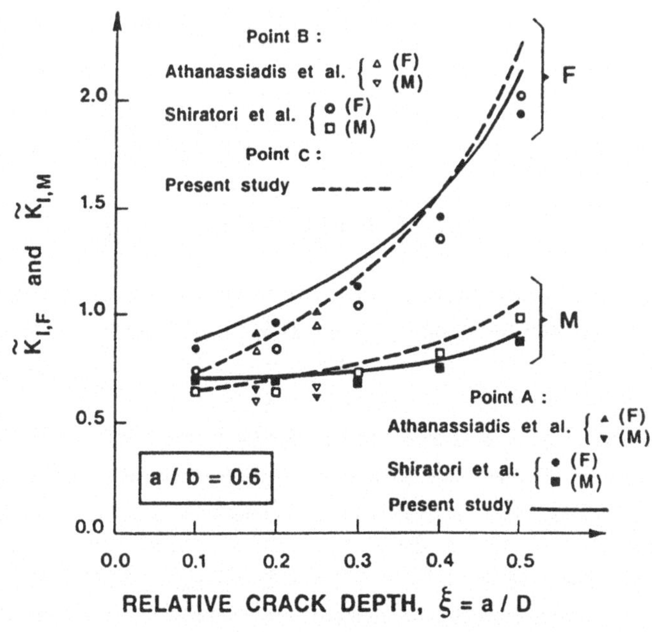

Figure 10. **Dimensionless stress-intensity factors (under tension and bending) at points *A*, *B* and *C*, for *a* / *b* = 0.6 (results taken from Refs [5,27,30,35,37]).**

presented in Refs [35,37] vary by approximately 10-15% from those deduced by other authors. Note that the results by Athanassiadis et al. [27] and by Shiratori et al. [5,30] are related to points A and B, while the other values in Fig.10 are related to points A and C, with point C ($\zeta / h = 0.1$) very close to point B but not coincident with it.

3.1.2. *Three-parameter model for stress analysis*

An additional geometrical parameter can be considered to improve the previous model : the ellipse shifting $s = a_{el} / a$, where a_{el} and a are the ellipse semi-axis along the Y-axis and the maximum crack depth, respectively (Fig.11) [41]. In this way the centre of the ellipse can be assumed to be external to the bar cross section ($s > 1$), and the generic crack configuration is defined by the three parameters s , $\xi = a / D$ and $\alpha = a_{el} / b_{el}$.

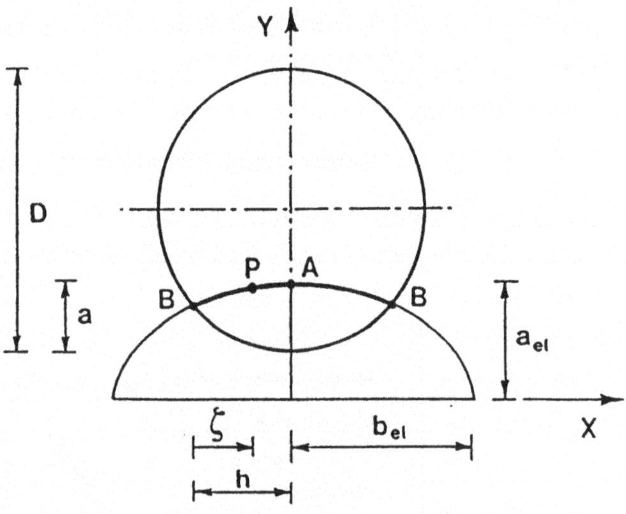

Figure 11. Ellipse shifting $s = a_{el} / a$.

Figure 12 shows the stress-intensity factors at point A ($\zeta / h = 1.0$) against the ellipse shifting s, for $\xi = 0.5$ and different values of aspect ratio α, in the case of both tension and bending. The results obtained for $\alpha = 0.0$ are obviously constant, since the straight-fronted crack configuration is independent of s. On the other hand, the curves for $\alpha \neq 0.0$ asymptotically tend to the horizontal straight lines determined for $\alpha = 0.0$, and are almost coincident with the asymptotes already for $s \geq 10$. The more the parameters ζ / h and ξ decrease, the stronger this tendency is. The configurations related to $s = 1$ have been analysed in the previous sub-section, while the results deduced in the case of $\alpha = 0.0$, which is equivalent to the case of $s = \infty$, can also be assumed for $s \geq 10$, according to the above

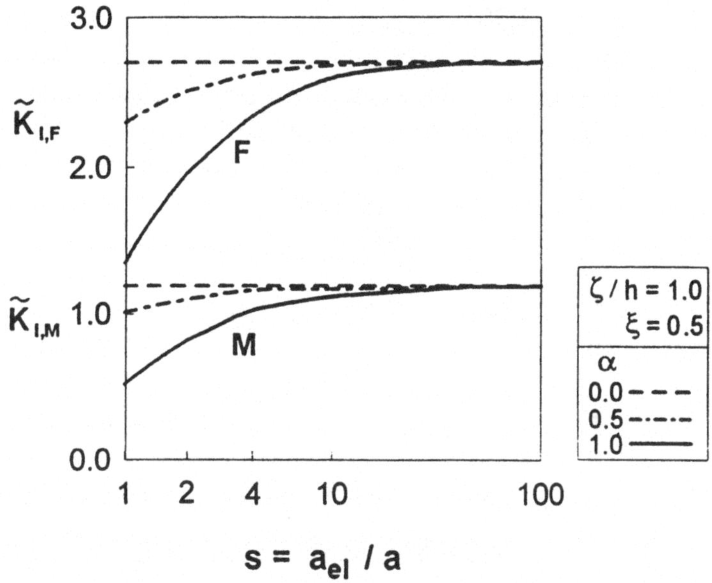

Figure 12. Dimensionless stress-intensity factors at point A (ζ / h = 1.0) against ellipse shifting s, for $\xi = a / D = 0.5$ and different values of crack aspect ratio $\alpha = a_{el} / b_{el}$ [41].

remarks. Therefore, the influence of the ellipse shifting has to be examined in the range $1 < s < 10$.

The dimensionless stress-intensity factor for tension is displayed in Fig.13a for $\xi = 0.4$ and different values of α, in the case of $s = 2$. The slope of the curve $\widetilde{K}_{I,F}$ becomes negative for α greater than or equal to about 0.85, that is to say, the maximum stress-intensity factor is attained at point A for $\alpha \leq 0.85$ and near point B for $\alpha \geq 0.85$. This transition phenomenon is analogous to that pointed out for $s = 1$ (see Fig.9) and for $s = 4$ (Fig.13b). For the latter case, the change of slope sign occurs in correspondence to a value of transition aspect ratio, α_t, equal to about 1.15; that is, α_t increases by increasing the parameter s. Moreover, α_t decreases by increasing the relative crack depth ξ. Diagrams analogous to those for tension have been obtained for bending, with the transition aspect ratio lower than that determined for tension at the same values of s and ξ.

On the basis of the above remarks, it can be concluded that the most critical parts along the crack front are the deepest point and the ends of the flaw. As an example, the stress-intensity factors under tension at point A (ζ / h = 1.0) and point C (ζ/ h = 0.1) for s = 2 are plotted in Fig.14. Analogous results have been obtained for several values of ellipse shifting in the range $1 < s < 10$.

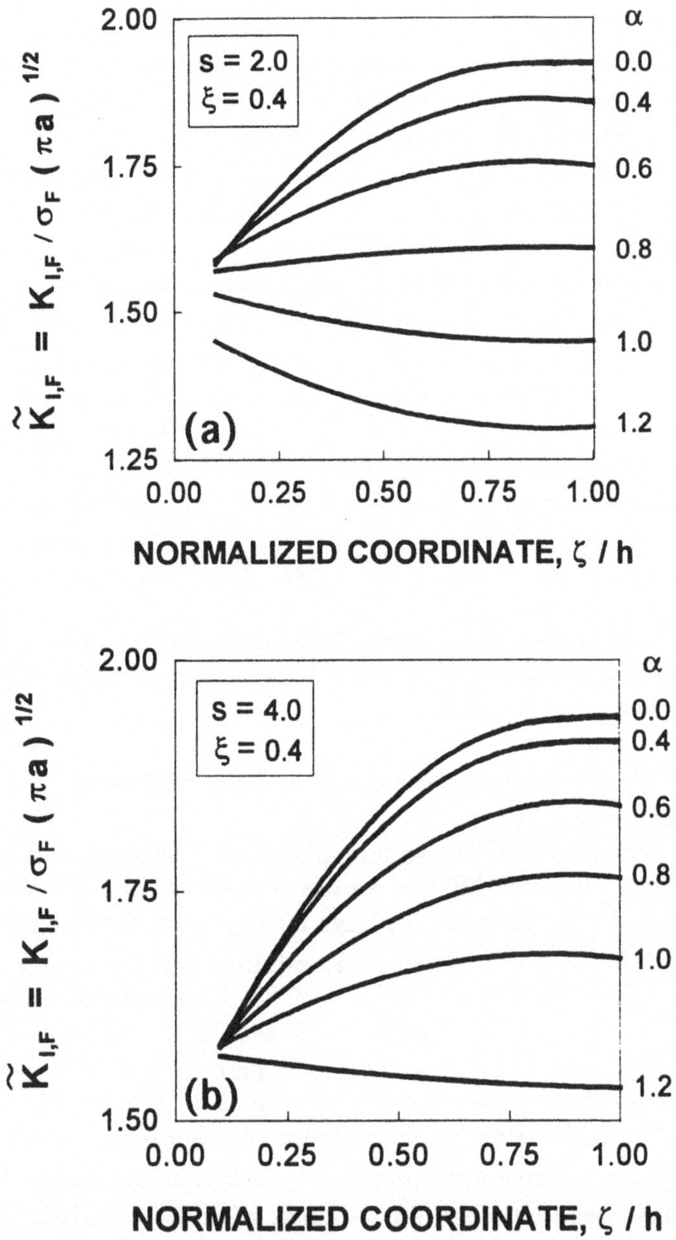

Figure 13. **Transition phenomenon dependent on the aspect ratio α in the case of tension and $\xi = a/D = 0.4$, for ellipse shifting $s = 2.0$ (a) and $s = 4.0$ (b).**

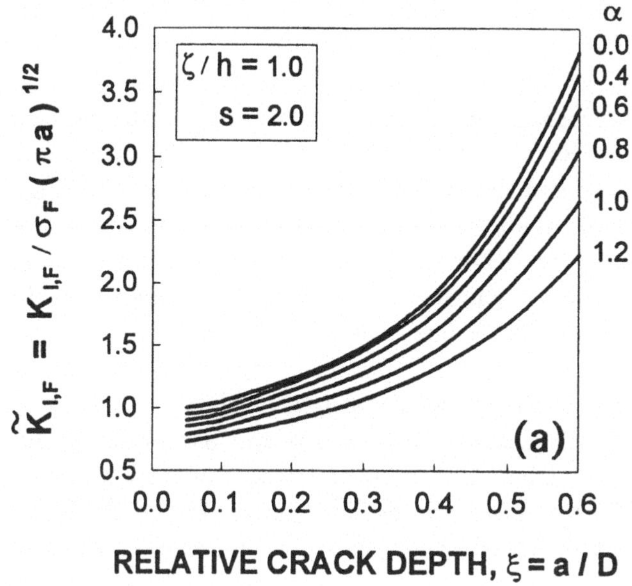

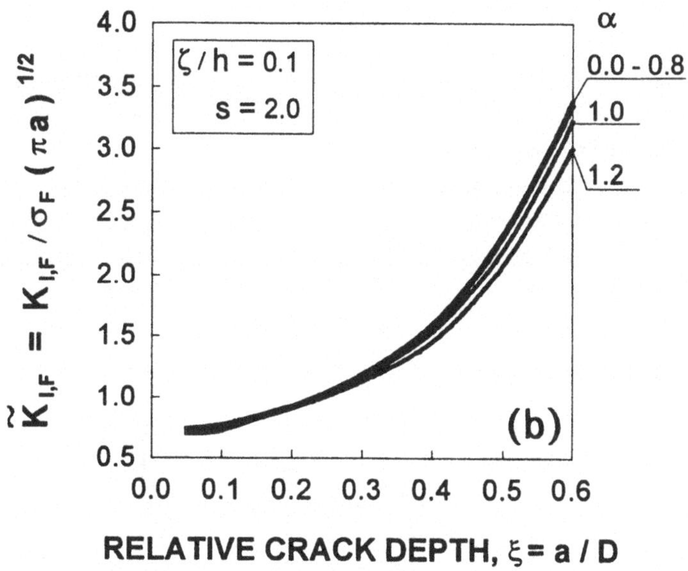

Figure 14. **Dimensionless stress-intensity factor against crack depth, for tension loading and s = 2 : (a) at point A ($\zeta / h = 1.0$); (b) at point C ($\zeta / h = 0.1$).**

3.2. FATIGUE PROPAGATION OF PART-THROUGH CRACKS IN ROUND BARS

The fatigue propagation of surface flaws in metallic round bars has been examined in Refs [3,27,28,31,37-44]. Crack growth patterns have experimentally and theoretically been analysed, by pointing out that these defects present an elliptical-arc shape and tend to follow preferred fatigue growth paths, for different initial flaw sizes, material properties and loading conditions. Similar conclusions have also been drawn for part-through cracks in plates (see previous sections).

3.2.1. *Two-parameter model for fatigue analysis*

Consider an elliptical-arc flaw in a round bar under cyclic loading with constant amplitude. The generic crack front with semi-axes a and b (Fig.15) will grow after one cyclic loading step to the new configuration described by the following equation ([37] and chapter 18 in Ref.[3]) :

$$\frac{x^2}{(b^*)^2} + \frac{y^2}{(a^*)^2} = 1 \tag{11}$$

where the two unknowns a^* and b^* can be determined through the condition that the coordinates of the points A^* and C^*, obtained from eqn (5), must satisfy eqn (11).

The propagation paths (thick lines) for surface flaws under cyclic bending with stress range $\Delta\sigma_M$ = 100 N mm $^{-2}$ are presented in Fig.16 [40]. Seven initial crack configurations are examined, with relative crack depth ξ

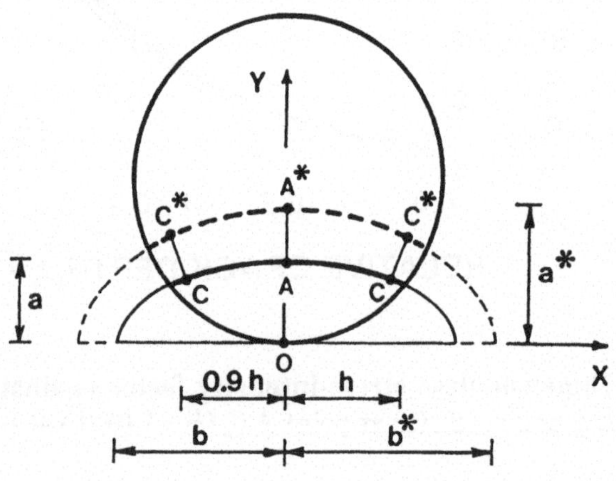

Figure 15. Fatigue crack propagation : two-parameter model [37].

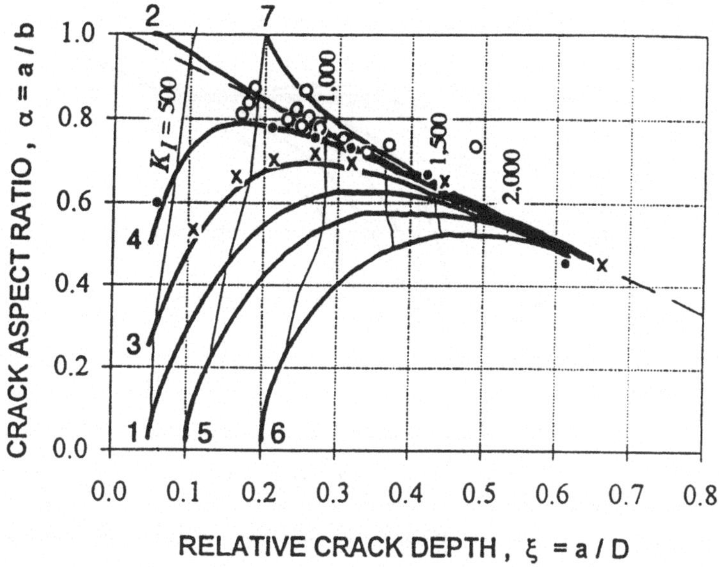

Figure 16. Fatigue propagation paths (thick lines) for different initial crack configurations under bending [40]. Some iso-K curves (thin lines) are also plotted, with K_I ranging from 500 to 2,000 N mm$^{-3/2}$. Experimental data from Ref.[31] (o) and Ref.[43] (x and •).

equal to 0.05, 0.10 or 0.20, and crack aspect ratio $\alpha = a/b$ equal to 0.001 (straight front), 0.25, 0.50 or 1.00 (circular front). All the propagation paths tend to converge to an inclined asymptote in the diagram of α against ξ (dashed line in Fig.16). In particular, α is equal to about 0.5 for $\xi = 0.6$. Several experimental data reported in the literature [31,43] lie on the propagation curves determined by the above theoretical model.

Some iso-K curves (thin lines) are also displayed in Fig.16, that is, the crack configurations where the maximum stress-intensity factor attains a given value are connected in this diagram. The crack growth phenomenon will become unstable, i.e. catastrophic failure will occur, when the fatigue fracture toughness K_{fC} is attained along the crack front. In other words, even if the theoretical propagation patterns in Fig.16 have been analyzed for ξ up to 0.65, an instantaneous failure can take place for relative crack depth smaller than this value.

The fatigue propagation paths for axial loading tend to converge to an inclined asymptote higher than that obtained for bending loading [37,40].

The intersection angle β (Fig.8) against the parameter ξ is plotted in the case of bending loading(Fig.17), for the seven initial crack configurations examined above. As has already been discussed in the present paper, some authors have theoretically deduced that the crack growth process requires a

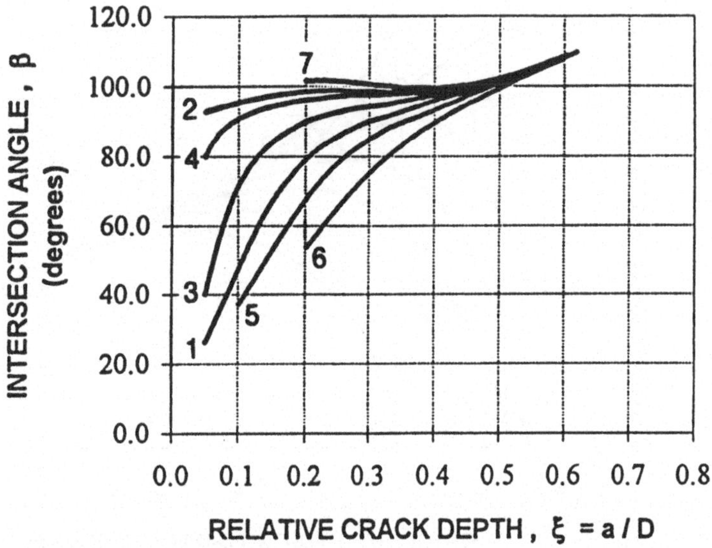

Figure 17. Crack front intersection angle β against relative crack depth ξ for the propagation paths plotted in Fig.16 (cyclic bending loading) [40].

square-root singularity and, consequently, β must be equal to about 100° for $\nu = 0.3$ and Mode I (Fig.3). The intersection angle numerically obtained (Fig.17) tends to values very close to the theoretical result.

The same conclusions on the fatigue crack growth paths and the intersection angle β can also be drawn for different initial crack sizes, material properties and loading conditions. In particular, the influence of the material constants A and m and of the stress range have been analyzed in Refs [37,40].

3.2.2. *Three-parameter model for fatigue analysis*

A three-parameter model is proposed to theoretically analyse the fatigue propagation of elliptical-arc surface flaws in round bars. The ellipse centre is assumed to lie out of the bar cross-section (Fig.18), and the generic crack front after one cyclic loading step can be expressed by the following equation [41]:

$$\frac{x^2}{\left(b_{el}^*\right)^2} + \frac{\left(y + y^*\right)^2}{\left(a_{el}^*\right)^2} = 1 \tag{12}$$

where the three unknowns a_{el}^*, b_{el}^* and y^* are deduced from the least squares method. More precisely, eqn (12) can be re-written as

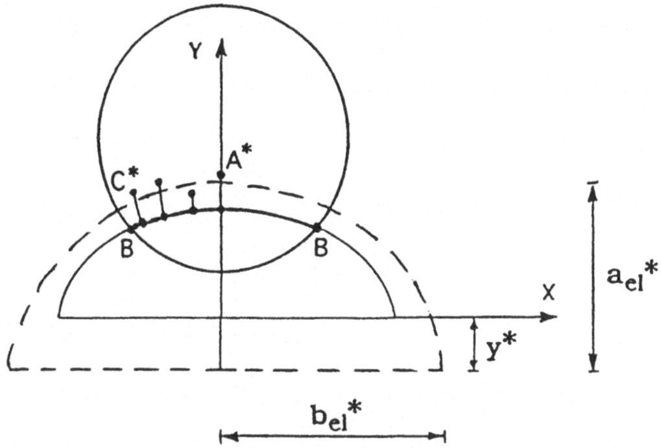

Figure 18. Fatigue crack propagation : three-parameter model [41].

$$x^2 = f_0 + f_1\, y + f_2\, y^2 \tag{13}$$

with f_0, f_1 and f_2 being functions of the above-mentioned three unknowns (and vice versa). The values of f_0, f_1 and f_2 can be calculated by assuming that the coordinates of several points along the new crack front (A*, C*, etc., Fig.18), obtained from eqn(5), must minimize the summation of the squared distances of the above points from the curve described by eqn(13) (least squares method). Successively, the three unknowns in Fig.18 can easily be determined.

As an example, the propagation patterns for surface flaws under cyclic bending with stress range $\Delta\sigma_M = 100$ N mm^{-2} are presented in Fig.19. The initial flaws examined have relative crack depth ξ_0 equal to 0.1 and crack aspect ratio α_0 equal to 0.001 (straight front, cases 1 to 3) or 1.0 (circular front, cases 4 to 6). The ellipse shifting s_0 is made to vary from 1.0 to 10.0, but only some values from 1.0 to 2.5 are considered in Fig.19 for better intelligibility of the diagram. For each initial defect configuration analysed (s_0, ξ_0, α_0), the least squares method is applied by considering five points along half the crack front, with the normalized coordinate equal to 0.10 (point C), 0.25, 0.50, 0.75, 1.00 (point A), respectively.

Preferred patterns in the diagram of $\alpha = a_{el} / b_{el}$ against $s = a_{el} / a$ and $\xi = a / D$ are followed during fatigue crack growth. For example, the equations of some propagation paths in the range $0.1 \le \xi \le 0.6$ can be written as follows :

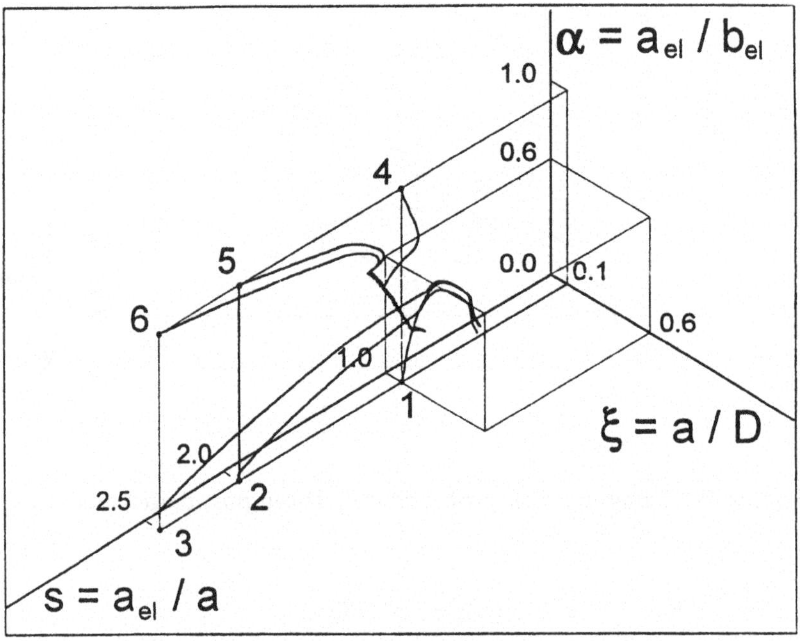

Figure 19. Fatigue propagation paths for different initial flaw configurations under bending loading ($\Delta\sigma_M$ = 100 N mm^{-2}) [41].

For point 1 with coordinates (s_0, ξ_0, α_0)=(1.0; 0.1; 0.001)

$$\alpha = -0.251 + 4.317\,\xi - 4.900\,\xi^2, \tag{14}$$

$$s = 1.026 - 0.240\,\xi + 0.460\,\xi^2, \tag{15}$$

For point 3 (2.5; 0.1; 0.001)

$$\alpha = -0.086 + 3.517\,\xi - 4.004\,\xi^2, \tag{16}$$

$$s = 2.973 - 10.517\,\xi + 12.860\,\xi^2, \tag{17}$$

The propagation paths tend to converge to an inclined asymptotic plane, with the parameter s equal to about 1.0 - 1.5 for $\xi = 0.6$. Therefore, even if the initial flaw presents the ellipse centre very distant from the bar cross-section ($s_0 \gg 1.0$), this centre tends to shift toward the bar circumference during fatigue crack propagation. The same conclusions can be drawn for cyclic axial loading. Moreover, it can be remarked that, for $\xi = 0.6$, the crack aspect ratio α is equal to about 0.7 - 0.8 for axial loading and 0.6 - 0.7 for bending loading, and that is consistent with previous experimental and theoretical results [37,40].

The intersection angle β (Fig.8) against the parameter ξ is plotted in the case of bending loading (Fig.20), for the six initial crack configurations examined above. The intersection angle numerically obtained (Fig.20) tends to values very close to the theoretical results previously discussed (Fig.3).

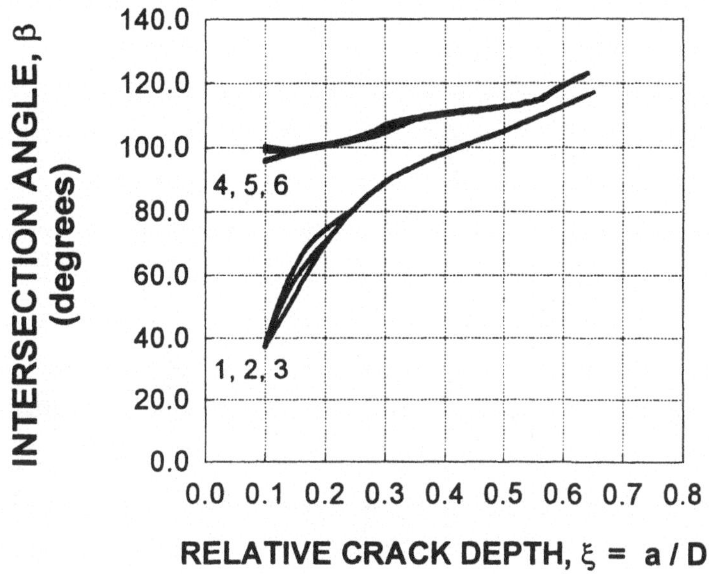

Figure 20. Crack front intersection angle β against relative crack depth ξ for the propagation paths plotted in Fig.19 (cyclic bending loading).

4. Part-through Cracks in Round Pipes

Several investigations related to part-through cracks in metallic pipes have been carried out [31,32,45-54]. A circumferential external surface flaw in a round pipe under constant amplitude cyclic axial and bending loading is analysed in the following. The stress-intensity factor variation along the crack front is determined through a finite element analysis and the fatigue flaw propagation paths are theoretically obtained by employing a two-parameter model analogous to that used in Section 3.2.1. for round bars.

4.1. STRESS CONCENTRATION IN PART-THROUGH CRACKED ROUND PIPES

The surface defect is assumed to present an elliptical-arc shape (Fig.21) with aspect ratio $\alpha = a_{el} / b_{el}$, while the relative depth ξ of the deepest point on the crack front is equal to the ratio between the maximum crack depth a and the pipe wall thickness t; the ellipse shifting $s = a_{el} / a$ defines the distance of the ellipse centre from the external perimeter of the pipe cross-section. The parameter R / t , with R = internal radius of the pipe, ranges from 1 (thick wall thickness) to 10 (thin wall thickness).

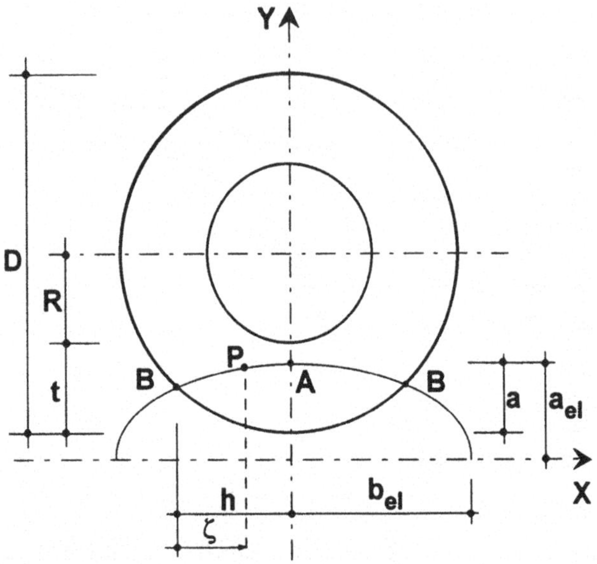

Figure 21. Circumferential external surface flaw in a round pipe.

The dimensionless stress-intensity factors for tension and bending are calculated from eqns (3) and (4) , where

$\sigma_F = F / [\ \pi (D^2 - 4\,R^2) / 4\ ]$ = applied uniform tensile stress,

$\sigma_M = M / [\ \pi ((D/2)^4 - R^4) / D\ ]$ = maximum bending stress.

The influence of the crack aspect ratio α on the stress-intensity factor distribution is displayed in Fig.22 for $\xi = 0.3$, $s = 1$ and $R / t = 1$. The transition phenomenon pointed out in Section 3.1. for round bars is also observed for round pipes. In particular, the change of the slope sign occurs for $\alpha_t = 1$ in the case of tension and for $\alpha_t = 0.9$ in the case of bending. For other values of the geometrical parameters (ξ, s, R / t), different values of the transition aspect ratio α_t can be determined, with α_t for tension greater than that for bending at the same values of ξ, s and R / t .

Figure 23 shows the stress-intensity factors in the case of both tension and bending, for several values of ellipse shifting s. The influence of this parameter is remarkable since the maximum is attained either in the middle point of the crack front or near the intersection point between crack front and external surface , and that depends on the value of s .

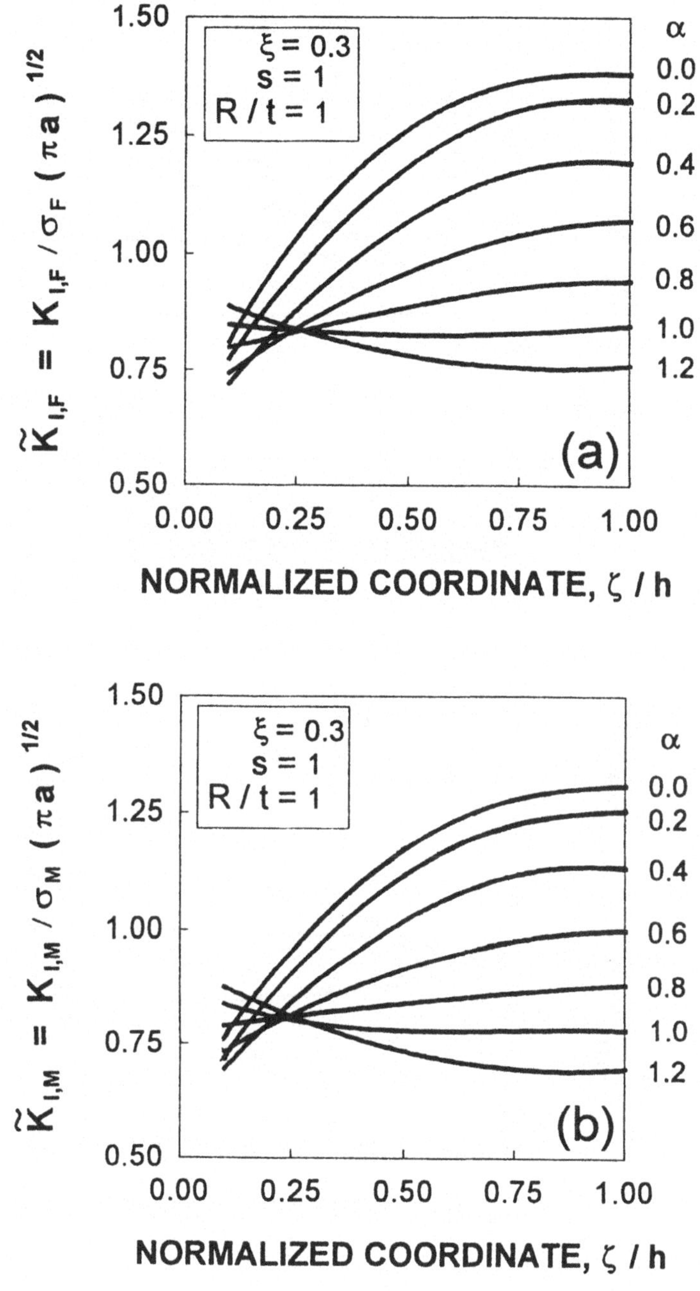

Figure 22. Influence of the crack aspect ratio α for both tension (a) and bending (b), in the case of $\xi = 0.3$, $s = 1$ and $R / t = 1$.

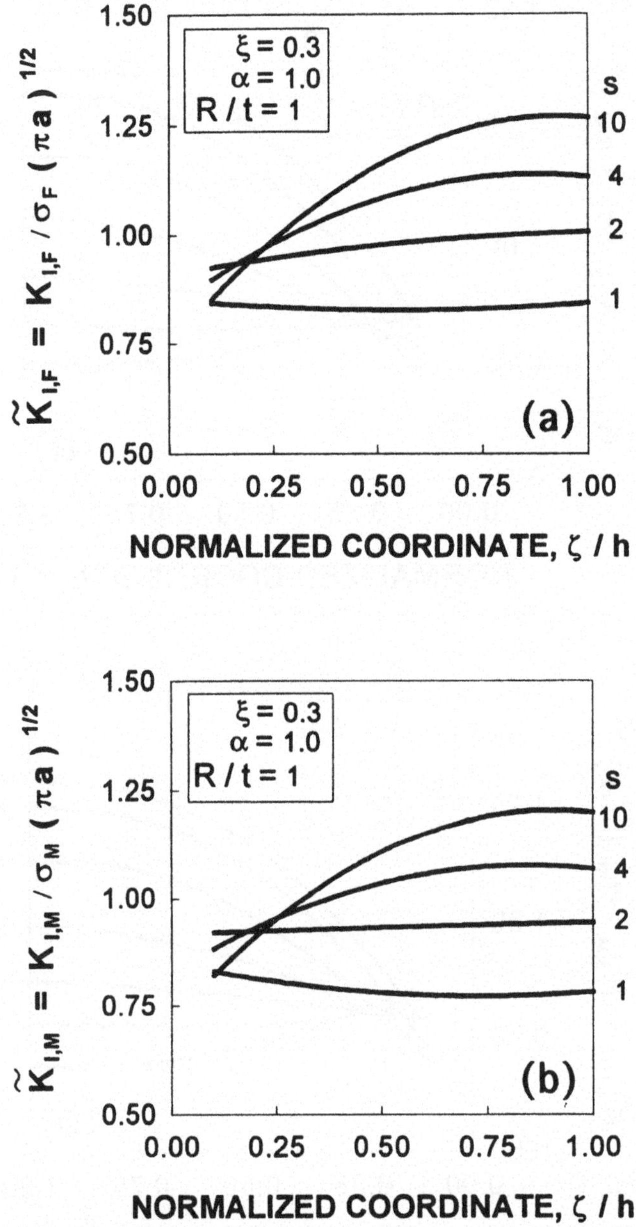

Figure 23. Influence of the ellipse shifting s for both tension (a) and bending (b), in the case of $\xi = 0.3$, $\alpha = 1.0$ and $R/t = 1$.

From the above remarks we can conclude that the most critical parts on the front are the deepest point A and the ends of the defect, as in the case of surface flaws in round bars. Diagrams analogous to those in Fig.24 have been obtained for different values of the geometrical parameters being considered.

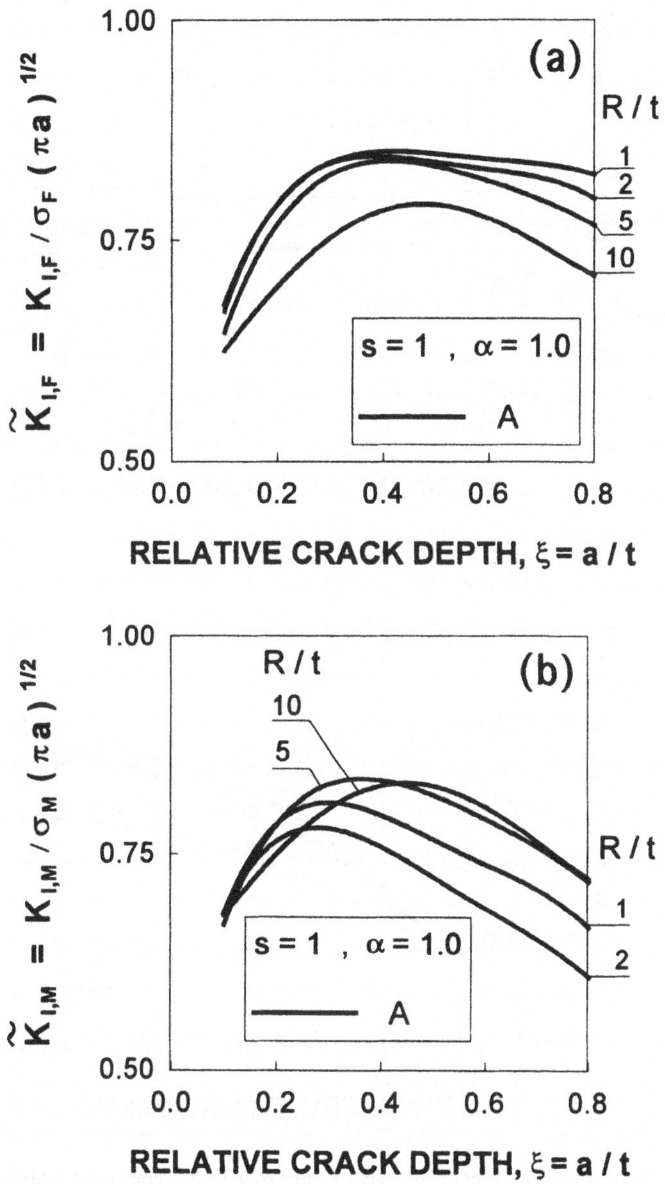

Figure 24. Stress-intensity factor at point A for both tension (a) and bending (b), in the case of $s = 1$, $\alpha = 1.0$ and different values of ξ and R/t.

Some results of the present study are compared with those from other authors [32,52] (Fig.25). The agreement between the values plotted is quite satisfactory.

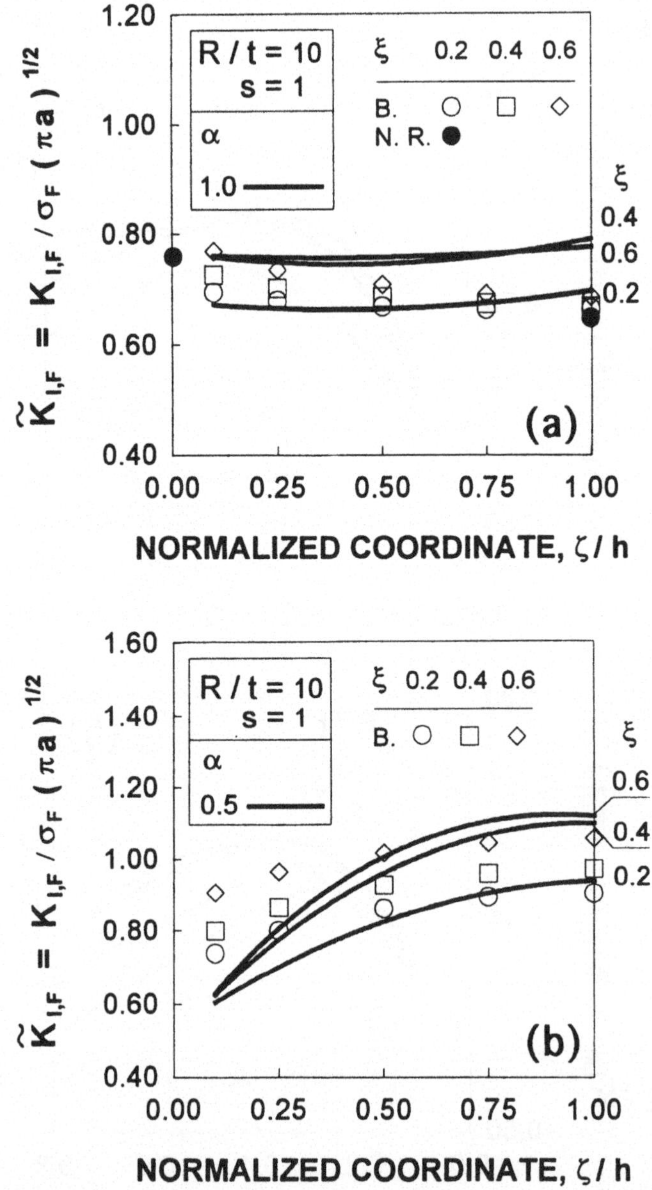

Figure 25. Comparison between the results of the present study and those taken from Refs[32,52] for $\alpha = 1.0$ (a) and $\alpha = 0.5$ (b), in the case of tension, $R / t = 10$, $s = 1$ and different values of ξ.

4.2. FATIGUE PROPAGATION OF PART-THROUGH CRACKS IN ROUND PIPES

The fatigue growth of circumferential external surface flaws in pipes is theoretically examined, by assuming the ellipse centre to lie on the external perimeter of the pipe cross-section during the whole propagation (Fig.26). By employing a two-parameter model analogous to that used in Section 3.2.1. for round bars, the generic crack front after one cyclic loading step can be expressed by the following equation:

$$\frac{x^2}{\left(b_{el}^*\right)^2} + \frac{y^2}{\left(a_{el}^*\right)^2} = 1 \tag{18}$$

where the two unknowns a_{el}^* and b_{el}^* can be obtained from the condition that the coordinates of the points A* and C*, deduced from eqn (5), must satisfy eqn (18).

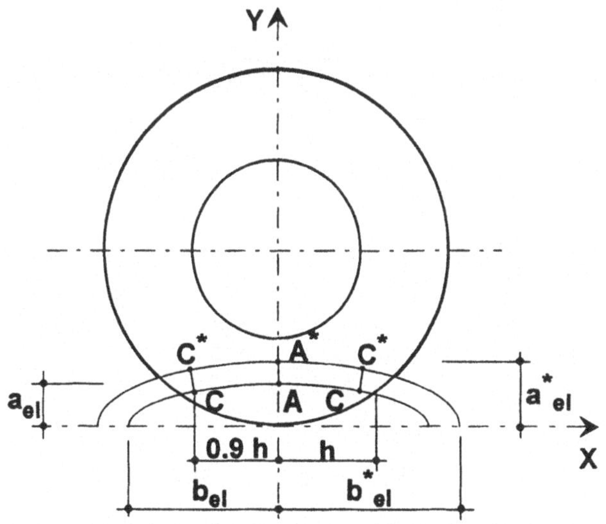

Figure 26. Fatigue crack propagation : two-parameter model.

Figure 27 shows the propagation paths for both axial and bending loading, in the case of thick wall thickness of the pipe (R / t = 1). Six initial flaw configurations are analysed : the crack aspect ratio α is equal to 0.0 for the cases 1 to 3, while the relative crack depth ξ is equal to 0.05 for the

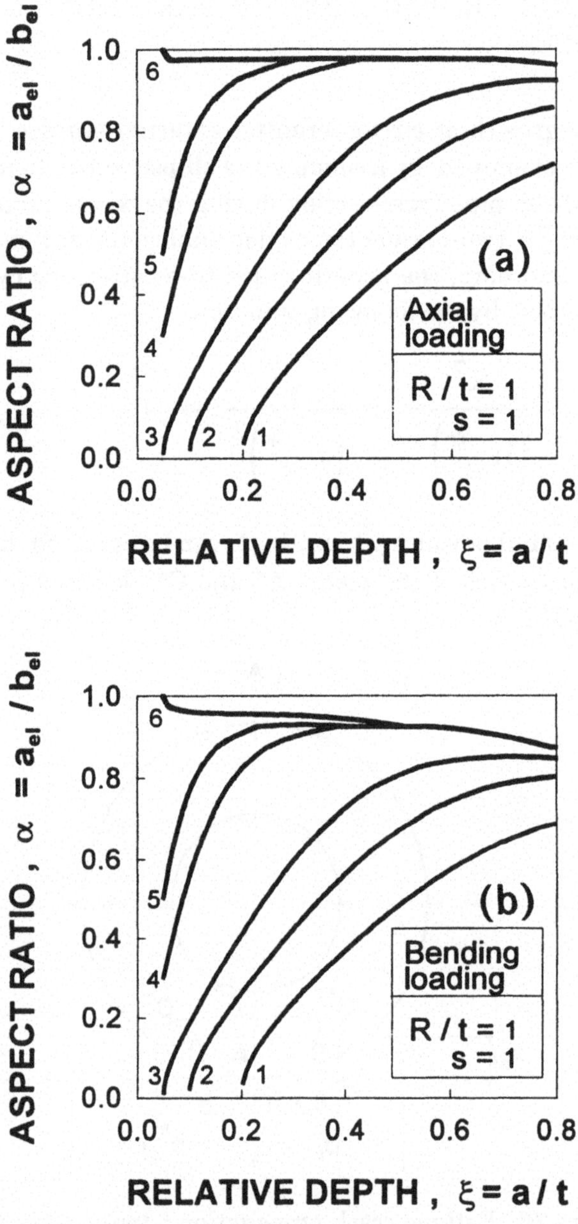

Figure 27. Fatigue crack propagation paths for axial loading (a) and bending loading (b).

cases 3 to 6. The fatigue behaviour for pipes is different from that observed for plates and round bars. As a matter of fact, the crack growth paths in the diagram of α against ξ converge to an inclined asymptote for plates and round bars, while they converge to a horizontal asymptote for pipes (Fig.27). Such asymptote almost corresponds to the straight line expressed by the equation $\alpha = 0.9 \sim 1.0$ in the case of axial loading (especially for low values of the initial relative crack depth), while it is little lower ($\alpha = 0.8 \sim 0.9$) in the case of bending loading.

Finally the intersection angle β between crack front and external surface, examined in Sections 2. and 3., is displayed against the parameter ξ (Fig.28), for the six initial crack configurations analysed above. Such angle, numerically obtained through the two-parameter model, tends to values very close to the theoretical result (equal to about 100°), discussed in Section 2.

5. Conclusions

The propagation of part-through flaws in metallic plates, round bars and pipes under cyclic axial and bending loading has been examined numerically. The following conclusions can be drawn :

(1) The influence of the flaw aspect ratio α on the stress-intensity factor distribution along the front of a surface defect is remarkable. The maximum stress-intensity factor is attained in correspondence to the deepest point on the crack front for low values of α, while it is attained near the external surface for $\alpha = 1$ (circular-arc front). A transition phenomenon is noticed for intermediate values of aspect ratio.

(2) Preferred propagation paths for elliptical-arc flaws in cyclically loaded round bars are obtained through numerical models. It is shown that the fatigue crack growth curves converge to an inclined asymptote in the diagram of flaw aspect ratio α against relative crack depth ξ of the deepest point on the front, for different initial crack sizes, material properties and loading conditions. Several experimental data reported in the literature lie on such theoretical paths. Moreover, for all cases considered, the intersection angle β between crack front and external surface tends to values very close to those theoretically deduced by means of fracture energy considerations.

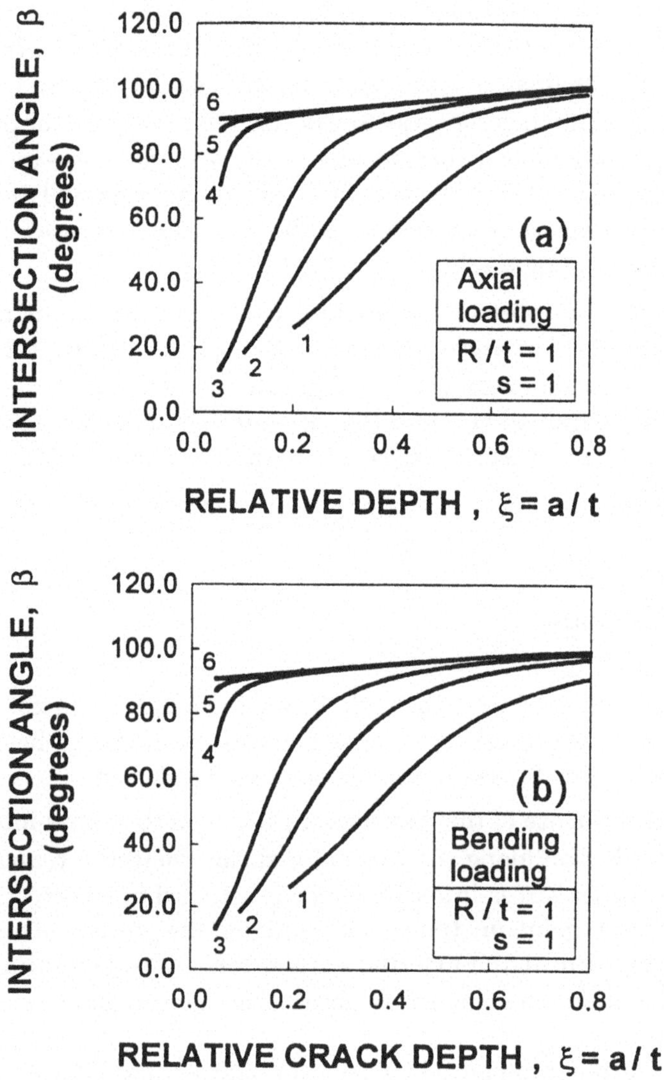

Figure 28. Crack front intersection angle β against crack depth ξ for the propagation paths in Fig.27 : axial loading (a) and bending loading (b).

(3) The conclusions at point (2) are analogous to those drawn by several authors for plates.

(4) The conclusions at point (2) are also analogous to those drawn in the present paper for round pipes. The only difference is that the crack propagation paths, obtained numerically for pipes, tend to converge to a horizontal asymptote in the diagram of α against ξ , with α equal to 0.9 ~ 1.0 for axial loading and 0.8 ~ 0.9 for bending loading.

Acknowledgements

The authors gratefully acknowledge the research support for this work provided by the Italian Ministry for University and Technological and Scientific Research (MURST) and the Italian National Research Council (CNR).

6. References

1. R.A. Smith (Editor), *Fatigue Crack Growth. 30 Years of Progress*. Proc. Conf. Fatigue Crack Growth, Cambridge, U.K. (1984); Pergamon Press, Oxford, U.K. (1986).
2. K.J. Miller, *Mater. Sci. Technol.* **9**, 453-462 (1993).
3. A. Carpinteri (Editor), *Handbook of Fatigue Crack Propagation in Metallic Structures*. Elsevier Science Publishers B.V., Amsterdam, The Netherlands (1994).
4. H. Tada, P.C. Paris and G.R. Irwin (Editors), *The Stress Analysis of Cracks Handbook*. Del Research Corp., Hellertown, Pennsylvania, U.S.A. (1973).
5. Y. Murakami et al. (Editors), *Stress Intensity Factors Handbook*. Vols. I and II, Pergamon Press, Oxford; U.K. (1987); Vol. III, The Society of Materials Science, Japan, and Pergamon Press, Oxford, U.K. (1992).
6. C. Ruiz and J. Epstein, *Int. J. Fract.* **28**, 231-238 (1985).
7. J.P. Benthem, *Int. J. Solids Struct.* **13**, 479-492 (1977).
8. Z.P. Bazant and L.F. Estenssoro, *Proc. 4th Int. Conf. Fract.*, Waterloo, Canada, 1977 (Edited by D.M.R. Taplin), Vol. 3a, pp. 371-385, Pergamon Press, Oxford, U.K. (1977).
9. Z.P. Bazant and L.F. Estenssoro, *Int. J. Solids Struct.* **15**, 405-426 (1979).
10. J.P. Benthem, *Int. J. Solids Struct.* **16**, 119-130 (1980).
11. S.X. Wu, *Engng Fract. Mech.* **22**, 897-913 (1985).
12. *ASME Boiler and Pressure Vessels Code*, Section XI, Appendix A, American Soc. Mechanical Engineers (ASME), U.S.A. (1981).
13. Published Document BS, PD 6493, *Guidance on Some Methods for the Derivation of Acceptance Levels for Defects in Fusion Welded Joints*. British Standards Institution, London, U.K. (1991).
14. A. Carpinteri, *J. Strain Anal. Engng Design* **28**, 117-123 (1993).
15. R.S. Barsoum, *Int. J. Fract.* **10**, 603-605 (1974).
16. R.D. Henshell and K.G. Shaw, *Int. J. numer. Meth. Engng* **9**, 495-507 (1975).

17. R.S. Barsoum, *Int. J. numer. Meth. Engng* **10**, 25-37 (1976).
18. I.S. Raju and J.C. Newman, Jr., *Engng Fract. Mech.* **11**, 817-829 (1979).
19. J.C. Newman, Jr. and I.S. Raju, *Engng Fract. Mech.* **15**, 185-192 (1981).
20. X.R. Wu, *Engng Fract. Mech.* **19**, 387-405 (1984).
21. P.C. Paris and F. Erdogan, Trans. American Soc. Mechanical Engineers (ASME), *J. Basic Engng* **85D**, 528-534 (1963).
22. J. Schijve and D. Broek, *Aircraft Engng* **34**, 314-316 (1962).
23. A. Carpinteri, *Fatigue Fract. Engng Mater. Struct.* **15**, 365-376 (1992).
24. A. Carpinteri, *Computers Struct.* **44**, 1317-1338 (1992).
25. X.B. Lin and R.A. Smith, *Fatigue Fract. Engng Fract. Mech.* **18**, 247-256 (1995).
26. M.A. Astiz and M. Elices, *Proc. 2nd Int. Conf. Numer. Meth. Fract. Mech.*, pp. 93-106 (1980).
27. A. Athanassiadis, J.M. Boissenot, P. Brevet, D. François and A. Raharinaivo, *Int.J.Fract.* **17**, 553-566 (1981).
28. W.D. Dover, *Fatigue Fract. Engng Mater. Struct.* **5**, 349-353 (1982).
29. H. Nisitani and D.H. Chen, *Trans. Japanese Society Mechanical Engineers* (1984); in *Stress Intensity Factors Handbook* (Edited by Y. Murakami et al.), Vol. 2, pp. 654-656, Pergamon Press, Oxford, U.K. (1987).
30. M. Shiratori, T. Miyoshi, Y. Sakai and G.R. Zhang, *Trans. Japanese Society Mechanical Engineers* (1986); in *Stress Intensity Factors Handbook* (Edited by Y. Murakami et al.), Vol. 2, pp. 659-665, Pergamon Press, Oxford, U.K. (1987).
31. R.G. Forman and V. Shivakumar, in *Fracture Mechanics:Seventeenth Volume*, ASTM STP 905, pp. 59-74 (1986).
32. I.S. Raju and J.C. Newman, in *Fracture Mechanics : Seventeenth Volume*, ASTM STP 905, pp. 789-805 (1986).
33. J. Llorca and V. Sánchez-Gálvez, *Fatigue Fract. Engng Mater. Struct.* **26**, 869-882 (1987).
34. A. Carpinteri, *Engng Fract. Mech.* **42**, 1035-1040 (1992).
35. A. Carpinteri, *Fatigue Fract. Engng Mater. Struct.* **15**, 1141-1153 (1992).
36. J. Toribio, *Int.J.Fract.* **53**, 367-385 (1992).
37. A. Carpinteri, *Int. J. Fatigue* **15**, 21-26 (1993).
38. A. Levan and J. Royer, *J.Fract.* **61**, 71-99 (1993).
39. A. Carpinteri and R. Brighenti, *Invited Keynote Lecture, Proc. 12th Spanish Conf. Fract.*, pp. 11-18, La Coruña, Spain (1995).
40. A. Carpinteri and R. Brighenti, *Int. J. Fatigue* **18**, 33-39 (1996).

41. A. Carpinteri and R. Brighenti, to be published on *Fatigue Fract. Engng Mater. Struct.* **19** (1996).
42. T.L. Mackay and B.J. Alperin, *Engng Fract. Mech.* **21**, 391-397 (1985).
43. T. Lorentzen, N.E. Kjaer and T. Henriksen, *Engng Fract. Mech.* **23**, 1005-1014 (1986).
44. M. Caspers, C. Mattheck and D. Munz, *Z. Werkstofftech.* **17**, 327-333 (1986).
45. F. Delale and F. Erdogan, *J.Appl.Mech.* **49**, 97-102 (1982).
46. B.K. Neale, A.B. Haines and A.G. Miller, *Fatigue Fract. Engng Mater. Struct.* **12**, 597-609 (1989).
47. Z.G. Chen and X. Zhang, *Engng Fract. Mech.* **32**, 639-652 (1989).
48. L.P. Pook, *Fatigue Fract. Engng Mater. Struct.* **13**, 135-143 (1990).
49. Y. Dai, M. Rödig and J. Altes, *Fatigue Fract. Engng Mater. Struct.* **14**, 11-23 (1991).
50. H. Gao and G. Herrmann, *Engng Fract. Mech.* **41**, 695-706 (1992).
51. M.K. Kassir, C.H. Hofmayer and K.K. Bandyopadhyay, *Engng Fract. Mech.* **43**, 807-813 (1992).
52. M. Bergman, *Fatigue Fract. Engng Mater. Struct.* **18**, 1155-1172 (1995).
53. A. Carpinteri, R. Brighenti and A. Spagnoli, *Invited Keynote Paper, Proc. 1st Int. Conf. Computational Meth. Testing Engng Integrity (CMT 96)*, pp. 3-10, Kuala Lumpur, Malaysia (1996).
54. A. Carpinteri, R. Brighenti and A. Spagnoli, *Proc. 11th Biennial European Conf. Fract.*, Poitiers-Futuroscope, France (1996).

MULTIAXIAL FATIGUE LIFE PREDICTION METHODS FOR ENGINEERING COMPONENTS

T. D. LIEBSTER and G. GLINKA
University of Waterloo
Department of Mechanical Engineering
Waterloo, Ontario, CANADA
N2L 3G1

Abstract

A method for calculating elasto-plastic notch tip strains and stresses in bodies subjected to multiaxial loading is presented. Two approximate formulae are derived based on the analysis of strain energy density in the notch tip region. They represent the lower and upper limits of the band within which the actual elasto-plastic notch tip strains can be found. All of the necessary relationships are derived for a general multiaxial stress state. Each method consists of a set of seven linear algebraic relations that can easily be solved for elastic-plastic strain and stress increments, given the hypothetical, elastic, notch tip stress history and the material stress-strain curve. Results of the validation show that the proposed methods compare well with experimental and finite element data.

A multiaxial fatigue parameter based on strain energy density which normalises fatigue data obtained under a variety of mean stress levels and loading combinations is presented. When combined with the notch analysis, it enables fatigue life predictions to be calculated. This parameter represents the proportion of the overall strain energy that is contributed by the stresses and strains on the critical (fracture) plane. It is shown that multiaxial fatigue life data may be accurately correlated by applying this parameter to the experimental results for SAE 1045 steel components tested under both independently and simultaneously applied tension and torsion.

Nomenclature

α, β	-	indices. $\alpha, \beta = 1,2,3$ summation is not implied
α_σ	-	angle of principal axis with respect to the specimen axis
$\Delta\varepsilon_{22}$	-	normal strain range in the critical plane

R.A. Smith (ed.), Reliability Assessment of Cyclically Loaded Engineering Structures, 101–136.

$\Delta\varepsilon_{ij}^{p}$	-	plastic strain increments
$\Delta\gamma_{21}$	-	shear strain range in the critical plane
δ_{ij}	-	Kronecker's delta, $\delta_{ij} = 1$ for $i = j$ and $\delta_{ij} = 0$ for $i \neq j$
ΔP	-	axial load range
$\Delta\sigma_{21}$	-	shear stress range in the critical plane
$\Delta\sigma_{22}$	-	normal stress range in the critical plane
$\Delta\sigma_{ij}$, $\Delta\varepsilon_{ij}$	-	increments in notch tip stress and strain components
ΔT	-	torsion moment range
ΔW^{e}	-	elastic strain energy density
ΔW^{p}	-	plastic strain energy density
E	-	modulus of elasticity
ε_{22}^{a}	-	amplitude of normal strain component in the critical plane
ε_{eq}^{p}	-	equivalent plastic strain
ε_{eq}^{pE}	-	equivalent plastic strain determined from the the ESED method
ε_{ij}^{a}	-	actual elasto-plastic strain components in the notch tip
ε_{ij}^{A}	-	strain components in fixed coordinate system co-axial with the specimen axis
ε_{ij}^{E}	-	elasto-plastic notch-tip strains obtained from the ESED method
ε_{ij}^{e}	-	notch tip strain components obtained from linear elastic analysis
ε_{ij}^{N}	-	elasto-plastic notch-tip strains obtained from the Neuber method
ε_{ij}^{n}	-	nominal strain tensor components
ε_{ij}^{p}	-	plastic components of the notch-tip strain tensor
ε_{n}	-	nominal strain
ESED	-	equivalent strain energy density
F	-	axial load
γ^{*}	-	multiaxial fatigue parameter
γ_{21}^{a}	-	amplitude of shear strain component in the critical plane
H	-	plastic modulus
K'	-	cyclic strength coefficient
k, K	-	material constants
K_{ε}	-	strain concentration factor
K_{σ}	-	stress concentration factor
K_{t}	-	theoretical elastic stress concentration factor
ν	-	Poisson's ratio
n	-	exponent of power law stress-strain curve
n'	-	cyclic strain hardening exponent
ν_{e}, ν_{p}	-	elastic and plastic Poisson's ratios respectively
ν_{eff}	-	effective Poisson's ratio
N_{f}	-	number of cycles to failure or to the initiation of a 1 mm crack
R	-	radius of a cylindrical specimen
S	-	equivalent nominal stress
σ_{21}^{a}	-	amplitude of shear stress component in the critical plane
σ_{21}^{max}	-	maximum absolute value of shear stress in the critical plane

$\sigma_{22}{}^{a}$	-	amplitude of normal stress component in the critical plane
$\sigma_{22}{}^{max}$	-	maximum value of normal stress in the critical plane
S_{eq}	-	equivalent nominal stress
σ_{eq}	-	equivalent stress
$\sigma_f{}'$	-	axial cyclic fatigue strength coefficient
S_{ij}	-	deviatoric stress components
σ_{ij}	-	stress components associated with the critical plane
$\sigma_{ij}{}^{a}$	-	actual stress tensor components in the notch tip
$\sigma_{ij}{}^{A}$	-	stress components in fixed coordinate system co-axial with the specimen axis
$\sigma_{ij}{}^{e}$	-	notch tip stress tensor components obtained from linear elastic analysis.
$\sigma_{ij}{}^{E}$	-	notch tip stress tensor components obtained from the ESED model
$\sigma_{ij}{}^{N}$	-	notch tip stress tensor components obtained from the Neuber solution
σ_n	-	nominal stress
$\sigma_n{}^{F}$	-	nominal (average) stress in the net cross section due to axial load F
σ_o	-	parameter of the material stress-strain curve (yield limit)
S_y	-	yield stress
T	-	torque
t	-	wall thickness
$\tau_f{}'$	-	torsional cyclic fatigue strength coefficient
τ_n	-	nominal shear stress in the net cross section obtained from the torsion formula
W	-	strain energy density
W^*	-	multiaxial fatigue strain energy density parameter
W^a	-	actual strain energy density at the notch tip
W^E	-	strain energy density at the notch tip according to the equivalent strain energy density (ESED) hypothesis
W^e	-	strain energy density at the notch tip obtained from linear elastic solution
W^N	-	strain energy density at the notch tip according to the generalized Neuber's rule

1. Introduction

Fatigue durability and strength analysis of machine components and structures subjected to multiaxial cyclic loads requires the determination of elasto-plastic strains and the accumulated fatigue damage at the point where the highest stress concentration occurs. This requires models that can efficiently simulate the notch tip stress-strain histories caused by externally applied cyclic loads and fatigue parameters that can account for damage caused by multiaxial stress states. The resulting local stress-strain histories can be used to determine the amount of fatigue damage at the notch, and to further estimate the fatigue life of the component.

Any fatigue life prediction procedure consists of three main areas that are used to analyse and input data into the calculation procedure:

- material properties;
- loading/stress history; and
- cyclic stress-strain analysis and damage evaluation.

The material and loading history inputs are similar in most methods; however, the local inelastic stress-strain analyses and the fatigue damage calculation methods may differ depending on the general philosophy used by an analyst. The main subject of this paper is the presentation and critical analysis of some of the contemporary approaches to multiaxial fatigue that are gaining some popularity among researchers and practising engineers.

2. Elasto-plastic notch tip stress-strain calculation methods

The most frequently used methods for calculating the notch tip stress-strain field due to cyclic loads are Neuber's rule [1], which has been extended to fatigue problems by Topper et al. [2], and the equivalent strain energy density (ESED) method [3]. An extension of Neuber's rule for multiaxial stress states has been proposed by Hoffman and Seeger [4] and recently by Barkey and Socie [5]. A more general extension of Neuber's rule and the ESED method for multiaxial loading has been proposed by Moftakhar *et al.* [6]. The method proposed in [6] is based on strain energy density considerations and is governed by the assumption that the multiaxial loads are applied in a proportional manner. A similar formulation, appropriate for notched bodies subjected to monotonically applied non-proportional multiaxial loads, can be found in reference [7]. The extension of those models to multiaxial cyclic loading histories is discussed below.

2.1 UNIAXIAL STRESS OR PLANE STRAIN STATES AT THE NOTCH TIP

If the dimensions and external loads of a body are such that a state of plane stress exists, the stress state at the notch tip is uniaxial (Fig. 1a). In such a case, four independent relations are required to define the one unknown notch tip stress and three unknown strain components. Similarly, if the notched body is in a state of plane strain (Fig. 1b), four relations are needed to determine the four unknown notch tip strain and stress components. Three independent relationships can be defined by the material constitutive equations and only one additional relationship is required.

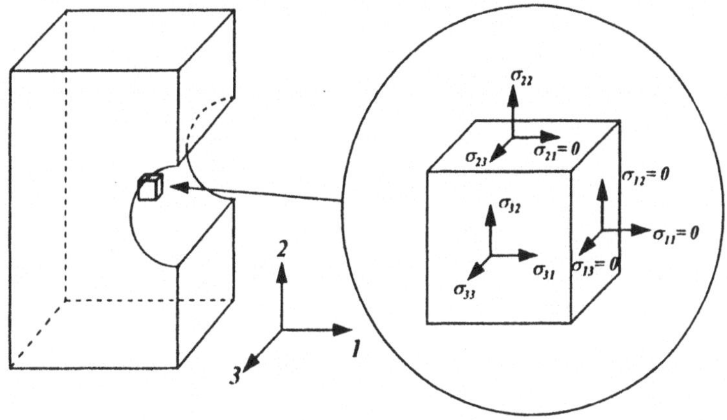

(a) Plane Stress

$$\sigma_{ij} = \begin{bmatrix} 0 & 0 & 0 \\ 0 & \sigma_{22} & 0 \\ 0 & 0 & 0 \end{bmatrix} \qquad \varepsilon_{ij} = \begin{bmatrix} \varepsilon_{11} & 0 & 0 \\ 0 & \varepsilon_{22} & 0 \\ 0 & 0 & \varepsilon_{33} \end{bmatrix}$$

(b) Plane Strain

$$\sigma_{ij} = \begin{bmatrix} 0 & 0 & 0 \\ 0 & \sigma_{22} & 0 \\ 0 & 0 & \sigma_{33} \end{bmatrix} \qquad \varepsilon_{ij} = \begin{bmatrix} \varepsilon_{11} & 0 & 0 \\ 0 & \varepsilon_{22} & 0 \\ 0 & 0 & 0 \end{bmatrix}$$

(c) General

$$\sigma_{ij} = \begin{bmatrix} 0 & 0 & 0 \\ 0 & \sigma_{22} & \sigma_{23} \\ 0 & \sigma_{32} & \sigma_{33} \end{bmatrix} \qquad \varepsilon_{ij} = \begin{bmatrix} \varepsilon_{11} & 0 & 0 \\ 0 & \varepsilon_{22} & \varepsilon_{23} \\ 0 & \varepsilon_{32} & \varepsilon_{33} \end{bmatrix}$$

Figure 1: Stress state at a notch tip (notation): (a) body in plane stress, (b) body in plane strain, (c) general biaxial stress state.

The additional relationship that is required for plane stress or plane strain states at the notch tip is generally defined by either Neuber's rule [1] or the ESED [3] relation. Both models relate the fictitious "linear elastic" stresses and strains at the notch tip $(\sigma_{ij}^{e}, \varepsilon_{ij}^{e})$ to the actual elasto-plastic stresses and strains $(\sigma_{ij}^{a}, \varepsilon_{ij}^{a})$ as shown in Fig. 2.

2.1.1 Neuber's Rule

The Neuber rule [1] was initially proposed for a notched body loaded in pure shear, but is most often used for notches under tensile or bending loads (Fig. 3). It relates the theoretical stress concentration factor, K_t, to the actual stress concentration, K_σ, and the strain concentration, K_ε, as follows:

$$K_t^2 = K_\sigma K_\varepsilon \tag{1}$$

where $$K_t = \frac{\sigma_{22}^{e}}{\sigma_n}; \quad K_\sigma = \frac{\sigma_{22}^{N}}{\sigma_n}; \quad K_\varepsilon = \frac{\varepsilon_{22}^{N}}{\varepsilon_n}; \quad \varepsilon_n = \frac{\sigma_n}{E}$$

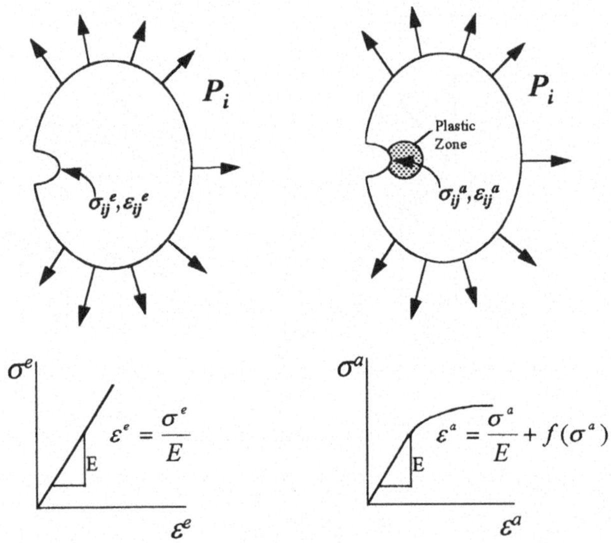

Figure 2: Geometrically identical elastic and elastic-plastic bodies subjected to identical boundary conditions.

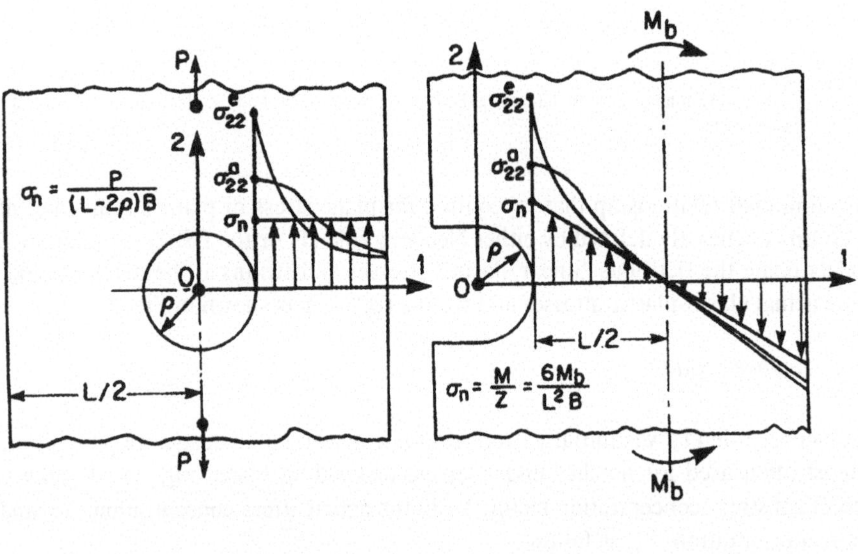

Figure 3: The nominal (σ_v), hypothetical elastic (σ_{22}^e), and the actual elastic-plastic (σ_{22}^a) stress distributions near notches for tensile loading (left) and bending load (right).

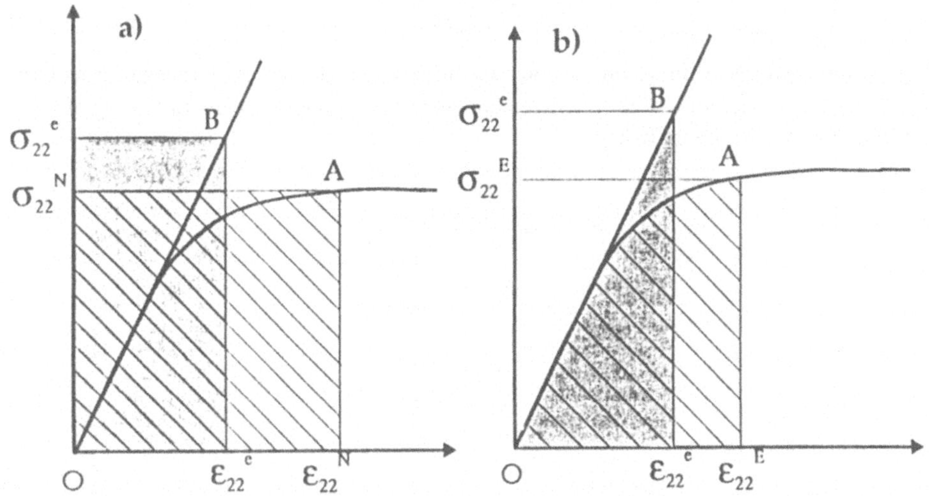

Figure 4: Graphical interpretation of (a) Neuber's rule, and (b) the equivalent strain energy density (ESED) method.

In the case of notched bodies in plane stress for which the stress state in the notch tip is uniaxial, Eq. (1) can also be written in a form that relates the elasto-plastic strain and stress components (ε_{22}^N and σ_{22}^N) to the hypothetical linear elastic notch tip strain and stress (ε_{22}^e and σ_{22}^e) that would occur in a purely elastic body with the same geometry an loading.

$$\sigma_{22}^{e}\,\varepsilon_{22}^{e} = \sigma_{22}^{N}\,\varepsilon_{22}^{N} \tag{2}$$

Thus, when the notch tip is subjected to a uniaxial stress state (as for a body in plane stress), Eq. (2) represents the equality of the total strain energy density at the notch tip, represented by the rectangles A and B in Fig. 4a. The total strain energy density is defined as the sum of the strain energy density and the complementary strain energy density. A relationship similar to Eq. (2) can also be written for notched bodies in plane strain.

The uniaxial constitutive stress-strain curves for engineering materials are most often given in the following general form:

$$\varepsilon = \frac{\sigma}{E} + f(\sigma) \tag{3}$$

For a uniaxial constitutive law in the form of Eq. (3), the set of four equations that is necessary to determine all of the stress and strain components at the notch tip of a body in plane stress are given below.

$$\begin{aligned} \varepsilon_{11}^{N} &= -\nu\frac{\sigma_{22}^{N}}{E} - \frac{1}{2}f(\sigma_{22}^{N}) \\ \varepsilon_{22}^{N} &= \frac{\sigma_{22}^{N}}{E} + f(\sigma_{22}^{N}) \\ \varepsilon_{33}^{N} &= -\nu\frac{\sigma_{22}^{N}}{E} - \frac{1}{2}f(\sigma_{22}^{N}) \\ \varepsilon_{22}^{e}\sigma_{22}^{e} &= \varepsilon_{22}^{N}\sigma_{22}^{N} \end{aligned} \tag{4}$$

2.1.2 Equivalent Strain Energy Density Rule

The equivalent strain energy density (ESED) relationship [3] was initially proposed for a notched body in plane stress. It is written as the equality of the strain energy densities at the notch tip:

$$\int_{o}^{\varepsilon_{22}^{e}} \sigma_{22}^{e}\, d\varepsilon_{22}^{e} = \int_{o}^{\varepsilon_{22}^{E}} \sigma_{22}^{E}\, d\varepsilon_{22}^{E} \tag{5}$$

Eq. (5) is a statement of the equality between the strain energy density, W^e , at the notch tip of a linear elastic body and the strain energy density, W^E, at the notch tip of a geometrically identical elastic-plastic body subjected to the same load (Fig. 2). This relationship is shown graphically in Fig. 4b, and represents the equality of the area under the linear-elastic curve and the area under the actual elastic-plastic σ^a-ε^a material curve.

For a uniaxial stress-strain constitutive law given in the form of Eq. (3), the relationships necessary to determine all of the stress and strain components at the notch tip of a body in plane stress are given in the form of four equations;

$$\varepsilon_{11}^{E} = -\nu\frac{\sigma_{22}^{E}}{E} - \frac{1}{2}f(\sigma_{22}^{E})$$
$$\varepsilon_{22}^{E} = \frac{\sigma_{22}^{E}}{E} + f(\sigma_{22}^{E})$$
$$\varepsilon_{33}^{E} = -\nu\frac{\sigma_{22}^{E}}{E} - \frac{1}{2}f(\sigma_{22}^{E}) \tag{6}$$
$$\frac{1}{2}\varepsilon_{22}^{e}\sigma_{22}^{e} = \int \sigma_{22}^{E} d\varepsilon_{22}^{E}$$

2.2 MULTIAXIAL STRESS STATES

For the case of a general multiaxial loading applied to a notched body, the state of stress near the notch tip is triaxial. At the notch tip, the stress state is biaxial because of the free surface (Fig. 1c). Since equilibrium of the element at the notch tip must be maintained, $\sigma_{23} = \sigma_{32}$ and $\varepsilon_{23} = \varepsilon_{32}$. Therefore, there are seven unknowns at the notch tip; three stress components and four strain components. A set of seven independent equations is required to completely define the stress-strain state at the notch tip. The material constitutive relationships provide four equations, leaving three additional relationships to be established.

2.2.1 Proportional Loading

The Constitutive Equations

The generalized elasto-plastic constitutive relationships are derived from the uniaxial stress-strain curve by using the principles of elasticity and plasticity theory. The simplest analysis uses a general form of Hencky's total deformation equations.

$$\varepsilon_{ij} = \frac{1+\nu}{E}\sigma_{ij} - \frac{\nu}{E}\sigma_{kk}\delta_{ij} + \frac{3}{2}\frac{\varepsilon_{eq}^{p}}{\sigma_{eq}}S_{ij} \tag{7}$$

where

$$\sigma_{eq} = \sqrt{\frac{3}{2}S_{ij}S_{ij}} \qquad \varepsilon_{eq}^{p} = \sqrt{\frac{2}{3}\varepsilon_{ij}^{p}\varepsilon_{ij}^{p}}$$

$$S_{ij} = \sigma_{ij} - \frac{1}{3}\sigma_{kk}\delta_{ij} \qquad \sigma_{kk} = \sigma_{11} + \sigma_{22} + \sigma_{33}$$

It is also assumed that

$$\varepsilon_{eq}^{p} = f(\sigma_{eq}). \tag{8}$$

The function $f(\sigma_{eq})$ is identical to the relationship between stress and plastic strain for a uniaxial tension or compression test. Therefore, Hencky's equations can be written in terms of stresses only;

$$\varepsilon_{ij} = \frac{1+\nu}{E}\sigma_{ij} - \frac{\nu}{E}\sigma_{kk}\,\delta_{ij} + \frac{3}{2}\frac{f(\sigma_{eq})}{\sigma_{eq}} S_{ij} \tag{9}$$

Note that the results from Hencky's equations are identical with those of the incremental theory of plasticity (which is presented later) when the deviatoric stresses remain in fixed proportions.

Generalised Neuber's and the ESEDRules

In a recent study, Moftakhar *et al.* [6] generalised both Neuber's rule and the ESED method to address multiaxial notch tip stress states in bodies subjected to proportional loading. They proposed that there exists a band within which the actual notch tip strains will always fall. They also found that the lower bound of this band is obtained by using the generalised ESED method based on Eq. (5) and the upper bound is given by Neuber's rule in the form of Eq. (2).

One of the three additional equations required to define the notch tip stress-strain field is based on the hypothesis that the equations of strain energy density for uniaxial loading should also hold for the case of multiaxial stress states, provided that the strain energy density is calculated by accounting for all of the stress and strain components at the notch tip;

$$\sigma_{ij}^{e}\,\varepsilon_{ij}^{e} = \sigma_{ij}^{N}\,\varepsilon_{ij}^{N} \tag{10}$$

On the principal stress axes, Eq. (10) can be reduced to a simple form consisting of the principal stress and strain components.

$$\sigma_1^{e}\varepsilon_1^{e} + \sigma_2^{e}\varepsilon_2^{e} + \sigma_3^{e}\varepsilon_3^{e} = \sigma_1^{N}\varepsilon_1^{N} + \sigma_2^{N}\varepsilon_2^{N} + \sigma_3^{N}\varepsilon_3^{N} \tag{11}$$

According to Eq. (10), the total strain energy density in the notch tip of an elasto-plastic body is the same as that in a geometrically identical elastic body subjected to the same external load

Analogously, the ESED method can be generalized for an arbitrary multiaxial stress state as,

$$\int_0^{\varepsilon_{ij}^{e}} \sigma_{ij}^{e}\,d\varepsilon_{ij}^{e} = \int_0^{\varepsilon_{ij}^{E}} \sigma_{ij}^{E}\,d\varepsilon_{ij}^{E} \tag{12}$$

On the principal stress axes, Eq. (12) is written as,

$$\int_0^{\varepsilon_{ij}^e} \sigma_1^e d\varepsilon_1^e + \sigma_2^e d\varepsilon_2^e + \sigma_3^e d\varepsilon_3^e = \int_0^{\varepsilon_{ij}^E} \sigma_1^E d\varepsilon_1^E + \sigma_2^E d\varepsilon_2^E + \sigma_3^E d\varepsilon_3^E \qquad (13)$$

Calculating the integral in Eq. (13) requires knowledge of the stress-strain relationship given by Eq. (9). The final expression can be presented in closed form by using all of the principal stress components and the equivalent stress. This is shown in reference [8].

In addition to Eq. (10) and Eq. (12), it has been observed that the contributions from each stress-strain component to the overall strain energy density at the notch tip is almost the same in both the hypothetical linear elastic body and the geometrically identical elastic-plastic body. This observation has resulted in the formulation of two additional equations that are related to the fractional contributions of the strain energy density. These equations are shown in the following sections.

Complete Set of Equations Based on the Generalised Neuber's Rule

The basic equation for the upper bound limit, which relates the linear elastic and the elasto-plastic notch tip strains and stresses, is the generalised Neuber rule (Eq. (10)). Since all stress components on the traction free notch tip surface must vanish, i.e. $\sigma_{11} = 0$, $\sigma_{12} = 0$, and $\sigma_{13} = 0$, the notch tip stress state can always be reduced to a biaxial stress state characterised by two principal stress components (σ_2 and σ_3) associated with three principal strain components (ε_1, ε_2 and ε_3). Thus, the whole notch-tip stress-strain problem can be reduced to the solution of a set of five nonlinear algebraic equations consisting of

- the generalized Neuber rule

$$\sigma_2^e \varepsilon_2^e + \sigma_3^e \varepsilon_3^e = \sigma_2^N \varepsilon_2^N + \sigma_3^N \varepsilon_3^N \qquad (14)$$

- the constitutive stress-strain relationships

$$\begin{aligned}
\varepsilon_1^N &= -\frac{\nu}{E}\left(\sigma_2^N + \sigma_3^N\right) - \frac{f(\sigma_{eq}^N)}{2\sigma_{eq}^N}\left(\sigma_2^N + \sigma_3^N\right) \\
\varepsilon_2^N &= \frac{1}{E}\left(\sigma_2^N - \nu\sigma_3^N\right) + \frac{f(\sigma_{eq}^N)}{2\sigma_{eq}^N}\left(\sigma_2^N - \frac{1}{2}\sigma_3^N\right) \\
\varepsilon_3^N &= \frac{1}{E}\left(\sigma_3^N - \nu\sigma_2^N\right) + \frac{f(\sigma_{eq}^N)}{2\sigma_{eq}^N}\left(\sigma_3^N - \frac{1}{2}\sigma_2^N\right)
\end{aligned} \qquad (15)$$

where $$\sigma_{eq}^N = \sqrt{(\sigma_2^N) - \sigma_2^N \sigma_3^N + (\sigma_3^N)^2}$$

- the fractional contribution of the total strain energy density

$$\frac{\sigma_2^e \varepsilon_2^e}{\sigma_2^e \varepsilon_2^e + \sigma_3^e \varepsilon_3^e} = \frac{\sigma_2^N \varepsilon_2^N}{\sigma_2^N \varepsilon_2^N + \sigma_3^N \varepsilon_3^N} \tag{16}$$

Equations (14)-(16) form a set of five equations from which all of the principal stress components (σ_2^N and σ_3^N) and the principal strain components (ε_1^N, ε_2^N and ε_3^N) can be calculated on the basis of data obtained from a linear elastic analysis, i.e. based on the stress components σ_2^e and σ_3^e and the strain components ε_1^e, ε_2^e and ε_3^e.

Complete Set of Equations Based on the Generalised ESED Hypothesis

The generalised ESED relationship of Eq. (5) is the principal equation for the lower bound formulation that relates the linear-elastic and the elastic-plastic notch-tip strains and stresses. Again, for the traction free notch tip surface the stress state can be represented by two principal stress components. The final set of equations can be written as follows:

- the generalized ESED

$$\frac{1}{2}\left(\sigma_2^e \varepsilon_2^e + \sigma_3^e \varepsilon_3^e\right) = \frac{1}{3E}(1+\nu)\left(\sigma_{eq}^E\right)^2 + \frac{1-2\nu}{6E}\left(\sigma_2^E + \sigma_3^E\right) + \int_0^{\varepsilon_{eq}^{pE}} \sigma_{eq}^E \, d\varepsilon_{eq}^{pE} \tag{17}$$

where $\varepsilon_{eq}^{pE} = f\left(\sigma_{eq}^E\right)$ and $\sigma_{eq}^E = \sqrt{(\sigma_2^E)^2 - \sigma_2^E \sigma_3^E + (\sigma_3^E)^2}$

- the constitutive equations

$$\begin{aligned}
\varepsilon_1^E &= -\frac{\nu}{E}\left(\sigma_2^E + \sigma_3^E\right) - \frac{f(\sigma_{eq}^E)}{2\sigma_{eq}^E}\left(\sigma_2^E + \sigma_3^E\right) \\
\varepsilon_2^E &= \frac{1}{E}\left(\sigma_2^E - \nu\sigma_3^E\right) + \frac{f(\sigma_{eq}^E)}{2\sigma_{eq}^E}\left(\sigma_2^E - \frac{1}{2}\sigma_3^E\right) \\
\varepsilon_3^E &= \frac{1}{E}\left(\sigma_3^E - \nu\sigma_2^E\right) + \frac{f(\sigma_{eq}^E)}{2\sigma_{eq}^E}\left(\sigma_3^E - \frac{1}{2}\sigma_2^E\right)
\end{aligned} \tag{18}$$

- the fractional contribution of the total strain energy density

$$\frac{\sigma_2^e \varepsilon_2^e}{\sigma_2^e \varepsilon_2^e + \sigma_3^e \varepsilon_3^e} = \frac{\sigma_2^E \varepsilon_2^E}{\sigma_2^E \varepsilon_2^E + \sigma_3^E \varepsilon_3^E} \tag{19}$$

Equations (17)-(19) make it possible to determine all of the stress and strain components in the notch tip provided that the hypothetical elastic notch tip stresses, σ_2^e and σ_3^e , and the material stress-strain curve are known.

2.2.2 Non-Proportional Loading

In the case of plastic deformation caused by non-proportional loading, the final stress-strain state is dependent on the loading path. As a result, relationships that define the local stress-strain state in a notched body subjected to multiaxial loading must be developed in an incremental form. Three stress increments and four strain increments have to be found at the notch tip for each load increment. Because the principal axes of the notch tip stress tensor rotate with respect to the notch frame of reference, all of the equation have to be written in the fixed, not principal axes.

The incremental constitutive relationship forms four independent equations. The remaining three equations that are necessary for a complete formulation of the notch tip problem can be determined by using the strain energy criteria that was discussed above, but formulated in incremental form.

Material Constitutive Model

The most frequently used material constitutive model for incremental plasticity is the Prandtl-Reuss relationship. For an isotropic body, the Prandtl-Reuss relationship can be expressed as:

$$\Delta \varepsilon_{ij} = \frac{1+\nu}{E} \Delta \sigma_{ij} - \frac{\nu}{E} \Delta \sigma_{kk} \delta_{ij} + \frac{3}{2} \frac{\Delta \varepsilon_{eq}^p}{\sigma_{eq}} S_{ij} \tag{20}$$

The multiaxial incremental stress-strain relation is obtained from the uniaxial stress-strain curve of Eq. (3) by relating the equivalent plastic strain to the equivalent stress such that

$$\Delta \varepsilon_{eq}^p = \frac{df(\sigma_{eq})}{d \sigma_{eq}} \Delta \sigma_{eq} \tag{21}$$

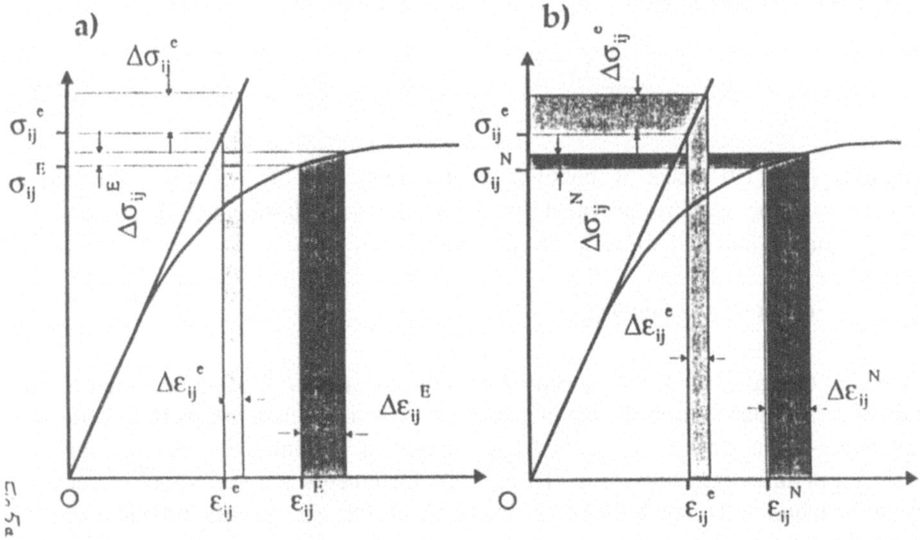

Figure 5: Graphical Representation of: (a) Incremental ESED Method, (b) Incremental Neuber's rule

Incremental Neuber's Rule

It is proposed that for a given increment of external load, the corresponding increment of the total strain energy density at the notch tip in an elastic-plastic body can be approximated by that which would be obtained if the body was to hypothetically remain elastic throughout the loading history. The total strain energy is defined as the sum of the strain energy density (as described above) and the complementary strain energy density. Mathematically, this can be written as:

$$\Delta\Omega^{e} = \Delta\Omega^{N}$$
$$or \tag{22}$$
$$\sigma_{ij}^{e}\,\Delta\,\varepsilon_{ij}^{e} + \varepsilon_{ij}^{e}\,\Delta\,\sigma_{ij}^{e} = \sigma_{ij}^{N}\,\Delta\,\varepsilon_{ij}^{N} + \varepsilon_{ij}^{N}\,\Delta\,\sigma_{ij}^{N}\,.$$

Eq. (22) is called the incremental Neuber relation since it reduces to Neuber's rule in its original form of Eq. (2) for uniaxial notch tip stress states. Furthermore, it represents a statement of equality between the increment of the total notch tip strain energy density that is obtained from a linear elastic solution and the corresponding increment that is obtained from an elastic-plastic analysis. A graphical representation of the incremental Neuber's rule is shown in Fig. 5b, where the horizontal and vertical rectangles represent the strain energy increment for an associated pair of stress and strain components.

The four constitutive relationships used in conjunction with the generalised Neuber's rule (Eq. (22)) will only be sufficient to describe five of the seven unknown notch tip stress-strain increments. Therefore, two more independent equations are required to completely define the notch tip stress and strain increments for a given increment of the applied load. Again, it is has been proposed [7] that the contribution of each elastic-plastic stress-strain component to the total strain energy density at the notch tip is equal to the contribution of the same increment of the stress-strain component when it is assumed that the body remains elastic during the loading history. This proposal can be expressed as:

$$\sigma^e_{\alpha\beta}\Delta\varepsilon^e_{\alpha\beta} + \varepsilon^e_{\alpha\beta}\Delta\sigma^e_{\alpha\beta} = \sigma^N_{\alpha\beta}\Delta\varepsilon^N_{\alpha\beta} + \varepsilon^N_{\alpha\beta}\Delta\sigma^N_{\alpha\beta}. \tag{23}$$

where α and $\beta = 1, 2, 3$ but summation does not apply.

In general, Eq. (23) represents six independent equations that result in Eq. (22) when they are added together side by side.

The three non zero relations implied by Eq. (23) and the four constitutive equations of Eq. (20) form a set of seven independent equations that are sufficient to determine the unknown increments $\Delta\sigma_{ij}^N$ and $\Delta\varepsilon_{ij}^N$, as shown below.

- incremental constitutive equations

$$\begin{aligned}
\Delta\varepsilon^N_{11} &= -\frac{\nu}{E}(\Delta\sigma^N_{22} + \Delta\sigma^N_{33}) - \frac{1}{2}(\sigma^N_{22} + \sigma^N_{33})\frac{\Delta\varepsilon^{pN}_{eq}}{\sigma^N_{eq}} \\
\Delta\varepsilon^N_{22} &= \frac{1}{E}(\Delta\sigma^N_{22} - \nu\Delta\sigma^N_{33}) + \frac{1}{2}(2\sigma^N_{22} - \sigma^N_{33})\frac{\Delta\varepsilon^{pN}_{eq}}{\sigma^N_{eq}} \\
\Delta\varepsilon^N_{33} &= \frac{1}{E}(\Delta\sigma^N_{33} - \nu\Delta\sigma^N_{22}) + \frac{1}{2}(2\sigma^N_{33} - \sigma^N_{22})\frac{\Delta\varepsilon^{pN}_{eq}}{\sigma^N_{eq}} \\
\Delta\varepsilon^N_{23} &= \frac{1+\nu}{E}\Delta\sigma^N_{23} + \frac{3}{2}\frac{\Delta\varepsilon^{pN}_{eq}}{\sigma^N_{eq}}\sigma^N_{23}
\end{aligned} \tag{24}$$

where $(\sigma^N_{eq})^2 = (\sigma^N_{22})^2 + (\sigma^N_{33})^2 - \sigma^N_{22}\sigma^N_{33} + 3(\sigma^N_{23})^2$

- energy equations

$$\begin{aligned}
\sigma_{22}^{e}\Delta\varepsilon_{22}^{e}+\varepsilon_{22}^{e}\Delta\sigma_{22}^{e} &= \sigma_{22}^{N}\Delta\varepsilon_{22}^{N}+\varepsilon_{22}^{N}\Delta\sigma_{22}^{N}\\
\sigma_{33}^{e}\Delta\varepsilon_{33}^{e}+\varepsilon_{33}^{e}\Delta\sigma_{33}^{e} &= \sigma_{33}^{N}\Delta\varepsilon_{33}^{N}+\varepsilon_{33}^{N}\Delta\sigma_{33}^{N}\\
\sigma_{23}^{e}\Delta\varepsilon_{23}^{e}+\varepsilon_{23}^{e}\Delta\sigma_{23}^{e} &= \sigma_{23}^{N}\Delta\varepsilon_{23}^{N}+\varepsilon_{23}^{N}\Delta\sigma_{23}^{N}
\end{aligned} \tag{25}$$

Incremental Equivalent Strain Energy Density (ESED) Relations

It has been proposed [7] that for a given increment of external load, the corresponding increment in the strain energy density at the notch tip in an elastic-plastic body can be approximated by that which would be obtained if the body was to remain elastic throughout the loading history. This hypothesis can be expressed as:

$$\begin{gathered}
\Delta W^{e}=\Delta W^{E}\\
or\\
\sigma_{ij}^{e}\Delta\varepsilon_{ij}^{e}=\sigma_{ij}^{E}\Delta\varepsilon_{ij}^{E}.
\end{gathered} \tag{26}$$

Eq. (26) represents a statement of equality between the increment of notch tip strain energy density obtained from a linear elastic solution and that obtained from an elastic-plastic analysis. A graphical representation of the incremental ESED method is shown in Fig. 5a, where the strain energy densities are represented by the vertical bars of the trapezoidal shape.

The four constitutive relations used in conjunction with the generalized ESED equation will only be sufficient to formulate a set of five equations. Therefore, two more independent equations are required to completely define the notch tip stress-strain increments for a given increment in the applied load. Based on observations of numerical and experimental data, the following hypothesis was proposed which states that the equality of energies also applies to all corresponding stress and strain components;

$$\sigma_{\alpha\beta}^{e}\Delta\varepsilon_{\alpha\beta}^{e}=\sigma_{\alpha\beta}^{E}\Delta\varepsilon_{\alpha\beta}^{E}. \tag{27}$$

Note that in Eq. (27), the indices $\alpha, \beta = 1, 2,$ and 3, but summation is not implied. The three non-zero relations given by this equation, and the four constitutive equations, are sufficient to determine the three unknown stress and the four unknown strain increments. The complete set of incremental equations based on the equality of strain energy increments (ESED) is given below.

- Incremental Constitutive Relations

$$\Delta\varepsilon_{11}^{E} = -\frac{\nu}{E}(\Delta\sigma_{22}^{E} + \Delta\sigma_{33}^{E}) - \frac{1}{2}(\sigma_{22}^{E} + \sigma_{33}^{E})\frac{\Delta\varepsilon_{eq}^{pE}}{\sigma_{eq}^{E}}$$

$$\Delta\varepsilon_{22}^{E} = -\frac{\nu}{E}(\Delta\sigma_{22}^{E} - \nu\Delta\sigma_{33}^{E}) + \frac{1}{2}(2\sigma_{22}^{E} + \sigma_{33}^{E})\frac{\Delta\varepsilon_{eq}^{pE}}{\sigma_{eq}^{E}}$$

$$\Delta\varepsilon_{33}^{E} = \frac{1}{E}(\Delta\sigma_{33}^{E} - \nu\Delta\sigma_{22}^{E}) + \frac{1}{2}(2\sigma_{33}^{E} - \sigma_{22}^{E})\frac{\Delta\varepsilon_{eq}^{pE}}{\sigma_{eq}^{E}} \tag{28}$$

$$\Delta\varepsilon_{23}^{E} = \frac{1+\nu}{E}\Delta\sigma_{23}^{E} + \frac{3}{2}\frac{\Delta\varepsilon_{eq}^{pE}}{\sigma_{eq}^{E}}\sigma_{23}^{E}$$

where $$(\sigma_{eq}^{E})^2 = (\sigma_{22}^{E})^2 + (\sigma_{33}^{E})^2 - \sigma_{22}^{E}\sigma_{33}^{E} + 3(\sigma_{23}^{E})^2$$

- Energy Equations

$$\sigma_{22}^{e}\Delta\varepsilon_{22}^{e} = \sigma_{22}^{E}\Delta\varepsilon_{22}^{E}$$
$$\sigma_{33}^{e}\Delta\varepsilon_{33}^{e} = \sigma_{33}^{E}\Delta\varepsilon_{33}^{E} \tag{29}$$
$$\sigma_{23}^{e}\Delta\varepsilon_{23}^{e} = \sigma_{23}^{E}\Delta\varepsilon_{23}^{E}$$

In order to determine the notch tip elastic-plastic strains and stresses at the end of the loading history, they must first be evaluated for each increment of the applied load. Initially, the first reference state is taken as the point at which yielding occurs at the notch tip since it can be found from an elastic analysis of the body. For each increment of external load, the increments in the elastic-plastic notch tip strains and stresses are computed from either Eq. (24) and Eq. (25), or Eq. (28) and Eq. (29) along with the hypothetical elastic notch tip stress-strain history and the material stress-strain curve. The stress and strain states at the end of given load increment are then computed using:

$$\sigma_{ij}^{n} = \sigma_{ij}^{o} + \sum_{k=1}^{n-1}\Delta\sigma_{ij} + \Delta\sigma_{ij}^{n}, \tag{30}$$

$$\varepsilon_{ij}^{n} = \varepsilon_{ij}^{o} + \sum_{k=1}^{n-1}\Delta\varepsilon_{ij} + \Delta\varepsilon_{ij}^{n}, \tag{31}$$

where n denotes the number of the load increment.

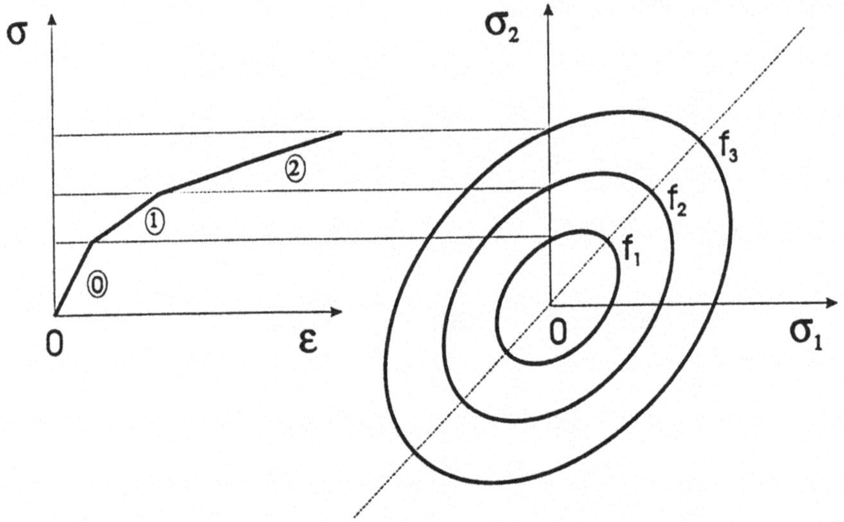

Figure 6: Piecewise linearisation of the material σ–ε curve and the corresponding work-hardening surfaces.

Incremental Cyclic Plasticity Model

The set of equations defined by the incremental Neuber's rule and ESED method can be solved only if the relationship between the equivalent plastic strain increment, $\Delta\varepsilon_{eq}^{p}$, and the equivalent stress increment, $\Delta\sigma_{eq}$, is known during the application of a given load increment. It is known that the current $\Delta\varepsilon_{eq}^{p}$-$\Delta\sigma_{eq}$ relationship depends on the previous load path. Therefore, the use of the incremental Neuber's rule or the incremental ESED method must be associated with a plasticity model that deals with path dependent material behaviour.

Several models are available in the literature. The model proposed by Mroz [9] and more recently improved by Garud [10] is the most popular. Mroz [9] proposed that the uniaxial stress-strain material curve can be represented by a set of work-hardening surfaces in three-dimensional stress space (Fig. 6). In the case of a two-dimensional stress state, such as that of a notch tip, the work-hardening surfaces are represented by ellipses on the coordinate plane for which the axes are defined by the principal stress components. The equation of each work-hardening surface is defined by

$$\sigma_{eq} = \sqrt{(\sigma_2)^2 - \sigma_2\sigma_3 + (\sigma_3)^2} \tag{32}$$

The load path dependency effects are modelled by prescribing a translation rule for the motion of the ellipses which move with respect to each other over distances given by the stress increments. The ellipses move within the boundaries of each other, but they do not intersect. If an ellipse comes in contact with another, they move together as one rigid body. The translation rule proposed by Garud [10] avoids the intersection of the ellipses that can occur with the original Mroz [9] model when the load path is defined by finite stress increments (as in the case of a numerical solution). The Garud translation rule is illustrated in Fig. 7 and can be described by considering only two work-hardening surfaces (two ellipses) as follows:

(1) The line of action of the stress increment, $\Delta\sigma$, is extended to intersect the next larger non-active surface, f_2, at point B_2.

(2) Point B_2 is connected to the centre, O_2, of the surface f_2.

(3) A line is extended through the centre of the smaller active surface , O_1, parallel to the line O_2B_2 to find point B_1 on surface f_1.

(4) The conjugate points B_1 and B_2 are connected by the line B_1B_2.

(5) Surface f_1 is translated from point O_1 to point O_1' such that vector O_1O_1' is parallel to line B_1B_2. The translation is complete when the end of the vector defined by the stress increment, $\Delta\sigma$, lies on the translated surface f_1'.

The translation rule described above assures that the two ellipses will be tangent to each other at the common point B_1B_2 without intersecting. Two or more tangent ellipses translate together as rigid bodies and the largest moving ellipse (Fig. 7) indicates which linear constitutive relationship should be used for a given stress-strain increment. In most publications, the plasticity models are described as algorithms for calculating strain increments that result from given stress increments or vice versa. In the case of the notch analysis described above, both the strain and stress increments are determined from the either the Neuber or the ESED methods. Therefore, the plasticity model is only needed to indicate which work-hardening surface is active during the next load increment.

The Mroz-Garud model [10] was chosen here as an illustration. Obviously, any other plasticity model can be associated with the incremental Neuber and ESED method.

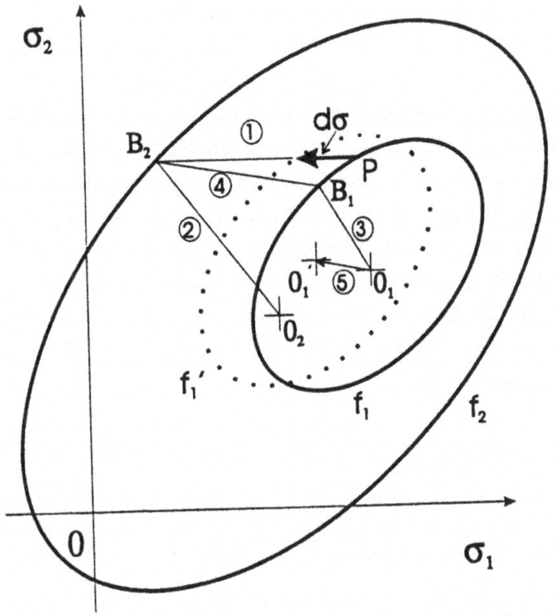

Figure 7: Geometrical Illustration of the Mroz-Garud Incremental Plasticity Model

3. Comparison with numerical data

3.1 PROPORTIONAL LOADING

The accuracy of the proposed method is demonstrated by comparison with the finite element data of Seeger and Hoffman [4], who analysed a cylindrical specimen with a circumferential notch subjected to simultaneous proportional tensile and torsional loading (Fig. 8). The load in the finite element calculations was applied in such a way that the ratio of the nominal notch tip shear to tensile stress was $\tau_n/\sigma_{nF} = 0.411$. The nominal stresses were determined from the net cross section dimensions.

$$\sigma_{nF} = \frac{P}{\pi(R-t)^2} \; ; \qquad \tau_n = \frac{2T}{\pi(R-t)^3} . \tag{33}$$

The stress concentration factors for tension and torsion were $K_F = 3.89$ and $K_T = 2.19$ respectively. The stress concentration factors were defined (Fig. 8) as:

$$K_F = \frac{\sigma_{22}^e}{\sigma_{nF}} \quad \text{and} \quad K_T = \frac{\sigma_{32}^e}{\tau_n} \tag{34}$$

The ratio of the notch tip hoop stress to axial stress in tension was $\sigma_{33}^e / \sigma_{22}^e = 0.2$. A bi-linear stress-strain relation was used for the calculations. Its mathematical form is given by

$$\varepsilon = \frac{\sigma}{E} \quad \text{for} \quad \sigma \leq \sigma_0 \tag{35}$$

and

$$\varepsilon = \frac{\sigma_o}{E} + \frac{\sigma - \sigma_o}{H} \quad \text{for} \quad \sigma > \sigma_o \tag{36}$$

where $E = 94400$ MPa, $H = (0.005)E$, $\sigma_o = 550$ MPa and $\nu = 0.3$.

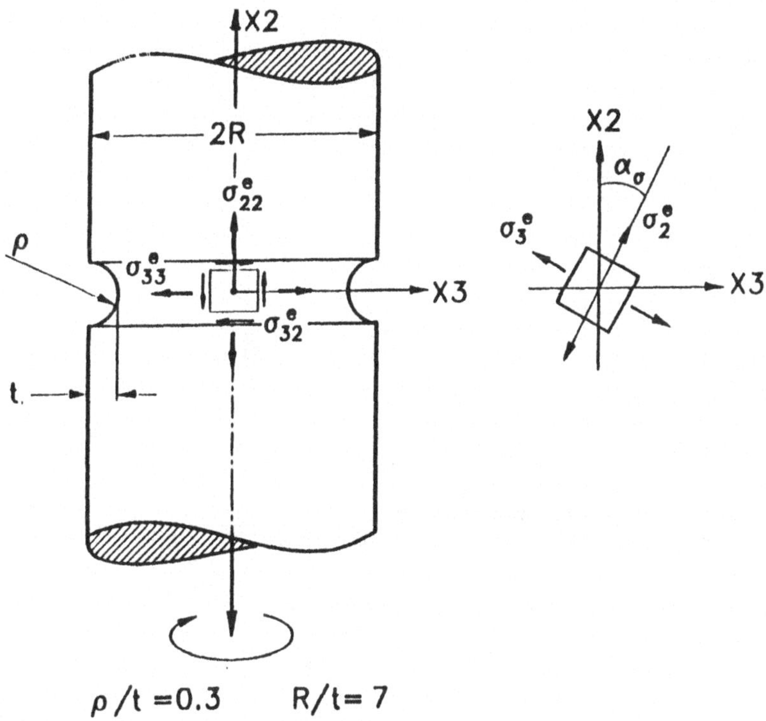

Figure 8: Geometry and dimensions of the cylindrical specimen tested by Hoffman and Seeger [4].

Both the Neuber (Eqns. (14)-(16)) and the ESED (Eqns. (17)-(19)) method were used for calculating the notch tip strains, ε_{ij}^{E} and ε_{ij}^{N}, and stresses, σ_{ij}^{E} and σ_{ij}^{N}. The solution method for both sets of equations is given in reference [7]. The calculated strains and stresses were compared with the Seeger-Hoffman [4] elasto-plastic finite element data (Fig. 9), where the nominal equivalent stress, S, was used as the reference.

$$S = \sqrt{\sigma_{nF}^{2} + 3\,\tau_{n}^{2}} \tag{37}$$

It has been found that the finite element data were within or close to the band predicted by the Neuber and ESED methods. It is believed that these methods give the upper and the lower bounds for the elasto-plastic strains and stresses in the notch tip. However, due to numerical errors, the actual stresses might sometimes be close to, but outside the band. This is especially true for the principal stress component with the smallest magnitude.

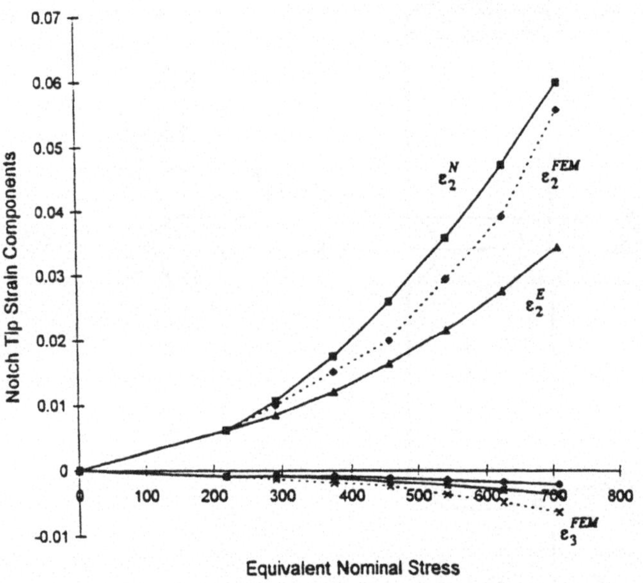

Figure 9: Comparison of the Neuber and ESED solutions with the multiaxial elasto-plastic finite element data Notch tip strain components.

3.2 NON-PROPORTIONAL LOADING

The accuracy of the proposed incremental Neuber's rule and incremental ESED method were assessed by comparing the calculated notch tip stress-strain histories to those obtained using the finite element method. The elastic-plastic results from the finite element analysis of reference [7], were obtained using the ABAQUS [11] finite element package.

The geometry of the notched element was that of the circumferentially notched bar similar to that one shown inFig. 8. The loads that were applied to the bar were torsion and tension according to the path shown in Fig. 10. The nominal torsional stresses, $\tau_{n.}$, and tensile stresses, σ_{nF}, were determined based on the net cross section according to Eq. (33). The basic proportions of the cylindrical component were $\rho/t = 0.3$ and $R/t = 7$ resulting in the torsional and tensile stress concentration factor $K_T = 3.31$ and $K_F = 1.94$ respectively. The ratio of the notch tip hoop to axial stress under tensile loading was $\sigma_{33}^e / \sigma_{22}^e = 0.284$.

The material for the notched bar was SAE 1045 steel with a cyclic stress-strain curve approximated by the Ramberg-Osgood relation:

$$\varepsilon = \frac{\sigma}{E} + \left(\frac{\sigma}{K'}\right)^{\frac{1}{n'}} \tag{38}$$

The material properties were: $E = 202$ GPa, $\nu = 0.3$, $S_y = 202$ MPa, $n' = 0.208$, and K' = 1258 MPa..

The maximum applied load levels were chosen to be 50% higher than would be required to cause yielding at the notch tip if each load were applied seperately. Specifically, the maxima were $\sigma_{nF}^f = 103$ MPa, and $\tau_n^f = 90$ MPa. The final ratio of the nominal stresses was $\sigma_{nF}^f/\tau_n^f = 1.133$. The normalised nominal equivalent net sectional stress ratio was defined as

$$\frac{S_{eq}}{S_y} = \frac{\sqrt{\sigma_{n.s.}^2 + 3\,\tau_{n.s}^2}}{S_y} \tag{39}$$

and reached a value of 0.92 at the end loading path, a value that indicates almost general yielding of the net section.

The elastic stress histories at the notch tip that were caused by the load path are shown in Fig. 10. They were used with Eq. (24)-Eq. (25) and Eq. (28)-Eq. (29) to calculate the notch tip elastic-plastic strains and stresses. The calculated strains and stresses were subsequently compared to the elastic-plastic finite element results.

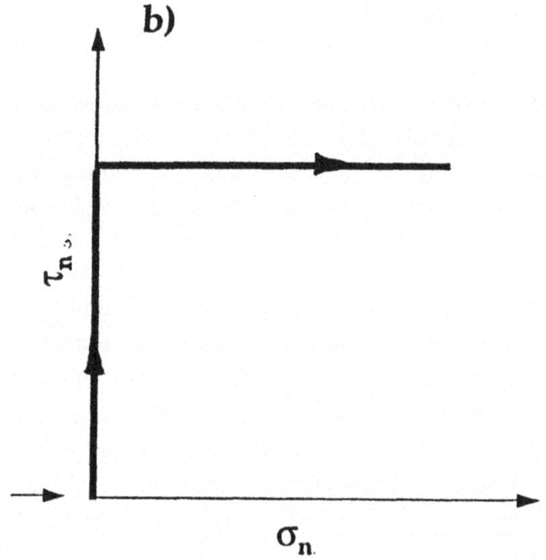

Figure 10: The torsion-tension load path

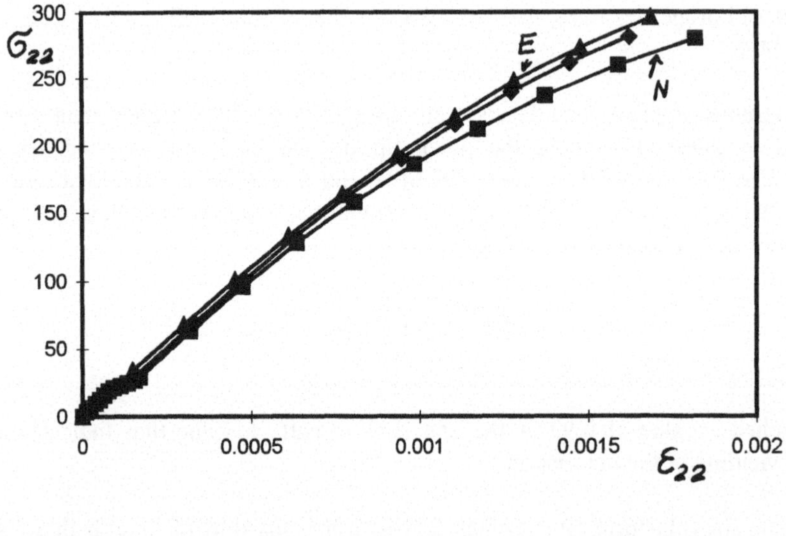

Figure 11: The history of the notch tip strains and stresses due to non-proportional torsion-tension loading: (a) axial stress and strain.

The strain components (ε_{22} and ε_{23}) and the stress components (σ_{22} and σ_{23}) that were calculated are shown in Fig. 11. Note that the results from both models and the finite element analysis are identical in the elastic range. This is expected since the models converge to the elastic solution in the elastic range. Just beyond the onset of yielding at the notch tip, the strain results that were predicted using the proposed models and the finite element data begin to diverge.

It can be concluded that the incremental Neuber's method predicts an upper bound, and the incremental ESED method, a lower bound approximation to the actual notch tip strains.

In order to predict the fatigue life of a notched component subjected to multiaxial cyclic loading, a parameter is needed to calculate the fatigue damage caused by multiaxial, cyclic stress states. This parameter should use the existing standard fatigue properties that have already been established for many engineering materials.

4. Multiaxial Fatigue Parameters

In general, most multiaxial fatigue theories can be divided into three categories: stress-based, strain-based and energy-based theories. The most popular stress-based criteria are extensions of classical failure theories such as the maximum principal stress, the maximum shear stress or the octahedral shear stress. These stress-based criteria are most often employed in high cycle fatigue problems where very little plasticity occurs. None of these theories, however, is sufficient to correlate various loading modes in situations where cyclic plasticity is involved.

The most popular strain-based theories are strain versions of the classical failure theories mentioned above, mainly concerning octahedral strain which has showed some capability of correlating various modes of cyclic loading. Nevertheless, it is often found that strain based criteria are not sufficient to model a wide variety of loading modes. In particular, the mean stress effect is not taken into account.

Energy based criteria have been proposed and tested by several researchers [10,12,13], but have not been generally accepted. The main criticism is that strain energy density, which is a scalar quantity, is unable to account for the loading history. However, the advantage of an energy criterion is that it incorporates both strain and stress components in the damage parameter.

In general, reasonably good correlations of multiaxial fatigue data have been obtained using critical plane approaches [14-16]. These theories postulate that fatigue cracks grow on certain planes and that only the shear and normal stresses and strains acting on these planes contribute to crack formation and growth. These theories are supported

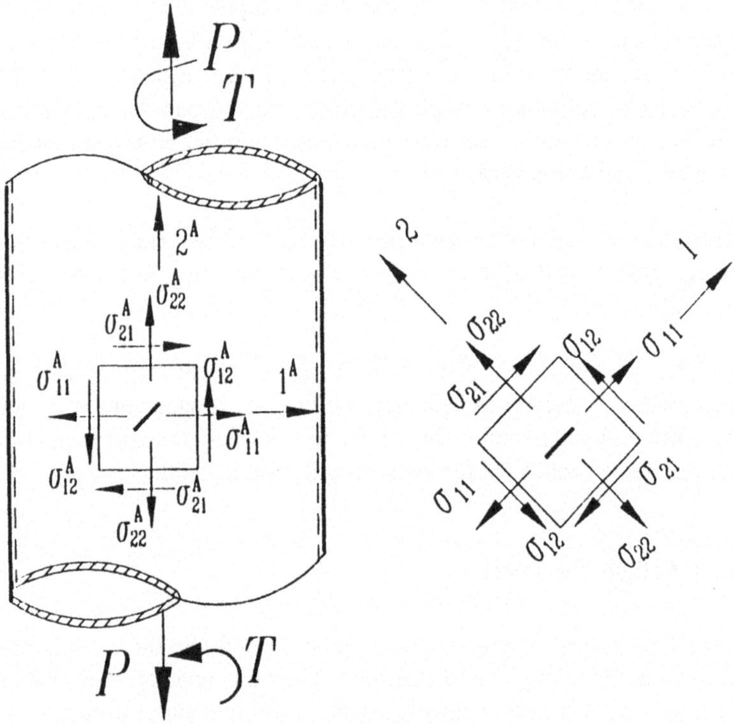

Figure 12: Co-ordinate system used in the multiaxial fatigue analysis (a) System of co-ordinates 1^A and 2^A associated with the main axes of the body and applied stress components σ_{ij}^A, (b) System of co-ordinates 1 and 2 associated with the critical plane and stress components σ_{ij} in the critical plane.

by various experimental observations of the physical mechanism of the fatigue damage process [16-19].

4.1 CRITICAL PLANE MODELS AND STRAIN ENERGY DENSITY CRITERIA

Critical plane models have been proposed by several researchers such as Findley [14], McDiarmid [15], Kandil *et al.* [16] and Lohr and Ellison [17]. Most of the proposed criteria are given in the form of expressions involving a combination of the stress [15] or strain [16] components associated with the critical plane. In general, the strain criterion can be given in the form of the critical plane shear and normal strain components, as proposed by Kandil *et al.* [16]

$$\gamma^* = \frac{\Delta\gamma_{21}}{2} + K\frac{\Delta\varepsilon_{22}}{2} \tag{40}$$

McDiarmid [15] gave an analogous stress criterion associated with the critical plane;

$$\tau^* = \frac{\Delta\sigma_{21}}{2} + k\frac{\Delta\sigma_{22}}{2} \tag{41}$$

The notation associated with Eq. (40) and Eq. (41) is shown in Fig. 12. These parameters they are sometimes criticised for a lack of formal correctness from the continuum mechanics viewpoint. The main difficulty concerns the interpretation of these equations, which represent the algebraic sum of the shear and normal strain or stress components acting in the critical plane. Also, the shear and normal components are weighted using an experimental parameter, K or k, hoped to be constant for a given material [16]. Fatemi and Kurath [18] have shown that the weighting parameter K in Eq. (40) may vary with fatigue life. It has been suggested [19] that in the case of multiaxial fatigue, life estimates should be based on a combination of both strains and stresses because they both contribute to cyclic damage.

With this in mind, the effects of stress and strain may be accounted for by applying a strain energy density relation [20] formulated on the critical plane, that is analogous to equation Eq. (40) and Eq. (41):

$$W^* = \frac{\Delta\gamma_{21}}{2}\frac{\Delta\sigma_{21}}{2} + \frac{\Delta\varepsilon_{22}}{2}\frac{\Delta\sigma_{22}}{2} \tag{42}$$

The novelty of the fatigue strain energy parameter W^* lies in the fact that it represents only that fraction of the strain energy contributed by the stresses and strains on the critical plane, not the overall strain energy. The parameter W^* is load path dependent, contrary to the criteria based on the overall strain energy density contributed by all stress and strain components at a given point. In addition, Eq. (42) is acceptable from the point of view of the formalism of continuum mechanics because the energy components $\Delta\sigma_{21}\Delta\gamma_{21}$ and $\Delta\sigma_{22}\Delta\varepsilon_{22}$ are scalars and can be added algebraically. Thus, the parameter W^* contains those important features often mentioned in the literature. In particular, it is

- associated with a favourable or critical plane;
- load path dependent;
- formally correct from the continuum mechanics viewpoint; and
- expressed without using any empirical fitting parameters.

However, the parameter W^* in the form expressed by Eq. (42) cannot account for the effects of mean stress. Therefore, further modifications are needed.

4.2 MEAN STRESS EFFECT IN MULTIAXIAL FATIGUE

Extensions to the critical plane strain criteria and the energy criteria which account for mean stress effects have been proposed by many researchers. These extensions are based on a few important observations [21, 22]. First, the alternating shear stress and strain on the critical plane are the primary cause of fatigue. Second, the mean or maximum values of the normal and shear stress and strain components on the critical plane are important secondary contributors to fatigue damage. Consequently, several versions of the critical plane strain criterion which account for the effect of the normal mean stress have been proposed. That of Fatemi and Kurath [18], and Socie *et al.* [22] has shown some capability of correlating various multiaxial fatigue data:

$$\gamma^{*} = \frac{\Delta \gamma_{21}}{2}\left(1 + k\frac{\sigma_{22}^{\max}}{S_y}\right) \tag{43}$$

Eq. (43) does not account for the mean shear stress effect. On the other hand, the formally correct strain energy criteria that account for mean stress effects often ignore the existence of the critical plane. One such criterion has been proposed recently by Ellyin *et al.* [23]

$$W^{*} = \Delta W^{e} + \Delta W^{p} = f(N_f) \tag{44}$$

It can be shown [6,7] that the overall strain energy density does not depend strongly on the load path because it depends only on the current stress state. For this reason Eq.(44) may not show a significant load path dependency despite the fact that one of the energy terms is indeed load path dependent. Nevertheless, the combination of plastic work per cycle and the elastic strain energy holds some promise.

A strain energy density criterion associated with the critical plane concept has been proposed by Chu *et al.* [24]. It combines both maximum stresses with the corresponding strain amplitudes on the critical plane:

$$W^{*} = 2\,\sigma_{21}^{\max}\gamma_{21}^{a} + \sigma_{22}^{\max}\varepsilon_{22}^{a} \tag{45}$$

The mean stresses (shear and normal) have been accounted for in Eq. (45), providing that both stress-strain components are alternating. Eq. (45) may ignore the mean stress effect of constant normal stress in the case of nonproportional loading when the amplitude $\varepsilon_{22}{}^{a} = 0$.

Based on the above considerations, a modified energy criterion associated with the critical plane concept has been proposed [25] in the following form:

$$W^{*} = \frac{\Delta \gamma_{21}}{2}\frac{\Delta \sigma_{21}}{2}\left[\frac{1}{1 - \dfrac{\sigma_{21}^{\max}}{\tau_{f'}}} + \frac{1}{1 - \dfrac{\sigma_{22}^{\max}}{\sigma_{f'}}}\right] \tag{46}$$

The form of Eq. (46) has been determined after considering the results of fatigue experiments carried out under constant amplitude proportional and non-proportional multiaxial loading [15-19], where it was concluded that the leading fatigue parameters were the alternating shear stress, $\Delta\sigma_{21}$, and strain, $\Delta\gamma_{21}$, on the critical plane. The maximum normal stress, σ_{22}^{max}, and the maximum shear stress, σ_{21}^{max}, are used in the equation to account for the mean stress effect. The normal stress on the critical plane assists in opening the crack and thus accelerates its growth. This correction is analogous to those used for crack tip closure in fracture mechanics. Similarly, the maximum shear stress effectively helps to overcome any sliding friction that takes place between the crack surfaces.

The advantages of Eq. (46), when compared with Eq. (42), are that the mean stresses have been taken into account and that only the alternating shear stress and strain are considered for the cyclic effect. This is in agreement with many experimental fatigue observations. Another advantage is that the parameter gives the orientation of the critical plane since its maximum coincides with the plane of maximum shear strain amplitude for the case of proportional loading.

In addition, the parameter W^* will predict different lives for the case of static compressive and tensile loads superposed on cyclic shear loading. This is in agreement with experimental observations [26].

4.3 PREDICTIVE CAPABILITIES OF THE MULTIAXIAL PARAMETER W^*

In order to assess the capability of the fatigue strain energy parameter, W^*, given by Eq. (46), to correlate fatigue lives corresponding to various loading modes, fatigue data obtained under different multiaxial loading conditions and different mean stresses were analysed and plotted in the form of W^* vs. N_f diagrams.

The set of experimental data was taken from Kurath *et al.* [27] who tested thin-wall tubular specimens (Fig. 13) made of hot rolled and normalized SAE 1045 steel. The specimens were thin-walled tubes with a 2.54 mm wall thickness, 25.4 mm internal diameter and 210 mm long with the gage length of 33 mm. The specimens were tested under three different loading conditions: tension, torsion, and simultaneous proportional tension and torsion. A constant ratio of applied shear to normal strain range $\Delta\gamma_{12}^{A}/\Delta\varepsilon_{22}^{A}$ was maintained during a given test. This ratio varied from 0 to ∞. The number of cycles, N_f, obtained from testing tubular specimens coincided with the creation of surface crack of 1mm long.

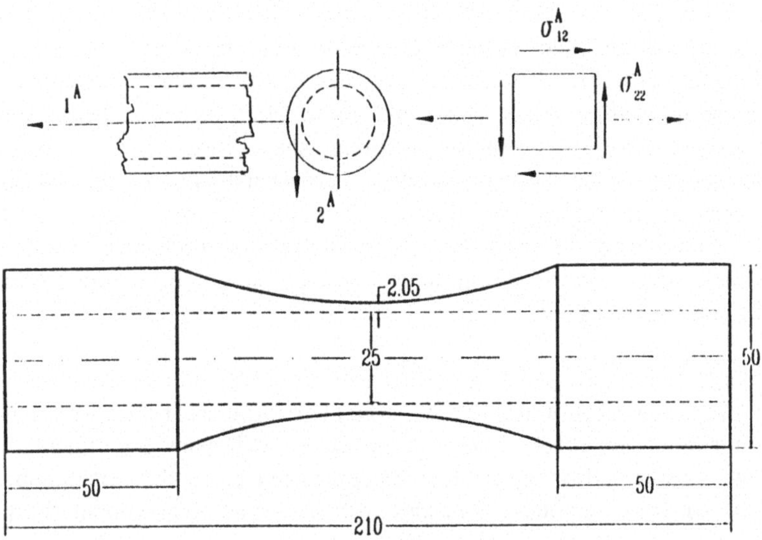

Figure 13: Geometry and dimensions of tubular specimen tested by Kurath et.al [16]. Coordinate system is shown (dimensions in mm).

The uniaxial fatigue data obtained from cylindrical specimens 6 mm in diameter that were tested under strain control, were also available in the form of the well known Manson-Coffin equation;

$$\frac{\Delta\varepsilon}{2} = \frac{\sigma_f{}'}{E}(2\,N_f)^b + \varepsilon_f{}'(2\,N_f)^c \tag{47}$$

First, the parameter W^* for the uniaxial tests involving solid cylindrical specimens was determined using Eq. (46). The plot W^* - N_f for the uniaxial data, denoted by the solid line is shown in Fig. 14. The parameter W^* in Eq. (46) was again plotted against the experimental fatigue life, N_f, obtained from tubular specimens and the results denoted by symbols are also shown in Fig. 14. It is apparent that all of the data, regardless of the type of loading and mean stress level, collapse onto one line which can be satisfactorily approximated by the standard curve determined from cylindrical specimens tested under fully reversed uniaxial loading (Eq. (47)).

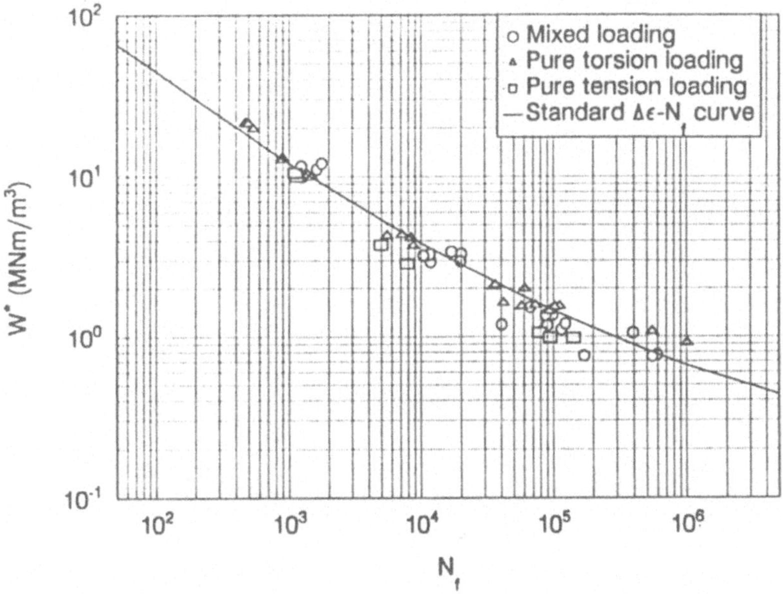

Figure 14: Strain energy density parameter W^* (Eqn (43)) versus experimental fatigue life N_f data for SAE 1045 steel tubular specimens tested under cyclic tension and torsion. (O mixed loading; Δ – pure torsion; □ pure tension; — standard $\Delta\varepsilon$ - N curve.)

Fig. 15 shows the variation of the parameter W^* with the orientation of the plane from the maximum shear strain plane for the SAE 1045 steel specimens at various levels and types of loading. All of the curves in Fig. 15 indicate that the plane of maximum damage coincides with the plane of maximum shear strain amplitude.

The second set of experimental data was taken from that reported by Fash [28]. Constant amplitude load controlled fatigue tests were conducted on shaft specimens (Fig. 16) made of SAE 1045 steel. The cracks initiated at the smaller shoulder radius and the fatigue life was defined as number of cycles, N_f, required to grow a crack to a length of 1 mm on the surface. The specimens were tested under fully reversed bending, torsion and simultaneous bending and torsion. The ratio of the bending to torsion moment M_b/M_t was kept constant during each test resulting in proportional cyclic loading. However, due to plastic yielding at the notch, the stress state at the notch tip did not remain exactly proportional. Therefore, the incremental Neuber approach to the notch-tip stress-strain analysis was adopted. The fatigue damage was determined using the fatigue parameter of Eq. (46). The fatigue life was calculated using the critical

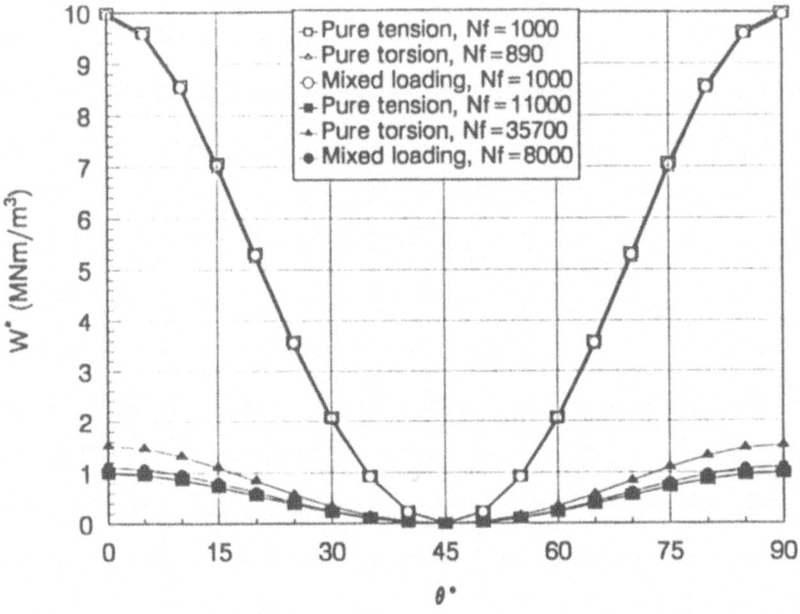

Figure 15: Variation of parameter W^* (Eqn (43)) with orientation of plane for SAE 1045 steel. (□ pure tension, N_f = 1137; Δ pure torsion, N_f = 1269; O mixed loading, N_f = 1258; ■ pure tension, N_f = 94525; Δ pure torsion, N_f = 102100; ● mixed loading, N_f = 115500.)

plane strain energy versus life relationship (W^* - N_f) obtained from uniaxial strain controlled tests. Comparison of the predicted and experimental lives is shown inFig. 17.

The whole procedure discussed above is capable of predicting fatigue lives for a variety of multiaxial loading modes using only the uniaxial cyclic stresses-strain and fatigue strain-life curves. However, the predicted lives tend to be conservative in the high-cycle fatigue regime. All the data that have been analysed up to this point were obtained under proportional multiaxial cyclic loading. It is yet to be determined whether or not the methods described above will be equally successful at predicting fatigue lives for non-proportional loading.

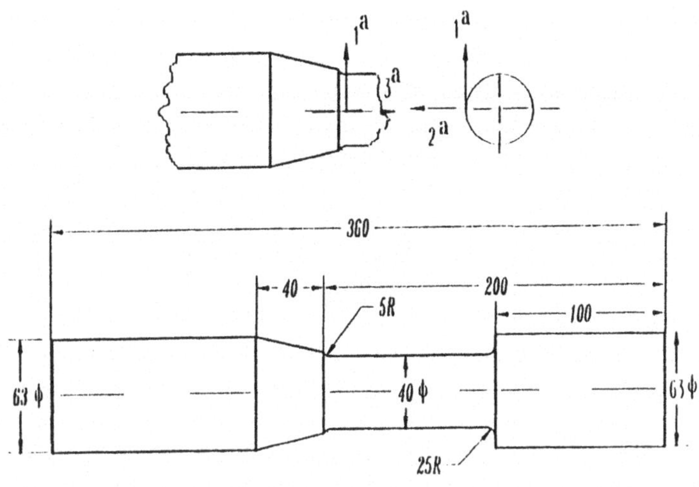

Figure 16: Geometry and dimensions of the notched shaft specimen [28] tested under bending and torsion

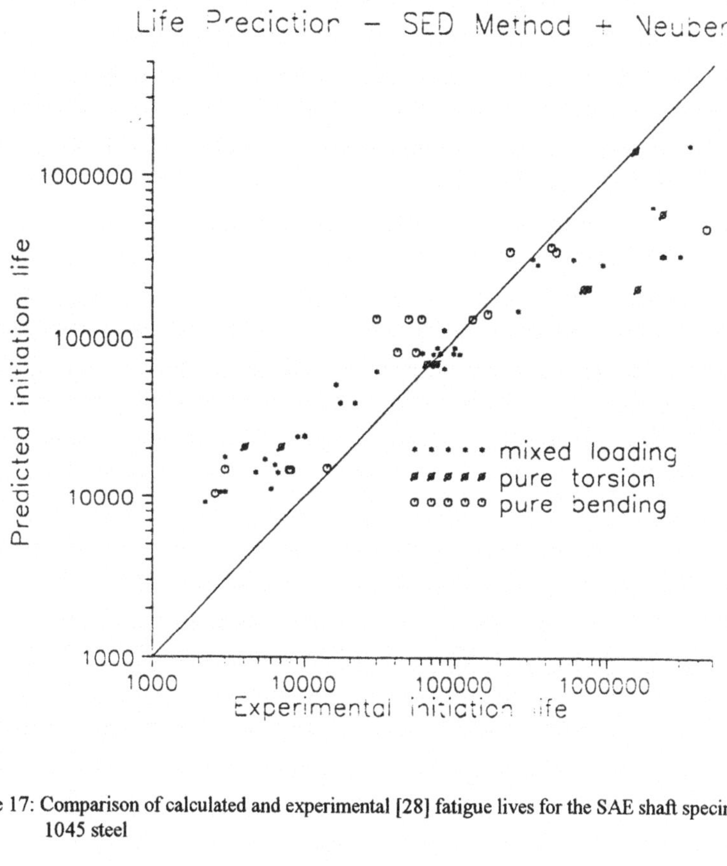

Figure 17: Comparison of calculated and experimental [28] fatigue lives for the SAE shaft specimens made of SAE 1045 steel

5. Conclusions

Two methods for calculating elasto-plastic notch tip strains and stresses have been proposed for general multiaxial loading. The methods have been formulated using both the total strain energy density and the strain energy density relationships in conjunction with with the Mroz-Garud cyclic plasticity model. It has been found that the generalised Neuber's rule, which represents the equality of the total strain energy density at the notch tip, gives an upper bound estimate for the elasto-plastic notch tip strains. The generalised equation of the equivalent strain energy density yields a lower bound solution for the notch tip strains and stresses. The method has been verified by comparison to numerical and experimental data for for various loading paths and material properties.

The calculated notch tip strains and stresses were subsequently used for calculating fatigue damage according to a proposed multiaxial fatigue damage parameter that associates strain energy density with the critical plane concept.

6. References

1. Neuber, H. (1961) Theory of Stress Concentration Shear Strained Prismatic Bodies with Arbitrary Non-Linear Stress-Strain Law, *ASME Journal of Applied Mechanics* **28**, 544-550.

2. Topper, T.H., Wetzel, R.M. and Morrow, J. (1969) Neuber's Rule Applied to Fatigue of Notched Specimens, *Journal of Materials* **4**, 200-209.

3. Molski, K. and Glinka, G. (1981) A Method of Elastic-Plastic Stress and Strain Calculation at a Notch Root, *Material Science and Engineering* **50**, 93-100.

4. Seeger, T. and Hoffman, M. (1986) The Use of Hencky's Equations for the Estimation of Multiaxial Elastic-Plastic Notch Stresses and Strains, Report No. FB-3/1986, Technische Hochschule Darmstadt.

5. Barkey, M.E., Socie, D.F. and Hsia, K.J. (1994) A Yield Surface Approach to the Estimation of Notch Strains for Proportional and Non-proportional Cyclic Loading, *ASME Journal of Engineering Materials and Technology* **116**, 173-180.

6. Moftakhar, A., Buczynski, A. and Glinka, G. (1995) Calculation of Elasto-Plastic Strains and Stresses in Notches under Multiaxial Loading, *International Journal of Fracture* **70**, 357-373.

7. Singh, M.N.K., Glinka, G. and Dubey, R.N., (1996) Elastic-Plastic Stress-Strain Calculation in Notched Bodies Subjected to Non-proportional Loading, *International Journal of Fracture*, to be published.

8. Reinhardt, W., Moftakhar, A., and G. Glinka (1995) An Efficient Method For Calculating Multiaxial Elasto-Plastic Notch Tip Strains And Stresses Under Proportional Loading, *27th ASTM Fracture Mechanics Conference*, Blacksburg, Virginia, to be published.

9. Mroz, Z. (1967) On the Description of Anisotropic Workhardening, *Journal of Mechanics and Physics of Solids*, **15**, 163-175.

10. Garud, Y. S. (1981) A New Approach to the Evaluation of Fatigue under Multiaxial Loading, *Journal of Engineering Materials and Technology*, ASME, **103**, 118-125.

11. Hibbit, H., Karlsson, B. and Sorenson, P.(1989), *ABAQUS Theory Manual*, Version 4.8.

12. V. M. Radakrishnan (1980) An Analysis of Low Cycle Fatigue Based on Hysteresis Energy, *Journal of Fatigue & Fracture of Engineering Materials and Structures*, **3**, 75-84.

13. Ellyin, F., and Kujawski, D. (1984) Plastic Strain Energy in Fatigue Failure, *Journal of Engineering Materials and Technology, ASME*, **106**, 342-347.

14. Findley, W.M. (1959) Theory for the Effect of Mean Stress on Fatigue of Metals under Combined Torsion and Axial Load or Bending, *Journal of Engineering for Industry*, **81**, 301-306.

15. McDiarmid, D.L., (1991) A General Criterion of High Cycle Multiaxial Fatigue Failure, *Journal of Fatigue & Fracture of Engineering Materials and Structures*, **14**, 429-453.

16. Kandil, F.A., Brown, M.W., and Miller K.J. (1982) Biaxial Low Cycle Fatigue of 316 Stainless Steel at Elevated Temperature, *The Metals Society*, London, **280**, 203-210.

17. Lohr, R.D. and Ellison, E.G. (1980) A Simple Theory for Low Cycle Multiaxial Fatigue, *Journal of Fatigue & Fracture of Engineering Materials and Structures*, **3**, 1-17.

18. Fatemi, A. and Kurath, P. (1988) Multiaxial Fatigue Life Predictions under the Influence of Mean-Stress, *Journal of Engineering Materials and Technology, ASME*, **110**, 380-388.

19. Bannantine, J.A. and Socie, D.F. (1992) Multiaxial Fatigue Life Estimation Technique, *ASTM STP 1122*, eds. M.R. Mitchell and R.W. Landgraf, American Society for Testing and Materials, Philadelphia, 249-275.

20. Glinka, G., Plumtree, A. and Shen, G. (1995) A Multiaxial Fatigue Strain Energy Density Parameter Related to the Critical Plane, *Journal of Fatigue & Fracture of Engineering Materials and Structures*, **18**, 37-46.

21. Findley, W.N. (1953) Combined Stress Fatigue Strength of 76S-T61 with Super-imposed Mean Stresses and Corrections for Yielding, *Technical Note 2924*, NACA-National Advisory Committee for Aeronautics.

22. Socie, D.F., Kurath, P. and Koch, J. (1989) A Multiaxial Fatigue Damage Parameter, *Biaxial and Multiaxial Fatigue-EGF3*, eds, M.W. Brown, and K.J. Miller, Mechanical Engineering Publications, London, 535-550.

23. Ellyin. F. and Golos, K. (1988) Multiaxial Fatigue Damage Criterion, *Journal of Engineering Materials and Technology, ASME*, **110**, 36-41.

24. Chu, C.C., Conle, F. and Bonnen J.J. (1992) Multiaxial Stress-Strain Modelling and Fatigue Life Prediction of SAE Axle Shafts, *Symposium on Multiaxial Fatigue, ASTM STP 1122*, eds. M.R. Mitchell and R.W. Landgraf, American Society for Testing and Materials, Philadelphia.

25. Glinka, G., Wang, G. and Plumtree, A. (1995) Mean Stress Effects in Multiaxial Fatigue, *Journal of Fatigue and Fracture of Engineering Materials and Structures*, **18**, 755-764.

26. Koch, J.L. (1985) Proportional and Non-Proportional Biaxial Fatigue of Inconel 718, *University of Illinois at Urbana-Champaign, Report No. 121/UILU-ENG, 85-3605.*

27. Kurath, P., Downing, S.D. and Galliart, D. (1989) Summary of Non-Hardened Notched Shaft Round Robin Program, *Multiaxial Fatigue, Analysis and Experiments SAE, AE-14*, eds. G.E. Leese and D. Socie, Society of Automotive Engineers, 13-32.

28. Fash, J.W. (1985) An evaluation of damage development during multiaxial fatigue of smooth and notched specimens, *University of Illinois at Urbana-Champaign, Report No. 123/UILU-ENG 85-3607.*

ASSESSMENT OF FATIGUE IN HIGH-DUTY ENGINEERING COMPONENTS

JOHN F. KNOTT
School of Metallurgy and Materials
The University of Birmingham, U.K.

1. Introduction

It is instructive to treat engineering components as constituent parts of *structures* or *machines*. *Structures* are, in principle, *functionally static* designs, such as bridges, whose *function* is to maintain a *static* connection from shore to shore, or offshore drilling platforms, whose *function* is to maintain drilling equipment in a *static* location. *Machines* are *functionally dynamic*, comprising an assemblage of components, designed to move relative to one another, often at high speed: automobile engines, aero-engines, turbines, alternators, high-speed manufacturing systems, robotic devices and so forth. Combinations are often found: for example, a power-crane, in which the engine/motor *machine* operating at many revolutions per minute (rpm) experiences a power surge every time it is called upon to operate, but in which the cantilevered jib-boom in the support *structure* experiences loading every time a weight is lifted. An automobile spring or rail spring is a component of interest: it is basically *structural* in nature and yet is subjected to cyclic frequencies akin to those of *machines*.

The *functional distinction* may be characterised also by the nature of the *service duty* to which a component is exposed and by the *fatigue lifing methodology* employed to address *lifetime integrity*. *Structures* contain fabricated components: girders, sections, tubes, plates, joined by processes such as welding, riveting, bolting or adhesive bonding. The *function* is *static*, but *cyclic loading* is experienced through factors such as traffic loading (for a bridge), variable wind gusts, wave loadings, collisions or temperature changes leading to thermal strains (of particular significance with respect to high-temperature plant or to space-platforms, plunged from intense heat to "black cold" as they rotate). The *duty* comprises *both* the *static* functional loads *and* anticipated *cyclic* loads. In terms of broad generalisation, *structures* are likely to contain meso/macro-scopic defects (upwards of 0.2mm in size), but the cyclic loads are small fractions of the yield/proof stress, unless *structural resonance* is induced by the duty. The *fatigue methodology* applied to structures *for design purposes* is *conceptually* that of the integration of crack-propagation-rate to predict life, although, *in practice*, resort may be made to ("S/N") *design-curves* such as those produced by TWI for various types of welded joint.

R.A. Smith (ed.), Reliability Assessment of Cyclically Loaded Engineering Structures, 137–164.

Components in *machines* are deemed to consist of high-quality ("defect-free") material, provided to a high degree of surface finish. Any defects are at the micro-scale. Heavily-worked bar-stock or forgings are commonly used; close control is employed in heat-treatment and attention is paid, first to surface finish, more critically, to "*surface engineering*". Thermal and/or mechanical treatments are used to create surface layers which are both hard and in a state of compressive residual stress. *Machine* components are generally subjected to a large number of (necessarily) low-amplitude cycles: e.g. small bending moments on a crankshaft in a car-engine, resulting from the wear of bearings. For an engine to last 150,000 km, running at 100 kmh^{-1} and 3500 rpm, it is necessary to sustain more than $3x10^8$ cycles. The *fatigue methodology* employed for *machine components* is traditionally the *S-N curve* (devised originally by Wohler for comparable cases of rotating bending). Again, *conceptually*, the principles of fracture mechanics might be recognised, and there is interest in pursuing "short-crack" crack-growth-rate formulations. In practice, however, these can only provide conceptual guidance, *not* engineering design data. *Integrity* will be assured by appropriate *"S-N" data* and the ability of the metallurgist to provide *consistent material* and *consistent surface engineering.*

2. Long-Crack Growth

2.1 FATIGUE-CRACK GROWTH-RATE AND 'LIFING'

Conventionally, the increment in fatigue crack growth per cycle, da/dN, is related to the instantaneous value of stress-intensity-factor range, $\Delta K = K_{max} - K_{min}$ (where K_{max} and K_{min} are respectively the maximum and minimum instantaneous values of stress-intensity-factor in the cycle) through the Paris expression [1].

$$da/dN = A\Delta K^m \qquad \text{..........1)}$$

where A and m are constants. In general, ΔK is given by:

$$\Delta K = Q\,\Delta\sigma(\pi a)^{1/2} \qquad \text{..........2)}$$

where Δσ is the applied stress range, a is the instantaneous crack length and Q is a geometrical factor, having a value of 1.12 for a through-thickness edge crack and 0.67 for a semi-circular edge crack. Equation (1) is a particular version of the more general form.

$$da/dN = B\,\Delta\sigma^v a^w \qquad \text{..........3)}$$

where B, v and w are constants. Equations (1), (2) and (3) are reconciled if m = v =2w.

The application of equation (1) or (3) to calculate a component's "lifetime" (or number of cycles to failure, N_f), may be illustrated, using equations (1) and (2), by separating variables, and integrating between limits, a_o (initial defect size) and a_f (final defect size) to give:

for m $\neq 2$ $$a_o^{1-m/2} - a_f^{1-m/2} = \{(2-m)A/2\}\, Q^m \Delta\sigma^m \pi^{m/2}\, N_f \qquad \text{..........4)}$$

for m = 2 $$\ln\,(a_f / a_o) = AQ^2 \Delta\sigma^2 \pi N_f \qquad \text{..........5)}$$

Similar results may be obtained using equation (3):

for w $\neq 1$ $$a_o^{1-w} - a_f^{1-w} = (1-w) B \Delta\sigma^v N_f \qquad \text{..........6)}$$

for w = 1 $$\ln\,(a_f / a_o) = B \Delta\sigma^v N_f \qquad \text{..........7)}$$

In each case, it is possible to calculate N_f for a given value of $\Delta\sigma$ if the constants in equations (1), (2) and (3) have been determined experimentally and if the values of the initial defect size, a_o, and final defect size, a_f, can be measured or calculated.

The initial defect size is a function of the material's original processing history (e.g. the distribution of pores in a casting, or laps in drawn bar) and of the details of fabrication (e.g. toe-cracks in a welded joint). It can be controlled/assessed in three main ways. The first of these is *process control*, usually combined with mechanical testing. For many high-duty alloys (destined for machine components) metallurgical processing is able to restrict defects to very small sizes, such that they can be detected only by (destructive) metallographic techniques. Clearly, this cannot be done for the actual component that is to enter service. What is done is to establish that a particular processing route gives satisfactory results in *specimen tests* or *model component tests* tests, (taken to failure) and the *processing details* are then "*sealed*" to assure that all components are of equally good quality. Statistically-based "design-curves" and safety-factors are produced to ensure that "design lives" can be achieved safely. An example of rather more quantitative "process control" is found in the forming of nickel-alloy aero-engine discs by consolidation of atomised powder, (see later). The mesh size of the sieve used to grade the powder provides a physical limit to the size of non-metallic particle that can be entrapped in the alloy powder.

A second method to control the initial defect size is the *proof-test* or *overspeed test*. Here, a component is subjected *either* to a single overload (e.g. a pressure-vessel is stressed to $0.8\sigma_Y$, when the normal design stress is $0.67\sigma_Y$) *or* to a sequence of high-amplitude fatigue cycles (e.g. a steam-turbine disc, before it enters service). If the component contains too large a defect, it will fail during the proof test: if it does not fail, the difference between the critical size at high stress in the proof stress and that at lower stress in normal operation provides a margin for crack growth. In many cases, this margin is not large, but the proof test also induces "overload" residual stresses which can give rise to a "*warm prestressing*" (WPS) effect, increasing fracture toughness at low temperature, or to a *retardation* of fatigue-crack growth. An extreme example of this is the *auto-frettage* of a gun-barrel.

The third method for assessing a_o is to use *non-destructive-inspection*, NDI. Fluorescent dyes or magnetic inks can be used to detect surface-breaking defects, but, to detect and size buried, *crack-like defects*, the only viable NDI technique is *ultrasonics*: X-rays can detect volumetric defects, but not narrow cracks. Typically, for steel structures, an ultrasonic probe would be of some 25mm in diameter, emitting a pulse of ultrasound at 2MHz: a wavelength of 3mm in steel. (This relates to the limits of resolution and sizing of defects). In material which is very clean in terms of inclusion content, it is possible to use a much higher frequency and smaller wavelength: for nickel alloy aeroengine discs, Rolls-Royce use 50MHz and can detect defects of 0.15mm in size.

The *final crack size*, a_f, in equations (4)-(7) may be determined by the length at which *fast fracture* occurs (when K_{max} becomes equal to the material's fracture toughness K_{IC}) by the length at which the remaining ligament undergoes *plastic collapse*, or by some other criterion: e.g. generation of such a *high compliance* (loss of stiffness) in the cross-section that the *component loses* its *structural function*. If the stress-range applied to a component is constant, a crack accelerates rapidly as it grows and the number of cycles associated with propagation from, say, $0.9a_f$ to a_f, can be insignificant compared with N_f. Under such conditions, variations in a_f can be ignored and, with a_0 fixed, equations (4) and (5) can be re-arranged to give

$$\log N_f = -m \log \Delta\sigma + const \qquad \text{..........8)}$$

Rearrangement of (6) and (7) gives (8) with m replaced by v. Equation (8) is the basic "S-N" curve (S=$\Delta\sigma$) for a material containing an initial defect. If a block of material contains a distribution of defects of different lengths, orientations and degrees of mutual "shielding", a set of testpieces cut from the block will demonstrate a "scatter" in values of N_f which is, in principle, calculable from the "scatter" in values of a_0 [2].

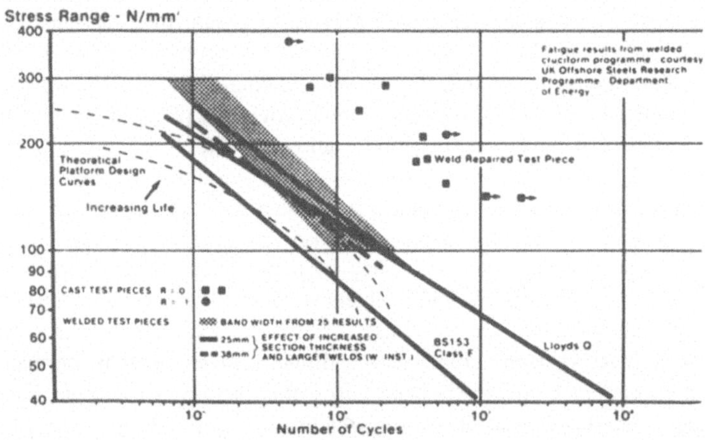

Figure 1. Fatigue performance of as-cast "node" (individual filled squares) compared with BS 153 Class F design curve and Lloyds Q design curve.

A practical example of the use of equation (8) is found in TWI "*Fatigue Design Curves for Welded Joints*". Here, archetypal welded features (simple butt welds, unstressed fillets, "T" or "K" tube-on-tube joints etc.) have been tested by laboratories all over the world and the data have been assembled in (log-log) "S-N" form. *Mean lines* are drawn through the data and *design curves* are constructed to lie two standard deviations (s.d) below the means (with corrections to produce straight lines). Welds may contain complicated combinations of stress-concentrators, residual stresses and fabrication defects, so that it is not necessarily a simple matter to derive the "S-N" curve by direct integration, but the conceptual links are clear. A simple butt-weld contains no stress-concentration, access is easy and defects should be of minimal size: a "K" tube-on-tube joint has a high stress concentration, access in a "crotch" of the K is difficult and it may be difficult to avoid defects. Here, the use of cast "nodes" to ameliorate stress-concentrations and to reduce defect size has led to marked improvements in fatigue performance. (see Fig. 1).

A final use of the crack integration approach is to treat *variable-amplitude loading*, assuming no interaction between the various loading "blocks". Suppose that the amplitude of the first block is $\Delta\sigma_1$ and that the crack grows from a_o to a_1 during the N_1 cycles for which $\Delta\sigma_1$ is applied. Let the integral corresponding to the LHS (left-hand-side) of equations (4) to (7) be represented by I_{01}: the RHS contains N_1. For failure at $\Delta\sigma_1$, let the LHS integral be I_{of}, corresponding to $N_{f(1)}$. Then $N_1/N_{f(1)} = I_{01}/I_{of}$. Let the second block have amplitude $\Delta\sigma_2$, and let the crack grow from a_1 to a_2 during N_2 cycles. The integration leads to $N_2/N_{f(2)} = I_{12}/I_{of}$. Pursuit of the argument for 'n' blocks which eventually lead to failure gives

$$\sum_{r=o}^{r=n} \frac{N_r}{N_{f(r)}} = \frac{1}{I_{of}} \sum_{r=o}^{r=n-1} I_{r,r+1} = (1/I_{of})\,(I_{01} + I_{12} + I_{23}....I_{n-1,n}) \qquad ....9)$$

Now, the last integral corresponds to the last stage of growth to a_f, and the summation of all the individual integrals $I_{r,\,r+1}$ must be equal to I_{of}, since they simply represent incremental stages of crack growth from a_o to a_f. Hence, the RHS of equation (9) becomes $I_{of}/I_{of} = 1$, so that (9) becomes

$$\sum_{r=o}^{r=n} \frac{N_r}{N_{f(r)}} = 1 \qquad ....10)$$

This is the Palmgren-Miner Law, which is commonly used in conjunction with "S-N" curves. With the assumption of no interaction between different loading blocks, and the second assumption that small changes in a_f are insignificant, it can been seen that equation (10) is compatible with equation (8), since both are derived from a common integration of crack growth.

2.2 THE FATIGUE-CRACK GROWTH-RATE CURVE.

Equation (1) is based on a fracture-mechanics characterisation of fatigue-crack growth-rates, whereas equation (3) is more general. Both have been shown to be of equal value in the calculation of lifetimes and which is to be preferred depends *either*, on which provides the better fit to empirical data, *or* on which possesses better theoretical justification. One body of early results tended to support equation (3) with $v = 3$, $w = 1$, but, more recently, there has been general acceptance of equation (1). Two types of model have been proposed: one, based on the value of *crack-tip opening displacement*, suggesting $m = 2$: the other, relating to stored *strain-energy* and suggesting $m = 4$ [3] . The former model is summarised in [4] and in the following paragraphs. The basis for the latter model appears to be fallacious, since it is observed that intermediate annealing causes a crack to accelerate (as a result of decreases in yield-stress and in plastic-wake closure forces), whereas damage-accumulation/stored energy models would predict a decrease (because stored energy has been removed by the annealing). Verification of the form of equation (1) has been sought by plotting log (da/dN) vs. log ΔK: if it holds, the graph should be a straight line of slope m. Early measurements were made on a rather coarse scale and linear forms with a wide variety of m-values were reported. Ritchie [5] made measurements, using the sensitive potential drop (P.D.) technique and showed that the general (log:log) curve could be divided into three regions: a central "Paris Law" region and two steeper curves: at low ΔK, approaching a "threshold" value ΔK_{th}; at high ΔK, approaching catastrophic fracture, as K_{max} approaches K_{IC}. (fig. 2).

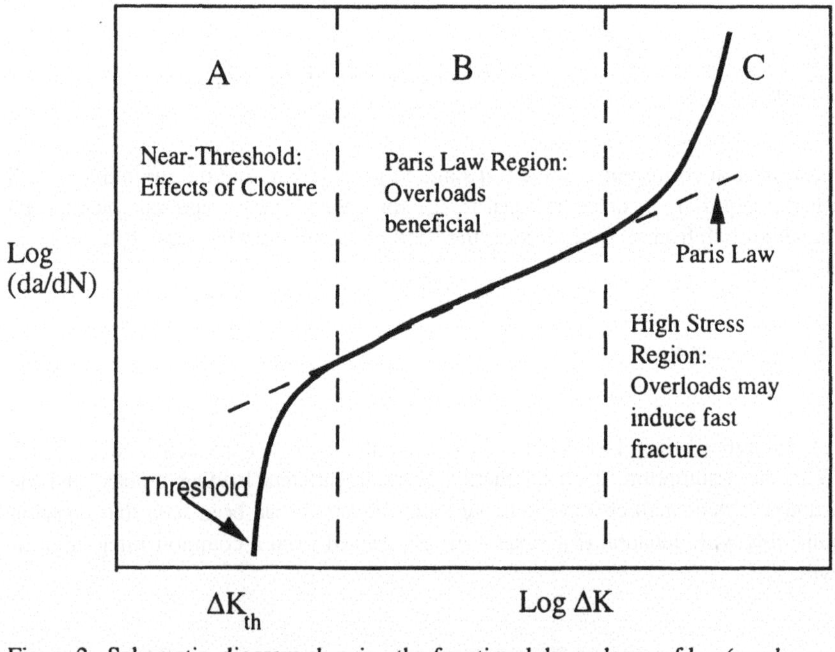

Figure 2. Schematic diagram showing the functional dependence of log (crack growth-rate, da/dN) on log (stress-intensity-range, ΔK)

By examining the behaviour of deliberately embrittled steel, it was possible to elucidate the role of brittle, "static" or "monotonic" modes of fracture at high K_{max} and then to analyse data previously reported in the literature to show [5,6] that high "m-values" were associated with brittle behaviour (low K_{IC} values). The values of "m" for tough materials tended to be close to 2, in general support of a CTOD model (with some allowance for differences in the effects of cyclic hardening on the cyclic flow stress, σ_{cf} in the region of the crack tip), see fig. 3.

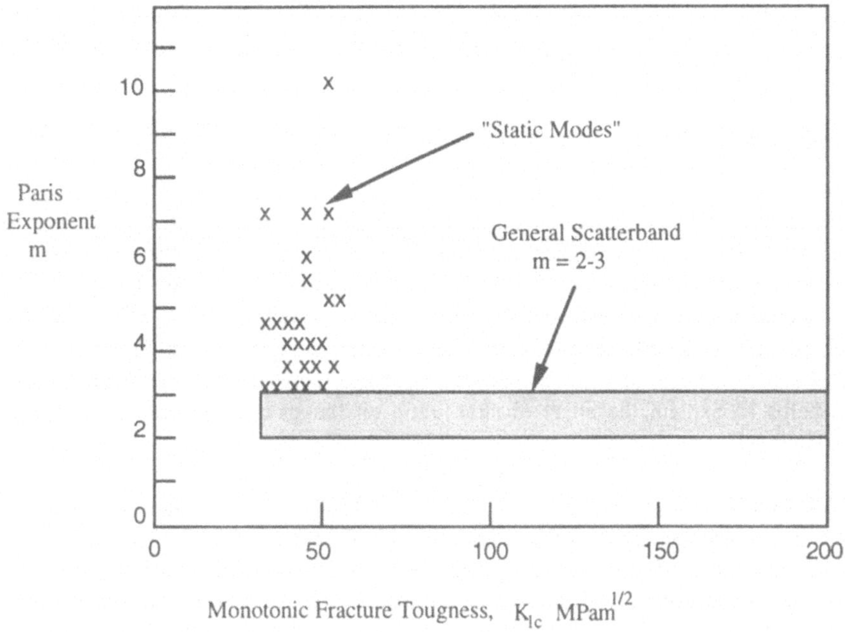

Figure 3. Schematic "dependence" of Paris exponent, m, on Monotonic Fracture Toughness (after [5])

Using the general expression, $\delta = K^2/2\sigma_Y E$, where δ is the CTOD, σ_Y is the yield stress and E is Young's modulus, it is possible to define a CTOD range, $\Delta\delta = \delta_{max} - \delta_{min}$, where δ_{max} and δ_{min} correspond to K_{max} and K_{min}. Defining the stress-ratio R as K_{min}/K_{max}, we can write.

$$\Delta\delta = \{\Delta K^2/(4\sigma_{cf}E)\}\{(1+R)/(1-R)\} \qquad ....11)$$

where the reversed flow stress is taken as $2\sigma_{cf}$. If it is assumed that (da/dN) is proportional to $\Delta\delta$, it follows that (da/dN) should be a function of (1+R)/(1-R) at constant ΔK. There is experimental evidence to support this relationship [4]. The proportionality between (da/dN) and $\Delta\delta$ is a function of the lack of rewelding in each crack-tip cycle and hence depends in a sensitive manner on the environment. Growth rates in vacuum are much lower than those in air.

At high ΔK (K_{max} approaching K_{IC}), it is found that the area fraction of brittle cleavage facets, A_c, in a number of steels rises rapidly with K_{max}. As K_{IC} is approached, it is not unreasonable to express the term $(1-A_c)$ as a linear function of the parameter $(K_{IC}-K_{max})$: in essence, the graph of A_c vs. K_{max} is *rotated* through 180° and the new origin is located at $K_{max} = K_{IC}$, $A_c = 1$. Now, consider that an area is being traversed by a crack comprising both fatigue striations and cleavage. Since the cleavage component is composed of a number of facets, each produced in a quarter of a cycle, it may be assumed that the cleavage occurs virtually instantaneously. The fatigue component then has to "make good" only an area $(1-A_c)$, so that the "fatigue" growth-rate, $(da/dN)_{app}$, is apparently enhanced by a factor $(1-Ac)^{-1}$. Since $(1-A_c)$ is linearly proportional to $(K_{IC} - K_{max})$, $(1-A_c)^{-1}$ is proportional to $(K_{IC} - K_{max})^{-1}$ and $(da/dN)_{app}$ may be written

$$da/dN = A' \{\Delta K^2/4\sigma_{cf}E\}\{(1+R)/(1-R)\}(K_{IC} - K_{max})^{-1} \quad ....12)$$

This form is quite similar to that proposed by Forman [7], recalling that K_{max} may be written as $\Delta K/(1-R)$, but has a physical basis [4]. The argument is somewhat less sharp for "static" modes involving inclusion - nucleated voids, because the voids are likely to take more than a quarter-cycle to grow, but the general principle still holds. It is of interest that brittle SiC particles in a particulate Al-matrix metal-matrix composite (MMC) are observed to crack at a critical value of K_{max} (dependent on particle size) and that the "static mode" approach helps to explain the effect of stress-ratio on fatigue-crack-growth behaviour in these materials [8].

At low ΔK values, a "*threshold*" ΔK_{th} is observed, below which the value of (da/dN) is virtually zero (experimentally, less than approximately 0.1nm per cycle). The "threshold" is, however, very much a feature of "long-crack" tests and the manner in which the threshold is approached, by "winding-down" the stress-amplitude and allowing the crack to grow (more and more slowly, over an increasingly greater number of cycles) through the prior plastic zone. The dislocation arrays in such zones behind the growing crack-tip produce compressive residual stresses which act to provide "*clamping forces*" in the "*plastic wake*". Premature "closure" of a crack is further enhanced by environmental effects, such as oxide-film formation [9] and by surface roughness [10]. The threshold value, ΔK_{th}, is greatly reduced if a high tensile mean stress (high R-ratio) is associated with the fatigue cycle, because the tension helps to counteract the compressive "clamping". Again, if a crack is propagating (at modest R-ratio) at low ΔK, just above ΔK_{th}, and part of the plastic wake is removed, by spark-machining from behind the crack-tip, the value of (da/dN) is found to increase sharply, because clamping forces have been removed [11].

These results have direct relevance to the behaviour of short cracks: small processing or fabrication defects which are not associated *a priori* with a compressive residual stress field. A significant defect is needed to ensure that fatigue cracks start to propagate at stresses markedly below yield: microstructurally short cracks, initiated by slip-bands, require high stresses, which may often be unrealistically high with respect to conventional fatigue design stresses. It is vitally important to ensure that *small-scale specimens* are tested in a manner such that they *fully reflect* the *characteristic features* of the *engineering application* to which they are supposed to relate. Very high applied stresses, close to yield, can produce ambiguities (and ill-defined residual stresses in the unloading cycle) but the general principle is that it is quite possible for a residual stress-free, small defect to propagate at a value of ΔK lower than ΔK_{th} at modest R-ratio. It is possible to reconcile "high-R" ΔK_{th} values with limiting values of ΔK at which residual-stress-free small defects begin to propagate, but these "limit" ΔK values, ΔK_l, are extremely low: $\leq 3MPam^{1/2}$ for structural steels [12]. Although there are versions of equation (12) which incorporate a $(\Delta K-\Delta K_{th})^2$ term, prudent, practical application of the equation takes ΔK_{th} as zero, to give slightly conservative predictions.

The residual-stress distribution created when a (long) crack grows through a region in which the prior crack-tip strain-amplitude (due to a higher ΔK-value) is higher than the current strain-amplitude produces "clamping" effects which can be effective at moderately-high ΔK as well as at low ΔK. An *overload* in the fatigue cycle is often beneficial in terms of *retarding* the subsequent crack growth-rate. However, the benefit (or otherwise) of the effect depends critically on the region in the (log:log) da/dN vs. ΔK curve associated with the application of an "overload" ΔK (K_{max}). If all processes are controlled by simple plasticity ($\Delta\delta$), any overload ΔK produces compressive residual stresses and retards cracks growth in a manner similar to that experienced in the near-threshold region. *However*, in the high ΔK (K_{max}) region, an overload may produce a "burst" of fast fracture, giving an *increase*, rather than a *decrease*, in overall crack growth-rate [13].

2.3 APPLICATION OF FATIGUE-CRACK GROWTH-RATE CURVES TO ENGINEERING APPLICATIONS.

Conventional fatigue-crack growth-rate curves are determined, using specimens containing single long, through-thickness cracks, e.g. a/W values ≥ 0.25 in specimens with W = 20mm, for which compliance functions and potential-drop calibrations are well-characterised. In contrast, defects in engineering components may be located near stress-concentrators, are often of irregular shape (which can, perhaps, be approximated to circumscribing ellipses or semi-ellipses) and may be found in clusters. *Interaction* between adjacent defects needs to be considered. As explained above, the *residual stresses* associated with service defects are often quite different from those generated by the plastic-wake behind a long-crack tip in the "wind-down" sequence. Small service defects can be free of compressive stress, and, for such defects, use must be made of "closure-free" near-threshold data. In contrast, *surface treatments* such as case-hardening or shot-peening can produce high levels of compressive stress, which provide a closure component much greater than that experienced in the long crack threshold test.

In this section, attention is paid to geometrical effects, to the interaction of fatigue cracks and to the behaviour of "natural", crack-like defects. Effects of residual stress are treated in a later section.

The propagation of a defect which is located ahead of a macro-scale concentrator may be treated elastically as a sharp crack emanating from the tip of the major axis of an elliptical (or semi-elliptical) notch, which, in one limit is a circle (or semi-circle) and, in the other limit, is also a sharp crack. It is possible to derive stress-intensity factor solutions for "notch + crack" configurations: in mode I, these solutions are usually numerical, but, in mode III, it is possible to obtain an analytical solution (using a sequence of conformal mapping transformations). Solutions can also be obtained by loading the small crack with the stress distribution ahead of the notch (in the absence of the crack) using a suitable weight function: formal equivalence may be established for mode III, using the mapping transformations [14].

If the "root radius" at the tip of the stress-concentrator is ρ and the length of crack ahead of the tip is d, the stress intensity factor, K, rises from zero at $d/\rho = o$, to a value equal to that for a sharp crack of length "notch + crack", for $d/\rho \sim 1$, in mode III, but to a slightly higher value in mode I. Consider a large circular stress concentrator (such as a nozzle in a pressure-vessel) with a small (radial) crack ahead of it lying normal to the maximum principal stress. As $d \rightarrow o$, the situation approximates to that of an edge crack, for which the form is $K = (3\sigma) \times 1.12\ (\pi d)^{1/2}$, since the circumferential stress is 3σ. As d increases, the local stress decreases from 3σ, but the maximum "notch + crack" value of K is approximately 5% greater than that for the sharp crack of equivalent length.

A second treatment of the "notch + crack" configuration is to envisage that a *plastic* zone is set up ahead of the notch and that the plastic strain amplitude within this zone is responsible for the initiation of, and early stages of propagation of, a "small crack". There is an argument that such a crack may start to grow, but then arrest, because the (notch-field) plastic strain amplitude decreases. The effect of plastic strain amplitude on "initiation" has been demonstrated in an aluminium alloy, heat-treated to give the *same initial* flow stress, but *different cyclic* flow stresses [15]: however, it remains an open question as to whether the role of *notch-root plasticity* is one of *plastic strain amplitude* or *closure* (due to clamping forces from the "notch-root" dislocation arrays). It is necessary to carry out both annealing experiments and experiments at high mean stress to substantiate an appropriate model. In general, the "*non-propagating*" *fatigue cracks* found in notched-bar tests are associated with materials which *cyclically strain-age*: this provides a mechanism to increase the flow stress ahead of a growing defect, thereby increasing σ_{cf} (c.f. equation 11). or 12).) and reducing $\Delta\delta$ and da/dN (possibly to zero).

The propagation of single, semi-elliptical cracks and the *interaction* between two such cracks has been studied by Soboyejo [16,17]. Using "top-hat" ("ridged") specimens it is possible, in bend, to introduce *semi-elliptical cracks* having the same characteristics as those for long, through-thickness cracks: the ridges are then machined off (with care) before testing. Propagation of such cracks in bending-fatigue leads to a "preferred crack profile", particularly at ΔK values significantly above threshold. The interaction of two such cracks in bending is clear. The cracks grow individually until the "inside" edges touch. There is then a very rapid increase in growth-rate until the crack propagates as a single defect. These studies lead to a "re-characterisation" procedure for clusters of defects which has a physical basis, and which is rather less conservative than the present standard procedures. It is possible to also investigate the effect of crack "shielding" for small elliptical cracks in a linear, or staggered array.

A similar study, by Cowling [18], has examined the behaviour of small, hydrogen-induced cracks, designed to simulate "toe" cracks in the heat-affected-zone (HAZ) of a welded joint. Bulk specimens of HY80 steel were first, given a heat-treatment which produced a microstructure corresponding to that of an as-welded HAZ. The specimen was then subjected to a small bending moment in a jig and the great majority of its surface was "stopped-off" with an electrically-insulating lacquer. Hydrogen was charged electrolytically and, because the HAZ was hard, small (intergranular) cracks could be produced. The specimens were then unloaded, de-lacquered and stress-relieved/tempered before being subjected to fatigue loading in the near-threshold region. Tests were carried out at both a low R-ratio (R=0.2) and a high R-ratio (R=0.7).

The results showed good agreement with those obtained by Soboyejo on the same steel and heat-treatment, but the *threshold values* were *higher* at R=0.2 than for the "long" (fatigue-induced) semi-elliptical crack. This was attributed to the fact that the intergranular, hydrogen-induced crack exhibited greater surface roughness than the transgranular, fatigue-induced crack. For a stress-ratio, R=0.7, threshold values were similar for both types of crack. These findings point up some of the difficulties in relating results obtained on *specimens* to the behaviour of *type-tests* or *features*, such as those used to derive TWI design curves for welded joints. A welded joint may be associated with different levels of residual stress, depending on subtle changes in parent/weld chemistry, heat input, thermal cycle and details of stress relief. Hydrogen-induced, HAZ "toe cracks" in notionally equivalent joints may then experience different stress-ratios, which can modify behaviour at low ΔK values and hence give rise to scatter in the type-test. The information obtained from specimens provides the basic understanding which eventually will help to improve processing to obtain better fatigue properties, but design at present needs to be based on the *design curves*. The replacement of welded *nodes* in offshore oil rigs by *cast nodes* led to significant improvements in performance, but the design stresses for the cast nodes had to be based on *freshly-acquired data* for cast features, see fig. 1.

3. Machine Components

3.1 THE "S-N" CURVE AND "INTERMEDIATE" COMPONENTS

The traditional approach to treat the fatigue-lifing of *smooth* machine components assumes that fatigue is *initiation-controlled* and, for load-controlled cycles, makes use of the "S-N" curve to relate fatigue life to stress-amplitude (*via* a suitable *safety-factor*). For ferritic steels and other strain-aging materials, the "S-N" curve may exhibit a sharp *fatigue-limit*, but, for other materials, it is conventional to define an *endurance limit* corresponding to a fixed number of cycles, e.g. 10^8. Stress-concentrating features, such as holes, keyways or radii, are treated by factoring down the fatigue design limit by the *elastic stress-concentration factor* (ESCF), modified by a fatigue sensitivity index.

In pure, defect-free metals, fatigue is initiated by *slip processes*, which, naturally, respond to the plastic *strain amplitude* in the cycle, see [4]. The plastic strain amplitude assumes high significance in high temperature fatigue, for which the cycle is often "driven" by thermal strains as temperatures change. These are *total* strains, but the *yield stresses*, and hence *elastic strains*, at high temperatures are small: additionally, cyclic hardening may be of low significance, because recovery processes can occur at high temperature as cycling proceeds. At low temperatures, it may be necessary to analyse the duty cycle in detail, to extract the cycle-by-cycle plastic strain component, which correlates directly with slip processes.

This is the *rationale* for what may be described as a "science-based" procedure, but, under load-control, it may be argued that the simple, experimental "S-N" curve contains all the relevant subtleties of strain-partitioning and cyclic hardening in the data that it embodies. The argument is reinforced when surface-hardened material is tested: load cycles of appropriate amplitude enable the performance of the hardened component to be assessed *as it would operate in service*; insistence on subjecting the material to a significant plastic strain-amplitude can "wipe-out" the beneficial compressive stresses, which are vital to maintain integrity in service. *It is crucial to ensure that tests are designed to reproduce service conditions.*

For carefully polished pure metals, which are "smooth" at the micro-level, it is necessary to operate slip processes, so that crack-like *intrusions* can be developed to *initiate Stage I* fatigue cracks in slip bands. These cracks later propagate as *Stage II* cracks, normal to the tensile stress. For single crystals of copper, an apparently stable *wall and ladder* structure is associated with a plastic strain amplitude of *approx. 0.9%*: this value of strain then corresponds to an *intrinsic fatigue limit*, [19].

The relevance of such observations to engineering alloys is questionable. Alloys are polycrystalline, with different slip systems in different grains, and may contain a variety of stress-concentrators, such as inclusions or other second-phase particles, even if the surface is scratch-free, and free of residual stresses, deriving from machining processes. Additionally, there are microstructural features such as precipitate-free-zones (PFZ), which may be softer than the grains and allow slip to localise, producing intergranular initiation. If the surface is scratched or is "rough" (with "intrusions" and "extrusions") as a result of machining, the need to generate crack-like intrusions by slip is removed and slip can be concentrated at the scratch or "intrusion" from the very first cycle. The concept of an *intrinsic fatigue limit* is then *irrelevant* to fatigue *design* in these alloys.

For engineering alloys, it is, in principle, more relevant to regard an "S-N" curve as the result of integrating fatigue-crack growth-rates, *c.f.* equation 8). whilst bearing in mind the *caveats* described in section 2.2. It is possible to identify certain *machine components*, as defined by *function* and *duty*, which are conventionally manufactured from material which is assumed to be *homogeneous* (at the *macro*-scale) by the design engineer, but which actually contains a distribution of *meso/micro*-scale defects, which can be identified, and characterised.

The first component is a *marine propeller*, which, for a large, ocean-going vessel, such as an oil-tanker, may be well over 12m from blade-tip to blade-tip. The component is basically "one-off" (surprisingly, twin-propellers are not always the norm) and there is no option to carry out "spin-rig" or type-tests of anything approaching full-size. Fatigue duty is experienced by bending moments exerted on near-hub sections as a result of differential drag as the propeller rotates and the vessel moves forward. The propellers are fabricated by casting (in a copper-based) nickel-aluminium bronze, NAB, (chosen because it has good corrosion-resistance in sea-water) and castings are prone to contain defects, such as shrinkage cavities, which are, unfortunately, most likely to be located in (thick) near-hub-sections.

Using replica techniques, combined with fractography, Taylor has characterised the growth of fatigue cracks from shrinkage porosity in NAB and the distributions of such pores [20]. The growth of a single crack in this *machine component* is precisely equivalent to the results obtained by Soboyejo [16,17] and Cowling [18] for defects relevant to structures and the size-scales are not dissimilar: Taylor found that the "S-N" curves corresponded to (casting) defects of roughly 0.5mm in size, see fig. 4. Using this figure, based on experiment, it is possible to *predict design limits* for propellers: such limits are remarkably close to those employed in current propeller design codes, but indicate that some of these codes are only marginally conservative.

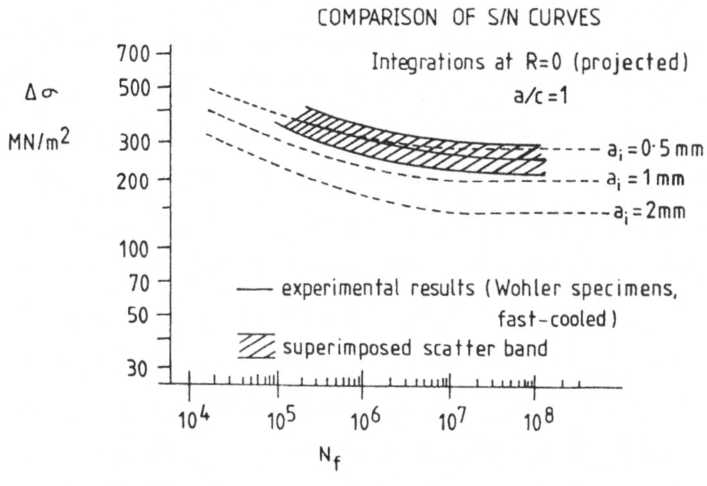

Figure 4. Comparison of "S-N" Curve for Fatigue Failure in Nickel-Aluminium-Bronze (hatched area) with Predictions of Fatigue-Crack Growth Rate Integrations (after [20])

A second use of Taylor's data is not to take just a single initial defect size, but to incorporate the full initial distribution and to use this to predict the scatter associated with the *lifetime* on the *"S-N" curve* [2]. Consider, for simplicity, equation 5). and its integration to derive equation 8). The assumptions employed in this integration (apart from the absence of any interaction effect) are that both a_0 and a_f are constant. Now assume that variations in a_f are insignificant, but that *values of* a_0 *are distributed*: following Taylor's results, the form of that distribution approximates to a negative exponential. It is then possible to define a *test volume* in a specimen and to carry out a *series of integrations* which give the *corresponding distribution* for N_f in equation 8). As the values of a_0 decrease, the *small crack* regime is involved and the *integration* tends to become *conceptual* rather than *quantitative* (refer to section 2.2). Given limitless resource, it would be possible to *compare* the distribution in N_f at high $\Delta\sigma$ (small a_0) with that at lower $\Delta\sigma$ (larger a_0) to deduce the differences in growth-rate law that small cracks entail, but there is little point in doing this until processing has been improved to eliminate the larger defects.

The marine propeller, although by design a machine component, should perhaps be regarded as *intermediate* in that the fatigue-life is clearly able to be explained in terms of the integration of growth-rates of a set of *inherent defects*. Similar principles hold for new design concepts for aluminium alloy cast "nodes" as critical parts in *space-frames* for possible use in light-weight automotive vehicles. Here, the *function* is *structural* but the duty (highly dependent on terrain and suspension system) may be classed as *intermediate*: in a 200,000km lifetime (at, say, 80kmh^{-1}) 10^6 cycles correspond to one *significant jolt* every 200m or every 9s. The role of different microstructural features is currently being investigated by Green and Campbell at Birmingham. From *tensile tests*, *four* levels of defects can be identified, in order: i). casting porosity (the worst) ii). entrained oxide films, iii). iron-containing intermetallics, iv). fine-scale microstructure. These features should hold equally for fatigue, and experience gained on commercial die-castings indeed demonstrates that fatigue cracks are associated with pores or entrained oxide films. Only when these have been eliminated is there significant value in reducing iron content or controlling fine-scale microstructure. Contrast this with high-strength, wrought, 7000 series aluminium alloys, in which control of iron content is critically important. The degradation in fatigue properties resulting from the presence of pores and films is the reason why "safety factors" for castings are much greater than those for wrought material. Research by Jiang at Birmingham is establishing the scientific basis for such factors, which, in turn, makes it possible to focus on those features of processing which must be addressed if the fatigue resistance of the castings is to be improved to levels associated with wrought material. Figs 5 and 6 show fatigue-crack initiation from a pore in a cast aluminium alloy.

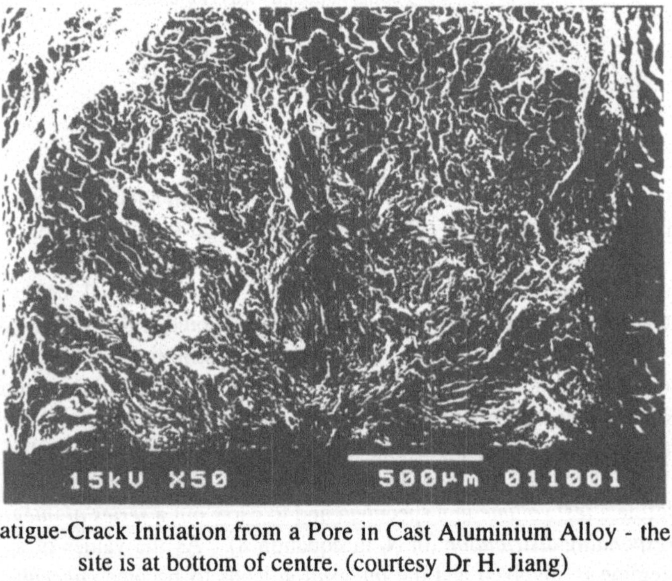

Figure 5. Fatigue-Crack Initiation from a Pore in Cast Aluminium Alloy - the initiation site is at bottom of centre. (courtesy Dr H. Jiang)

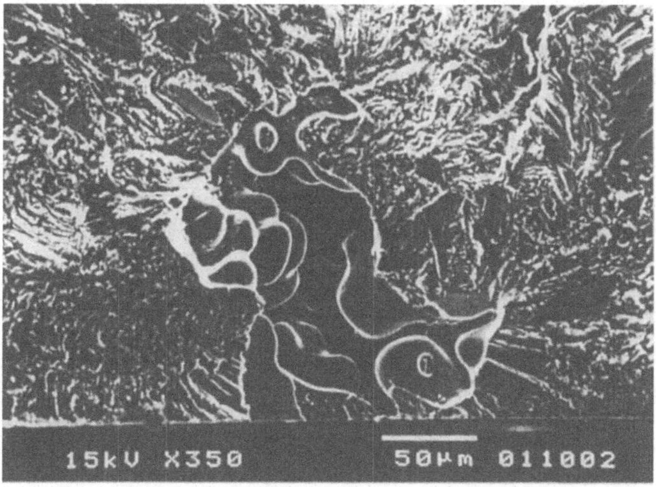

Figure 6. Higher magnification photograph of Initiation Site - seen to be a near-surface shrinkage cavity. (courtesy Dr. H. Jiang).

Another machine component in which it is possible to characterise inherent defects is the high-pressure (high-temperature) turbine disc in a gas-turbine aero-engine. This is approximately 0.8m external diameter with a central bore of some 0.2m diameter and a thickness of 0.1-0.2m. The disc carries a set of turbine blades, spaced around its rim. In a modern disc, the rim temperature is upwards of 650°C and the bore temperature may rise to 400°C. The maximum stress (at the bore) is of order 1GPa. To meet this duty, discs are fabricated from nickel-base superalloys, which have been developed to sustain increasingly higher temperatures and hence to resist deformation at high temperatures. This has involved the addition of high melting-point, transition metal, alloying elements such as Nb and Mo.

Such developments have implications with respect to the processing route. Traditionally, a disc was made by casting a blank and forging this to shape. The necessary changes in composition have: first, given rise to marked segregation in the cast blank (with Mo and Nb perhaps forming carbides); second, made it difficult to forge the alloy (because it has been designed to be resistant to deformation at high temperatures). The result has been to produce discs which exhibit segregation and hence spatial variations in resistance to deformation and oxidation. To eliminate such segregation, *high-duty discs* are produced by *powder processing*. Here, the alloy is melted, in a refractory alumino-silicate crucible, and is then poured, *via* a refractory-lined tundish, through a circular nozzle, to emerge as a single metal stream. This is *atomised*, by lateral impingement (from a circumferential array of orifices) of a gas jet which breaks up the continuous stream into a set of droplets. These solidify as powder particles, which are subsequently sieved, hot-isostatically pressed

(HIPped) and forged to produce the final disc. Conventionally, the atomising gas is argon, which is *inert* and might produce (microscopic) gas bubbles (of the size of inter-dendritic spacings) during HIPping. An alternative is to use an atomising gas such as nitrogen, which could be *reactive* and form sold particles. From the point of view of fatigue "lifing" in these discs, however, the main concern at present is that aluminosilicate refractory particles in the crucible or tundish may be scoured off by the dense metal stream and enter the solidified powder. It is then possible for such particles to pass through all the subsequent processing steps and end up in the final disc, where they may act as fatigue-crack initiators, see figs. 7 and 8.

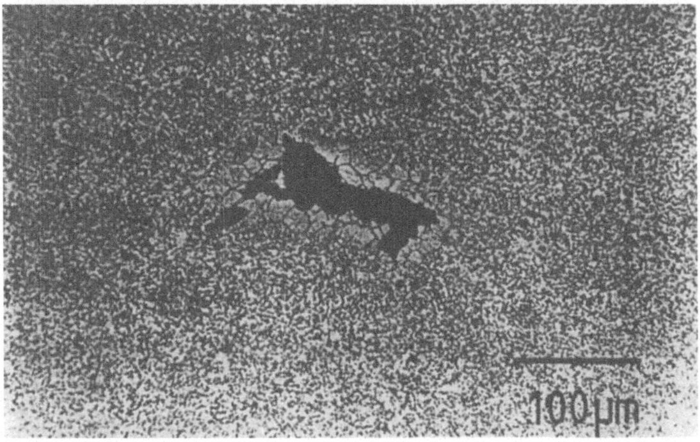

Figure 7. Alumino-silicate inclusion in "doped" nickel-base superalloy. Note that chemical interaction at high temperature gives rise to a "halo" around the inclusion. (courtesy Dr. P. Woollin).

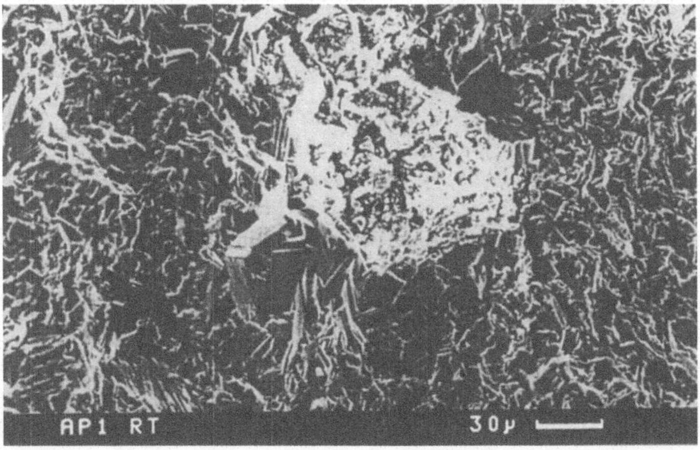

Figure 8. Fatigue fracture surface showing initiation from an alumino silicate inclusion (bright patch in centre). Note "faceted" growth in "halo" region. (courtesy Dr P. Woollin)

It is possible to use equation 4). to estimate the size of initial *defect*, which in combination with the service duty is associated with a specified *design life*. For through-thickness cracks the specific form of equation 1). for nickel-base superalloys is [21]:

$$da / dN = 4 \, x \, 10^{-12} \, \Delta K^{3.3} \qquad \text{.........13)}$$

This leads to guideline figures for 10,000 major flight cycles as follows: $\Delta\sigma$ = 800MPa (e.g. Waspaloy), a_0 = 0.22mm; $\Delta\sigma$ = 950MPa (e.g. Astroloy), a_0 = 0.1mm; $\Delta\sigma$ = 1200MPa (e.g. René 95), a_0 = 0.04mm, see fig. 9. These precise values are clearly affected by the value of Q (equation 2). assumed, but give quantitative guidelines to process control, in that they can be related to the mesh size of the sieves used to screen powder: a larger mesh will allow larger alumino-silicate particles to pass through. Some particles may not be equiaxed, and arguments concerning the "angle of attack" of a particle's major axis with respect to the mesh then have to be considered, but it is instructive to use the guideline figures to discuss technical, economic and safety factors.

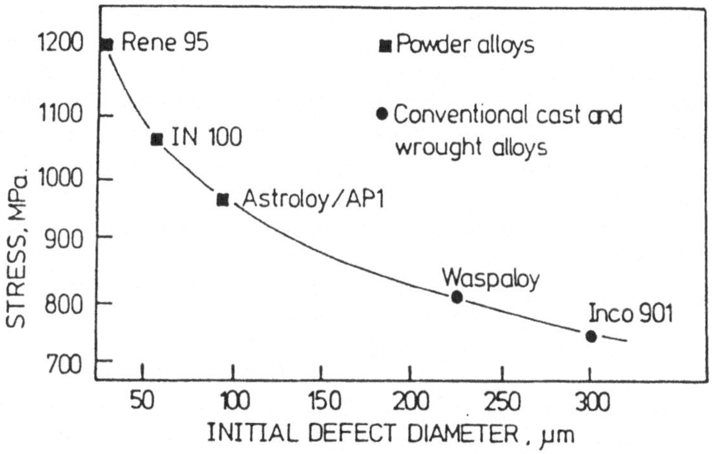

Figure 9. Size of initial defect associated with 10,000 flight cycles in nickel-alloy discs.

It is clear that *higher duty requires finer sieving*. This not only *reduces powder yield*, and hence directly makes the alloy much more costly, but, for finer particle sizes (e.g. 0.04mm) also necessitates investment in more sophisticated (and expensive) powder-handling equipment and the employment of a more highly-trained workforce. The balance between economic and technical efficiency, on the one hand, and safety, on the other, is delicately poised. The public wants cheap air-fares, hence high thrust/weight, combined with good fuel consumption. This dictates high temperatures and high working stresses. The safety issues require 10,000 (or now, more commonly, 20,000) major flight cycles without failure and these can be achieved only with more stringent controls on initial defect size, which, in turn, give rise to greatly increased component costs. It may well be that 1GPa represents the highest *stress* which it is economically viable to utilise, although further alloy development may enable discs to operate under stresses of this order at higher *temperatures*.

There is a strong suggestion that the LEFM calculations, based on equations 1). 2). and 4). are unduly pessimistic in their predictions of fatigue lives for actual discs: experimental results on specimens and in spin-rig tests show that discs can sustain higher stresses than those predicted by LEFM integration [22], see fig. 10. There are two possible reasons for this. The first is based on the probability that a test sample does not happen to contain a large defect. This is more tenable for fatigue specimens (although an occasional poor result should be obtained), but less so for a model disc, because the volume of highly-stressed material in a disc is quite large. A second reason is that specimens/model discs contain compressive residual stress distributions, *either* at the *meso-level* (arising from processes such as grinding or shot-peening) *or*, at the *micro-level* (due to thermal-expansion-mismatch strains between particle and matrix).

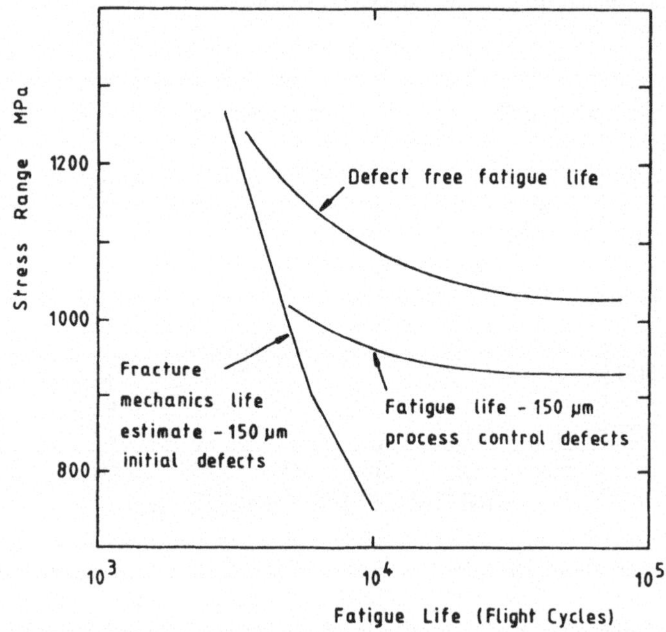

Figure 10. Comparison of Fracture Mechanics estimate with Fatigue Life in a Nickel-based disc alloy (after [22]).

The general theme of residual stress is addressed in the following section, but it is relevant here to note the work of Woollin [23,24], who studied fatigue in a nickel-base superalloy, deliberately *doped* with a high volume fraction of *alumino-silicate particles*. Within the limits of broad scatter-bands, he showed that short cracks grown from alumino-silicates at *room temperature* were able to propagate at ΔK values associated with the threshold region in long-crack tests, but *propagated more slowly* than short cracks initiated in slip bands, see fig. 11. This was attributed to a "clamping" effect, arising from compressive thermal strains in the particles. At 600°C in vacuum, however, the relative effect was much reduced, and this could be explained in terms of a reduction in the magnitude of the clamping stress.

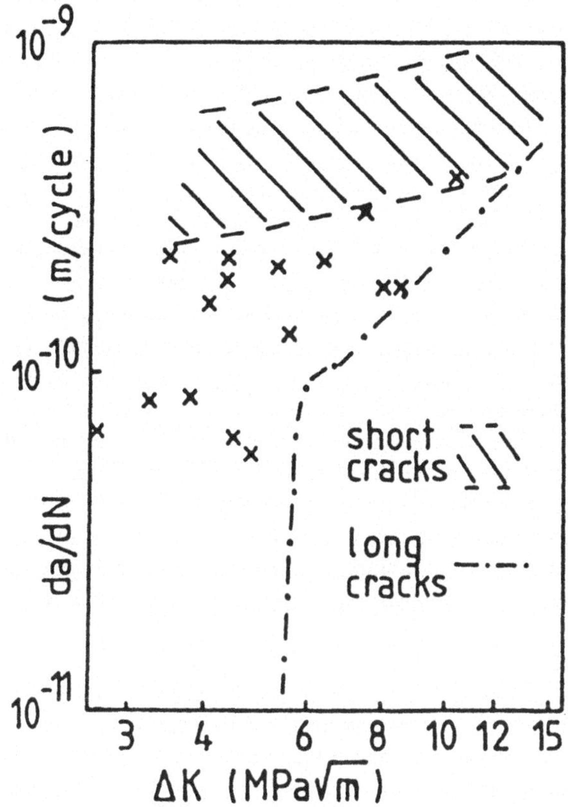

Figure 11. Comparison, in nickel-based disc alloy API, tested at room temperature of long-crack growth, growth from short cracks initiated in slip-bands (hatched region) and short cracks initiated at inclusions (crosses). (see [23], [24])

3.2 RESIDUAL STRESSES AND 'SURFACE ENGINEERING'

Residual stress distributions are associated with both structures and machine-components. In general, they are more likely to arise fortuitously in structures, as a result of fabrication process and to be produced deliberately in machine components, primarily to prolong "initiation" and/or early stages of crack growth. The *scale* of residual stresses may be at the *macro/meso*-level(e.g. as a result of thermal and transformation strains induced when a large, steel weldment cools), at the *meso/micro*-level (as a result of reaming holes, or case-hardening surfaces), or at the *micro*-level (associated with strains around precipitates or particles). Analyses of such stress distributions are available, but they are not yet fully incorporated into fatigue-lifing procedures.

The sources of residual stress may be thermo-chemical of mechanical. Both sources may be encountered in structures and in machine components. *Fusion welding* of steel and aluminium structures entails the filling of the welding-preparation gap with liquid metal, which undergoes a major contraction when it solidifies. In steels, the actual values of thermal mismatch strains between weld metal, HAZ, and "parent" wrought material are modified by phase transformation on (rapid) cooling, but the weld-metal contraction usually dominates, resulting in tensile residual stresses. It is generally held that the maximum value of residual stress cannot exceed the yield stress because, if it does, stresses will be relieved by yielding. This provides the *rationale* for *stress-relief heat-treatments*, in which temperature is increased slowly (in steel, to *circa* 650°C) thus reducing the yield stress and hence the magnitude of the (weld-metal contraction) residual stress. The stress-relief argument does not apply if the residual stress field is pure hydrostatic tension, because yielding is induced by the deviatoric component of the stress field (as is the plastic deformation associated with fatigue processes). High values of hydrostatic stress are encountered only in the deep interior of a thick-section joint, whereas fatigue is traditionally associated with near-surface regions (in "defect free" material). Even if the deep interior contains a pore, or a particle which fractures under load, the "interior surface" might be regarded as operating *in vacuo*, so that interior initiation/early-growth would be much retarded compared with that from exterior surfaces, exposed to air or to a corrosive environment. An interesting possibility is the presence of hydrogen in a internal pore subjected to a combination of high hydrostatic tension and cyclic loading.

Thermal cycles similar to those associated with fusion welding are encountered in *cladding* processes. A good example is to be found in BWR and PWR pressure vessels. Here, to prevent excessive amounts of corrosion products entering the coolant, the ferritic steel vessel is clad on interior surfaces with a layer of austenitic steel, some 5-7mm thick. The *underclad* region experiences a thermal cycle similar to that of the HAZ in a fusion weld, but thermal mismatch strains persist during stress relief, because the thermal expansion coefficients of austenitic steel and ferritic steel are significantly different. Residual stresses in fabricated structures may also be produced by mechanical means: by the reaming of holes or the punching of rivets, which both give plastic expansion to an inner annulus, which is left in a state of residual compression on unloading. The beneficial compressive stress is partially offset by the roughening of "bore" surfaces induced by the reaming or punching operations. Also the residual *tensile* stresses in welded joints may be "relieved" (suppressed) by heavy, local plastic deformation, induced in a semi-empirical manner by shot-peening or hammer-peening. Factors such as this relate to the comments made earlier on the "generally, *fortuitous* generation of residual stresses in *structures*".

In *machine components*, compressive residual stresses in *surface* layers are *engineered* by thermo-chemical or mechanical means. In steels, *case-hardening* processes include *case-carburising*, *case-nitriding* and *flame*-or *induction-hardening*. The first two of these processes involve the diffusion into surface layers of either carbon or nitrogen atoms. Carburising is carried out at ca. 880°C, in the austenitic state, so that phase changes on cooling take place *after* the production of *graded material.* If the final product is not uniformly of the precise dimensions required, correction by *local machining* presents a problem, because readjustment of internal stresses gives rise to further changes in external dimensions, followed by a necessity for further "corrective" machining, which, in turn produces further unanticipated dimensional changes.

Case-hardening of precision components in *steels* is effected by *case-nitriding.* This is carried out on a component which has *already been quenched and tempered,* typically at 600-650°C. The nitriding is carried out at ~480°C, either in "cracked ammonia" (a mixture of NH_3 and H) or in a "plasma": the aim is to produce *atomic nitrogen* which diffuses readily into surface layers to induce high hardness and a high level of surface compressive stress. In medium (0.25-0.4%) carbon steels, surface compression can be induced by a combination of thermal cycles and phase transformation. Typical processes are *flame-hardening* and *induction-hardening.* For both, *surface layers* are taken into the austenite region, and the carbon equivalent is such that on cooling (by metallic conduction) the outer layers transform to martensite whilst the core remains unaffected (e.g. as a tempered martensite), because only the surface layers have been reaustenitised. Transformation of the outer layers to martensite would normally involve a volume expansion (whose magnitude is a function of the carbon content), but this expansion is resisted by the untransformed core, so that the surfaces are both hard and in compression, which helps to guard against fatigue-crack initiation (by minimising plastic strains) and short-crack propagation (since high compression induces large closure forces). The work of Clark [25] on notched bars, subjected to induction-hardening, case-nitriding, and case-carburising treatments clearly demonstrates that the residual stresses may be treated as uniaxial compressions which can be counteracted by employing high stress ratios.

Another way in which the formation of martensite retards fatigue crack propagation may be observed in so-called TRIP (transformation-induced plasticity) steels. Here, the combination of shears and tensions in the stress field ahead of the growing crack promote the formation of martensite, which has a higher volume than the matrix from which it forms, and so generates compressive, "clamping" forces. Martensite is also effective in giving high "threshold" stress-intensity factors in dual-phase steels [26].

In addition to residual strains/stresses produced by phase transformations, it is possible to generate residual stress distributions as a result of differences in the thermal expansion coefficients of constituents. A good example is that of the powder-formed nickel superalloy gas-turbine disc, in which the metal matrix contains refractory alumino-silicate inclusions. The inclusions and matrix may be deemed to be in (strain-free) equilibrium at the forging temperature of *ca* 1150-1180°C. On cooling, because the thermal expansion coefficient of the matrix is much greater than that of the inclusion the inclusion is left in a state of *compression* with the surrounding matrix in a state of *tension.* The magnitude of these stresses is a function of thermal expansion mismatch and elastic constants and may be expressed as the *uniform* compression in the particle p. If a crack grows out from the particle, it is still subjected to compressive "clamping" forces over the central part of its length. For a crack of (half) length equal to twice the particle radius, r_i, the "negative" (compressive) stress intensity factor due to residual stress is:

$$K_{res} = 0.058p\,(2\pi r_i)^{1/2} \qquad \text{.........14)}$$

Surface compression may also be induced by mechanical means. Local plastic deformation is produced by pre-loading, which produces permanent deformation in an elastic matrix. When the pre-load is removed, the "spring-back" from the elastic matrix generates compressive stresses in the plastically-deformed region. Examples are found in the processes of shot-peening and hammer-peening (deformation produced by the impact of hard shot or by hammer blows), in the "ballizing" of lugs (an over-size steel ball is forced through the lug), in the cold-expansion of a bush in a bolt-hole or a rivet-hole, or in the *autofrettage* of a gun-barrel. Here a gun-barrel is over-pressurised before entering service, so that a significant fraction of the wall deforms plastically and residual compression is induced on unloading. This compression is effective under normal operation.

The issue of fatigue-crack propagation in an autofrettaged gun-barrel has been treated quantitatively by Clark [27] who has examined the effects of different amounts of yielding: fractions of 32%, 58% and 80% wall-thickness. For 58% yielding, the peak compression is -500MPa at 10% wall-thickness (there is a Bausclinger effect which induces reversed plastic flow at the inner surface), zero stress at 40% wall thickness and a peak tension of +250MPa at the plastic/elastic interface (58% wall thickness). The steel's yield stress was 1050MPa. Initial cracks are produced at the corners of rifling grooves during the first few firing cycles, which austenitise material close to the surface, and then allow this to transform to martensite, rather as in induction-hardening, but now with sharp stress-concentrators present. "Heat checked" cracks of this sort are a little longer than the depth of a rifling groove, say, 3mm. The heating effects are local to the surface, so that longer-range residual stresses are not annealed-out.

Clark's analysis uses a *weight function* approach to load a crack with the pre-existing compressive stress distribution to calculate a residual (negative) stress intensity factor K_{res}, which must be subtracted from the applied stress intensity factor, to give an "effective" value, which can be related to da/dN data. For a barrel yielded to 58% wall thickness (t), K_{res} for a 3mm crack (approx. 0.1t) is -48MPam$^{1/2}$, and for a 6mm crack (0.2t) it is -60MPam$^{1/2}$. Such values reduce the fatigue-crack propagation rate under normal firing pressures by an order of magnitude, which gives a substantial increase in the life of a gun-barrel. The autofrettage may also be accomplished by drawing an over-sized mandrel through the barrel, which closely simulates the ballizing process, and Clark has carried out an analogous analysis of the benefits of residual compressive stress with respect to the fatigue life of a lug in an aircraft structure [28].

In principle, similar principles may be applied to calculate the benefit to fatigue life of shot-peening, which is commonly used to enhance the performance of machine components from automotive coil springs or leaf springs to aero-engine turbine discs. Small steel shot subjects to plastic deformation which gives rise to compressive stresses in unloading. The shot, however, is spherical, making circular indents and impacts on the surface in a random fashion. Some 200-300% "coverage" may be employed to ensure that surface layers are "uniformly" covered. The surface is roughened by the peening process and, in practice, not all regions may be deformed plastically.

A two-dimensional model of the effect of a single shot can be constructed by indenting surfaces with cylindrical rollers. For rigid-plastic material, the slip-line field beneath such an indentation has the form of a logarithmic spiral and, again, compressive stresses are generated on unloading. Recently, Li at Birmingham has studied a system in which a nickel-base superalloy of 0.2% P.S. 1100MPa and U.T.S. 1340MPa was indented by a silicon carbide roller of diameter 4mm. For a load of 15kN (in a bar with depth W=8mm, and thickness B=8mm) finite-element analysis indicated that the maximum compressive stress was approx. -650MPa at a depth of 0.45mm below the indented surface. Zero stress was obtained at a depth of approx. 1.5mm and a maximum tensile stress of approx. +250MPa at a depth of 3.0 to 3.3mm. These values are of similar character to those obtained by Clark for the auto-frettaged gun barrel.

Li has proceeded to study short crack growth in plain specimens, by growing long cracks and then machining away the specimen behind the crack tip. At ΔK values below 15MPam$^{1/2}$, these short cracks were observed to propagate more rapidly than long cracks, as expected. The next stage was to take a "short crack" specimen and indent it with the cylindrical roller under a load of 15kN. Such indented cracks then grew *much more slowly*, even more slowly than long cracks, and a dark band of fretted oxide could be observed on the fracture surface to a depth of approx. 1.5mm (i.e. in the region predicted to contain residual compressive stress). The observations can be rationalised in terms of an effective ΔK value which incorporates closure contributions from both the plasticity associated with short crack growth in the absence of "applied" residual stress and the compressive stress distribution induced by indentation. Further work is examining effects of indentation load, stress ratio and location of the short crack. An issue of particular interest is a short crack located midway between two roller indentations, to explore what happens if a defect in a shot-peened component is located in a region which has been "missed" (not plastically-indented) by the shot. Work carried out by Braid, described in [29], suggests that beneficial compression should still be present.

In absence of a full mechanics treatment, involving closure, recourse is made (e.g. by Rolls-Royce) to the concept of an "equivalent initial flaw size" EIFS. The test specimens are blocks of substantial size, which have been subjected to *precisely the same* (shot-peened) surface treatment as that experienced by a service component, such as a turbine disc. The total number of cycles to failure is recorded, and the fatigue life surface is examined fractographically to enable striation spacings to be measured. If the stress intensity factors are such that the striation spacing is equal to da/dN, it is possible to invert the da/dN vs ΔK relationship, to derive dN/da as a function of a at constant $\Delta\sigma$. Knowing $\Delta\sigma$ and N_f, and assuming that all the life is taken up by propagation, it is possible to extrapolate back to the EIFS. This value can then be used in design calculations for the service components.

The EIFS method is excellent in pragmatic terms, because, for fracture design, it compares like with like, but it is rather less satisfactory in terms of the quantitative correlation of EIFS with the actual defects present in the material and observed, on the fatigue surface, to be the "initiation" sites. In castings, where the sites are typically voids, correlation is good, because macroscopic residual stress fields are not large, or are tensile (as in welds), so that closure effects are small. Even in the tests on doped nickel alloys (with no macroscopic residual stress field), correlation is reasonable and can be taken as a lower bound, because the range of the microscopic residual stress field is small. For shot-peened, or other "engineered" surfaces, however, the sizes of real defects and EIFS will be widely different and the benefit, in terms of fatigue life, of reducing the defect content and size distribution cannot be quantified until the insights gained by the work of the type carried out by Clark and Li are incorporated into sensitivity analyses. For powder-processed aero-engine discs, such sensitivity analyses have direct bearing on the mesh sizes required for powder sieving and hence on the critical balance between economic factors and guaranteed fatigue life.

The work on surface-treated components and especially those subjected to mechanical treatments gives rise to a number of issues. The first generic issue concerns the location and nature of a defect with respect to the residual stress distribution. This is usually regarded as balance between high compression but eventual access to the external environment for a near-surface defect, contrasted with high tension but vacuum conditions for a sub-surface defect. Brittle "static" modes occur with equal facility in air and vacuum, and there is always the possibility that the sub-surface crack is filled with hydrogen or some other reactive gas. Sub-surface initiation has been observed, but the conditions for this need to be established in detail. In commercially shot-peened springs, the findings of Todinov at Birmingham are that defects in incoming bar stock are still of significance in controlling fatigue life.

A second issue relates specifically to the shot-peening of components, such as aero-engine discs which will subsequently operate at high temperature. Although the (shot-peened) bores do not exceed 400°C, there are (shot-peened) bolt-holes towards the periphery of the disc which experience higher temperatures. The melting point T_m of Ni is 1453°C (1726K) so that $0.5T_m$ is 863K i.e. 590°C. This is generally held to be the temperature above which self-diffusion is rapid. A lower fraction of T_m may be significant, in the long term, for significant creep relaxation, and this must be put in the context of 2×10^4 flight cycles being equivalent to 8×10^4h at temperature (averaging 4h for European and Transatlantic flights). There are two concerns. *Either*, the residual stresses are not relieved and there is extensive exposure to high, sub-surface stress at high temperature (leading perhaps to creep damage/cracking) *or* the stresses are relieved, simply leaving the surface roughness engendered by the original shot-peening. Alternative mechanical forms of inducing residual stress are subject to the same types of problem, but there may be further scope for thermo-mechanical treatments.

The third issue is of great potential significance to the more consistent and better response of machine components to service duty in the absence of consciously "engineered" surfaces. This is concerned with machining schedules and the effect of these on fatigue life. A machining process, be it grinding, turning, milling or whatever, simultaneously gives rise to two effects. The first is *surface roughness*. The produces "natural" intrusions, which allow plastic strain to be concentrated from the first cycle. The second is *plastic deformation*, which, after machining, leaves the surface in a state of residual compressive stress. The balance between these two is of critical importance to the performance of as-machined components under fatigue loading, yet neither parameter is generally well-characterised and neither features explicitly in general design rules. The work of Taylor [30] in this area should be highly regarded and its extension to general engineering practice, to close the gap between "machinists" and "stressers", is extremely exciting. Wealth creation and economic success are often based on higher precision in engineering design. What we have sight of here is the ability to "fine-tune" fatigue design, not just in terms of stress-concentration factors for gross features, but also in terms of machining schedules tuned to individual materials.

4. Conclusions.

The paper has addressed issues of fatigue in components of *structures* and *machines*, almost exclusively at ambient temperature and in non-aggressive environments. In this broad field, it is possible to recognise a number of different inputs: *pragmatic design* against fatigue failure, *initial defect content* and the sensitivity of fatigue life to metallurgical control or the control of fabrication processes, quantification of the value of *surface engineering* for machine components. For all these inputs, it is necessary to recognise the *relevant endpoint*. This can only be the prevention of *premature failure* due to fatigue, which requires *quantification* of *fatigue damage/crack growth rate* as a function of *duty cycles in service*. *Pragmatic design* takes test-assemblies designed to simulate service components and subjects them to simulations of service duty. In many cases, the simulations are close copies (perhaps, even duplicates) of service components. *Design data* given by such a pragmatic procedure are of immense value, but the tests are expensive and it is not often easy to offer similar resource to "test out" new material compositions, new processing routes, new "operational transients" which may, or may not be significant with respect to lifing. Detailed (fracture mechanics/fractography) analysis of test data and specimens is not always encouraged.

In test specimens, much has been made of the Kitagawa plot [31]. Arguing that a constant fatigue crack threshold level, ΔK_{th}, implied a simple linear relationship between $\Delta\sigma$ and $a^{-1/2}$, Kitagawa produced bounding lines on a plot of log$\Delta\sigma$ vs log a which gave a linear relationship of slope -0.5 between $\Delta\sigma$ and *a* for large *a* ("structures"), but possessed a cut-off for small *a* ("machines") equal to the fatigue life limit/endurance limit. Much has been made of the fact that some "short crack" results give failures at values of $\Delta\sigma$ lower than predicted by either limit, but a more rigorous examination [12] of the Kitagawa data shows that, once a correction has been made for a closure-free threshold, ΔK_0, all data points are, indeed, bounded by these two straight lines (on a log; log plot). Aero-engine discs are stressed to particularly high levels, and the required life is only 2 x 10^4 cycle but shot-peening is employed, so that whereas it is arguable that it might be necessary to incorporate short-crack growth-rate data in design: in practice, use is made of (da/dN) growth-rates, determined experimentally.

The *initial defect content* is an extremely important factor with respect to fatigue. *Defects* may be present *in the starting material*: examples have been given of *pores in castings* and *inclusions in powder-formed alloys*; or they may be *introduced during fabrication*: *hydrogen cracks in weldments*, *surface-defects* resulting from the reaming of holes. *Further "initial" defects* may be introduced during the lifetime of a component, either as a *result of an accident*, or by incautious disassembly and re-assembly during *scheduled maintenance* operations: such defects are likely to be scratches or gouges. In all these situations, it is important to realise that a *defect is not just a geometrical discontinuity*, but is associated with an *idiosyncratic residual stress-field*, resulting from the means by which it has been produced. In particular, the long-crack "threshold", generated by a "wind-down" sequence, is associated with closure forces in the plastic wake, and is often not conservative when used to design against the propagation of short cracks. In general, weldments where residual tensile stresses exist in critical areas will show better agreement with high-R long crack tests than will situations where the *residual stresses* are *compressive*. Here an *"effective" stress-intensity range* will have to be calculated. To ensure that test specimens generate data of value to design, the *defects* that they contain must *closely simulate those likely to be encountered in service.*

The final issue is that of *quantification of the value of surface engineering*. Stress analyses and experimental confirmation of residual stress fields *are* available and the *methodologies* used to treat fatigue crack growth is some selected case-studies are *highly plausible* but they have not yet been pursued sufficiently to treat the general range of problems found in the fatigue of machine components. The *pragmatic approach*, based on *EIFS*, produces convincing *design data*, but the size of the *EIFS* does *not correspond* to an *identifiable defect* on the fatigue surface. Until an appropriate micro-mechanical model has been properly established, the value of reducing the size of a defect in a component which is shot-peened or otherwise *surface-engineered* before it enters service, is unclear. It is of paramount importance to establish such links in a quantitative manner so that *design* against fatigue failure can become both *reliable* and *economic*.

The use of fracture mechanics to characterise fatigue crack in terms of *applied stress intensity factor range*, ΔK, has been established for over 30 years. During the last 10-15 years, there has been a growing appreciation of the importance of *residual stress* and how this affects, for example the threshold, the growth of short cracks, or the growth of cracks in a deliberately imposed residual stress field. The principles apply *equally* to components in *structures* and to those in *machines* and the goal is to be able to provide *quantitative assessments* of the relative values of initial *defect size* and of nature and magnitude of *residual stress* field in determining the precise *fatigue-life*. The paper has attempted to illustrate the principles required to do this.

5. *Acknowledgements.*

I would like to thank those who have worked with me in Cambridge and in Birmingham, on aspects of fatigue in engineering alloys. Some have been referenced in this paper, but there are several others who have contributed knowingly or unknowingly, to the views expressed in this paper. They should not be held responsible for anything other than what shine out as obvious truths.

Research into fatigue has been one of Birmingham's strengths over the years and is continuing strongly: my main interests are in conventional alloys, whilst Professor Paul Bowen is concerned with "advanced" alloys: titanium aluminides and metal-matrix-composites. The fatigue tradition was, however, set by Dr Jim Beevers, whose death in 1992 was such a great tragedy. This is the first time since then that I have written an overview on fatigue, and I would very much like to have it associated with a memory of my erstwhile colleague and friend.

REFERENCES.

1. Paris, P. and Erdogan, F. (1963) Trans. ASME Jnl. Basic. Eng. **85** p528.
2. Knott, J. F. "Near -Threshold Growth of Defects in Engineering Materials" in "Fatigue 87" Proc. 3rd Intl. Conf. on Fatigue, Ed R. O. Ritchie and E. A. Starke Jnr. EMAS, 1987, **1**, pp 497-515
3. Weertman, J. 1969 "Theory of Rate of Growth of Fatigue Cracks Under Combined Static and Cyclic Stresses" Intl. Jnl. Fracture Mechanics **5**, p 13.
4. Knott, J. F. "Models of Fatigue Crack Growth" in "Fatigue Crack Growth - 30 Years of Progress" ed R. A. Smith Pergamon 1986 pp 31-52.
5. Ritchie, R. O. and Knott, J. F. 1973 "Mechanics of Fatigue Crack Growth in Low Alloy Steel" Acta Met **21** p 639.
6. Ritchie, R. O. and Knott, J. F 1974 "Microcleavage Cracking during Fatigue Crack Propagation in Low Strength Steel" Mater. Sci. and Eng. **14** p 7.
7. Forman, R. G, Kearney, V. E. and Engle, R. M. 1967 ASME Jnl Basic Eng. 89 pp 459.
8. Kumai, S, King, J. E. and Knott, J. F. 1990 "Short and Long Fatigue-Crack Growth in a SiC-reinforced Aluminium Alloy" Fat. and Fract. Eng. Matls and Structures **13** pp 511-524.
9. Suresh, S, and Ritchie, R. O. 1984 in "Fatigue Crack Growth Threshold Concepts" ed D. L. Davidson and S. Suresh TMS-AIME, p227.
10. Kendall, J. M and Knott, J. F. 1988 "Near-Threshold Fatigue-Crack Growth in Air and Vacuum" ASTM STP **924 II** pp 103-114
11. James, M. N and Knott, J. F. 1985 "An Assessment of Crack Closure and the Extent of the Short Crack Regime in QIN (HY80) Steel Fatigue Fract. Eng. Mater. Struct **8** pp 177-191.
12. Kendall, J. M, James, M. N. and Knott, J. F. 1986 "The Behaviour of Short Fatigue Cracks in Steels" Proc. Europ. Conf. on Short Cracks Sheffield 1985. Ed K. J. Miller and E. R. de Los Rios, MEP pp 141-148.
13. Damri, D and Knott, J. F. 1993 "Fracture Modes Encountered following the Application of a Major Tensile Overload Cycle" Intl. Jnl. Fatigue **15** pp 53-60.
14. Clark, G "Fatigue-Crack Growth from Notches" PhD thesis, University of Cambridge Nov. 1975
15. Livesey, V. B. and Knott, J. F. 1981 "Fatigue-Crack Initiation at Notches" Res. Mechanica letters **1** pp 7-12.
16. Soboyejo, W. O, Kishimoto, K, Smith, R. A and Knott, J. F. 1989 "A Study of the Interaction and Coalescence of Two Coplanar Cracks in Bending" Fatigue Fract. Eng. Mater. Struct. **12** pp 167-174.
17. Soboyejo W. O, Walsh M. J, Cropper, K. R. and Knott, J. F. 1990 "Fatigue Crack Propagation of Co-planar Semi-elliptical cracks in Pure Bending" Eng. Fract. Mech. **37** pp 323-340.
Induced Cracks in QIN Steel" Fatigue Fract Eng. Mater. Struct. **12** pp 585-595.
19. Brown, L. M. 1977 "Dislocation Substructures and the Initiation of Cracks by Fatigue" Metal Science **11** pp 315-322.
20. Taylor, D. and Knott, J. F. 1982 "Growth of Fatigue Cracks from Casting Defects in Nickel-Aluminium-Bronze" Metals Technology **9** pp 221-232.
21. Brown, C. W and Hicks, M. A. 1982. "Fatigue Growth of Surface Cracks in Nickel-based Superalloys" Intl. Jnl. Fatigue pp 73-81.
22. Wildgoose, P, Turner N. G, Davies, H. F, Helliwell, B. J, Ubank, R. and Harrison, H. 1981 "Powder Metallurgical Innovations for Improved Hot-Section Alloys in Aero-Engine Applications" Powder Metallurgy pp 75-86.
23. Woollin, P. and Knott, J. F. 1987 "The Effect of Temperature on Fatigue Crack Initiation from Alumino-Silicate Inclusions" in "Fatigue 87" Proc. 3rd Intl. Conf. on Fatigue ed. R. O. Ritchie and E. A. Starke Jnr. EMAS pp 1087-1099.
24. Kendall, J. M, King, J. E, Woollin, P and Knott, J. F. 1987 "Short Fatigue Crack Growth in AP1: Roles of Microstructure and Powder Defects" Proc. Conf. "PM Aerospace Materials" (Luzern) Metal Powder Report Services Ltd. pp 7.1- 7.12.
25. Clark, G and Knott, J. F. 1977 "Effects of Notches and Surface-Hardening on the Early Growth of Fatigue Cracks" Metal Science **11** pp 345-353.
26. Minakawa, K, Matsuo, Y and McEvily, A. J. 1982. "The Influence of a Duplex Microstructure in Steels on Fatigue-Crack Growth on the Near-Threshold Region Metall. Trans. **12A** pp 439-445.

27. Clark, G. 1983 "Fatigue Crack Growth in Autofrettaged Thick-Walled Cylinders" in "Fracture Mechanics Technology Applied to Material Evaluation and Structure Design" ed G. C. Sih, Ryan, N. E and Jones, R. Martinus Nijhoff pp 417-430.

28. Clark,G. 1991. "Modelling Residual Stresses and Fatigue Crack Growth at Cold-Expanded Fastener Holds". Fat. Fract. Engng. Mater. Struct. **14**, pp 579-590

29. Knott, J. F. 1987 "Effects of Residual Stress on Fatigue-Crack Propagation and Fracture" Proc 4th Intl. Conf. on "Numerical Methods in Fracture Mechanics" ed A. R. Luxmoore, D. R. J. Owen, Y. P. S Rajapakse and M. F. Kanninen, Pineridge Press pp 607-626.

30. Taylor, D and Clancy, O. M 1991 "The Fatigue Performance of Machined Surfaces" Fatigue Fract. Eng. Mater. Struct. **14** pp 329-326.

31. Kitagawa, H, Takahashi, S, Suh, C. M and Miyashita, S 1979. "Quantitative Analysis of Fatigue Process - Microcracks and Slip Lines under Cyclic Strains" ASTM STP **675** pp 420-449.

MATERIAL CHARACTERIZATION REQUIRED FOR THE RELIABILITY ASSESSMENT OF CYCLICALLY LOADED ENGINEERING STRUCTURES

Part 1: Fatigue and Failure of Materials

A.J. KRASOWSKY* and L. TOTH**

* *Professor, Institute for Problems of Strength, National Academy of Science of Ukraine, Timiryazevskaya str. 2, 252014 Kiev, UKRAINE*

** *Professor, University of Miskolc, Department of Mechanical Engineering, 3515 Miskolc-Egyetemvaros, HUNGARY*

Abstract

Main stages of fatigue failure: i) damage of material's structure during cyclic loading; ii) fatigue microcrack nucleation; iii) subcritical fatigue crack growth up to the critical size; iv) collaps. The number of cycles required for each of this stages depending on practical situation. Some examples: pressure vessels, storage tanks, pipe lines, turbine blades etc. Main approaches to the engineering structure reliability assessment depending on the functional task of structure.

Cyclic inelastic deformation as the main factor responsible for fatigue of engineering materials. Standard requirements for fatigue testing at constant load amplitude and for strain-controlled fatigue testing. Plastic strain amplitude-controlled testing. Mechanical hysteresis loop and its transformations during cycling under mentioned conditions. Cyclic hardening, cyclic softening and cyclic stable materials. Dependence of stress, strain and plastic strain amplitudes as well as the accumulated plastic deformation on the number of cycles.

The most important equations characterizing the cyclic stress-strain behavior and the fatigue-life relationships. Six material parameters describing fatigue behavior of engineering materials. Relations between these parameters reducing the number of independent parameters down to four. Fatigue properties of five groups of engineering materials: unalloyed steels, low-alloy steels, high-alloy steels, aluminium and titanium alloys. Established linear correlations between material's parameters within each of these groups. Formal and physical sources of these correlations are analysed. The mentioned correlations cannot be explained only by the algebraic structure of power function of the Wohler's or Coffin-Menson's relations. Established correlations reduce the number of material's parameters required for the reliability assessment of cyclically loaded engineering structures.

Constant load amplitude fatigue testing and fatigue limit definition. Relation between the fatigue limit and the threshold value of the stress intensity factor (SIF) range. Threshold size of fatigue microcrack. Non-propagating fatigue microcracks and

R.A. Smith (ed.), Reliability Assessment of Cyclically Loaded Engineering Structures, 165–223.

their size for some engineering materials. Effect of different factors on the fatigue life relationship and on the fatigue limit: stress concentration, stress and strain states, stress ratio, frequency, temperature, corrosion environment, etc. Creep-fatigue interaction at high and at low temperatures. Cyclic creep and its relation to fatigue. High-temperature cyclic creep of two molybdenum alloys as compared with conventional creep at the same conditions.

Fatigue crack growth (FCG) rate diagram. Standard requirements for the FCG-diagram design . Some examples: low-carbon steel, nodular cast iron, nickel, nickel based superalloy. Three regions of FCG-diagram and main parameters describing it. Relation between the FCG rate and the plastic zone size and shape visible on the polished surface of specimen.

Some important formulas for the FCG-diagram description. The middle region of the FCG-diagram: Paris-Erdogan relation. Empirical parameters of this relation and some experimental evidences of strong linear correlation between them. This correlation cannot be explained only by the algebraic structure of power function. The most important fracture mechanics parameters governing fatigue crack propagation rate: stress intensity factor range, crack tip opening displacement range, J-integral range, etc. Fatigue crack closure phenomenon and its explanation. Effective stress intensity factor range and its definition. Crack retardation after single overload. Effects of different factors on the fatigue crack propagation rate: frequency, stress ratio, temperature, environment, etc.

SIF distribution along the front of part-elliptical crack. Weight-function method for calculation of this distribution. Fracture mechanics parameters required for the description of the part-elliptical fatigue crack growth rate: averaged, local and effective SIF along the crack front. Experimental correlation between the FCG rates of standard straight crack front and part-elliptical crack front specimens. Part-through crack growth under inhomogeneous stress field. Stabilized shape of the part-through crack under inhomogeneous stress field. Experimental verification on the four-point bend specimens with different kinds of stress concentrators and with the surface and quarter-elliptical cracks. Prediction of the fatigue crack growth in the structural elements under service conditions.

Fatigue and non-fatigue failure mechanisms accompanying fatigue failure. The role of non-fatigue mechanisms: dimples, cleavage facetts, intergranular facetts. Fatigue striation formation as a basic mechanism of fatigue failure. One-to-one "striation-loading cycle" correspondence. On the correlation between the macroscopic FCG rate and the striation space. The important guidelines for the retrospective analysis of fatigue crack propagation.

1. Introduction

Fatigue and corrosion fatigue is the most commonly encountered cause of failure of parts and structural elements in operation. Much attention has been paid to the study of this phenomenon during a period of over 100 years. As evidenced by the investigations, it is complicated and subject to the influence of many factors (temperature, environment, loading frequency, surface conditions, stress concentration, structural state of the material, etc.). The process of the material fatigue can be divided into at least four stages, each having its own specific features. These are:

i) fatigue damage accumulation in the material related to the changes in its structure and primarily dislocation structure [1];
ii) nucleation of a fatigue crack [2];
iii) fatigue crack growth from a microcrack nuclea to a critical size;
iv) instantaneous final fracture.

Depending on the circumstances, some of those stages may not be present. For instance, in many large-scale structures, such as pressure vessels, pipelines, metalworks, etc., cracks are present initially, particularly, if welding or riveting were used in their assembly. In this case the first two stages are absent and a structure is operating under conditions of a continuous crack growth. Other components, e.g. gas turbine engine blades, steam turbine casings, etc., operate under conditions of thermal cycles which induce thermal fatigue. Sometimes in this case, as a result of changes in the compliance of a component occurring as the cracks grow, the latter extend only to a certain length after which their growth may stop thus eliminating the final stage. In very brittle materials the critical size of a crack can be comparable to the size of a fatigue crack nucleus. In this case stage iii) can be absent and the last stage can immediately follow the second one. True, the investigations of many years into the fatigue process revealed that the first three stages are unfailingly related to cyclic inelastic strains. Therefore, it is difficult to imagine the nucleation of fatigue cracks in very brittle materials without such strains. And finally, in fatigue testing of smooth specimens at stress equal to or somewhat exceeding the fatigue limit, over 95% of the lifetime is related to the stage of crack nucleation [3]. For this reason, methods for the assessment of reliability of structures subject to cyclic loading may differ depending on the function and operating conditions of the structure. In this connection a need may arise in different characteristics of the material resistance to cyclic loading. When testing materials it is also very important to define clearly what event is taken as the specimen failure. If, for instance, this is the separation of a specimen into two halves, then the lifetime includes the stage of crack growth which requires a different description.

2. Mechanical fatigue of materials

In this section we shall mainly consider classical approach to the assessment of fatigue of materials.

2.1. FATIGUE OF SMOOTH SPECIMENS

The process of fatigue of materials is directly related to that of cyclic inelastic deformation. Therefore, we shall connect the consideration of the main types of mechanical fatigue testing of materials with specific features of inelastic deformation. Present-day standards for fatigue test [4,5] envisage the following main test regimes:

i) the tests which involve constant amplitude of cyclic stress, σ_a, (or load); ii) the tests with a constant amplitude of total strain, ε_a, i.e. strain-controlled fatigue testing. A variation of the latter regime is often used wherein the amplitude of inelastic (plastic) strain , ε_{ap}, is maintained constant during the test.

In the process of the material testing involving one of the above three regimes (σ_a = const; ε_a = const; ε_{ap} = const), structural changes occur which change the material properties. With some materials cyclic hardening occurs, with others cyclic softening, whereas the third ones remain neutral, i.e. cyclically stable.

Generally, the regularities of those changes are such that as the number of the load cycles in the regimes σ_a = const or ε_a = const grows, the intensity of the processes in cyclically hardening materials deminishes and the properties are gradually stabilized, saturation occurs. It is assumed with sufficient reliability that by the time half of the number of cycles to failure are realized, a complete stabilization of properties occurs. Within the test regime ε_a = const no accumulation of inelastic strains is observed and failure is typically fatigue. Within the regime σ_a = const, unilateral accumulation of strains takes place and fracture is typically quasistatic.

Let us use a cyclic hardening material to illustrate main features of the fatigue test process [6]. Figure 1 shows two mechanical hysteresis loops (stress vs strain relation in a loading-unloading cycle) for each of the above testing regimes. The loop denoted N_1 corresponds to the first load cycle, N_2 to a cycle with a stabilized hysteresis loop. It also presents the curves of the variation of other characteristics of a cycle which do not remain constant during the given test regime. For a homogeneous material an apparent relation holds among σ_a, ε_a and ε_{ap} at any instant of time:

$$\varepsilon_a = \frac{\sigma_a}{E} + \varepsilon_{a,p} \tag{1}$$

where E is the Young's modulus.

It is seen from Fig.1 that in the course of fatigue testing of a cyclically hardening material with the σ_a = const regime, the width and the area of the histeresis loop decrease, whereas in tests with the ε_{ap} = const regime they grow.

The behaviour of different materials under cyclic loading at σ_a = const regime is shown in Fig.2 [16]. As is seen from this Figure, for a cyclically hardening material (b) the hysteresis loop width decreases in the course of the test, and the total plastic strain tends to some limiting value. For this behaviour of the material a fatigue mechanism of failure is most probable. With a cyclically stable material (c), whose hysteresis loop has a constant width during the test, a unilateral accumulation of strains occurs (provided the loop's width in an even-numbered halfcycle is bigger than in an odd-numbered one). In this case quasistatic failure is common. And finally, with a cyclically softening material (d), both the loop's width and the accumulated plastic strain increase during the test.

One should note some conditionality of dividing materials into cyclically hardening, stable and softening. In fact, the behaviour of the given material under cyclic loading may depend on the regime of heat treatment, on the level of cyclic stress, and at constant stress amplitude - on the number of load cycles.

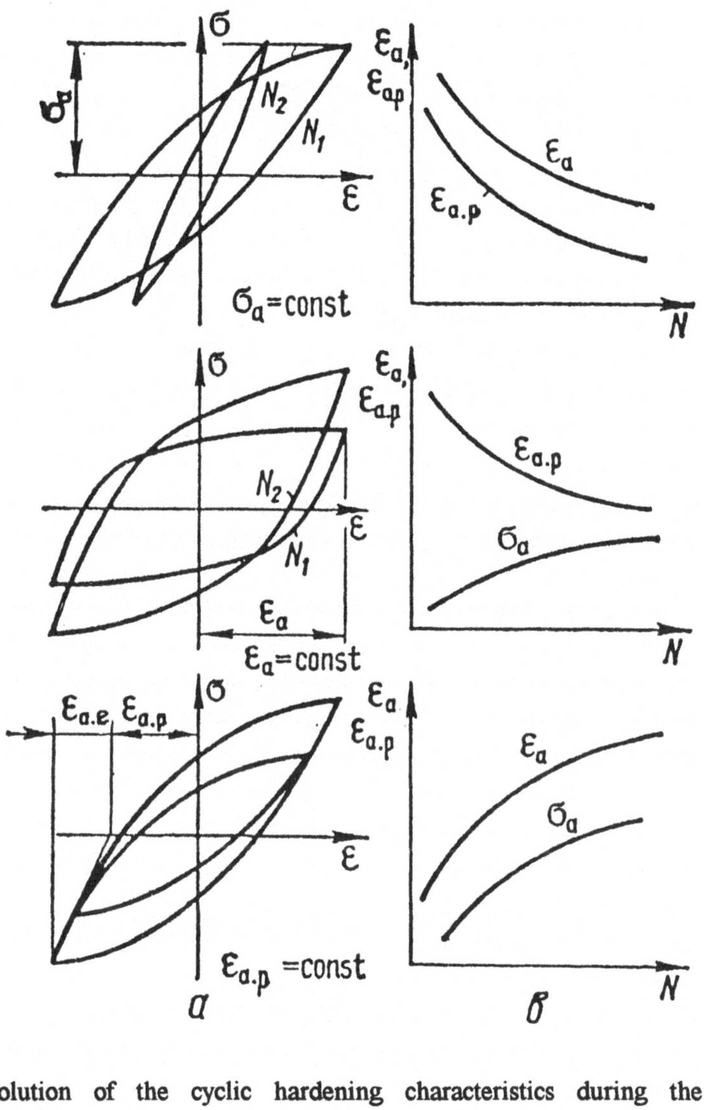

Figure 1. Evolution of the cyclic hardening characteristics during the cyclic deformation process for three loading regimes [6]: a) the shape of hysteresis loop; b) cyclic stress-strain curves.

It is the practice to write the cyclic hardening curve (stress vs plastic strain amplitude relation) in the form:

$$\sigma_a = K'\left(\varepsilon_{a,p}\right)^{n'} \tag{2}$$

where K' and n' are the parameters of cyclic hardening, the material constants by

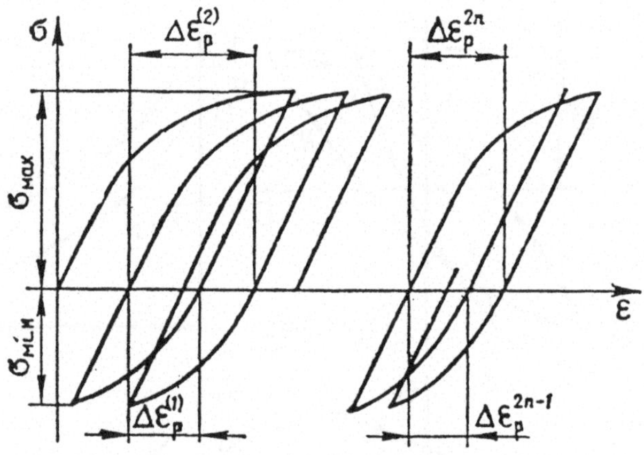

Figure 2A. The character of the mechanical hysteresis loop within the neighbouring 1st and 2nd, then in the odd, (2n-1), and even, (2n), loading halfcycles.

analogy with the power approximation of the static strain hardening curve:

$$\sigma = A\varepsilon^{n}, \tag{3}$$

where A and n are the strain hardening parameters under monotonic loading.

From Eqs(1) and (2) one can obtaine a constitutive relationship for cyclic deformation:

$$\varepsilon_a = \frac{\sigma_a}{E} + \left(\frac{\sigma_a}{K'}\right)^{\frac{1}{n'}}. \tag{4}$$

Standard [5] also recommends for use the following relationships for the lifetime:

$$\sigma_a = \sigma_f'\left(2N_f\right)^b \tag{5}$$

$$\varepsilon_{a,p} = \varepsilon_f'\left(2N_f\right)^c \tag{6}$$

$$\varepsilon_a = \frac{\sigma_f'}{E}\left(2N_f\right)^b + \varepsilon_f'\left(2N_f\right)^c \tag{7}$$

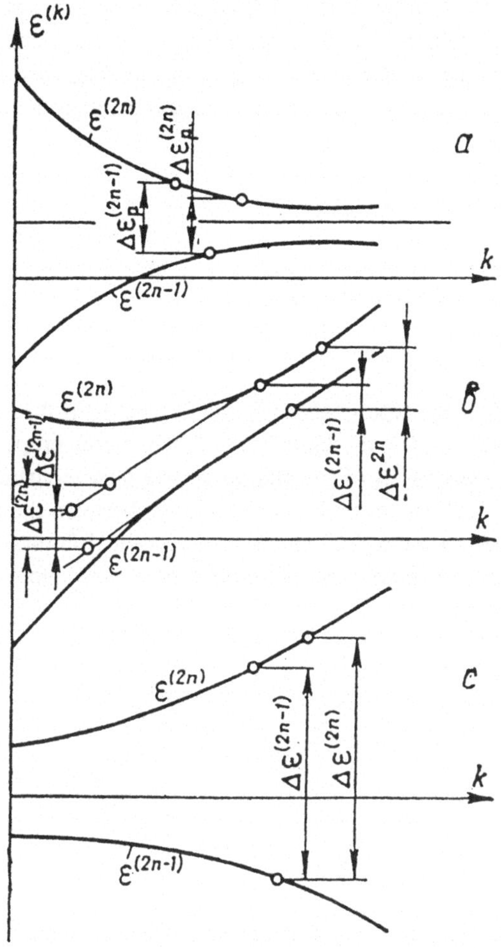

Figure 2B. The value of the accumulated plastic deformation within the mentioned in Fig.2A halfcycles during the fatigue testing (regime σ_a = const) of cyclic hardening (a), cyclic stable (b) and cyclic softening (c) materials. k is the number of the loading halfcycles.

Equation (5) reflects the main contribution to the lifetime in high-cycle fatigue, whereas Eq.(6) dominates in low-cycle fatigue. In those formulas N_f is the number of load cycles to failure, σ_f', ε_f', b and c are material constants determined experimentally. On the basis of Eqs (1), (2), (5) and (6), one can get the following relations between the constants for each material:

$$b = n' \cdot c \tag{8}$$

$$K' = \sigma_f' \Big/ \left(\varepsilon_f'\right)^{n'} . \tag{9}$$

These relationships were verified by experiments [7]. With their help it is possible to reduce the number of constants needed to characterize the material properties under cyclic loading from 6 to 4. The possibility of correlation among those constants extends further. As was shown in refs [8,9], there is a linear correlation between the parameters of power dependences of type (5) and (6) which describe the lifetime of smooth specimens under cyclic loading. In those works fatigue curves were approximated by power functions of the form:

$$N = C_{3e}\,\sigma_a^{\,n_{3e}}, \qquad n_{3e} = \frac{1}{b}; \qquad C_{3e} = \frac{1}{2}\left(\frac{1}{\sigma_f}\right)^{1/b} \tag{10}$$

$$N = C_{3p}\,\varepsilon_{a,p}^{\,n_{3p}}, \qquad n_{3p} = \frac{1}{c}; \qquad C_{3p} = \frac{1}{2}\left(\frac{1}{\varepsilon_f'}\right)^{1/c} \tag{11}$$

It turns out that for materials of the same class there are linear correlations between n_{3e} and lgC_{3e}, as well as between n_{3p} and lgC_{3p} . These correlations are shown in Figs 3 through 6. The plots were constructed using the reference data for unalloyed steels [10], low-alloy steels [11], high-alloy steels [12], as well as for aluminium and titanium alloys [13]. All the results correspond to strain-controlled fatigue testing regime, $\varepsilon_a = \text{const}$.

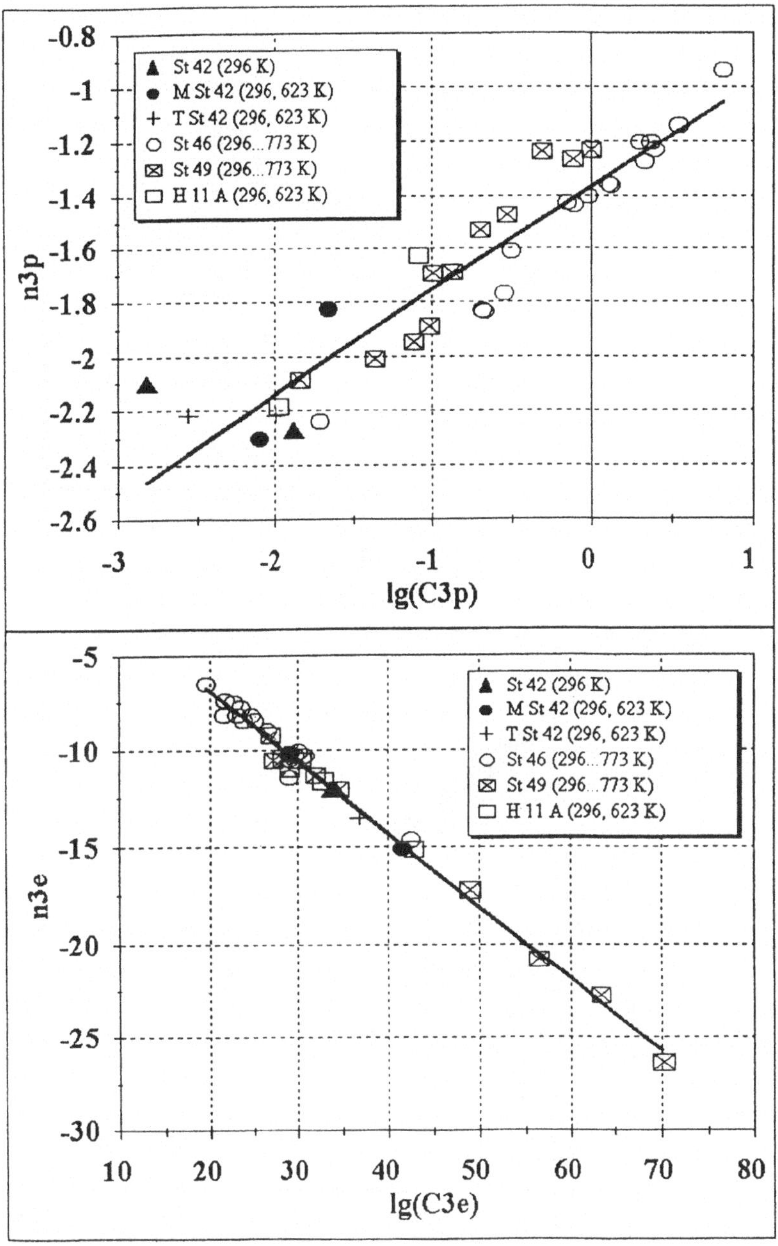

Figure 3. Correlation between n_{3e} and *lg* C_{3e}, Eq. (10), between n_{3p} and *lg* C_{3p}, Eq.(11), for unalloyed steels. The stress amplitude measure is [MPa], the life measure is [cycle].

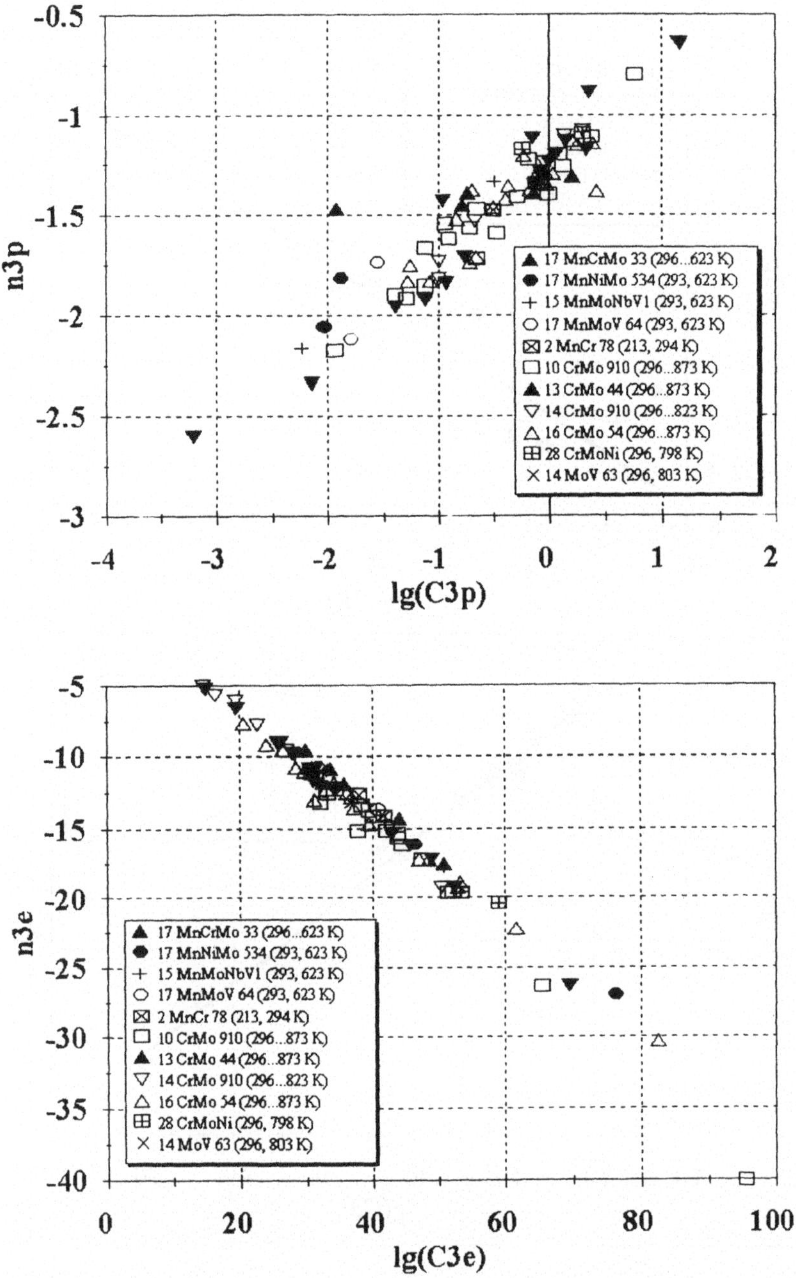

Figure 4. Correlation between n_{3e} and $lg\ C_{3e}$, Eq.(10), and between n_{3p} and $lg\ C_{3p}$, Eq.(11), for low-alloy steels. The measures are the same as for *Figure 3*.

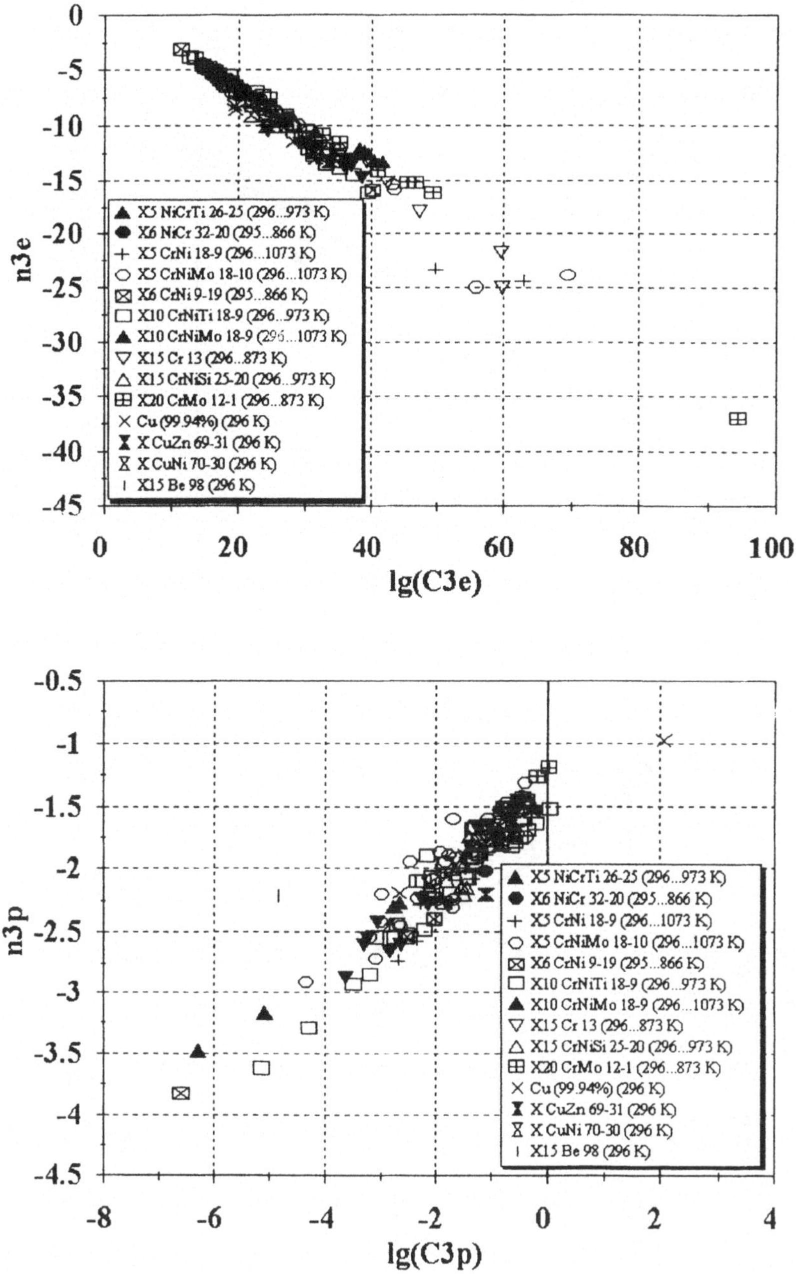

Figure 5. Correlation between n_{3e} and $lg\ C_{3e}$, Eq.(10) and between n_{3p} and $lg\ C_{3p}$, Eq.(11), for high-alloy steels. The measures are the same as for *Figure 3*.

a

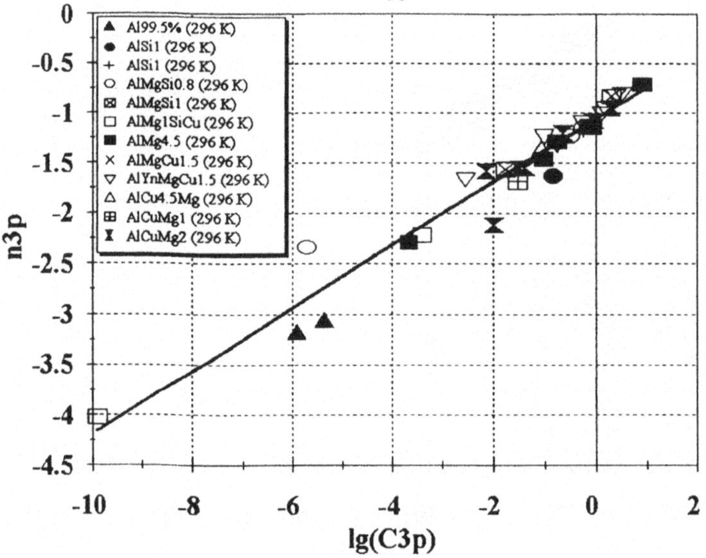

Al99.5% (296 K)
AlSi1 (296 K)
AlSi1 (296 K)
AlMgSi0.8 (296 K)
AlMgSi1 (296 K)
AlMg1SiCu (296 K)
AlMg4.5 (296 K)
AlMgCu1.5 (296 K)
AlYnMgCu1.5 (296 K)
AlCu4.5Mg (296 K)
AlCuMg1 (296 K)
AlCuMg2 (296 K)
n3p
lg(C3p)

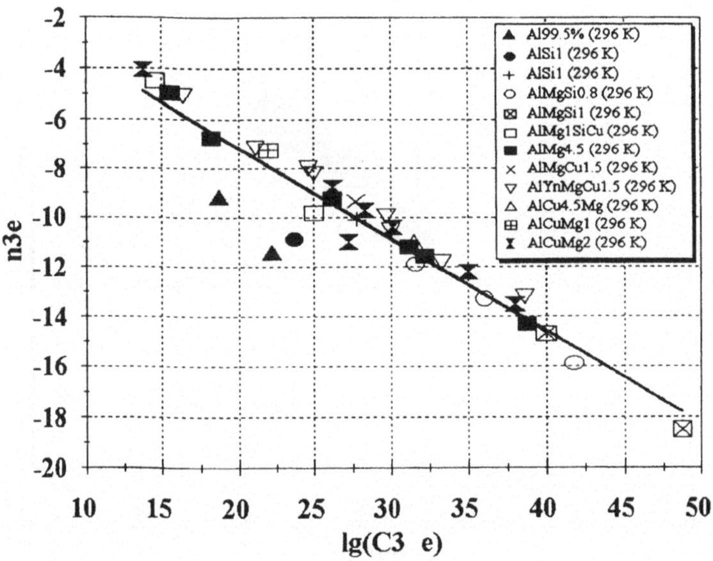

Al99.5% (296 K)
AlSi1 (296 K)
AlSi1 (296 K)
AlMgSi0.8 (296 K)
AlMgSi1 (296 K)
AlMg1SiCu (296 K)
AlMg4.5 (296 K)
AlMgCu1.5 (296 K)
AlYnMgCu1.5 (296 K)
AlCu4.5Mg (296 K)
AlCuMg1 (296 K)
AlCuMg2 (296 K)
n3e
lg(C3 e)

Linear correlations shown in Figs 3 through 6 can be present as strainght lines

$$n_{3e} = A_e \cdot \lg C_{3e} + B_e \tag{12}$$

$$n_{3p} = A_p \cdot \lg C_{3p} + B_p \tag{13}$$

TABLE 1. Quantitative characteristics of the linear correlation, Eqs(12), (13), for five groups of metallic materials

Group of materials	High-cycle fatigue			Low-cycle fatigue		
	A_e	B_e	*correlation factor* R_e	A_p	B_p	*correlation factor* R_p
Unalloyed steels	- 0.3752	0.6722	0.9890	0.3872	- 1.3689	0.8775
Low-alloy steels	- 0.3884	1.1219	0.9774	0.4185	- 1.2536	0.8539
High-alloy steels	- 0.3979	1.1051	0.9576	0.3619	- 1.4156	0.8615
Aluminium alloys	- 03706	0.2468	0.9010	0.3129	- 1.0649	0.9502
Titanium alloys	-0.3267	0.5096	0.9972	0.4275	- 1.1118	0.9669

Generalized data, which correspond to linear correlations, Eqs (12) and (13) for five classes of materials presented in Figs 3-6, are listed in Table 1. With the use of this Table and Eqs (12)-(13), the number of constants needed to describe the behaviour of the indicated groups of materials subjected to cyclic deformation can be reduced still more. It should be noted that Figs 3-6 present the data not only for different materials but also for different temperatures (with the exception of aluminium and titanium alloys for which only the ambient temperature data is available). Nevertheless, within each group of materials a stable correlation is observed between n_{3e} and $\lg C_{3e}$ (and consequently, between b and σ_f'), as well as between n_{3p} and $\lg C_{3p}$ (and consequently, between c and ε_f'). A sufficiently big quantity of experimental data was used for each group of materials. This can be used to characterize a material which was not studied provided that it is known that it belongs to one of the five groups mentioned. In this case it is possible to reduce appreciably the number of fatigue tests in order to obtain a complete set of the above characteristics. This can be done by both reducing the number of specimens and the test temperatures.

Fatigue limit. In the standards mentioned [4,5] no special attention has been drawn to the determination of a fatigue limit, σ_r. Nevertheless, this characteristic is often used in fatigue strength calculations. The magnitude of fatigue limit is generally related to the given number of cycles to failure (fatigue limit) and to the given probability of nonfailure (fatigue limit for p % survival) [14] on the *S-N* curve. Present-day investigations into fatigue of metallic materials reveal [6, 15] that fatigue crack nucleation can also occur at stresses below fatigue limit. Such cracks remain nonpropagating. Their threshold size, l_{th}, can be evaluated approximately by the methods of linear elastic fracture mechanics making some not quite obvious assumptions [6] of the relationship between σ_r, l_{th} and the threshold SIF value, $K_{a\,th}$ (the latter will be considered below). These estimations give the following value of the threshold crack size at the fatigue limit:

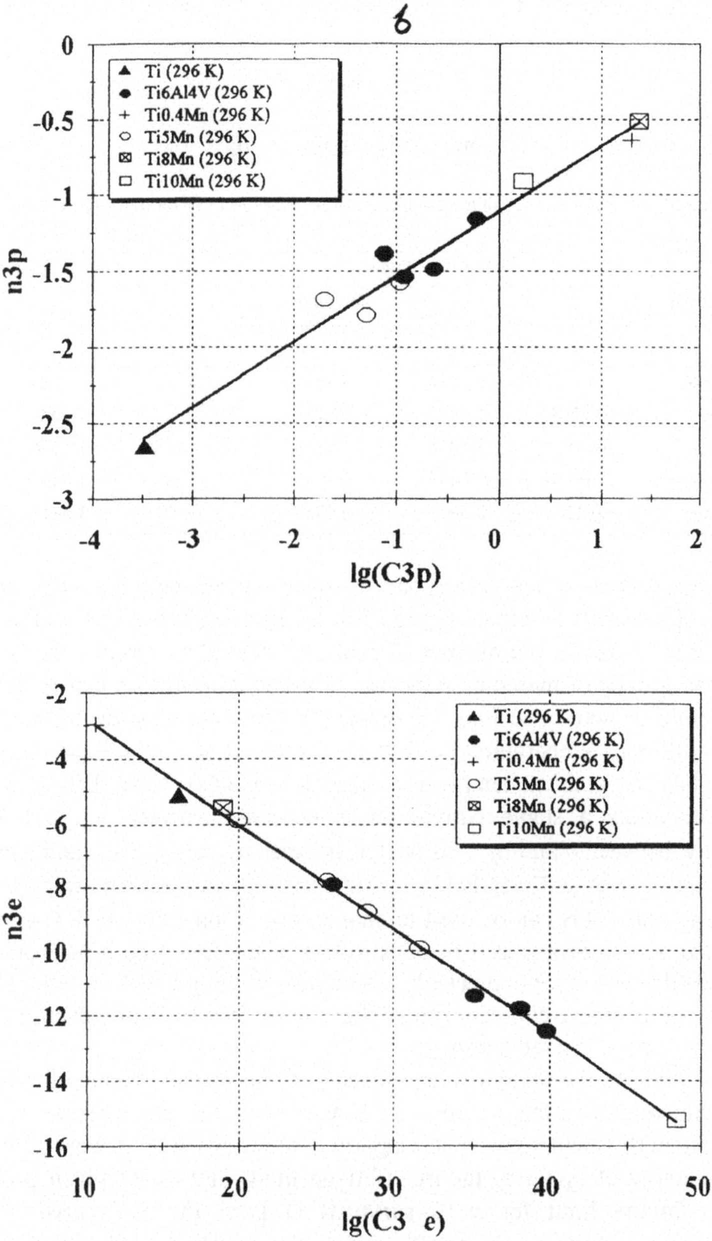

Figure 6. Linear correlations between n_{3e} and *lg* C_{3e}, Eq.(10), between n_{3p} and *lg* C_{3p}, Eq.(11), for aluminium (a) and titanium (b) alloys. The measures are the same as for *Fig.3*.

$$l_{th} \approx \mathrm{f} \cdot \left(\frac{K_{a\,th}}{\sigma_r} \right)^2 , \tag{14}$$

where f is a constant factor of an order of $1/\pi$ which varies within a narrow range and depends on the microcrack shape. For real $K_{a\,th}$ and σ_r values of metallic materials Eq.(14) yields the l_{th} magnitudes ranging from one-hundredth of a millimeter for high-strength materials to several tenths of a millimeter for low-strength ones. Those estimates yield the l_{th} value of a correct order of magnitude and give a qualitatively reasonable prediction of fatigue strength for notched specimens.

2.2. THE INFLUENCE OF VARIOUS FACTORS ON THE WÖHLER CURVE

Fatigue of materials is a very complicated phenomenon which depends on many factors. Those factors affect the Woehler curve determined on laboratory specimens in different ways. The effect of those factors on fatigue failure of structural elements under service conditions may not always be unambiguous. Here we shall briefly consider the effect of various factors on the process of materials fatigue to give the reader a general idea of this effect.

2.2.1. *Stress Concentration.*

The influence of stress concentration on fatigue is of mechanical nature. This influence is determined by three main causes: a local increase of the stress level; a change of the stress state in the stress concentration zone; nonuniformity of the stress and strain fields in the concentrator zone. Those circumstances can complicate appreciably the fatigue behaviour and its analysis, particularly in those cases when plastic deformation zones appear in response to local stress concentration.

The effect of stress concentration on fatigue is described with the use of the effective stress concentration coefficient:

$$\beta = \frac{\sigma_r}{\sigma_{rn}} , \tag{15}$$

where σ_{rn} is the fatigue limit of a notched specimen. It is known from the experiment that the value of σ_{rn} is always higher than that predicted on the basis of the theoretical stress concentration factor, α, which is equal to the ratio between the local stress peak and the nominal stress. That is, the effective concentration always turns out to be lower than the theoretical one, $\beta < \alpha$. A satisfactory explanation to this fact can be obtained [6] on the basis of the notion of nonpropagating fatigue microcracks of length l_{th}. This length is assumed to be the same for smooth and notched specimens. Writing, by analogy with Eq.(14), an expression for a crack of threshold length for the case when it is located at the notch root, and comparing it with Eq.(14), one can get the following expression for the fatigue limit of a notched specimen:

$$\sigma_{rn} = \frac{\sigma_r}{\alpha} \left[1 + \frac{1.14}{\rho} \left(\frac{K_{ath}}{\sigma_r} \right)^2 \right]^{1/2} , \tag{16}$$

where ρ is the radius of the elliptical notch vertex. The predictions made with Eq.(16) for different l_{th} values and different stress concentrators are shown in Fig.7. Those predictions are in agreement with the available experimental data [17]. The problems of the consideration of stress concentration in the fatigue assesment have been studied in detail by Heywood in his book [18].

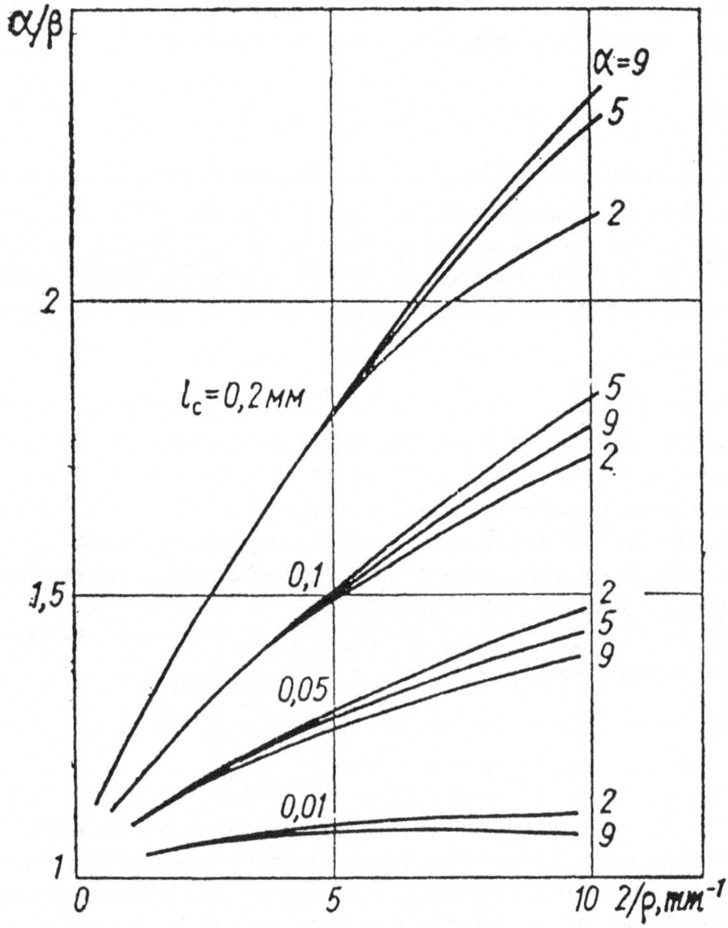

Figure 7. Dependence of the ratio α/β characterizing the relation between the theoretical and effective stress concentration factors as predicted by Eq.(16) for different notch acuities [6].

2.2.2. *Stress and Strain States*

These factors have a strong impact at all stages of fatigue. Here, considering the nature of fatigue phenomenon and its direct relation to cyclic plastic strains, one can suggest that the material strain state defines its fatigue more precisely than the stress state. Although it is clear that both these states are mutually related due to the existence of a constitutive equation for cyclic deformation, Eq.(4). According to Miller [15], the criterion of maximum shear strains (analog of Tresca-St. Venant's criterion for stresses) offers some advantages over other classical criteria. It allows one to describe a unified cyclic stress-strain curve for various ratios of strain tensor components during their in-phase variation. When the strains along different axes are in out-of-phase condition, the situation becomes even more complicated.

The study of the impact of the stress-strain state upon the material fatigue is of great practical importance. This importance is attributed to both local changes in the stress state under the effect of stress concentrators, and real complex stress-strain conditions which accompany the operation of structural elements (e.g. turbine discs and blades, pipe-lines and pressure vessels, shafts and contact pairs, etc.). A more detailed information concerning this problem can be found in specialized literature [19].

2.2.3. *Stress Ratio*

To take into account the mean stress, σ_m, which governs the cycle stress ratio, the most commonly used are the diagrams of either limiting stress in a cycle (in the σ_{max} - σ_m coordinates), or of limiting stress amplitudes in a cycle (in the σ_a - σ_m coordinates). Both diagrams are considered in detail in the courses of mechanics of materials. For this reason, we shall confine ourselves to only some analytical relationships which take into account the influence of the cycle stress ratio upon fatigue limit [20]:

$$\sigma_{a_r} = \sigma_{-1} - \xi\sigma_m , \tag{17}$$

where σ_{-1} is the fatigue limit at a fully-reversed cycle, $r = \dfrac{\sigma_{\min}}{\sigma_{\max}} = -1$, ξ is the coefficient of the material's sensitivity to a stress ratio, σ_{a_r} is the fatigue limit at an oscillating cycle. For carbon steels it is recommended to assume $\xi = 0.1 - 0.2$, whereas for alloy steels $\xi = 0.2 - 0.3$. Different expressions are used to describe analytically the diagrams of limiting amplitudes. The most simple of them are [21]:

$$\sigma_{a_r} = \sigma_{-1}\left[1 - \frac{\sigma_m}{\sigma_u}\right] \tag{18}$$

$$\sigma_{a_r} = \sigma_{-1}\left[1 - \left(\frac{\sigma_m}{\sigma_u}\right)^2\right] \tag{19}$$

where σ_u is the ultimate strength. Equation (18) is best suited to brittle materials of the cast iron type, whereas Eq. (19) to ductile materials. A detailed analysis of the influence of mean stresses in a cycle upon metals fatigue is presented elsewhere [18].

2.2.4. *Loading Frequency*

The impact of this factor on fatigue is related to the material sensitivity to the strain rate and to the self-heating of the specimen during the test. The available standards [4,5] allow a rise in the specimen temperature in the course of the test of no more than 2°C. Generally a loading frequency of no more than 170 Hz is sufficient for this purpose, though for some materials it should be lower. Fatigue limit versus loading frequency curves have a characteristic dome-shaped form [21]. In the load frequency region up to about 1000 Hz the fatigue limit of metals grows with the frequency. The influence of strain rate on the cyclic stress-strain curve and on the cyclic yield stress manifests itself therein. With further increase in frequency, the specimen self-heating leads to a competing effect of the reduction of cyclic yield stress, and therefore of the fatigue limit. The maximum of the fatigue limit versus loading frequency relation can be shifted towards higher frequencies by a better cooling of the specimen.

To take into account the frequency effect upon the fatigue curve, Coffin [22] introduced a correction factor for frequency into Eq.(11), after which it took the following form:

$$\left(N_f \cdot \nu^{k-1}\right)^{\beta} \cdot \varepsilon_{ap} = C_{3p}' . \tag{20}$$

Comparison of this equality with the experimental data obtained in the quite wide loading frequency range revealed fair agreement. In Eq.(20) $\beta = 1/n_{3p}$ and k reflects the sensitivity of the material to the frequency (at $k = 1$ the material is insensitive to the frequency).

2.2.5. *Temperature*

The character of high temperature effect upon Woehler's curve for nickel-based alloy ЭИ 437Б (19% Cr, 0.6%Si, 0.4%Mn, 2.4%Ti, Ni- bal.) under cyclic loading at $\sigma_a = const$ is shown in Fig.8. It is seen from the Figure that an increase in temperature leads to an appreciable reduction of the material resistance to cyclic loading. At temperatures of 1073K and 1173K there appears a region of drastic drop in the fatigue strength (at $N_f > 10^8$ cycles and at $N_f > 3 \cdot 10^7$ cycles , respectively). It was also noticed that the slope of the main part of Woehler's curve, $d\sigma_a / dN_p$, increases which is explained by an increase of the material softening rate with temperature.

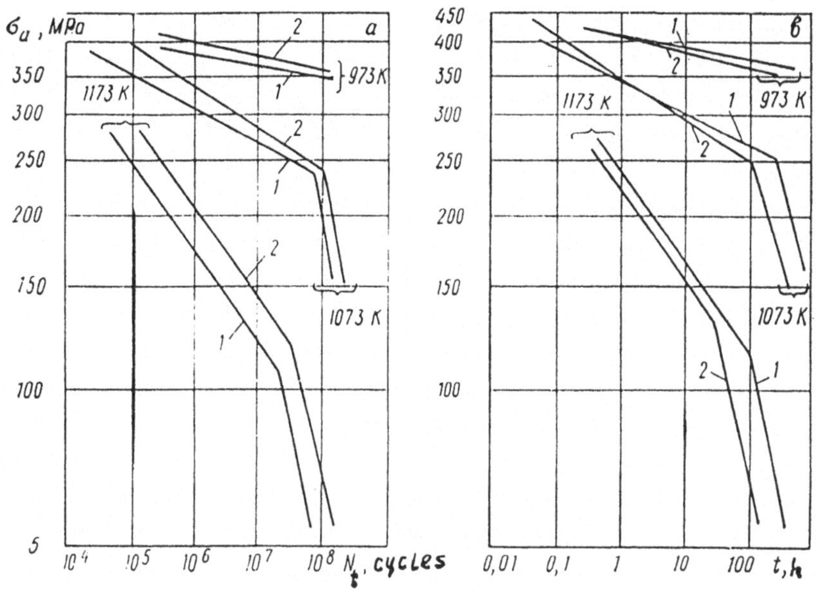

Figure 8. Dependence of the limit stress amplitudes on the loading cycles number (a) and on the time (b) for nickel-based alloy ЭИ 437Б [20]. Legend: 1- frequency 50 Hz ; 2- frequency 200 Hz.

This is the feature of the stress-controlled loading (σ_a = *const*) regime, at which unilateral accumulation of strains, i.e. creep process, occurs. Previously it was shown [23] that if the creep power law is obeyed, the sensitivity of both the creep rate, $n_1 = \left(\frac{\partial \ln \dot{\varepsilon}}{\partial \ln \sigma}\right)_T > 0$ and creep life $n_1' = \left(\frac{\partial \ln t_t}{\partial \ln \sigma}\right)_T < 0$ to stress varies inversely with the temperature. The n_1 and n_1' values are the slopes of the creep rate and creep life versus applied stress logarithmic relations. In fact, the value $n_2 = \left(\frac{\partial \ln N_f}{\partial \ln \sigma_a}\right)_T < 0$ is the slope of the main part of the Woehler curve which is as much related to the n_1' value as the creep proceses contribute to specimens failure at the cyclic loading regime σ_a = *const*. This can be shown by processing the results [24, 25] of creep and cyclic creep tests at σ_a = *const* regime for the molybdenum alloys ЦМ-6 (0.12% Zr, 0.0024%B, Mo-bal.) and ЦМ-10 (0.005%Al, 0.0021%B, Mo-bal.), Figs 9 and 10.

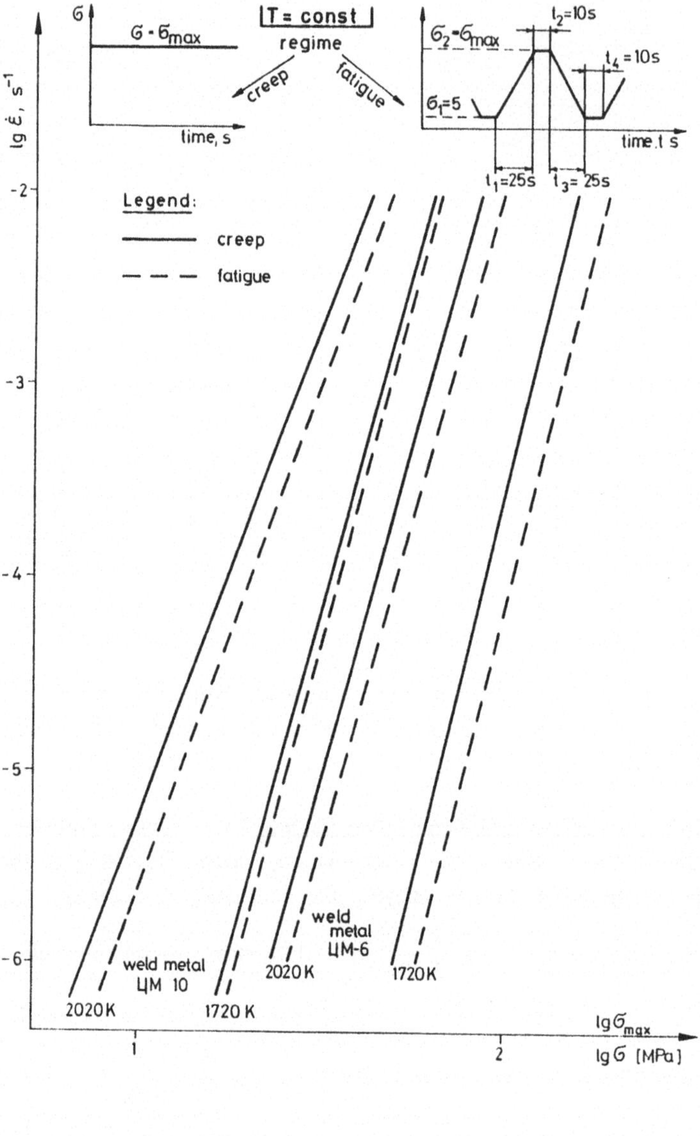

Figure 9. Dependence of creep rate and cyclic creep rate on the maximum stress for two molybdenum alloys.

The straight lines in the plot were obtained by averaging the experimental data by the least-squares method. The loading regimes are indicated in the upper parts of the Figures. For cyclic creep the time was measured in seconds rather than in the number of cycles. As is seen from the Figs 9 and 10, the behaviour of the creep and cyclic creep

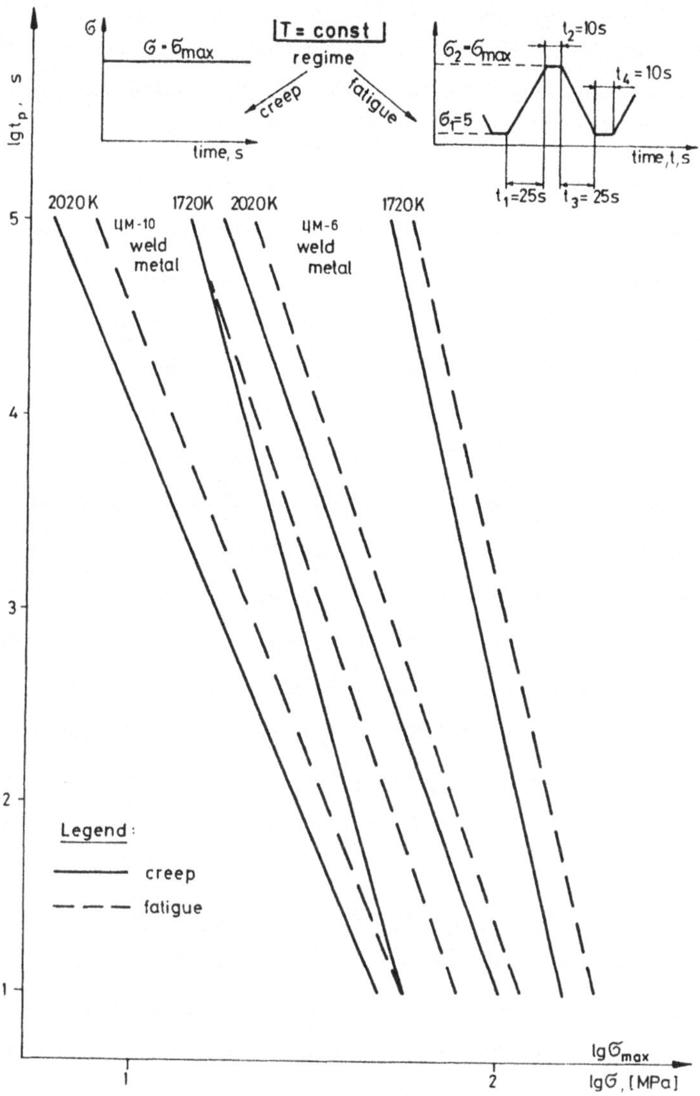

Figure 10. Dependence of life on the maximum stress at creep and cyclic creep for two molybdenum alloys.

plots is identical and corresponds to the power creep law. The slope of the lines diminishes with a rise in temperature. Since no creep occurs under minimum loads of cycle, it is reasonable that at the same σ_{max} value the cyclic creep rate is lower and the real time to rupture is longer than in the case of creep. As in the case of creep power law for different materials [23], similar linear correlation is observed in the given case between n_1 and lgC_1, as well as between n_1' and lgC_1' for creep and between n_2 and lgC_2, and between n_2' and lgC_2' for cyclic creep (Fig. 11).

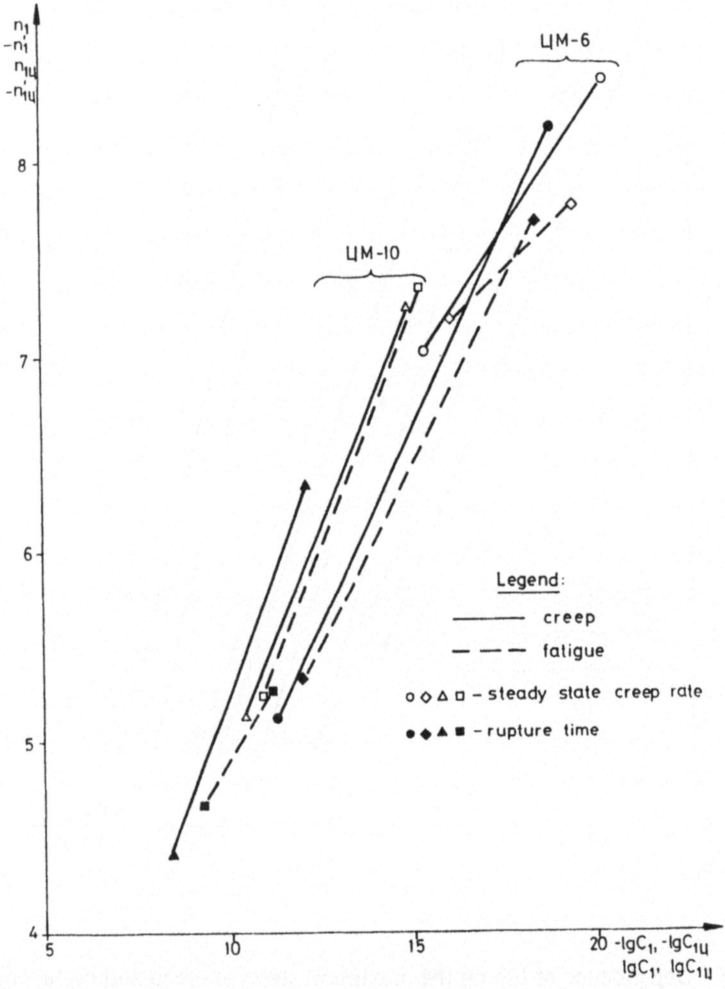

Figure 11. Linear correlation between empirical parameters n_1 and $lg\ C_1$ (power law of creep) and between n'_1 and $lg\ C'_1$ (creep rupture time) as well as between parameters n_2 and $lg\ C_2$ (power law of cyclic creep) and between n'_2 and $lg\ C'_2$ (cyclic creep rupture time) for two molybdenum alloys.

In order to represent the data for creep rate (n_1, n_2, lgC_1, lgC_2) and for time to rupture (n_1', n_2', lgC_1', lgC_2') on the same plot of Fig. 11, we have changed the signs for n_1', n_2', lgC_1 and lgC_2. It is seen from this Figure that the correlations presented correspond to each other fairly well. It can be shown [23] that if a correlation is observed between parameters n_i and lgC_i in the power law of the type

$$y = C_i x^{n_i} = C_i \left(\frac{x}{1} \right)^{n_i} \tag{21}$$

it cannot be explained by formal causes associated with the algebraic structure of Eq. (21). This correlation is related to specific features of thermodynamics of the plastic deformation process. It follows from the thermo-activation relationship for the process rate:

$$\dot{\varepsilon} = \dot{\varepsilon}_{0i} \cdot \exp\left[-U_i(\sigma^*)/kT \right] \tag{22}$$

and from the dependence of the activation energy U_i (σ*) on the effective stress in the form

$$U_i(\sigma^*) = U_{0i} \ln \frac{\sigma_i^*(0)}{\sigma^*} \quad . \tag{23}$$

In these formulas $\dot{\varepsilon}_0$ is the preexponent, k is Boltzmann's constant, T is the absolute temperature, U_{0i} is a constant with the dimensions of energy, $\sigma_i^*(0)$ is the yield stress at the absolute zero temperature (i.e. in the absence of thermal fluctuations).

From the Eqs (21), (22) and (23), after logarithming one can get

$$\lg C_i = \lg \dot{\varepsilon}_{0i} - \frac{U_{0i}}{kT} \lg \sigma_i^*(0); \quad n_i = \frac{U_{0i}}{kT} \tag{24}$$

Equations (24) reveal a linear correlation between n_i and lgC_i shown in Fig. 11, which is defined by thermodynamics of the process of inelastic deformation. The formula for n_i in Eq. (24) agrees well with the experimental data for various materials (Figs 12, 13).

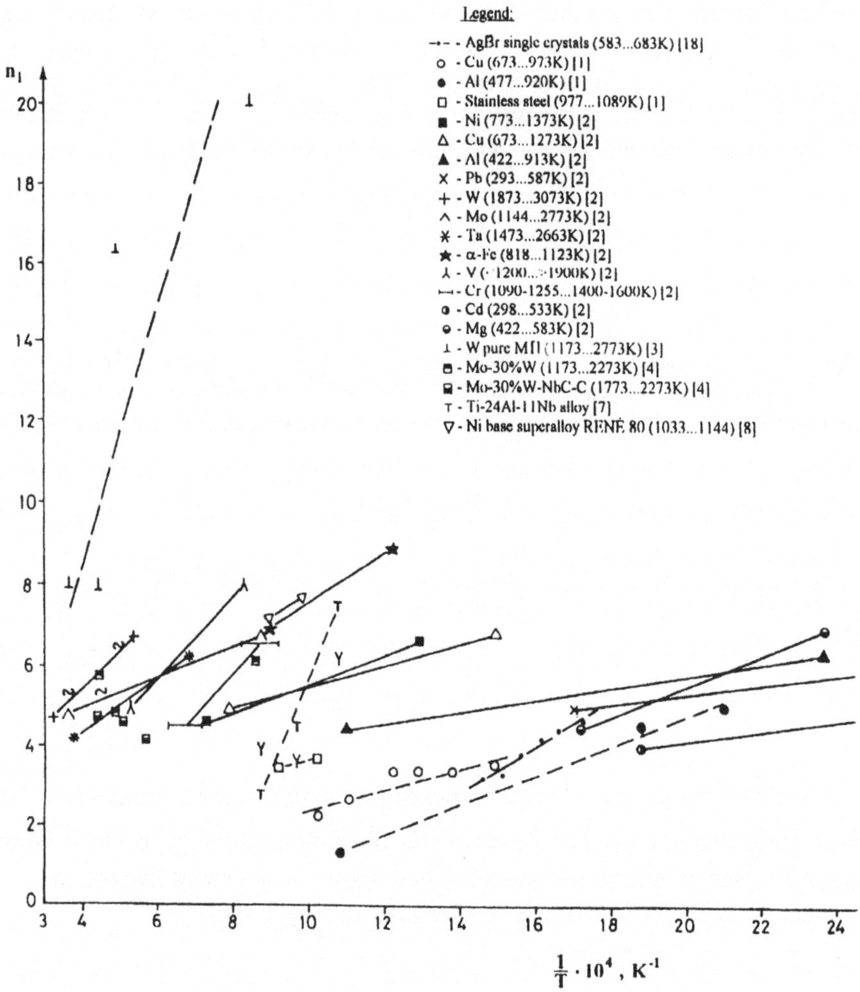

Figure 12. Dependence of the parameter n_1 of the creep power law on the reciprocal of test temperature for different materials.

The creep and cyclic creep results for alloys ЦМ-6 and ЦМ-10 shown in Fig. 14 do not contradict those data. Basing on the formula for n_i from Eq.(24), the slope of the

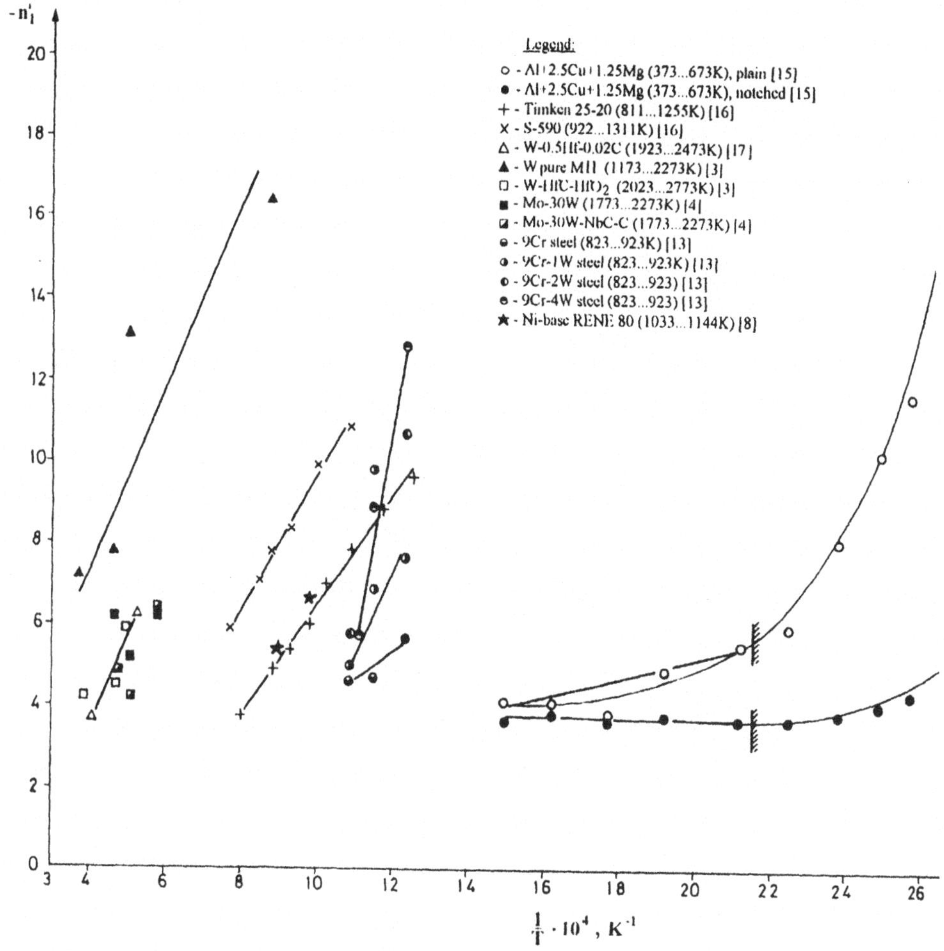

Figure 13. Dependence of the parameter $n`_1$ of the rupture time-stress power relation on the reciprocal of the test temperature for different materials.

plots $n_i(1/T)$ (Figs 12-14) is defined as

$$\frac{dn_i}{d\left(\frac{1}{T}\right)} = \frac{U_{0i}}{k} \quad , \tag{25}$$

i.e. it depends only on the magnitude of the constant U_{0i}. In ref. [23] the constant U_{0i} is shown to be in direct relation to the interatomic potential and for each material it has

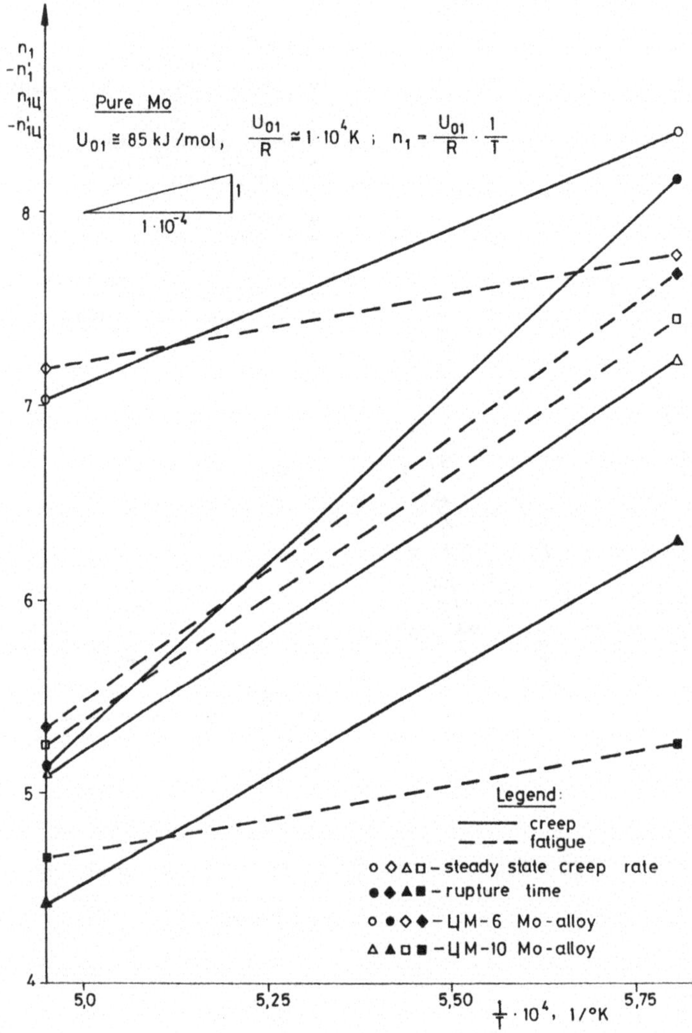

Figure 14. Dependence of the exponents n_1 and n_2 of the creep power law and the exponents n'_1 and n'_2 of the rupture time-stress power relation on the reciprocal of test temperature for molybdenum alloys at creep (n_1 and n'_1) and at cyclic creep (n_2 and n'_2).

the meaning of cohesive energy. This conclusion follows from the comparison of the values $k\,n_1T$ and $3kT_m$ (T_m is the melting point) presented in Fig. 15. A simple formula follows from this relation

$$n_1 \approx -n'_1 \approx 3\frac{T_m}{T} \tag{26}$$

for determining slopes n_1 and n_1' in the creep power law and of the rupture time curve:

$$\dot{\varepsilon} = C_1 \sigma^{n_1}, \; n_1 > 0; \tag{27}$$

$$t_r = C_1' \sigma^{n_1'}, n_1' < 0 \tag{28}$$

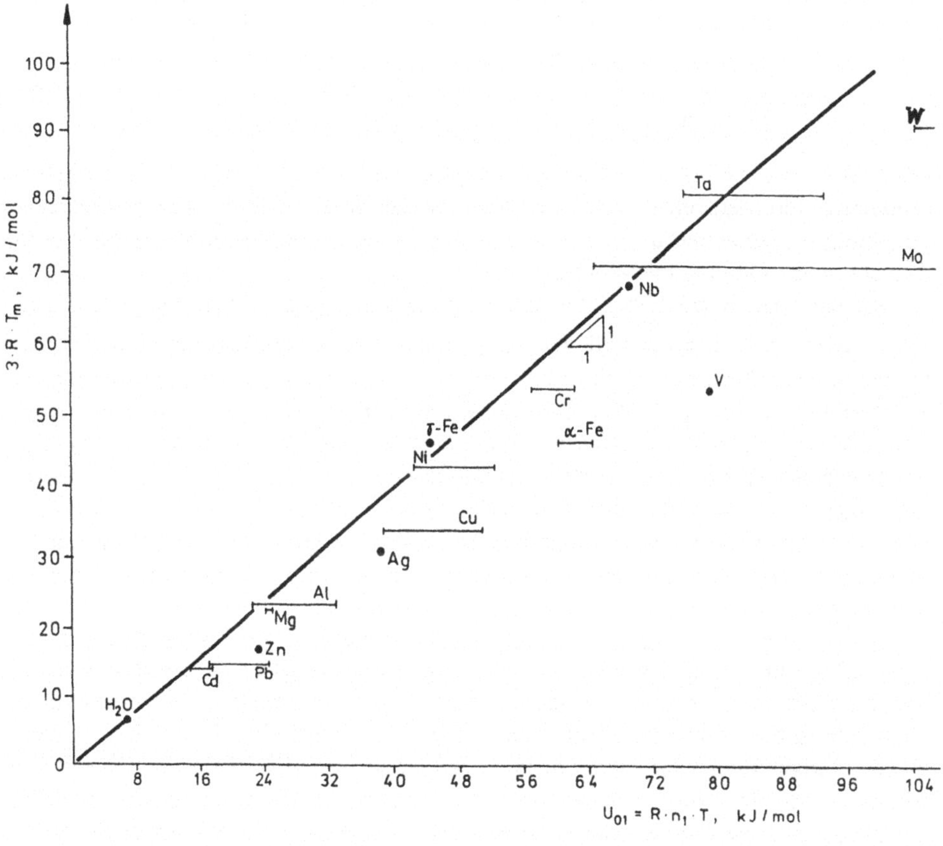

Figure 15. Relation of the constant U_{0_1} to the melting temperature for different materials.

The Equations (26),(27) and (28) can be used to evaluate cyclic creep and its contribution to fatigue failure of materials at elevated temperatures. In particular, an obvious advantage of Eq. (26) which has been validated for pure metals [23] is in the possibility of *a priori* assessment of slopes n_1 and n_1'. After this, it is sufficient to test only one specimen at creep in order to assess constants C_1 and C_1' in Eqs (27) and (28) from the $\dot{\varepsilon}$ and t_r values and in this way to describe both curves. Judging from the data of Figs 9 and 10, there are temperature regions and loading regimes for which slopes n_1 and n_2, as well as n_1' and n_2' are similar. Consequently, Eq. (26) can also be used to evaluate cyclic creep. However, these regions and regimes have to be defined. As for the equality of slopes n_1 and n_2, as well as n_1' and n_2', it makes it possible to reduce the problem of describing cyclic creep using the creep data to the development of such criteria of equivalent damage, which allow parallel translation of cyclic creep lines until they coincide with creep lines. Such a translation can be accomplished with the help of the introduction of either equivalent time or equivalent stress. For instance, in ref. [26] a criterion is proposed for the regime of cyclic loading indicated in Figs 9 and 10:

$$\sigma_{eq}=\left[\frac{\left(\sigma_2^{\beta+1}-\sigma_1^{\beta+1}\right)\left(t_1+t_3\right)}{\left(\sigma_2-\sigma_1\right)\left(\beta+1\right)\left(t_1+t_2+t_3+t_4\right)}+\frac{\sigma_2^{\beta}\,t_2+\sigma_1^{\beta}\,t_4}{t_1+t_2+t_3+t_4}\right]^{\frac{1}{\beta}} \tag{29}$$

where $\beta = -n_1'$ and other definitions are given in the Figs 9 and 10. This criterion revealed a fair agreement with the experimental data. It allows the possibility of predicting the behaviour of a material under cyclic creep conditions on the basis of the data obtained in the regime of ordinary creep.

As has been noted above, no unilateral accumulation of inelastic strains takes place under cyclic loading with the ε_a = const strain-controlled regime and typical fatigue failure takes place. In this case, even at elevated temperatures the conditions for a favourable occurring of creep deformation are absent. Though in this case, as the results of Figs 3-6 indicate, a linear correlation is observed among the parameters of the power dependences of strain or stress amplitudes on lifetimes, it is difficult to relate it to the thermodynamics of plastic flow as simple as in the case of creep. For instance, in Figs 3-5 there is much experimental data for elevated temperatures. Yet, the attempt to establish a correlation between the slopes n_{3p} or n_{3e} and the reciprocal of the test temperature fails.

With a drop in temperature the cyclic yield stress and fatigue limit increase. This is observed for all metals. But whereas for metals, which do not reveal a brittle-to-ductile transition behaviour (e.g. fcc-metals, austenitic steels), this leads to an increase in the resistance to both crack nucleation and growth, for embrittling materials (e.g. bcc - metals, carbon steels) the situation is more complicated. For them at temperatures below critical transition temperatures, the material fracture toughness decreases drastically and the critical crack size reduces. In this case, the fatigue crack growth zone on fracture surfaces decreases drastically and the zone of final brittle fracture increases. These peculiarities will be considered in the second part of the lecture dedicated to the fatigue crack growth.

2.2.6. *Corrosive Medium*

This extremely important and almost always present (with the exception of tests in high vacuum and in high-purity inert media) factor influences the lifetime at all the stages of fatigue failure and almost always reduces it. The mechanism of this influence is diverse. It depends on both the corrosive medium, and the material resisting it. For this reason, characterization of the material for the operation under cyclic loads in the presence of corrosive medium calls for the consideration of specific pairs “material-environment” and the conditions of their loading. Since the surveying of this subject might lead us far away from the main direction of this lecture, we refer the readers to specialized literature: the effect of corrosive media upon fatigue crack nucleation in refs [27, 28], whereas the effect of corrosive media on the fatigue crack growth in refs [29, 30].

2.2.7. *Other Factors*
Among these are the state of the surface layer of a component or a specimen, microstructure and the material state, heat treatment, variable strain or load amplitudes, fretting-corrosion, residual stresses, the shape of a cycle and other factors which sometimes can have a determining effect on the process of fatigue. Those readers who interested with this topic we recommend to refer to specialized books [6, 15, 18, 31].

3. Fatigue crack growth.

A study into the process of fatigue crack growth is of independent scientific interest since many structural elements operate during the greater, if not all, part of their lives under conditions of the existing fatigue crack growth. It has been noted above that these are, generally, those structural elements in which cracks were present originally (for example, in the region of a welded joint) or appeared shortly after the beginning of their operation due to a high level of stresses (for example, in the stress concentrator zone). The process of the fatigue crack growth, from the standpoint of its analysis , is even more complex as compared to the fatigue process in smooth specimens discussed above. The reason is that a nonuniform three-dimensional stress-strain state exists in front of the fatigue crack. Each microvolume of the material along the growing crack trajectory as the tip is approaching it is subject to cyclic elastic-plastic deformation with an increasing amplitude. Let us consider the main relationships of this process.

3.1. EXPERIMENTAL STUDY INTO FATIGUE CRACK GROWTH.

The methods for the experimental investigation into fatigue crack growth (FCG) have been standardized [32,33] which facilitates the comparison of the results obtained by different investigators. The main result of the investigation is presented in the form of a diagram of FCG rate versus the range of the coresponding parameter of fracture mechanics. Figure 16 illustrates the common view of such a diagram with the indicated crack growth rate per cycle, dl/dN, depending on the range of the stress intensity factor (SIF). The range of crack growth rates given in the diagram corresponds to commonly observed levels of the FCG rates. Such minimal rates were also observed in the experiment under loading in the ultrasonic frequency range at a low amplitude of SIF. The ranges of structural element dimensions typical for metals are given on the diagram for purposes of comparison. The diagram has three parts. The middle (**II**) part corresponds to the power dependence of the crack rate upon ΔK (on logarithmic coordinates this is a stright line). The first part **I** is limited from the left by the threshold value of SIF, ΔK_{th}. This is the maximum value of the SIF range at which the crack growth is not yet observed at a rather large number of load cycles. The third part

III is limited from the right by the critical value of SIF, $K_{max} = K_{I_{fc}}$. Under cyclic loading this value can be equal to the material fracture toughness, K_{Ic}, but can also differ from it. The reasons behind this we shall discuss below. The lower scale $\lg\sigma/\sigma_y$ corresponds to realistic crack length. This scale shows that a crack can have a definite propagation rate at quite low load levels.

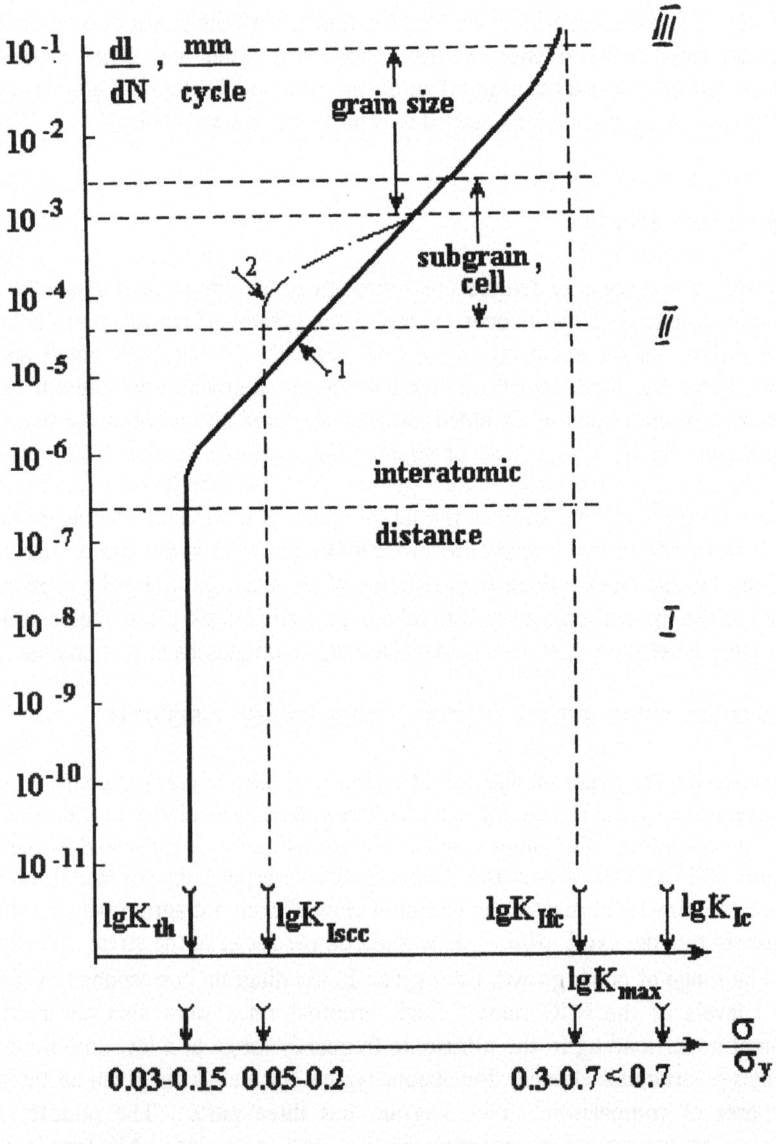

Figure 16. Schematic representation of the fatigue crack growth rate dependence on the stress intensity factor range.

3.1.1. Some Fatigue Crack Growth Rate Diagrams.

Figure 17 illustrates the FCG rate diagram for a high-strength nodular cast iron. It is typical for this material to nucleate a fatigue microcrack during (0.04-0.11) part of the specimen fatigue life. These standard FCG data correspond to a frequency of 16.7 Hz,

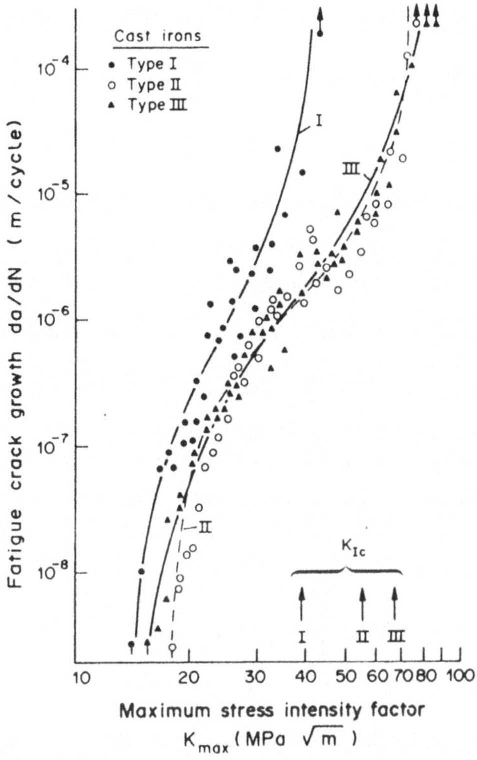

Figure 17. Fatigue crack growth rate diagram for three nodular cast irons.

stress ratio R = 0.17, for three-point bend specimens tested at the laboratory air. Quantitative characterization of Fig. 17 is presented in Table 2. Here the threshold values of SIF , K_{th} , are almost the same and their values are rather high which is typical of cast irons. The form of the FCG rate diagram depends upon how brittle or ductile the material under test is. The general trend is such that ΔK_{th} is usually increases and $K_{I_{fc}}$ decreases with an increase in brittleness, i.e. the diagram itself becomes steeper. A property of BCC metals to undergo a brittle-to-ductile transition with temperature makes it possible to observe this change in the FCG diagram on the same material. The FCG diagrams for the steel with 0.08% C obtained on 3 mm thick specimens are given in Fig.18. A reduction in temperature resulted in a considerable increase in K_{th} (the experiments were carried out at K_{min} =0.45 MPa, i.e.

TABLE 2. The main quantitative characteristics of the fatigue crack growth for three different nodular cast irons.

	Charactristic	I	II	III
	Average globule diameter, mm	50	110	40
	Globule as sphere, %	65-75	70-85	90-95
	Pearlite content, %	50	0	50
	Average distance between globules, mm	85	202	97
	K_{th}, MPa$\sqrt{m}$	11.5	14.6	12.5
	K_{fc}, MPa$\sqrt{m}$	33.6	63.2	67.0
	K_{1-2}, MPa$\sqrt{m}$	13.9	18.8	16.7
	K_{2-3}, MPa$\sqrt{m}$	27.9	49.9	50.4
Eq.(34)	$C_4\times10^{15}$, mm/cycle	0.96	330	32.0
	n_4	6.9	4.5	5.2

at the cycle stress ratio $R=\sigma_{min}/\sigma_{max}\approx 0$, therefore $\Delta K\approx K_{max}$) and a decrease in $K_{I_{fc}}$. Each of the experimental points indicated is the result of averaging of 8 or 10 measurements of the crack rate for the given level of load. Figure19 illustrates the dependences of characteristic dimensions of the plastic zone at the crack tip at 293 K (the plastic zone was observed well on the specimen polished side surface; at 113 K plastic zones were not observed) upon the level of cyclic load. With an increase in the load level, the zone shape changed from that characteristic of the plane strain to that elongated along the direction of the crack propagation (typical for plane stress). Characteristic dimensions of the zone (width, b, and length, d) are described by the following formulas:

$$b=b_0\left(\frac{K_{max}}{\sigma_y}\right)^{n_1},\quad d=d_0\left(\frac{K_{max}}{\sigma_y}\right)^{n_2}. \tag{30}$$

At $b\approx d\approx B$ (B is the specimen thickness), the maximum load

$$K_{max}\approx\sqrt{K_{th}K_{fc}} \tag{31}$$

which approximately corresponds to the intersection point of the curves in Fig.18. It follows from Fig.19 that the known Irwin's expression for the plastic zone radius is valid in a wide range of loads K_{max} if the greater dimension, b or d, is taken for the plastic zone diameter $2r$. The inclination of the normal of the fracture surface plane to the maximum tensile stress is consistent with the plastic zone evolution in the process of the load increase, Fig. 20. The angle of this inclination α is equal to zero (i.e. the crack plane coincides with the plane of the maximum tensile stress) only at low rates of the crack growth, less than10^{-5}mm/cycle, when the plastic zone is small as compared to the specimen thickness. As the load increases, this angle increases too, especially rapidly at the crack rate higher than 10^{-3} mm/cycle (Fig. 20, curve 1). At low temperatures the angle α is rather small in the whole range of the crack rates.

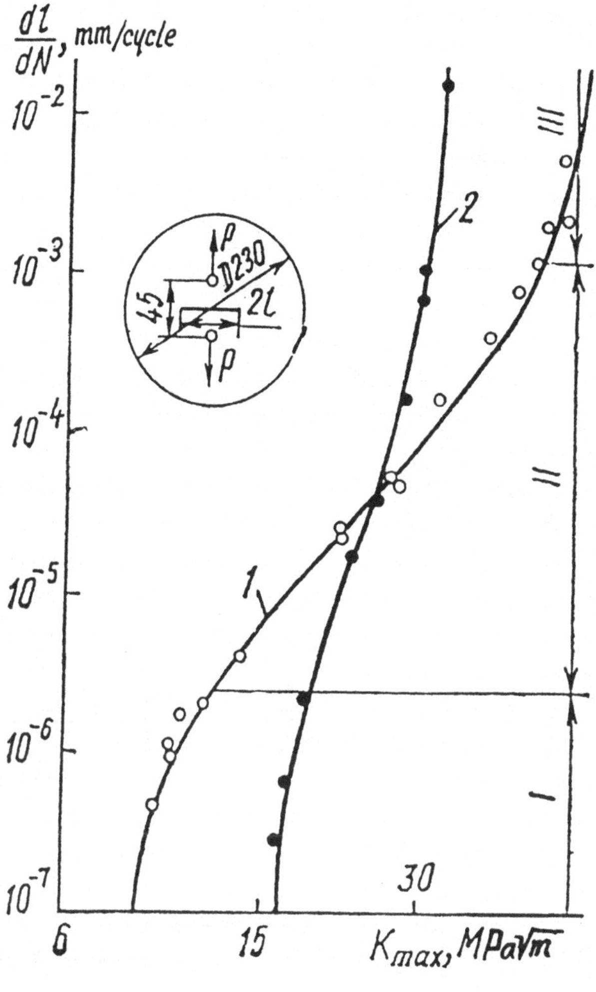

Figure 18. Fatigue crack growth rate diagrams for the low-carbon (0.08 % C) steel at temperatures 293 K (1) and 113 K (2). Here and in *Figure 19* the roman figures are related to the typical parts of the FCG-diagram.

Apart from the above mentioned two extreme temperatures, experiments were also carried out at intermediate temperatures. The temperature dependences of the fatigue crack growth rate at different values of K_{max} are presented in Fig. 21. At low load levels (curves 1 and 2) the crack growth rate decreases monotonically with temperature down to its complete arrest. The lower the K_{max} value, the higher the temperature of such an arrest. At $K_{max} \geq 22.8$ MPa$\sqrt{m}$ (curves 3 through 6), the dependences become nonmonotonic: with a decrease in temperature the crack rate first decreases but at temperatures below 160K it increases again, and increases the faster, the higher is the

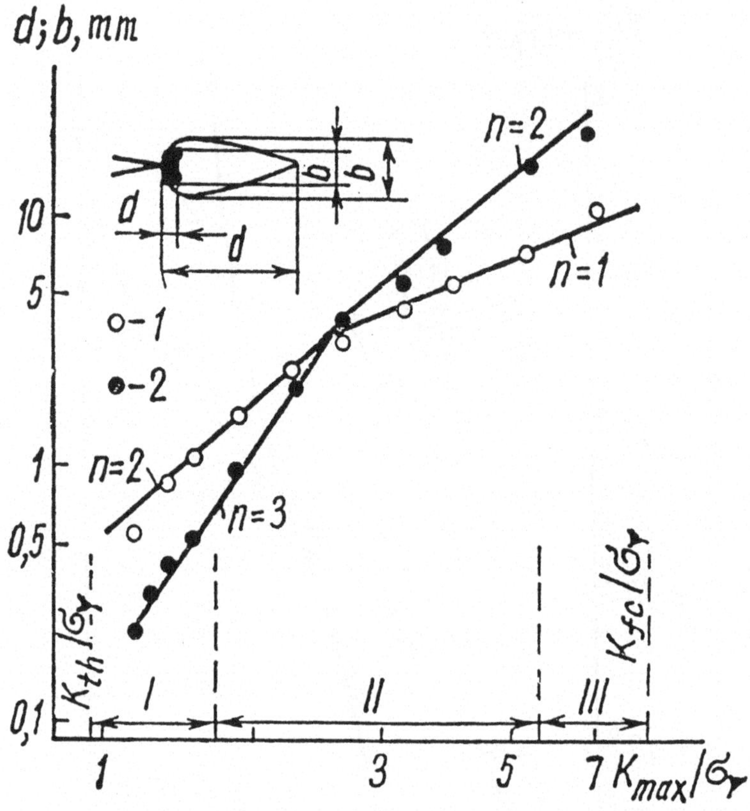

Figure 19. **Dependence of the width, b, (1) and the length, d, (2) of the plastic zone at the fatigue crack tip on the loading parameter K_{max}/σ_y at temperature 293 K.**

K_{max} value. At this temperature the 54 MPa$\sqrt{m}$ load (curve 6) results in an instantaneous failure of the whole specimen. All these peculiarities will be explained below when describing the mechanism of the fatigue crack growth.

3.1.2. Some Analytical Approximations of the FCG Diagram.

According to Forsight the FCG process is divided into two stages. The 1-st stage is associated with the crack nuclea propagation in the slip plane, the 2-nd one with its growth in the plane close to or coinciding with the plane of the maximum tensile stress.

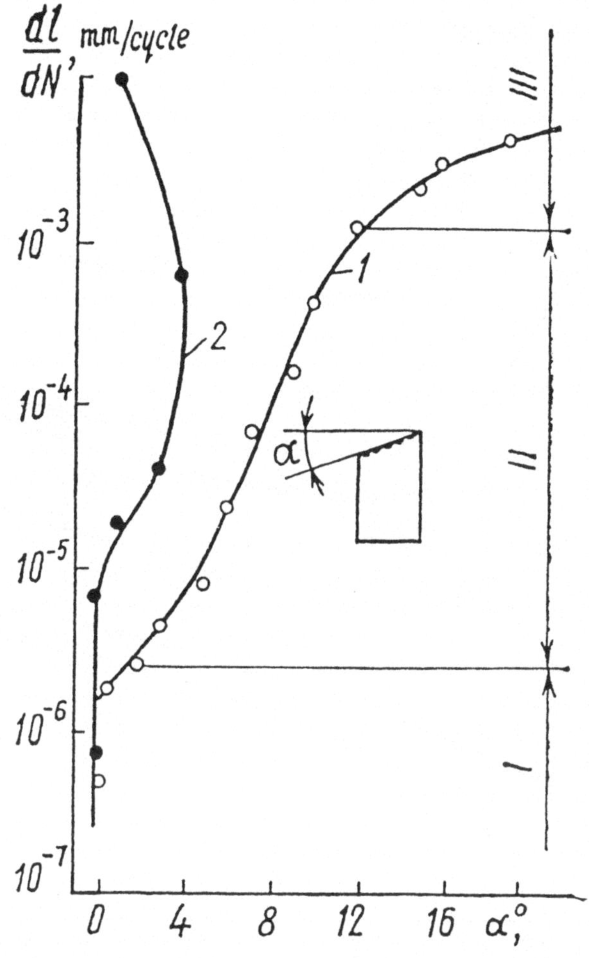

Figure 20. Dependence of the angle α, characterizing the slope of the fracture surface to the maximum tensile stress plane on the fatigue crack growth rate for low-carbon steel at temperatures 293 K (1) and 113 K (2).

According to Ref. [34], simple assumptions that the crack propagation at stage 1 is governed by the energy dissipated during the cyclic reverse shear in the slip plane whereas at stage 2 by the elastic strain energy being released, give two extreme assessments of the FCG rates. In one case, the energy must be proportional to ΔK_{II}^2 (ΔK_{II} is the SIF range of the mode II crack), and to ΔK_I^6 in the other one. These two

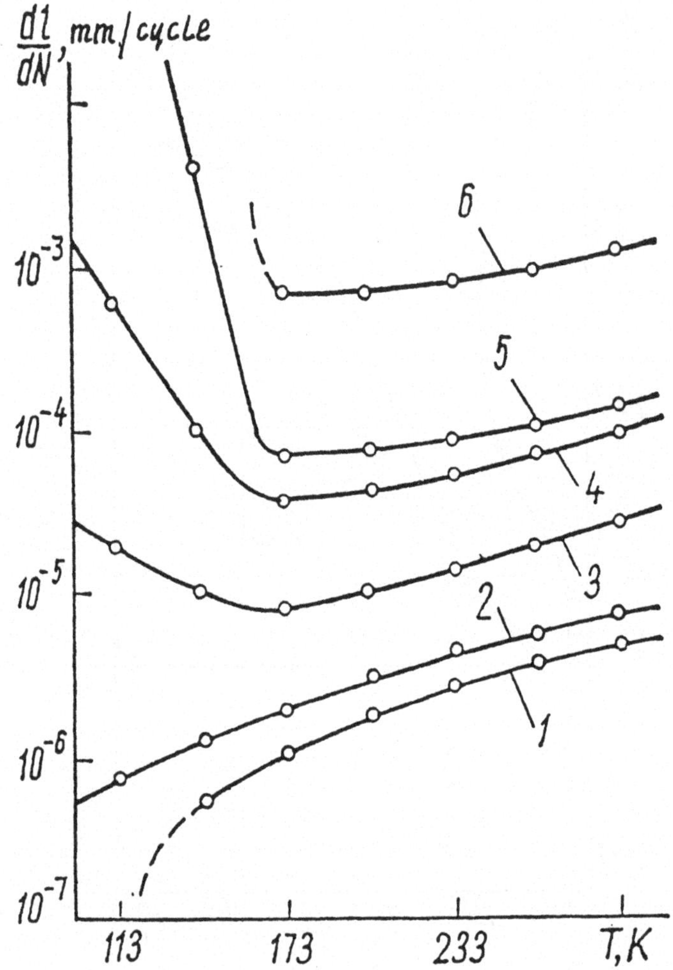

Figure 21. Relation between the fatigue crack growth rate and the test temperature for low-carbon steel at different levels of the maximum stress intensity factor, K_{max}, MPa√m: 1-15.2; 2-17.7; 3-23.6; 4-30.6; 5-34.4; 6-54.0.

simplified assessments give the expected interval of the crack growth rates:

$$C_1 \Delta K_I^2 \leq \frac{dl}{dN} \leq C_2 \Delta K_I^6 . \tag{32}$$

The well known power dependence to describe the middle part of the FCG diagram

$$\frac{dl}{dN} = C_3 \Delta K_I^4 \tag{33}$$

was proposed by Paris and Erdogan. Since then a great deal of experimental data have been obtained on the fatigue crack growth. As a rule, the middle portion of the diagram is well described by the power dependence but the exponent values can differ from four. Therefore now this formula is used in the following form:

$$\frac{dl}{dN} = C_4 \Delta K_I^{n_4} \; . \tag{34}$$

As was shown earlier by the authors of [35], the values of n_4 for different materials can differ greatly from 4. However, and what is all the more interesting, a rather strong linear correlation between n_4 and $\lg C_4$ for different classes of materials was found in that work. As an example of such a correlation we refer to Fig. 22. As was shown in Ref.[23], such a correlation cannot be explained by the structure of Eq.(34). A physical reason which should be clarified must exist for it. Similar correlation for the power law of creep [23] was explained from the standpoint of thermodynamics of the plastic flow process.

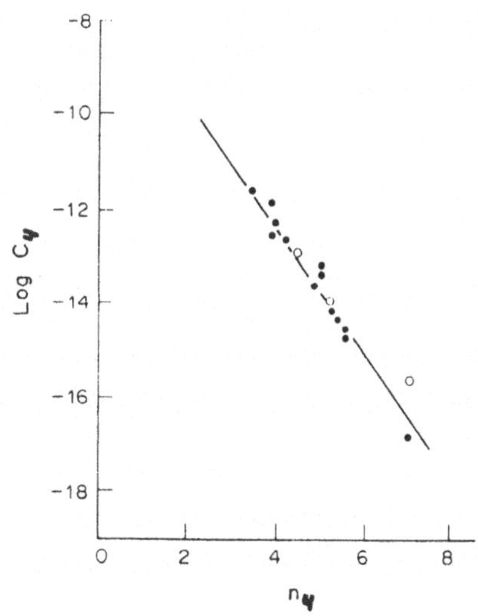

Figure 22. Linear correlation between parameters n_4 and $lg\ C_4$ of Eq.(34) for three nodular cast irons.

One of obvious disadvantages to Eq.(34) is the neglect of the stress ratio. This disadvantage can be eliminated to some extent if the effective SIF value is used in it instead of the SIF range, the former being defined as:

$$\Delta K_{I_{ef}} = K_{I_{max}} - K_{I_{op}} \tag{35}$$

where $K_{I_{op}}$ is the value of the SIF corresponding to the moment of the crack opening in a loading half cycle. At present, several hundreds of formulas have been proposed to describe the FCG diagram. Their review is outside the scope of this lecture. Here we merely present two simple formulas which claim description of a wider range of the crack rates. These are the Forman's formula which includes the stress ratio, R:

$$\frac{dl}{dN} = A \frac{\Delta K_I^n}{(1-R)K_c - \Delta K} \tag{36}$$

where A and n are the material constants, and the formula by Yarema - Mikitishin:

$$\frac{dl}{dN} = B \left(\frac{\Delta K_I - \Delta K_{th}}{K_{fc} - \Delta K_I} \right)^q \tag{37}$$

where B and q are the material constants. Equation (37) is recommended for positive values of the stress ratio, $R \geq 0$. By its use, the approximation of the experimental data was made in Fig.18 (solid lines) for the following values of parameters of Eq.(37) (the first figure refers to a temperature of 293K, the second one to a temperature of 113K): $K_{th} = 7.8$ and 15.4 MPa$\sqrt{\text{m}}$; K_{fc}=69.4 and 36.9 MPa$\sqrt{\text{m}}$; $B = 1.90 \times 10^{-4}$ and 6.84×10^{-5} mm/cycle; q =1.69 and 2.32.

Originally, in the above and other formulas by the range $\Delta K_I = K_{I_{max}} - K_{I_{min}}$ was meant one of the characteristics of the load cycle, i.e. the SIF range. However, the situation here is more complicated: at some moment of unloading, the crack surfaces can come into contact after which the crack no longer exists and further unloading has no influence on the situation at the crack tip. In other words, after the crack closed at the value of $K_I = K_{cl}$ prior to the K_{min}-value and then at a repeated load up to the moment of the crack opening at $K_I = K_{op}$, this part of the cycle has no effect on the situation at the crack tip. Elber [36] proposed to exclude it from the consideration by introducing the effective range of the SIF according to Eq.(35):

$$\Delta K_{ef} = (0.5 + 0.4R)\Delta K \tag{38}$$

This formula is not quite exact since ΔK_{cl} and ΔK_{op} were not calculated by Elber exactly. The problem of the ΔK_{ef} definition was discussed in many works whose survey is outside the scope of this lecture. It follows from those works that the phenomenon of the crack closure is complex by itself and does not exhaust the whole problem of the ΔK_{ef} definition. Thus, Broek [37], when analyzing the fatigue crack growth, considers it physically incorrect to take into account only that part of the cycle during which the crack is open.

The effect of crack closure is related to crack retardation after a single overload. This phenomenon is rather interesting but even more complex. Its practical importance is also obvious. The survey of works on the effect of a single overload on the crack retardation is given is Refs [37] through [39].

3.1.3. Fracture Mechanics Parameters Governing the FCG.

It was shown above, Fig. 18, that Eq.(36) wherein the SIF range is the main parameter of the fracture mechanics, approximates well experimental data on the FCG. However, this can be considered as a good fit. In fact, as is seen from Figs. 13 and 20, the use of the ΔK value has a rigorous substantiation only for a temperature of 113 K since in this case the crack is propagating in the plane of maximum tensile stress and the plastic zone is negligible. For a temperature of 293 K neither of these conditions is met, therefore the values of ΔK or $K_{I_{max}}$ should be considered here as conventional ones. In similar cases, the parameters of nonlinear fracture mechanics such as the crack opening displacement range, $\Delta\delta$, or the J - integral range, ΔJ, should be used. The use of these criteria to describe the FCG in the range of high loads or small crack lengths yields better results [40].

3.1.4. The Impact of Different Factors on the FCG.

In section 2.2 we have considered the influence of different factors on the conventional fatigue of materials. The same factors also influence the FCG. It is well known that even humidity of the laboratory ambient air can accelerate the fatigue crack growth by an order of magnitude as compared to its growth rate in inert medium or in vacuum (see Fig. 16). For this reason, the control over the environment wherein the FCG test are carried out is of great importance. However, the environment has different effects on different materials, therefore, in practice such influence is difficult to predict. One of the main characteristics of the material connected with its sensitivity to the environment is the threshold value, K_{Iscc}, as a maximum value of the SIF at which the crack growth is not yet observed under conditions of stress corrosion. This characteristic is also of great importance for the corrosion fatigue. Depending on the material and the type of the corrosive environment it can vary over wide limits. Figure16 illustrates the case when $K_{I_{scc}} > K_{th}$ and the crack acceleration begins in humid air at $K_{\max} \geq K_{I_{scc}}$. Another important factor is the specimen thickness. It has a pronounced effect on the stress-strain state at the crack tip. Depending upon the ratio of the plastic zone size to the specimen thickness, the situation at the crack tip can approach either the plane strain (as at 113 K in Figs. 18-20) or the plane stress (as at 293 K in the same Figures).

The effect of temperature on the FCG rate is illustrated by Fig.21. As will be seen below when considering the fractographic pictures, this influence is also manifested itself ambiguously and depends upon such factors as the load level, strain rate, the material structure, etc. At high temperatures the processes of creep and void growth can contribute significantly to the fatigue crack growth rate [41].

As in the case of smooth specimens the effect of loading frequency shows itself through the rate sensitivity of the material and through the material local heating at the crack tip. The influence of frequency on the FCG rate also manifests itself in combination with the effect of corrosive environment as well. Since corrosion is a time-dependent process, the effect of frequency on this process is dependent on the duration of the crack tip exposure to maximum load in each cycle.

3.2. PART - THROUGH CRACK GROWTH

This problem has been considered in Ref. [42] with the main results as follows.

Consider a mode I part-through crack of area S with a curved front Γ. Let R_i, θ_i be polar coordinates in a crack plane which determine the position of the arbitrary crack front point Q_i. The stress state near the crack front is characterized by the stress intensity factor distribution.

One can write the fatigue crack growth equations as follows:

$$dR_i/dN = F\left(\Delta K_i^*; C_{m,i}^*\right) \tag{39}$$

where ΔK^* is the fracture mechanics parameter taken to be a fatigue crack growth criterion, ΔK_i^* is the ΔK^* value at point Q_i, $C^*_{m,i}$ are the empirical constants and F is the function defined from test results for standard specimen with a through-the-thickness crack of length l:

$$dl/dN = F(\Delta K; C_m) \ .$$

The crack front location can be approximately characterized by a finite number of degrees of freedom. For instance, known works cited in [42] deal with semi-elliptical and quarter- elliptical cracks. In such cases one has a two-parametric crack front model outlined by a system of equations

$$\begin{aligned} da/dN &= F\left(\Delta K_A^*; C_{m,A}^*\right), \ i = 1, 2, \ldots \\ db/dN &= F\left(\Delta K_B^*; C_{m,B}^*\right), \ i = 1, 2, \ldots \end{aligned} \tag{40}$$

where a and b are the ellipse half-axes (the half-axis a is the crack depth), and A and B are the corresponding half-axis vertices.

The use of Eq.(40) for the part-through crack growth prediction brings up two matters of principle. First, the correctness of the actual flaw approximation by a part-elliptical crack has to be verified, as well as errors brought up by such approximation have to be estimated. Second, it should be found which linear-elastic fracture mechanics parameter ΔK^* is preferable as the fatigue crack growth criterion.

There are weighty enough reasons for using part- elliptical cracks in fracture analysis of structural components. Thus in well-known recommendations [43] the rules of real flaw replacement by part-elliptical cracks are worked out since such schematization simplifies appreciably the stress intensity factor computation procedure. Moreover, it is known that a part-through crack comes to a regular form during cyclic loading. For example, from experimental studies of part-through fatigue crack propagation one can see that the shape of a surface crack in a plate and a corner crack at a plate edge or at stress concentrators is close to being semi-elliptical and quarter-elliptical, respectively. The results of predicting the rate and shape changes of semi- and quarter- elliptical cracks during cyclic loading obtained on the basis of the model, Eq.(40), are usually in good agreement with the experimental data [42].

The ΔK^* parameter of Eqs (40) can be calculated in different ways. Furthermore, the three most commonly used criteria which can be applied for part-through fatigue crack growth prediction are studied.

(a) The local stress intensity factor range, ΔK . Using this criterion, the crack growth rate at an arbitrary front point depends on the stress intensity factor range at that point.

(b) The averaged stress intensity factor range, $\Delta\overline{K}$. Using this criterion, the crack growth rate is a function of the averaged energy release rate with the crack front translation. The $\Delta\overline{K}$ values at the vertices of the part- elliptical crack are defined as follows:

$$\begin{aligned} \overline{K}_A &= \left[(\delta S_A)^{-1} \int_\Gamma K^2 d(\delta S_A) \right]^{1/2} \\ \overline{K}_B &= \left[(\delta S_B)^{-1} \int_\Gamma K^2 d(\delta S_B) \right]^{1/2} \end{aligned} \tag{41}$$

where δS_A and δS_B are the crack surface increments due to the corresponding half-axis length variations.

(c) The effective stress intensity factor range, ΔK_e. This value is mostly related to the crack closure phenomenon. The experimental data for a surface flaw show the influence of the crack closure on the crack front point location. At the same time, the relationship between the ΔK_e value, on the one hand, and the micro- and macro-features of the material behaviour near the crack front (the variation in strain constraint and in stress asymptotics along the crack front, the change of the material crack growth resistanse over the specimen thickness and the redistribution of stresses due to plastic strains, etc.), on the other hand, is not found. Here we do not discuss the physical nature of the ΔK_e value.That is why within the framework of the macro-mechanics of fracture one can take the following definition of the effective stress intensity factor range at the crack front point Q_i:

$$\Delta K_{e,i} = \gamma_i \, \Delta K_i , \tag{42}$$

where coefficients γ_i are determined on the basis of experimental data from the condition that through-the-thickness and part-through crack growth rates have to coincide with equal effective stress intensity factor ranges.

Below, on the basis of the experimental data of Ref. [42], we describe a comparative analysis of the applicability of the local, averaged and effective stress intensity factor ranges for predicting the fatigue growth of mode I surface and corner cracks.

Crack propagation was studied on specimens made of low-alloy pressure vessel steel 15X2MΦA ($\sigma_{0.2}$ = 584 MPa, σ_u = 700 MPa, δ = 21%, ψ = 74.6%). The experiments were carried out in four-point bending fatigue at a frequency of 10 Hz and a stress ratio R = 0.32, in air at room temperature. The range of crack growth rate analysed varied within the Paris region (middle part of the fatigue crack growth diagram).Four types of part-through cracked specimen were used: a plate with a surface crack; a plate with corner cracks; a plate with a corner crack at a hole; a plate with corner cracks at a part-circular side notches. The maximum nominal stress value in a cycle for plates with part-through cracks varied from 0.38 to 0.9 of the yield stress.

The ΔK, $\Delta \bar{K}$ and ΔK_e parameters are determined in accordance with Egs.(41) and (42) from the known stress intensity factor distribution along the crack front. The SIF analysis of surface- and corner-cracked specimens was performed by using the weight function method. This method makes it possible to calculate the stress intensity factor of a part-elliptical crack without solving a boundary-value problem if the configuration function , f, and nominal stress distribution, σ, in the crack plane are known.

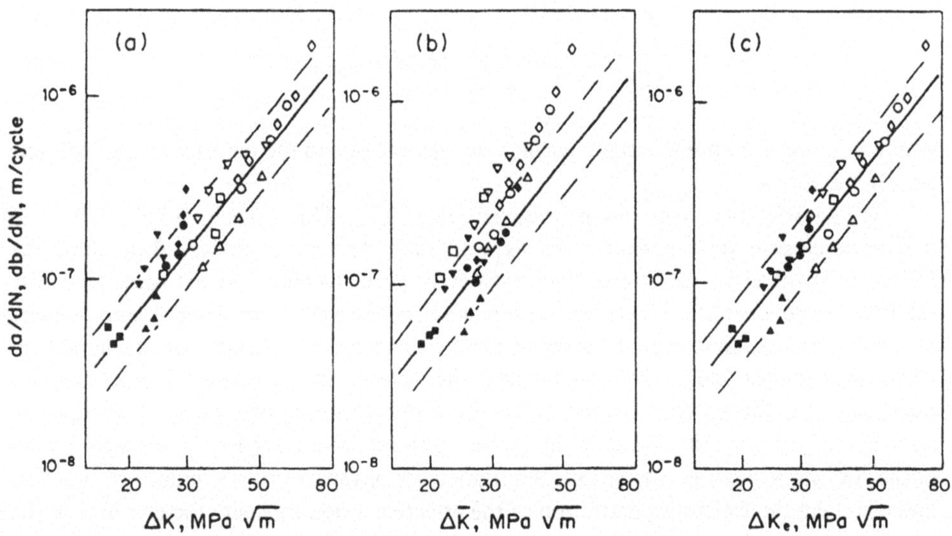

Figure 23. The surface crack growth rate *da/dN* (filled symbols) and *db/dN* (open symbols) versus the range of local (a), averaged (b) and effective (c) stress intensity factor values. Dashed and solid lines are the scatter band of the results for standard specimens and their approximation by Eq.(34) [42].

Figures 23 and 24 present fatigue crack growth rates at the deepest point, da/dN, and surface point, db/dN, for surface and corner flaws plotted versus the ΔK, $\Delta\overline{K}$ and ΔK_e values. Different symbols in Fig. 23 correspond to five surface-cracked specimens and in Fig. 24 to different types of corner-cracked specimens; solid and open symbols are the crack growth rates da/dN and db/dN, respectively. The dashed solid lines represent a scatter band of experimental results for standard specimens and their approximation by the Paris formula. The dl/dN versus ΔK data for standard specimens in the range of ΔK = 20-50 Mpa$\sqrt{\text{m}}$ are approximated by Eq.(34) in the form

$$dl/dN = C(\Delta K)^n \text{, m/cycle} \tag{43}$$

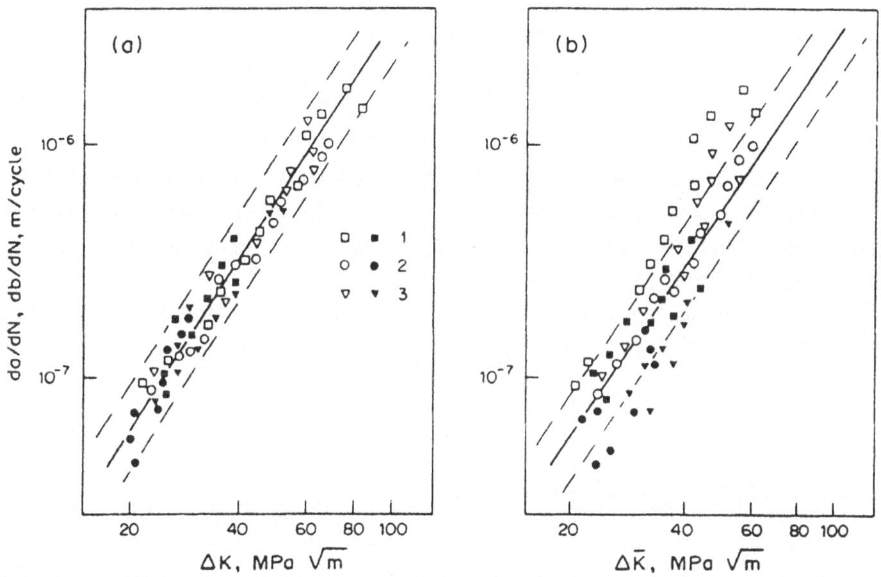

Figure 24. The corner crack growth rates *da/dN* (filled symbols) and *db/dN* (open symbols) versus the range of local (a) and averaged (b) stress intensity factor values. Dashed and solid lines are the scatter band of the results for standard specimens and their approximation by the Paris formula. 1, the crack at the plate edge; 2, the crack at a hole; 3, the crack at a side notch [42].

where l is the edge crack length, C = 2.96 $\times 10^{-11}$ m/cycle and n = 2.54. Note that the standard specimen test results revealed no anisotropy of the material fatigue crack growth properties.

The analysis of the experimental results on the surface crack growth shows that the db/dN versus ΔK_B data (Fig.23a) virtually coincide with the standard specimen test results. At the same time, with equal stress intensity factor ranges, the rate of surface crack propagation into the specimen depth has a larger value than the one over the specimen boundary.

If the averaged stress intensity factor range is used as the fatigue crack growth criterion, the majority of experimental points lie beyond the scatter band of the standard specimen data and the da/dN , db/dN versus ΔK and da/dN , db/dN versus $\Delta\overline{K}$ approximation curves have appreciably different slopes (Fig. 23b). With the above interpretation, the application of the averaged stress intensity factor range leads to essential errors when predicting the surface crack growth rate.

Consider the possibility of employing the effective stress intensity factor range as the surface crack growth criterion. To determine the γ_A and γ_B coefficients of Eq.(42), let us write the crack growth equations which correspond to the two-parametric crack front model:

$$da/dN = C_A\left(\Delta K_A\right)^n = C\left(\gamma_A \Delta K_A\right)^n \tag{44}$$

$$db/dN = C_B\left(\Delta K_B\right)^n = C\left(\gamma_B \Delta K_B\right)^n \tag{45}$$

Similarly to most known works, we assume the exponent n to be independend of the crack front point position and to be equal to a corresponding value found from standard specimen test results; this assumption is confirmed above when analysing the fatigue crack growth data in Fig. 23a.

Taking account of Eqs (44) and (45), a condition of the coincidence of surface crack growth rates in the depth and length directions has the form

$$C_A/C_B = \left(\gamma_A/\gamma_B\right)^n$$

from which the following relation is obtained:

$$\eta = \gamma_A/\gamma_B = \left(C_A/C_B\right)^{1/n}. \tag{46}$$

From the data in Fig. 23a we have $\gamma_B = 1$. The coefficient η, which characterizes

TABLE 3. The values of the coefficient η for the specimens with surface cracks [42]

Specimen number	Load level, $\sigma_{nom}/\sigma_{0.2}$	η	Average value, $\bar{\eta}$
1.2	0.65	1.089	
1.4	0.78	1.096	1.100
1.5	0.70	1.101	
1.6	0.90	1.113	

the difference in the surface crack growth rates da/dN and db/dN at $\Delta K_A = \Delta K_B$, was determined from Eq.(46) with the use of Eqs (44) and (45) obtained preliminarily by the least-square technique. The η values found in such a way for each specimen tested are listed in Table 3. The η values vary from 1.089 to 1.113 and are actually independent of the nominal stress level, while the average value is $\bar{\eta} = 1.1$.

Figure 23c presents da/dN , db/dN versus ΔK_e data obtained with the γ_A and γ_B coefficients found above. All the experimental points in Fig.23c are within the scatter band of the data for standard specimens and they group arround the line described by Eq. (43).

Thus the use of the effective stress intensity factor range in the prediction of the surface crack propagation under cyclic loading is preferable. The difference in surface crack growth rates in the depth and length directions with equal stress intensity factor values is taken into account by the factors γ_A and γ_B . In accordance with the work [42], the ratio of the values γ_A and γ_B is equal to $\eta = 1.1$. Table 4 gives a comparison of the $\bar{\eta}$ value obtained by the authors with the known results, as well as the estimates obtained on the basic of experimental data [42] using the technique described above. The results in Table 4 were obtained for the crack propagation rates corresponding to the Paris region of the fatigue crack growth curve.

Figure 24 presents fatigue crack growth data for three types of corner-cracked specimens. Experimental results for da/dN versus ΔK_A and db/dN versus ΔK_B (Fig. 24a) show an insignificant deviation and coincidence within a scatter band with the corresponding data for standard specimens. According to Fig. 24a, in the definition of the effective stress intensity factor range for a quarter-elliptical crack, one has to take

$$\gamma_A = \gamma_B = 1$$

Hence the fatigue crack growth data da/dN, db/dN versus ΔK and da/dN, db/dN versus ΔK_e for a corner flaw coincide. Similar to a surface flaw, the application of the averaged stress intensity factor for predicting the corner crack propagation leads to significant errors in the estimation of the fatigue crack growth rate (see Fig. 24b).

Thus the use of the effective stress intensity factor range as a fatigue crack growth criterion is advisable in predicting part-through crack propagation under cyclic loading.

TABLE 4. Estimates of the coefficients for different materials [42]

Material	Loading	Stress ratio, R	η
2024-T351	tension	0.1	1.098
12ГН2МФАЮ Steel	bending	0.5	1.131
PMMA	bending	0.1	1.270
Titanium alloy, nickel steel, aluminum alloys	tension, bending	0.0	1.111
BS 4360 50B	eccentric tension	0.05	1.133
15X2МФА Steel	bending	0.32	1.100

Within the middle part of the fatigue crack growth curve and for the ranges of thenominal stress and stress ratio studied (see Tables 2-4) the coefficients γ_A and γ_B for metallic materials are related by $\gamma_A = 1.1\gamma_B$ for a corner flaw. For the case when thethickness of a standard specimen and a part-through-cracked plate are close, in the present paper γ_B was found to be equal to 1.

Figure 25 shows the surface and corner crack aspect ratio (a/b) as a function of the dimensionless crack depth (a/h) for four types of specimens tested. Different points on each diagram of Fig.25 correspond to test results for specimens of the same configuration but with different nominal stress levels (see Table 2) and initial flaw dimensions (a_0, b_0). As one can see from Fig.25, irrespective of the initial crack configuration defined by the dimensionless parameters $(a/b)_0$ and $(a/h)_0$, the crack shape tends to a stable state during fatigue growth; this fact is in agreement with the known results [42]. Note from Fig.25 that the crack shape parameter corresponding to the stable front position is independent of the load level for the σ_{nom} range studied.

Experimental and theoretical analyses of the part-through crack shape variation and stable crack growth patterns have been of great interest for many authors. On the basis of the experimental data a number of analytical expressions have been derived for the semi-elliptical crack stable shape under cyclic tension and bending. Some results for the bending load are summarized in Table 5 [Eqs.(47)-(53)] and in Fig.26; Eq.(53), which approximates the data of the authors [42] (Fig. 25a), is also given. Note that all the presented results are valid within the middle, i.e. II, region of the fatigue crack growth curve. The scatter of the reviewed expressions for the stable crack shape is apparently stipulated by the differences in the material properties and test conditions.

TABLE 5. Stable shape equations for the semi-elliptical crack under cyclic bending [42]

Material	Parameter n of the Eqs (44),(45)	Stress ratio, R	Load level, $\sigma_{nom}/\sigma_{0.2}$	Equation
steels M-41, HT-60, HT-80	3	0	0.47-0.73	a/b = 1.05a/h (47)
E 36 steel	3-4	0.1-0.5	-	a/b = 0.84-0.86 a/h (48)
low alloy steel	3.3	0.04; 0.19	0.45-0.53	a/b = 0.97-1.29 a/h (49)
-	-	-	-	a/b = 1.07-1.05 a/h (49a)
12ГН2МФАЮ Steel	4.6	0.5	0.48-0.72	a/b = 0.85-0.83 a/h (50)
09G2 steel	-	0; 0.37	0.55-0.88	a/b = 0.94-0.89 a/h (51)
Cr-Ni-Mo low alloy steel	-	0	0.60-0.80	a/b = 0.96-0.78 a/h (52)
15X2MФA Steel	2.54	0.32	0.38-0.90	a/b = 1.03-0.98 a/h (53)

In view of the peculiarities of the reference solutions for part-through crack stable shapes, the problem of predicting the crack shape change, with account taken of the material properties and loading conditions, turns out to be important. The prediction of the part-through crack growth has to be based on the use of the substantiated crack growth model and criterion, the results of the stress intensity factor computation and the material fracture resistance properties found in testing standard specimens.

A two-parametric crack front model, Eq.(40), has been used in predicting the part-through flaw propagation under cyclic loading. For the case when the surface and corner crack development at the points A and B is described by the Paris expression, Eqs (40) lead to

$$
\begin{aligned}
da/dN &= C_A^*\left(\Delta K_A^*\right)^n \\
db/dN &= C_B^*\left(\Delta K_B^*\right)^n
\end{aligned}
\qquad (54)
$$

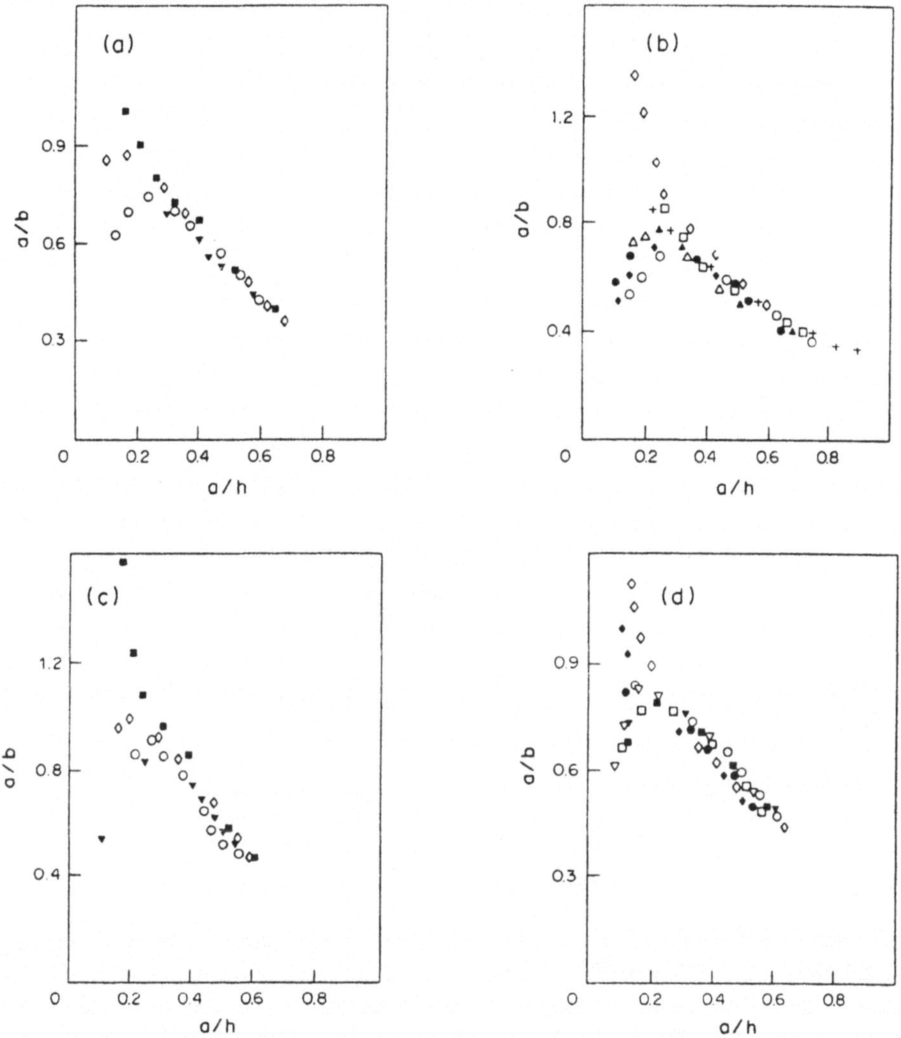

Figure 25. **Part-through crack shape variation under cyclic loading: (a) a plate with a surface crack; (b) a plate with corner cracks; (c) a plate with a corner crack at a hole; (d) a plate with corner cracks at part-circular side notches.**

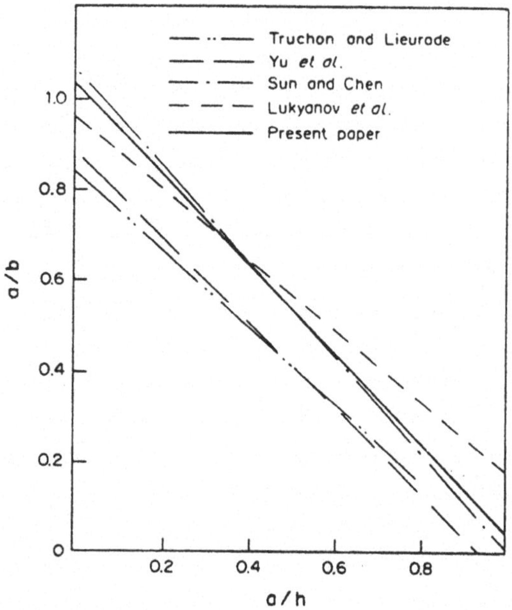

Figure 26. Stable crack growth patterns for the surface flaw in a plate under cyclic bending.

Having eliminated parameter N from Eq.(54), an equation is obtained,

$$\frac{d(a/b)}{d(a/h)}=\frac{(a/b)}{(a/h)}\left[1-(a/b)\frac{C_B^*}{C_A^*}\left(\frac{\Delta K_B^*}{\Delta K_A^*}\right)^n\right] \tag{55}$$

which describes the change in the surface and corner crack shapes under cyclic loading. The solution of Eq.(55) depends in particular upon the value of the exponent n; this fact explains the difference in the known experimental data on stable crack shapes (see Table 4 and Fig. 26). Figure 27 shows predicted curves of the change in surface and corner crack shapes for the four types of specimens studied in Ref. [42]. The results were obtained by the numerical integration of Eq.(55) under different initial conditions and with the use of the local, averaged and effective stress intensity factor ranges as fatigue crack growth criteria; the exponent in the Paris expression was taken to be $n = 2.54$ [see Eq.(43)]. Note that for a corner flaw an equality $\Delta K = \Delta K_e$ is observed, and the corresponding solutions of Eq.(55) found with the use of the local and effective stress intensity factor ranges coincide. Figure 27 demonstrates that with an increase in the parameter (a/h) the solutions of Eq.(55) for different initial conditions tend asymptotically to a limiting curve which corresponds to a stable crack shape. Thus the stable crack shape can be defined analytically as an asymptotic curve of the solutions of Eq.(55) under various initial conditions.

In Fig.28 the experimental data on the surface and corner crack shape variation in specimens of four types are compared with the stable shapes obtained by solving Eq.(55) with the use of the local, averaged and effective stress intensity factor ranges as ΔK^*. (In plotting the stable pattern for the corner crack, the symmetry condition as $a/h \rightarrow 0$ has to be considered; that is why a/b was taken equal to 1 with $a/h = 0$). It is evident from Fig.28 that the use of the ΔK_e criterion gives the best results when predicting a part-through crack shape. In this case, the experimental data actually coincide with the calculated curve up to $a/h = 0.6$. In the case when $\Delta\overline{K}$ was used as

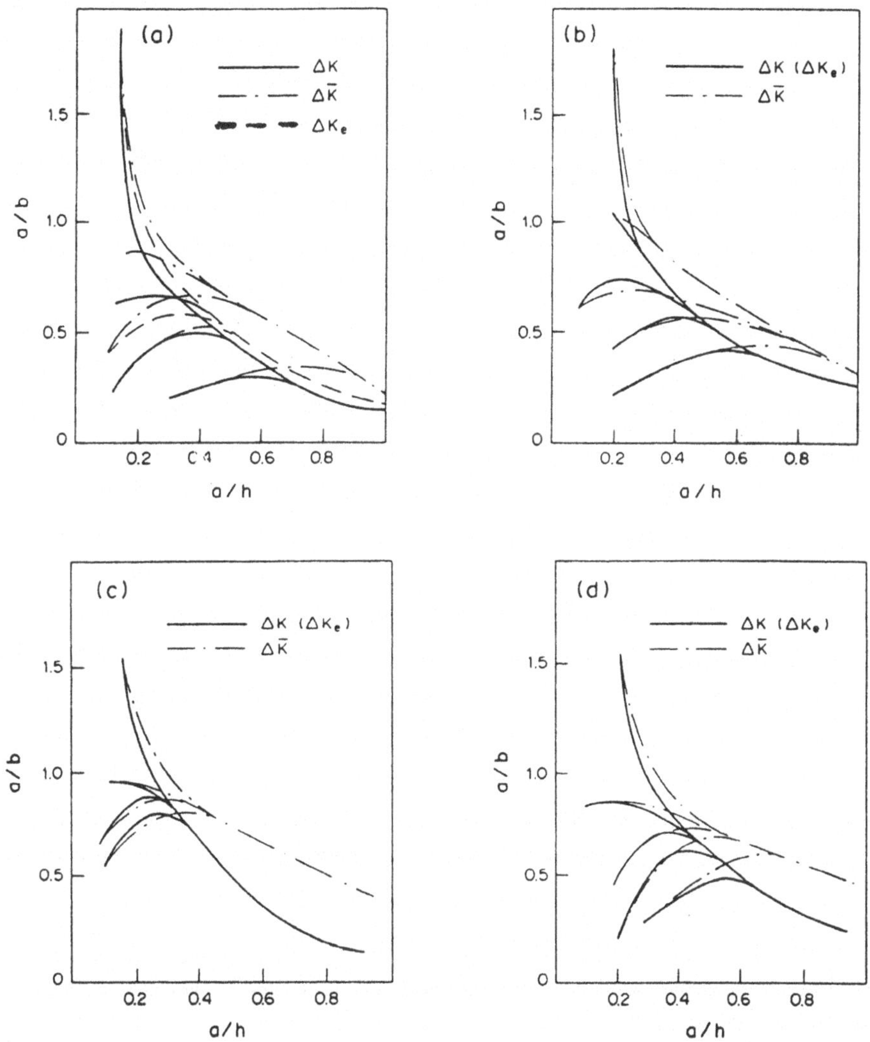

Figure 27. Predicted curves of the part-through crack shape variation. (a)-(d) are the same as in Fig.25.

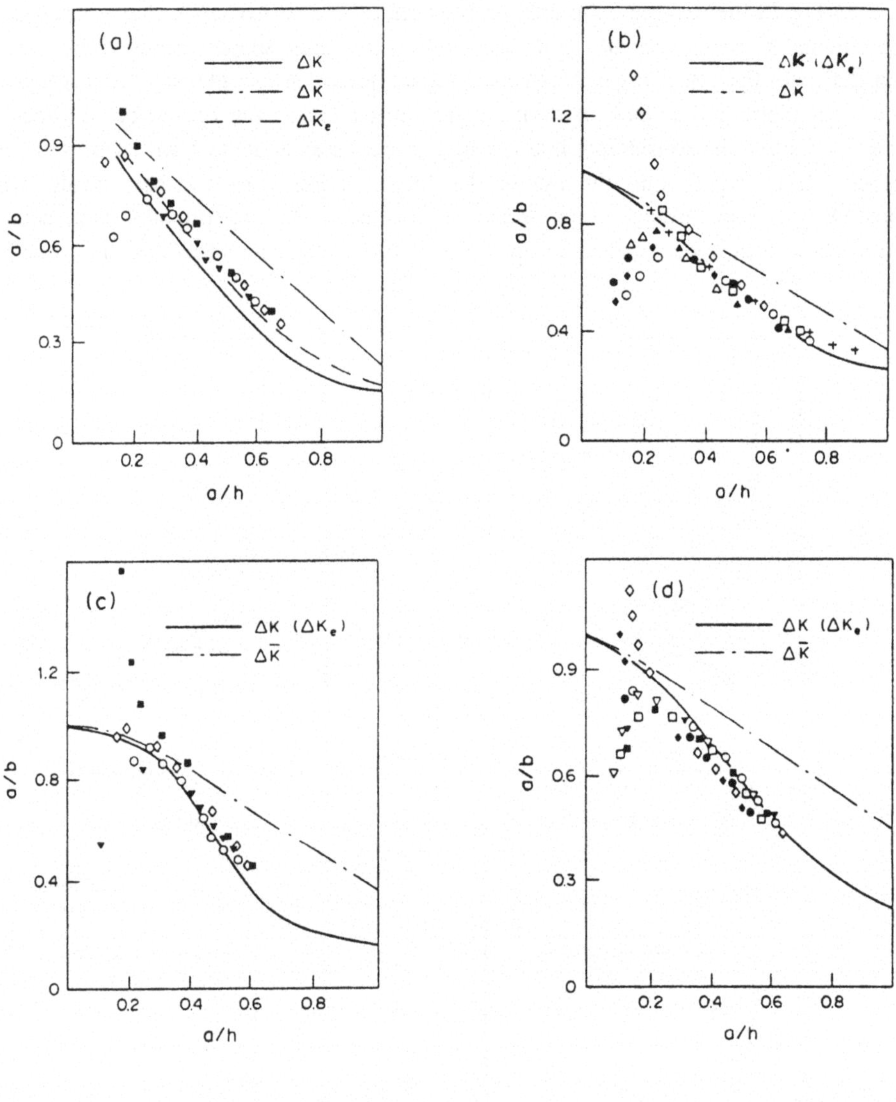

Figure 28. Comparison of the predicted stable crack growth patterns with the experimental data on the part-through crack shape variation. (a)-(d) are the same as in Fig.25.

the fatigue crack growth criterion, the predicted value of the crack aspect ratio turned out to be overestimated, while in the case when $\Delta\overline{K}$ was used (a surface crack), it was somewhat underestimated.

Similar conclusions on the applicability of various fracture mechanics parameters in predicting part-through crack growth under cyclic loading have been made above when analysing fatigue crack growth data. Therefore, the crack shape study may be used in estimating the validity of fatigue crack growth criteria. As compared to the fatigue crack growth rate analysis, the latter approach is more evident and less tedious.

Figure 27 shows an increase in the (a/b) versus (a/h) curve gradient with decreasing initial crack depth. This result is evident when analysing the behaviour of the solution of Eq.(55) with $(a/h)_0 \rightarrow 0$ and reflects the features of the part-through crack shape variation which were observed experimentally (see Fig.25). Herefrom, within a short period of time the shape of a macrocrack with a small initial depth becomes stable and it is determined by the stable shape solution. Thus, the data on the part-through crack shapes make it possible to determine the ranges of configuration parameters for a growing crack, the fact which is of practical importance. The use of stable shapes in design allows us to improve the analysis substantially and to obtain the most realistic estimates of structure safety.

When solving Eqs (54), one can find the crack growth life as a function of the crack depth, i.e. $N = N(a)$. With the use of this relationship and an appropriate failure criterion (e. g., $a = a_c$, $K_I = K_{Ic}$, $K_I = K_{Ifc}$) one can estimate the fatigue life of a structure, the duration of the in-between repair period, requirements to the NDT-control and others.

TABLE 6. Comparison of the predicted (N_{pre}) and true (N_{act}) part-through crack growth lives [42]

Specimen number	Initial and final crack dimensions (mm) a_0	a_c	Crack growth lives N_{pre}	N_{act}	N_{pre} / N_{act}
1.2	8.8	17.7	92,100	119,000	0.77
1.4	3.9	18.6	94,200	96,100	0.98
1.5	5.4	18.3	123,200	130,200	0.95
1.6	6.0	16.1	64,100	57,700	1.11
2.1	4.2	19.3	88,600	72,400	1.22
2.2	4.2	14.6	60,900	49,500	1.23
2.3	2.7	20.4	89,700	76,600	1.17
2.4	4.0	20.1	122,800	108,900	1.13
2.5	2.7	18.6	118,000	142,100	0.84
3.1	6.0	16.2	249,600	210,200	1.19
3.2	7.0	18.4	156,200	137,000	1.14
3.3	4.1	17.3	124,100	115,200	1.08
3.4	5.8	18.0	71,400	79,100	0.90
4.1	3.8	20.6	104,300	82,500	1.26
4.2	4.2	22.4	58,900	57,000	1.03
4.3	5.0	25.0	272,200	230,900	1.18
4.4	5.7	22.3	33,200	37,900	0.88

For the surface- and corner-cracked specimens, the functions $N(a)$ were determined with the use of the effective stress intensity factor range as the fatigue crack growth criterion. Table 6 compares the predicted (N_{pre}) and true (N_{act}) crack growth lives. The N_{pre} value was defined as the number of cycles for the crack growth from its initial size, a_0, to the final size, a_c, which corresponds to specimen failure or general yielding of the cracked section. The predicted crack growth lives are within 30% of the experimental results.

3.3. MECHANISMS OF THE FCG AND THE RETROSPECTIVE ANALYSIS OF FRACTURE SURFACES.

When studying the mechanisms of fatigue crack growth, the formations on the fracture surface which are specific to fatigue and referred to as fatigue striations attract the greatest attention (Fig.29). These striations are arranged normally to the local direction of the crack growth and, according to the common opinion are indicative of the crack front position at each load cycle. Fatigue striations are the most distinctive feature of fatigue failure. Their detection on the fracture surface is a documentary evidence of cyclic loading as a main cause of failure. However, the mechanism of the fatigue crack growth by the formation of fatigue striations is not the only one. As indicated by numerous fractographic investigations, depending on the material type and loading regime, the fatigue crack growth rate can also be influenced by "non-fatigue" mechanisms such as cleavage, intergranular cleavage, dimples, etc.

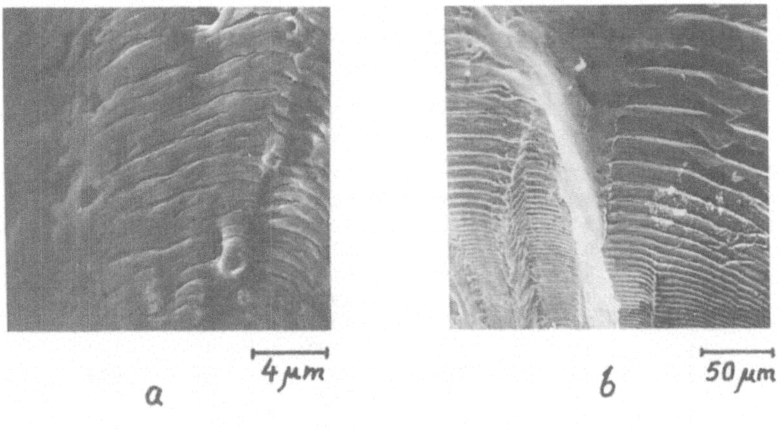

Figure 29. Fatigue striations on the fracture surface of steel (a) and nickel (b).

The authors of Ref. [44] concluded that for steels the following four types of crack growth mechanisms are generally realized: fatigue striations, microcleavage, void coalescence and intergranular rupture. The crack propagation by the fatigue striations mechanism is insensitive to the mean stress (except for low loading levels) and to the specimen thickness (except for high loading levels). The striation mechanism is usually observed at the crack growth rates corresponding to the middle part of the FCG diagram. Depending on the level of loading and the material structure, the striation mechanism is accompanied or replaced by one of the above mentioned "non-fatigue" mechanisms resulting in the crack acceleration. For example, the combination of the striation mechanism and cleavage results in the crack acceleration which depends upon the steel structure and the maximum stress. Similarly, the combination of the striation mechanism with the void coalescence mechanism results in the crack acceleration with an increase in the mean stress.

Another important observation noted in Ref. [44] is that low-strength steels and alloys show a stronger tendency to failure by the striation mechanism. The tendency toward departure from failure by the striation mechanism is usually observed with steels having low fracture toughness. The manifestations of other failure mechanisms, such as cleavage at low loading levels and ductile dimple failure at high loading amplitudes, were also observed in titanium alloys. Besides, the occurrence of the secondary cracking was observed which also resulted in the crack acceleration.

A nearly general rule for most of materials is that in the presence of corrosion environment, especially in the range of low loading levels, they show a tendency to integranular fracture. In the presence of the corrosion environment an increase in the loading frequency usually reduces the crack rate related to each loading cycle.

3.3.1. FCG Mechanisms in Low - Carbon Steel.

The results of fractographic examination of the specimens given in Figs 18 through 21 are in agreement with the above mentioned general rules. The main data of such investigation are presented in Fig.30. Here good correlation is seen between the slope of the fracture surface (Fig.20) and the portion of the fracture surface covered by dimples, particularly at 293 K. For every portion of the FCG diagram (Fig.18) there is a certain prevailing fracture mechanism which makes a major contribution to the crack growth rate. For the 1-st part of the FCG diagram ($\frac{dl}{dN} < 5 \times 10^{-6}$ mm/cycle) prevailing is a fracture surface which is specific to near-threshold crack rates; the fracture surface is quite plane with a smooth opaque appearance. For the second portion of the diagram ($5 \times 10^{-6} \leq \frac{dl}{dN} \leq 10^{-3}$ mm/cycle), the main mechanism is that of fatigue striations and at 113K it is cleavage.

The results of the fractographic examination at intermediate temperatures are presented in Fig.31. It is seen from Fig.31c that on the first part of the FCG diagram at low crack growth rates ($< 5 \times 10^{-6}$ mm/cycle) at all the temperatures, a typical near-threshold fracture mechanism dominates (we denote it a "lineage" structure) with a small number of cleavage sites, which disappear with the temperature, and striations detectable at near-room temperatures. With an increase in the crack growth rate (the second and partially the third portion of the FCG diagram in Figs 18 - 22 and Figs29 - 31(b) and (c)), the fracture microrelief of specimens tested at temperatures above and below 160 K differs considerably. In the first case prevailing are fatigue striations, in the second cleavage. From the comparison of Figs 21 and 25 it follows that at $K_{max} \geq 22.8$ MPa$\sqrt{m}$ and temperatures below 160 K prevailing on the surface are cleavage facets which is indicative of the relation between this temperature and the critical transition temperature under cyclic loading.

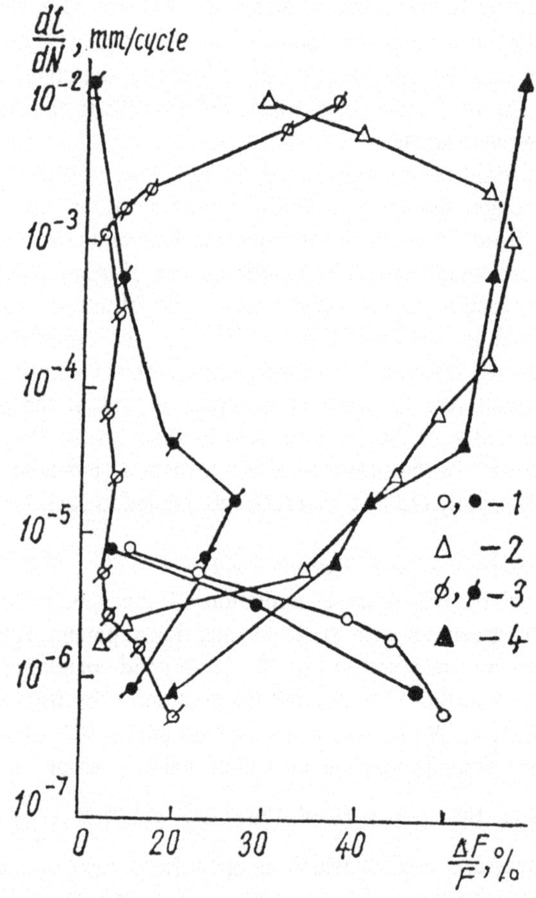

Figure 30. The fracture surface fraction, $\frac{\Delta F}{F}$%, exhibiting the appropriate failure mechanism appearance as dependent on the fatigue crack growth rate for low-carbon steel. Solid simbols are related to the temperature 113 K, open symbols - to the temperature 293 K. Failure mechanism appearance: (1) near threshold; (2) fatigue striations; (3) dimples; (4) cleavage.

The results of the measurement of mean striations spacing and its comparison with the actual crack growth rate are given in Fig.32 which shows that on the second and partially on the third part of the FCG diagram (Fig.18) a correlation is observed between the average striation spacing and the macroscopic crack growth rate. In this case, in the crack growth rate range between 10^{-5} and 10^{-3} mm/cycle, the greater part of the fracture surface is covered by fatigue striations. The results given in Figs 18

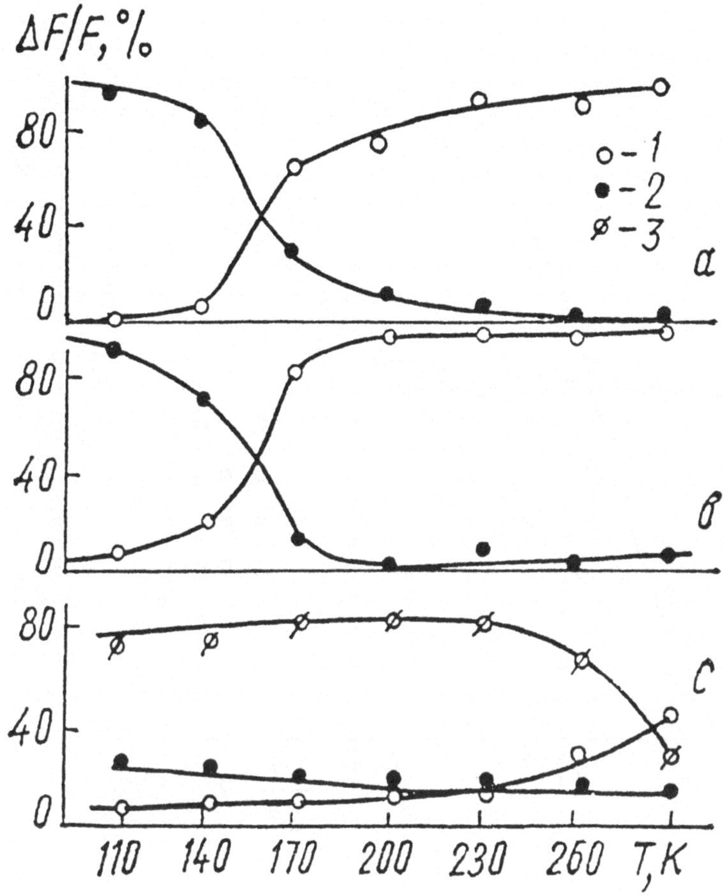

Figure 31. The fracture surface fraction, $\frac{\Delta F}{F}$%, exhibiting the appropriate failure mechanism appearance as dependent on the temperature at low loading levels, K_{max}, MPa$\sqrt{m}$: a) 34.4; b) 30.6; c)17.7. Legend: 1 - fatigue striations; 2 - cleavage; 3 - near threshold mechanism.

through 26 can be taken as guidelines when analysing the case hystories of structural elements made of such class of steels. One of important observations from these results is the conclusion that depending on temperature, loading level, material structure, corrosive environment, etc. several failure mechanisms can contribute

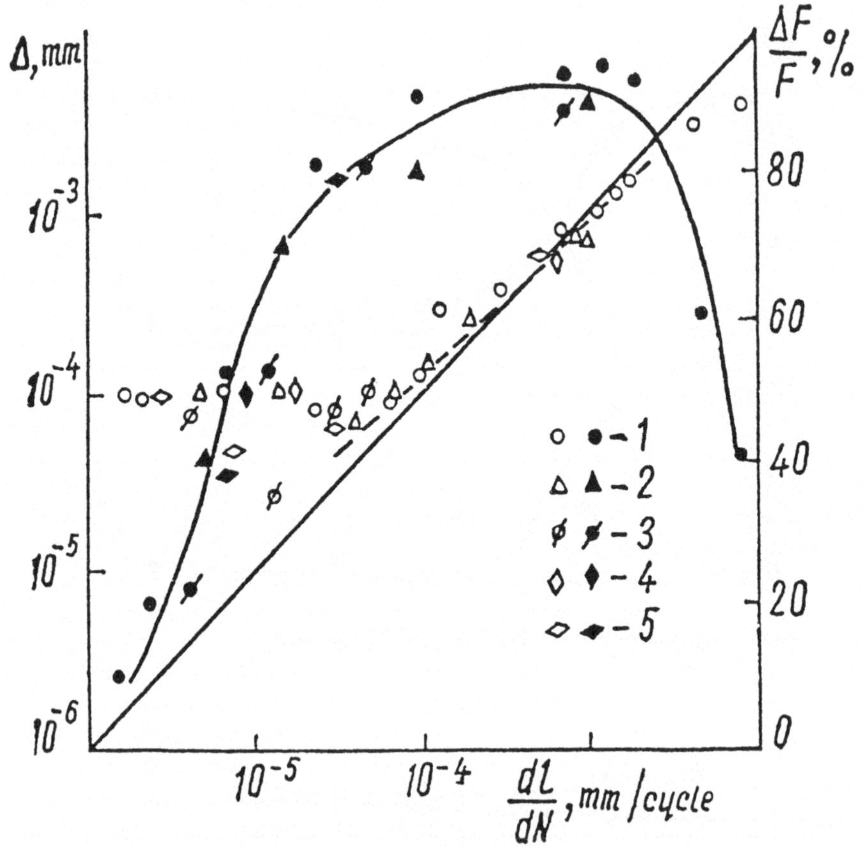

Figure 32. The comparison of the fatigue striation spacings, Δ (open symbols), and the fracture surface fraction, $\frac{\Delta F}{F}$%, exhibiting the fatigue striations (solid symbols), with the actual fatigue crack growth rate, *dl/dN*, at different temperatures, T° K: 1) 293; 2) 263; 3) 233; 4) 203; 5) 173.

simultaneously to the FCG rate. The fatigue striation mechanism gives the minimum FCG rate, whereas other mechanisms, being superimposed on the fatigue striation mechanism, accelerate the crack.

4.CONCLUSIONS

1.The linear correlation exists for engineering materials within each of five classes (unalloyed steels, low-alloy steels, high-alloy steels, aluminum alloys, titanium alloys) between the material parameters describing the power law relation of life to stress or strain amplitudes at high-cycle and low-cycle fatigue. This correlation reduce the number of material`s constants required to describe the behaviour of mentioned classes of materials subjected to cyclic deformation.
2. Similar linear correlation between above mentioned material parameters is observed at high temperatures and the constant load amplitude regime for two molybdenum alloys. There are much common behaviour observed under definite circumstances between steady state creep rate and cyclic creep rate as well as between the creep rupture time and cyclic creep rupture time. The reason for such correlation are the termally activated processes of the plastic flow and of the structural damage of materials.
3. The results of the surface and corner crack prediction under cyclic loading based on a two-parametric part-elliptical crack front model are in good agreement with the experimental data. When transforming the fatigue crack growth properties from the through-cracked standard specimens to structural components containing part-through flaws, use of the effective stress intensity factor range as the crack growth criterion is preferable. Within the middle region of the fatigue crack growth diagram, the effective stress intensity factor value at the deepest point of the semi-elliptical crack is 1.1 times the corresponding value at the surface point.
4. The stable crack shape analysis may be considered as an efficient way of estimating the validity of the part-through crack growth criteria. In this case only the specimen macrofracture surfaces are studied without producing fatigue crack growth rate experiments. The data on the stable crack shape may be used in fracture mechanics-based design of structures, in estimating the stress intensity factor and flaw depth from measurement s of the flaw length over the component surface, etc.

REFERENCES

1. Jin, N.Y. and Winter, A.T. (1988) The role of cross-slip of screw dislocations in fatigue behavior of copper single crystals, in *Basic Questions of Fatigue*: vol.1. ASTM Philadelphia, ASTM STP 924, pp.17-25.
2. Hunsche, A. and Neumann, P. (1988) Crack nucleation in persistent slip bands, in *Basic Questions of Fatigue*: vol.1. ASTM Philadelphia, ASTM STP 924, pp. 26-38.
3. Grosskreutz, J.C. (1971) Fatigue mechanisms in the sub-creep range, in *Metal Fatigue Damage*, ASTM Philadelphia, ASTM STP N495, pp.5-60.
4. *ASTM Standards* (1982) E 466-82, Standard Practice for Conducting Constant Amplitude Axial Fatigue Tests of Metallic Materials, Philadelphia, ASTM, vol. **03.01** (1993), pp.567-571.
5. *ASTM Standards* (1992) E 606-92, Standard Practice for Strain-Controlled Fatigue Testing of Metallic Materials, Philadelphia, ASTM,vol. **03.01** (1993), pp.632-646.
6. Troshtchenko, V.T., Khamasa, L.A., Pokrovsky, V.V., Klesnil, M., Polak, J., Lukas, P. (1985) *Cyclic Deformations and Fatigue of Metals*, Kiev: Naukova dumka,vol.1.(In Russian).
7. Lukas, P., Kesnil, M., Polak, J (1974) *Mat. Sci. Engng.*, N15, pp. 239-244.
8. Toth, L. *GEP*, (1989), **XLI**, N9, pp. 340-346 (in Hungarian).
9. Krasowsky, A.J. and Toth, L. (1994) A thermodynamic nature of power relations between some mechanical properties and temperature and time, *Strength of Materials*, N6, pp. 3-9 (in Russian).
10. Boller, C. and Seeger, T. (1987) *Materials Data for Cycling Loading*. Part A: Unalloyed Steels. Elsevier Ed., Amsterdam-Oxford.
11. Boller, C. and Selger, T. (1987) *Materials Data for Cycling Loading*. Part B: Low-Alloy Stells. Elsevier Ed., Amsterdam-Oxford.
12. Boller, C. and Seeger, T. (1987) *Materials Data for Cycling Loading*. Part C: High-Alloy Stells. Elsevier Ed., Amsterdam-Oxford.
13. Boller, C. and Seeger, T. (1987) *Materials Data for Cycling Loading*. Part D: Aluminium and Titanium Alloys. Elsevier Ed., Amsterdam-Oxford.
14. *ASTM Standards* (1987) E 1150-87, Standard Definitions of Terms Relating to Fatigue, ASTM Philadelphia,vol. **03.01** (1993), pp. 843-852.
15. Miller, K.J. (1982) Creep and Fracture. Mechanical and Thermal Behavior of Metallic Materials*: Proc. Intern. School of Phys. "Enrico Fermi"* Amserdam, N-Y., Oxford: North-Hol. Publ. Co.
16. Shneiderovich, R.M. (1968) *Strength at Static and Low-cycle Loadings*.-Moscow, Mashinostroyenie (in Russian).
17. Lukas, P. (1981) A model of critical microcracks at the fatigue limit and cyclic stress calculation.- *Mechanical Fatigue of Metals*. Nauk. dumka, Kiev, p.194.
18. Heywood, R.B. (1962) *Designing Against Fatigue*, London , Chapman a Hall Ltd.
19. High - Temperature Creep Fatigue (1988)., in R. Ohtani, M. Ohnami and T. Inoue (eds). Elsevier Appl. Edition.
20. Semsen, S.V., Kogayev, V.P. and Shneiderovich, R.M.(1963) *Limit Load and Strength of Structural Elements* 2-nd Ed., Moscow, Mashgis (in Russian).
21. Troshchenko, V.T. (1981) *Deformation and Failure of Metals at High-Cycle Loading*, Nauk. Dumka Ed. Kiev (in Russian).
22. Coffin, L.,Jr. (1974) Fatigue at high temperature-prediction and interpretation, *Proc. Instn. Mech. Engrs*. vol 188, 9/74, pp.109-127.
23. Krasowsky, A.J. and Toth, L. (1996) A thermodynamic analysis of the empirical power relationships for creep rate and rupture time, *Strength of Materials*, N2, pp. 5-22 (in Russian).
24. Bukhanovsky, V.V., Kharchenko, V.K., Polishtchuk, E.P. et al. (1992) Rupture stress and creep of the molybdenum alloys weldings at high temperatures, *Automatic welding*, N2 (467), pp. 15-20. (in Russian)
25. Bukhanovsky, V.V., Kharchenko, V.K., Kravchenko, V.S. et al. (1989) Cyclic creep and rupture of molybdenum alloys ЦМ-6 and ЦМ- 10 at high temperatures, *Strength of Materials*, N8, pp. 37-42 (in Russian).
26. Bukhanovsky, V.V., Borisenko V.A., Kharchenko, V.K. (1994) High-temperature strength and creep of the molybdenum alloys weldings at low-cycle fatigue, *Strength of Materials*, N11, pp. 32-44(in Russian).
27. Ford, F.P. (1983) Stress corrosion cracking in iron alloys at water environment, in C.L. Briant and S.K. Banerji (eds), *Treatise on Materials Science and Technology*, vol.25. Embrittlement of Engineering Alloys, Academic Press, New York, London.
28. Spahn, H. (1983) Hydrogen effect on toughness and crack growth. in W. Dahl and W. Anton (eds) *Werkstoffkunde Eisen und Stahl, Teil I: Grundlagen der Festigkeit, der Zahigkeit und des Bruchs*, Verlag Stahleisen mbH. Dusseldorf.

29. Romaniv, O.N. and Nikiforchin G.N. (1986) *Mechanics of Corrosion Cracking of Engineering Alloys*, Metallurgia, Moscow, 293 (in Russian).
30. Nelson, H.G. Hydrogen embrittlement, in C.L. Briant and S.K. Banerji (eds), *Treatise on Materials Science and Technology*, vol.25. Embrittlement of Engineering Alloys, Academic Press, New York, London.
31. Frost N.E., Marsh K.J. and Pook L.P. (1974) *Metal Fatigue*, Clarendon Press, Oxford.
32. *ASTM Standards* (1993) E 647-93, Standard test method for measurement of fatigue crack growth rates, Philadelphia, ASTM, vol. **03.01** (1993), pp.679-706.
33. *USSR Standards* (1982): РД 50-345-82 Methodical recommendations on the crack resistance determination at cyclic loading, Standards Ed., Moscow (in Russian).
34. Родней, М. (1976) Усталость высокопрочных материалов, in *Разрушение*, Мир, М. **3**, pp. 473-527.
35. Toth, L., Romvari, P. and Nagy, G. (1980) The fatigue crack growth laws, *Strength of Materials*, N12, pp. 18-28 (in Russian).
36. Elber, W. (1971) The significance of fatigue crack closure, in *Fatigue*. Philadelphia, ASTM STP N486, pp.230-242.
37. Broek, D. (1988) *The Practical Use of Fracture Mechanics*, Kluwer Academic Publishers. Dordrecht / Boston/ London
38. Krasowsky, A.J. (1980) *Brittleness of Metals at Low Temperatures*, Naukova dumka, Kiev, 340 (in Russian).
39. Fleck, N.A. (1988) Influence of stress state on crack growth retardation, in *Basic Questions in Fatigue*. Ed. by J. Fong and R. Fields, ASTM STP 924,vol. **1**, pp. 157-183.
40. *Handbook of Fatigue Crack Propagation in Metallic Structures* (1994) Ed. by A. Carpinteri, Amsterdam, Elsevier, vol.I, vol.II.
41. Kuwabara, K., Nitta, A., Kitamura, T. and Ogata, T. (1988) Effect of small-scale creep on crack initiation and propagation under cyclic loading, in *Basic Questions in Fatigue*. ASTM STP N924, vol.**2**, pp.41-59.
42. Varfolomeev, I.V., Vainshtok, V.A. and Krasowsky, A.J. (1992) Prediction of part-through crack growth under cyclic loading, *Engineering Fracture Mechanics*, **40**, N6, pp. 1007-1022.
43. *ASME Boiler and Pressure Vessel Code* (1977) Section 11.
44. Richards, C.E. and Lindley, T.C. (1972) The influence of stress intensity and microstructure on fatigue crack propagation in ferritic materials, *Engineering Fracture Mechanics*, **4**, N4, pp. 951-978.

MATERIAL CHARACTERIZATION REQUIRED FOR THE RELIABILITY ASSESSMENT OF CYCLICALLY LOADED ENGINEERING STRUCTURES
Part 2: FATIGUE APPLICATION

L. TOTH[1)] and A.J. KRASOWSKY[2)]

[1)] University of Miskolc, Department of Mechanical Engineering, Egyetemváros, MISKOLC 3515 HUNGARY

[2)] Institute for Problems of Strength, National Academy of Science of Ukraine, Timiryazevskaya St. 2, KIEV 252014 UKRAINE

1. Introduction

It has been said that no man is civilised or mentally adult until he realises that the past, the present, and the future are indivisible. That is true for any kind of sciences, so it is true for fatigue knowledge as well. This introduction tries to give an overview of the milestones which made possible to organise the Royal Society's Conference in 1979 under the title "***Fracture Mechanics in Design and Service - Living with Defects***".

The fatigue application is as old as fatigue knowledge itself because the development of engineering design against fatigue was initiated by failures and accidents. From this, it follows that the development is in close connection with history of engineering sciences. In other words it can be said that the most effective "engine" of development is accidents. If we try create periods in technical development of fatigue knowledge, then it can be a little bit arbitrarily divided into the following four parts:

- railway age;
- automobile period;
- aeroplane age;
- living with defects.

It is well known what is the difference between the "story" and "history". The paper focuses only for a brief story of fatigue, which was observed and detailed at first more than 150 years ago in 1838 [1].

The first period is closely connected to the opening of the first public railway between Stockton and Darlington on 24 September 1825 [2]. It was the first milestone in the quick development of material testing [3] and

R.A. Smith (ed.), Reliability Assessment of Cyclically Loaded Engineering Structures, 225–272.

fatigue. The failures, fractures without any remarkable plastic deformation caused catastrophycal accidents. One of the most remarkable happened in Versailles in 1842 [4] where the total loss of life was more than 60. In this period more and more unexpected breakages were observed [5] and analysed. This period of the industrial revolution can exactly be characterised by the lengths of railway network, shown in Fig. 1. for Germany and in Fig 2. for the world.

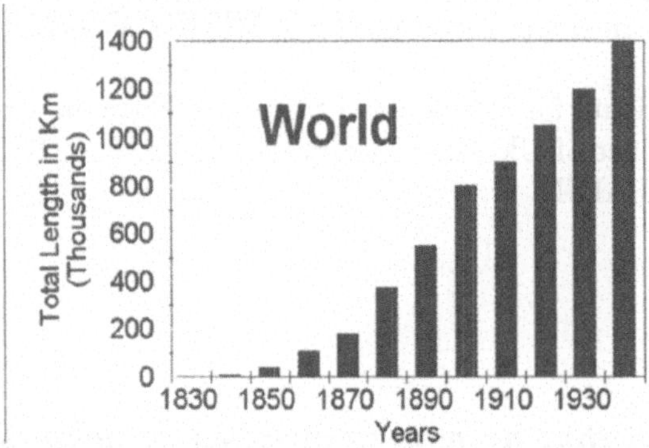

Figure 1. The length of railway network in the world.

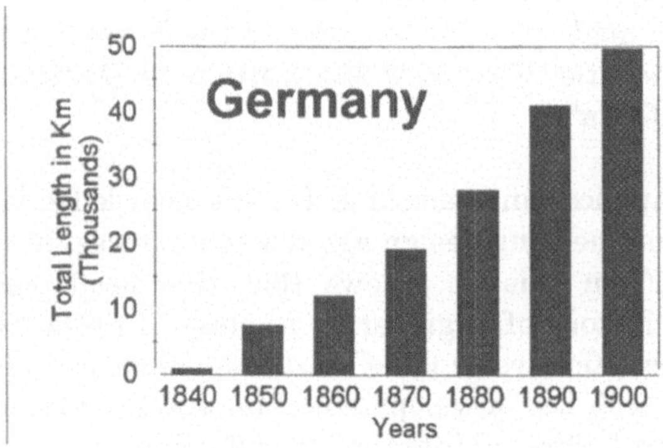

Figure 2. The length of railway network in in Germany.

The dominant part of research in England and Germany (these countries played dominant role in industrial revolution in the first and second period of the last century respectively) focused on elimination of these unexpected failures and fractures [10-13]. The results of experiences and new inventions [19-21, 23, 44, 46] were published and so the number of papers increased rapidly. This is summarised in Fig 3.

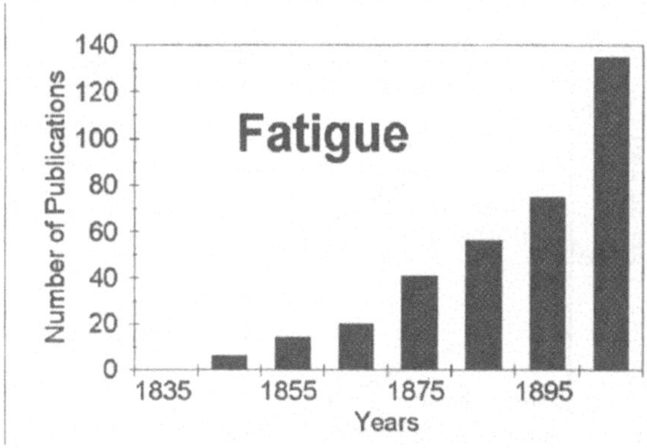

Figure 3. Total number of publication during the last century (on the basis of [8])

The most important milestones in this period can be summarised as follows:

- development of fatigue testing techniques;
- opening the research institutes and testing laboratories;
- observation of the nature of fatigue;
- determination of material properties which can be used for elimination (decrease of risk) of fatigue of railway axles.

The main results of experimental work carried out in England and especially in Germany are summarised in Wöhler's report on material selection and system design against fatigue of railway axles presented in 1870. The tests were performed on the equipment shown in Fig 4.

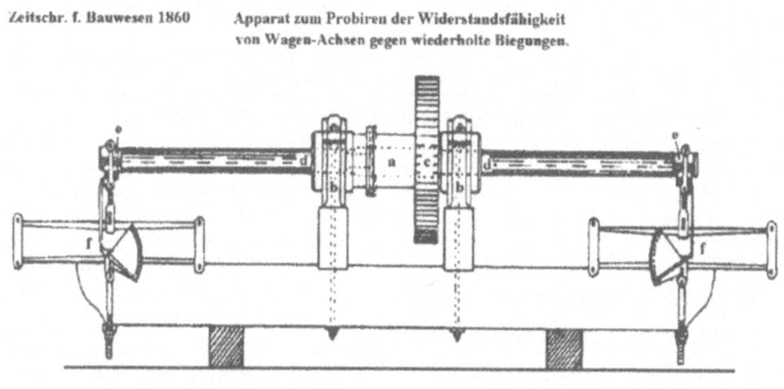

Figure 4. Wöhlers's machine for fatigue testing of railway axles.

One of the most cited results on smooth and notched axles is summarised in Fig. 5.

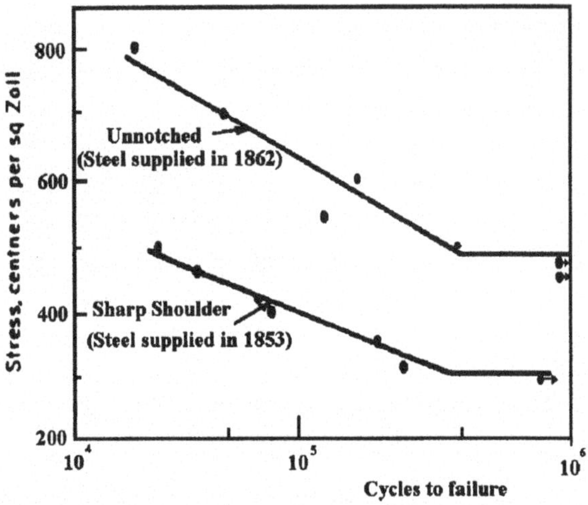

Figure 5. Fatigue curves for smooth and notched samples.

The most important milestones of this period are briefly summarised in TABLE 1.

TABLE 1. Milestones in fatigue of materials and material testing

Year	Event	Ref
1825	Opening of the first public railway (1824 September 27.)	[2]
1837	First publication on fatigue of driving rope	[1]
1842	Railway accident in Versailles with extensive loss of life	[4]
1843	First paper on "fatigue" tests of railway axles (York's paper)	[8]
1843	First paper on design of railway axles against fatigue (Rankie's paper)	[5]
1854	Introduction of term FATIGUE	[6]
1858	Opening the first Material Testing Laboratory (D. Kirkaldy)	[3]
1858	First Wöhler's paper	[7]
1860-1870	Wöhler's experiments on smooth and notched railway axles using axial-, bending and torsion loading condition, analysis of mean stress effect	[8]

1870	Wöhler's material selection and design system against fatigue of railway axles	[8]
≈1881	Bauschinger-effect ⇒ low cycle fatigue	[3]
1884	1st Bauschinger Conference on Material Testing in Münnich	[3]
1886	Martens-mirror, small displacement measurements	[3]
1895	International Society for Material Testing (L. Tetmajer, Zürich)	[3]
1895	Discovering of the X-ray by Wilhelm Conrad RÖNTGEN (in November)	[99]
1907	Stress distribution at the vicinity of sharp notches, crack (K. Weighard)	[25]
1908	New testing machine for reversals of stress (Smith, J.H.)	[9]
≈1910	Smith's safety diagram	[9]
1910	Analytical description of Stress vs. Life Time curve, Basquin σ_a=CN^n	[14]
1912	Solution of two-dimensional problem of elasticity, Kolosov-functions	[29]
1912	New testing machine for alternating load tests, (Haigh B.P)	[9]
1913	Stress distribution at the vicinity of sharp notches, crack (Inglis C.E.)	[26]
1914	Notch effect on fatigue limit (Heyn, E.)	[9]
≈1916	Haigh's safety diagram	[9]
1918	Fatigue testing of full-scale aeroplane components	[9]
1918	Introduction of the "dupuli-Tised film" (the first plastic base X-ray film)	[100]
≈ 1920	Repeated stress safety factors	[9]
1920	Energy balance concept for crack propagation (A.A. Griffith)	[27]
1924	Palmgren concept on fatigue damage at different stress level	[9]
1929	Load spectrum measurements for laminate (car) spring (Batson, Bradley J.)	[9]
1929	Patent of ultrasonic testing (S.J. Sokolov)	[3]
1930	Stress concentration and fatigue strength (Peterson R.E.)	[9]
1932	Load spectrum measurements for agricultural machine (Kloth, Stoppel)	[9]
1934	Magnetic testing (W. Gerhard)	[3]

1935	Introduction of "shape-strength" phenomena (Gestaltfestigkeit - A. Thum)	[9]
1935	Notch factor definition in fatigue (β_k - A.Thum)	[9]
1936	First crack-propagation law (De Forest A.V.)	[9]
1937	Automatic crack detection equipment (F. Förster)	[3]
1937	Neuber concept on notch theory	[9]
1937	Damage accumulation from stress cycles of varying amplitude (Langer B.E.)	[9]
1932-1938	Load spectrum measurements and publication for aircraft (Kaul H.W.)	[9]
≈ 1939	Introduction of "working-strength" phenomena (Betriebsfestigkeit - Gaßner)	[9]
1939	Introduction of strain-gauge technology in strain measurements	[3]
1939	Statistical nature of fatigue (W. Weibull)	[9]
1939	Introduction of cracks solutions for elastic body, Westrgaard-solutions	[30]
1943	Residual stress influence on fatigue (Horger O.J.)	[9]
1945	Miner concept in commulative damage in fatigue	[9]
1946	Crack solutions for elastic body for different loading conditions (Sneddon)	[31]
1951	Foundation of International Committee on Aeronautical Fatigue	[9]
1953	Low-Cycle-Fatigue (Manson S.S.-NACA≡NASA))	[9]
1953	Random fatigue (Freudenthal A.M., Gumbel E.J.)	[9]
1954	Low-Cycle-Fatigue (Coffin L:F. - General Electric)	[9]
1954	COMET accidents (Elba on 10/01 and near Naples on 8/04)	[28]
1956	Non-linear Corten-Dolan approach on commulative fatigue damage	[9]
1956	Introduction of "Crack extension force" by G. Irwin	[33]
1959	Introduction of the DGS diagram by J. Krautkrämer	[101]
1960	Electrohydraulic closed-loop testing machine	[3]
1961	Paris crack growth law (Paris, Gomez, Anderson)	[41]
≈ 1964	Computer aided (analogue) material testing system (Phil Mast)	[3]
1967	Computer aided electrohydraulic MTS system	[3]
1968	Introduction of ΔK_{eff}, Elber model for crack growth (W. Elber)	[88]
1970	The first standard for fracture mechanics testing (ASTM E 399-70)	[3]

1983	The first standard for fatigue crack growth testing (ASTM E 647-83)	[102]
1986	Application of RS232/V24 to ultrasonic testing equipment	[101]
1994	Digital ultrasonic testing equipment with built in DGS diagrams	[101]

An absolutely new age was began at the beginning of this century through the development of the automotive industry. The main results of this period can be summarised as follows:

- mean stress effect on endurance limit;
- creation of different safety diagrams;
- determination of notch stress factors using analytical solutions;
- experimental investigation of notch effect on fatigue behaviour;
- registration of the real load spectrum of different structural components (only observation);
- development of design methods against fatigue based on tests carried out on specimens;
- introduction of "shape-strength" (Gestaltfestigkeit) and "working-strength" (Betriebsfestigkeit) phenomena;
- recognition of statistical nature of fatigue [3, 80, 104],
- foundation of basic principles of fracture mechanics [15-18,24, 32, 34-39];
- invention of NDT testing methods;
- introduction of strain gauge technology into load spectrum registration (on line method).

The most important milestones of this age are also briefly summarised in TABLE 1.

The second and third periods (the automobile period and aeroplane age) of development of fatigue are a little bit overlapped. It follows from the nature of military research. These have always more or less secret characters. In spite these facts some main characteristic differences between these two periods an be summarised as follows:

⇒ main characteristic features of the automobile period of fatigue:

- fatigue behaviour of steel components;
- fatigue behaviour of welded (joined) components;
- notch effect on fatigue;
- fatigue tests on specimens;

- effect of machining technology on fatigue behaviour of steel components (heat treatment, surface treating i.e. surface deformation, corrosion media, roughness, etc.);
- full-scale tests are not general carried out,

⇒ main characteristic features of the aeroplane age of fatigue

- fatigue behaviour of light metals (aluminium alloys, lithium alloys, etc.), and materials are applied in engine technology (turbo- and usual engines);
- investigation of external effect on fatigue (temperature, corrosion media, etc.);
- presence of low-cycle fatigue, thermal fatigue [75, 76];
- fatigue behaviour of bolted joints;
- fatigue at real working loading conditions,
- measurement and analysis of real spectrum loads [107, 108];
- full-scale tests;
- fatigue behaviour of new advanced materials;
- development of electrohydraulic and computer aided testing systems;
- the fracture mechanics principles are basically created [53, 54, 63, 64, 82, 83, 85];
- the non-destructive testing methods for crack detection are basically developed [51, 52, 105].

Considering the above mentioned facts it can be said that the "aeroplane age" began at the end of the 1st World War and became a more intensive period from the middle of 30's. A new, international character of the fatigue research was developed by foundation of *International Committee on Aeronautical Fatigue* in 1951. A new intensive period was created by the appearance of new advanced materials used in astronautics and hypersonic aeroplanes. The most important milestones of this age are also briefly summarised in TABLE 1.

The common features of these above mentioned periods, is that in design of structural elements and components against fatigue the crack like defects (imperfections and flaws) are not allowed in spite of the fact that the behaviour of cracked components have been intensively investigated both experimentally and theoretically. This is caused on the one hand by some important accidents in military and human industrial history and on the other hand by rapid development of non-destructive methods used in supervision of structures and components. This last factor became more and more important and follows from the tremendous development of microelectronics. In this time two very important engineering areas interlocked: *non-destructive testing* (NDT) and *fracture mechanics* creating a new period of fatigue; the age of "living with defects". The main questions raised in this period are the following:

- what kind of material property(ies) can be used for characterisation of crack growth resistance of materials [45, 50, 66, 70, 73, 74, 81];
- how can it be determined experimentally using small specimens;
- how can it be transferred to full structural elements [48];
- what kind of NDT method(s) are needed;
- how accurately can size of detected flaws be determined;
- how good is the reliability and reproducibility of the NDT results, etc.?

These questions can be continued and always need to be answered in order to estimate the risk of decision about the reliability of cracked structural elements. The most important background for reliability estimation of cracked structural components were provided in the middle of 80's. They are the following:

- standard for experimental measurement of crack growth resistance of materials;
- the computer aided, servohydraulic mechanical testing systems for determination of crack growth resistance of materials, joints at very different working conditions;
- NDT testing methods for measurement of flaws on the surface and in bulk material;
- software packages, handbooks [56-60], databases [61, 62] for calculation of the stress intensity factor or local stress-strain field at the vicinity of crack-like defects;
- the measuring methods, data acquisition and process systems in order to experimental control of the assessment methods.

The way from the first detection of flaws in welded joints using X-ray techniques (in 1896 by Wight of Yale University) [99] up to the present day, can roughly be followed on the one hand by the facts summarised in Table 1. and on the other hand by the number of "fatigue" related publications. The last one is well demonstrated by the trend shown in Fig.6.

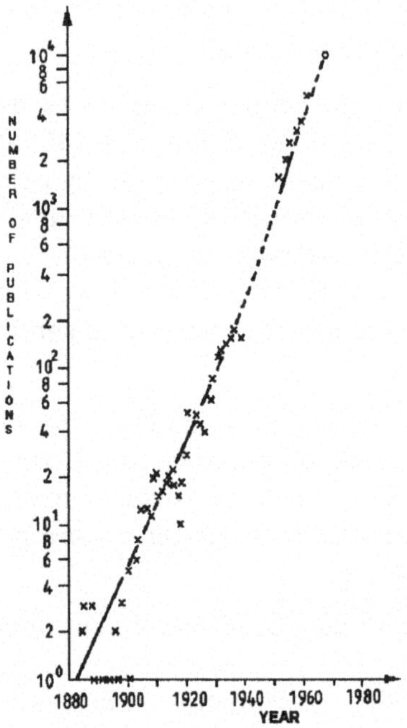

Figure 6. The number of publications related to fatigue

According to the compendium of fatigue bibliography of Man [8] up to 1950 the total number of fatigue related publications was 4080. This is not totally true because the publication of small countries or papers with obscure languages is not considered. The quick evaluation of this area can be followed by different databases. The DIALOGUE database (which contains app. 2-3 million records) includes 41,849 fatigue related papers. According the COMPENDEX database the following numbers were recorded:

- 3726 publications in 1988-89
- 3056 publications in 1990-91
- 3205 publications in 1992-93.

These data show that approx. 10 publications which deal with fatigue problems are prepared daily all round the word. In spite of this enormous number of publications the economic effect of fracture is remarkable in each country. Its ratio is approx. 4 % of the gross domestic product (GDP). The important role played by fatigue [47] is well represented by Figs. 7-11.

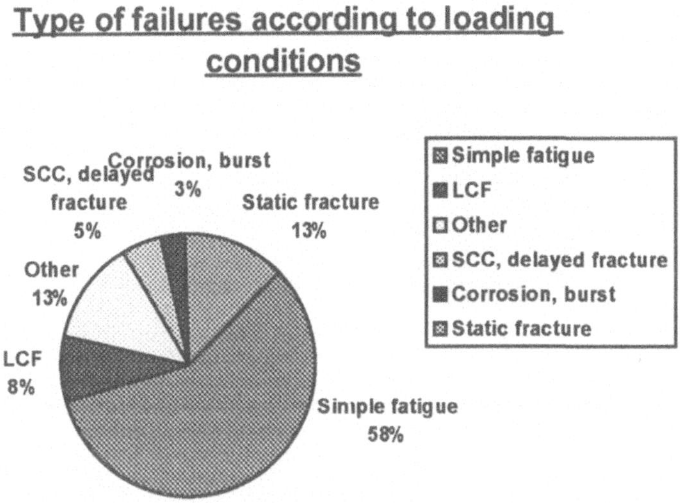

Figure 7. Effect of loading condition on frequency of failures (on the basis of 242 failures [47])

Fig 7. shows the remarkable importance of fatigue in failures. Its ratio is almost 80%! Considering that final fracture includes both crack initiation and propagation stages, it is important to emphasise the importance of the crack propagation period of life because this is the period in which all of the circumstances (NDT methods, crack growth resistance of materials, numerical methods of continuum mechanics) arise to allow remaining life time and reliability estimations.

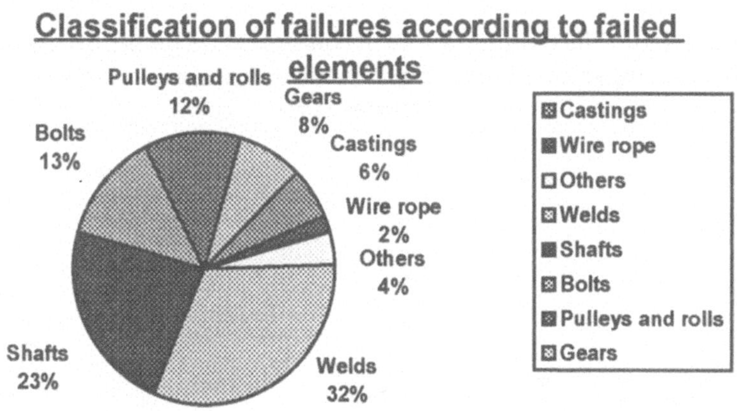

Figure 8. The frequency of failures of different type of structural components. (on the basis of 242 failures [47])

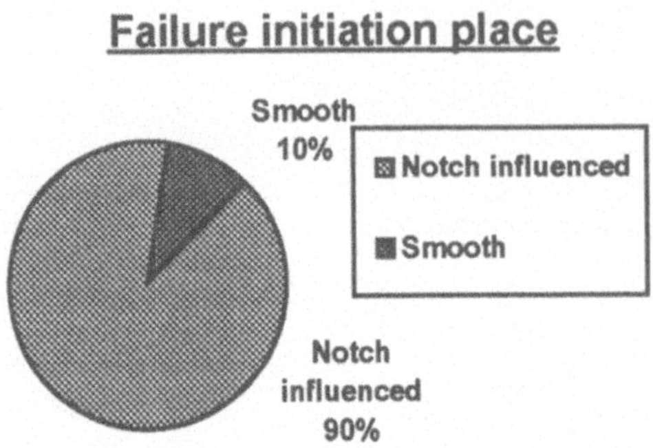

Figure 9. Effect of stress-strain (on the basis of 242 failures [47])

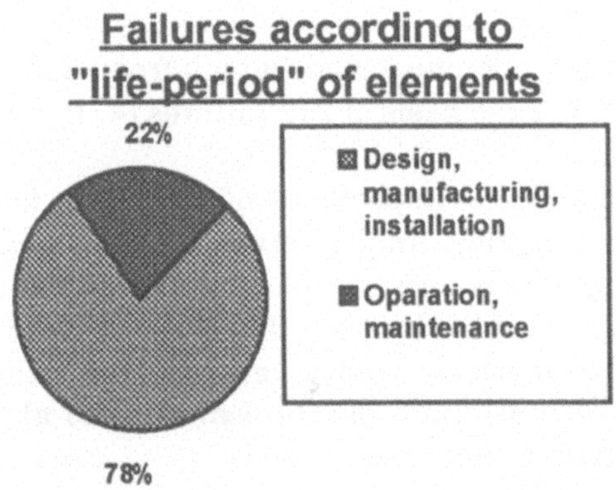

Figure 10. The dominant role of concentrators on failure design, manufacturing and installation problems in fracture. (on the basis of 242 failures [47])

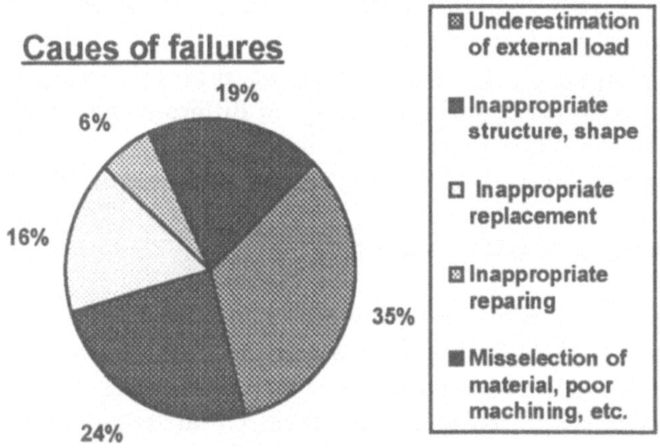

Figure 11. The importance of good estimation of real loading condition on reliability of structures.

Each of above figures represents the importance of practical use of fracture mechanics in reliability assessment of cyclic loaded engineering components having crack like defects. This concept includes three main, interconnected areas as shown in Fig.12.

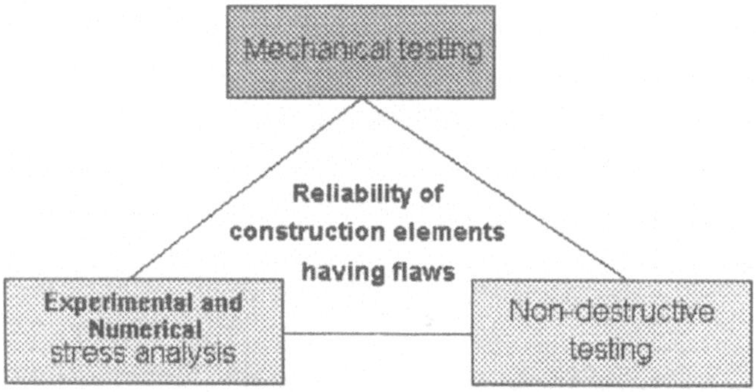

Figure 12. Basic knowledge in reliability assessment of structural components having crack like defects.

This paper tries to give brief overviews about the;

- main features of NDT methods special regarding to their reliability and application limits;
- material properties used for characterisation of fatigue crack resistance of materials (i.e. fatigue crack growth laws);

- the requirements of NDT testing results and safety factor of cracked elements;
- comparison of the danger of different cracks in different structural components made of different materials (by defining the crack propagation sensitivity index);
- practical use of the crack propagation sensitivity index.

2. Main features of NDT methods

The most important flaws in bulk material in structures can principally be detected from the beginning of this century. This goes back to Wilhelm Conrad Röntgen, discovering the X-ray technique. Surface flaws and cracks can be detected by visual observation. Great steps were realised by introduction of ultrasonic testing in 1929 and magnetic testing in 1934. The development of DGS (Distance-Gain-Size) diagrams (by Krautkrämer in 1959) in ultrasonic testing for determination of flaw sizes opened the possibility of practical use of fracture mechanics principles. This first DGS diagram can be seen in Fig.13.

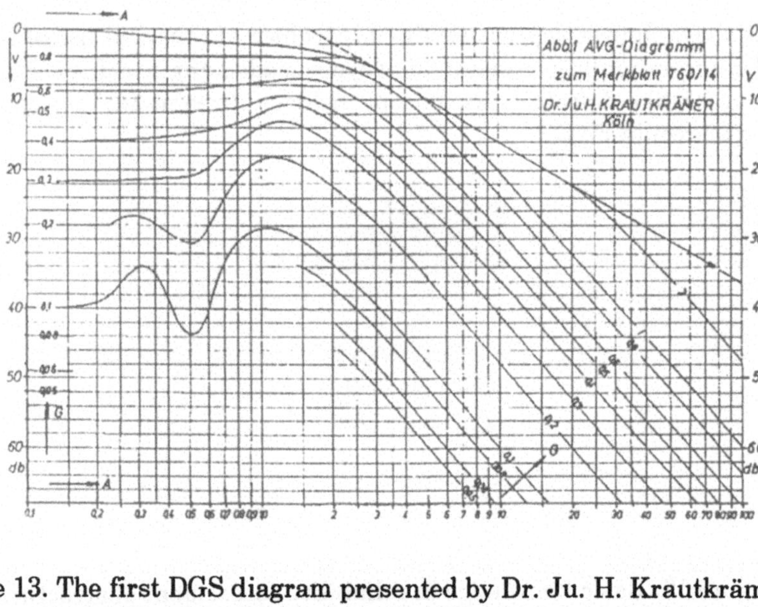

Figure 13. The first DGS diagram presented by Dr. Ju. H. Krautkrämer for determination of flaw size using ultrasonic examination

A new age was started with the appearance of digital techniques in ultrasonic testing from 1986. This greatly reduced human error in reproducibility of ultrasonic testing. The most important facts and milestones in this area are summarised in Table 1. as well. The main questions of NDT testing are the following:

- what is the reliability of testing;
- how much is the probability of flaw detection;
- how much is the reproducibility of the test results?

Knowing the answers for these questions, there is no difficulty in the use of fracture mechanics principles for reliability assessment of cracked elements. Unfortunately these questions are very complicated and can not easily be answered. The probability of flaw detection with widely used X-ray and ultrasonic testing can be seen in Fig.14. as a function of type of flaw and thickness of welded joints.

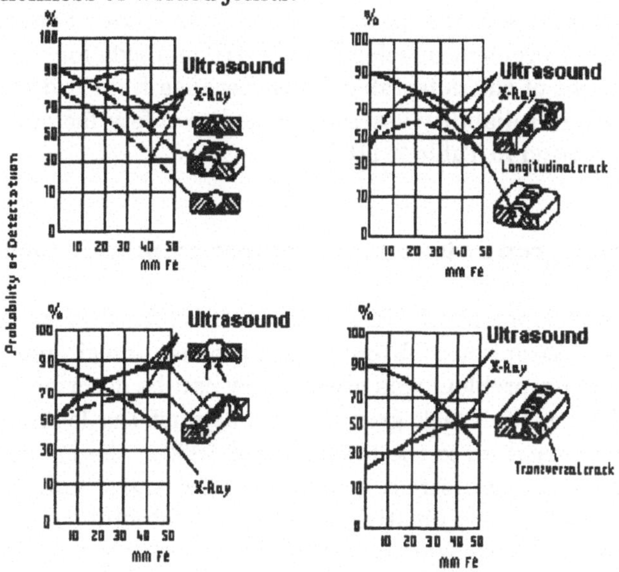

Figure 14. The probability of detection of different flaws in welded joints using X-ray and ultrasonic testing methods as a function of thickness.

Figure 13. shows that even the most dangerous flaws and longitudinal cracks in the heat affected zone cannot be detected more than 90% probability. It need to be emphasised the followings as well:

- X-ray examination is proposed for thin structural elements and ultrasonic method for thicker ones;
- the detection probability of crack like defects using X-ray methods is less than using ultrasonic examination;
- the probability of detection of crack-like defect increases with increasing thickness using ultrasonic examination,

The advantages, disadvantages and application fields of other NDT methods are discussed and summarised in the literature [51, 52, 105]. Following new publications in this area it can be stated, that a great effort is directed all round the world to determination of probability, reproducibility and reliability of testing results. It naturally follows from this fact, that knowing the flaw geometry and loading conditions, the danger of detected flaws can be estimated from the point of view of the reliability of structural components. Unfortunately there is no widely used approach(es) which interconnects the requirements of NDT testing results and reliability (uncertainties in NDT results, safety factor and remaining life time). It seems that one of the most important tasks in reliability assessment of cracked structural elements is to develop such approaches in which the these requirements are unambiguously interconnected as schematised in Fig.13. Introducing the crack growth sensitivity index of structural elements (see the paragraph 4.) this interconnection can be assured. The other problem in selection of NDT methods for a given task is the minimum sizes of the detectable flaws, because from it directly follows the uncertainties in reliability assessment and in safety factors. One of the results is summarised in Table 2. [104].

3. Fatigue crack growth laws for constanst amplitude loading

By the rapid development of NDT methods and fracture mechanics the requirement for description of fatigue crack propagation circumstances also increased. The main problem was always how can be describe the material response (da/dN i.e. the crack growth rate) against the loading condition. During the first period up to the beginning of 1960's the loading condition was characterised by the stress amplitude. The crack length in fatigue crack growth laws was explicitly included. An absolutely new age was started by Paul C. Paris, who proposed the stress intensity factor range (ΔK) for characterisation of the loading condition [41]. In this relation the crack length is implicitly involved. The acceptance of the Paris law was not to easy because ΔK is based on a totally elastic material response even at the crack vicinity, while the fatigue crack growth is the result of plastic deformation and degradation at the crack vicinity. These statements were clearly reflected in review of the presented paper of Paris and his co-workers. More than 20 years later in 1982 [43] P.C. Paris said the following:

"Ironically, the paper was promptly rejected by three leading journals, whose reviewers uniformly felt that 'it is not possible that an elastic parameter such as K can account for the self-evident plasticity effects in correlating fatigue crack growth rates'."

In spite of the above mentioned fact, the Paris law is the most widely used in our time. The most intensive research for description of fatigue crack growth circumstances under constant amplitude loading condition was performed in the 60's and 70's. An overview of proposed relationships is summarised in Table 3[67,68].

TABLE 2. A comparioson of minimum sizez of flaws detected by different NDT methods [104]

NDT method	Material and the detectability	Detectable flaws					
		Volume mm^3	Crack like defects, mm			Diameter	Relative size
			opening	depth	length	d, mm	$\bar{d}=d/H$, %
Optical	0,1, optical material		0,01	0,0001	0,1		
Capillary			0,001 -0,003	0,01	0,1		
color			0,001 -0,003	0,01 -0,05	0,1 -0,3		
fluorescent penetrant			0,001 -0,003	0,01 -0,05	0,1 -0,3		
Acoustic							
ultrasonic	1 -1,5 mm		0,001 -0,003	0,1 -0,3	1,0 -2,0		
acoustic emission					growth 0,04-0,09 mm		
Magnetic							
magnetic particle			0,001 -0,003	0,001 -0,005			
magnetic-fluorescent			0,25	2			
powder suspended in liquid				surface 0,01			
				embedded 0,5			
inductive				surface 0,22		surface 10 %	
Eddy current			0,0005 -0,01	0,1 -0,2	0,6 -2		
Electric							
using electron potential				0 - 25			
changing in electromagnetic field				0,1 - 120			
electric discharge				0,6			
electric resistance	0,1		0,15				10 %
Radio wave							
pore, slags to 15 mm depth		1					
to 15 - 40 mm depth						3	
cracks to 10 mm depth			0,1		10		
Temperature					0,02 - 0,06		
Radiation							
radiography							1-2 %
radioscopy						0,01	1 %
materials of (1,7 -1,8) g/cm^3 density with 100 - 1000 mm thickness		$30 - 3\cdot10^3$					
Pipes with 30 - 100 mm diameters	30						
Welded joints with 70 - 200 mm thickness		$1,5\cdot10^3$					
Castings with 450 mm thickness							
Rolled products with 250-300 mm thickness		$2\cdot10^3$					3 %
Plastics with 400 - 600 mm thickness						0,01	0,5 -1 %
X-ray topography		$1\cdot10^3$	0,001 -0,2			8	
Radioactive leak testing			0,0001				

TABLE 3. Some fatigue crack growth laws

No	Author	Year	Relationship	Parameters	
				Loading	Material
1	Stanley	1952	$\frac{da}{dN} = C\sigma_a^n a$	σ_a	C, n
2	Head	1953	$\frac{da}{dN} = \frac{C\sigma_a^3 a^{3/2}}{(R_{P0,2} - \sigma_n)^2 \sqrt{r_p}}; r_p = const;$ $C = \frac{1}{12} \frac{\lambda}{R_u - R_{P0,2}}$	σ_a	C, r_p $R_{P0,2}$
3	Forst, Irwin		$\frac{da}{dN} = \frac{C_a^2 a}{(R_{P0,2} - \sigma_a)^2}$	σ_a	C, $R_{P0,2}$
4	Weibull	1954	$\frac{d(2a/w)}{dN} = C\sigma_{nom}^n$	σ_{nom}	C,n
5	Forst, Dugd-ale	1958	$\frac{da}{dN} = Cf(\sigma_a^n) a\sigma_{mean} f(\sigma_a^n) = \sigma_a^2 \nu \sigma_a^3$	σ_{mean} ,$f(\sigma_a{}^n)$	C,n
6	McEvily Ilg	1958	$\lg\frac{da}{dN} = C_1 K_\alpha \sigma_{nom} - \frac{C_2}{K_\alpha \sigma_{nom} - \sigma_f} - C_3$	σ_{nom}, K_α	C_1,C_2, C_3, σ_f
7	Schijve	1960	$\frac{da}{dN} = C\sigma_a^{2,6} a^{1,5}$	σ_a	C
8	Liu	1961	$\frac{da}{dN} = Cf(\sigma_{mean}, \sigma_{max} - \sigma_{min})a$	$f(\sigma_{mean}$, σ_{max}- $\sigma_{min})$	C
9	Weibull	1961	$\frac{da}{dN} = Cw\sigma_{nom}^n$	σ_{nom}	C, n
10	Christ-ensen	1961	$\frac{da}{dN} = C\frac{(a/a_c)^m}{(1 - \frac{a}{a_c})^n}$	C, m, n	C, m, n
11	Valluri	1961	$\frac{da}{dN} = Ca\frac{[(\sigma_{max} - \sigma_f)/E]^2 (\Delta\sigma/\sigma_f)^2}{\ln[(\sigma_{max} - \sigma_f)/k]}$	σ_{max}, σ_{min}	C, k, E, σ_f

12	Paris, Gomez Anderson	1961	$\frac{da}{dN} = CK^4$ or $\frac{da}{dN} = C(\Delta K)^n$	K or ΔK	C or C and n
13	McEvily Boettner	1963	$\frac{da}{dN} = Ca^n \sigma^{2n}$	σ, n	C, n
14	Weibull	1963	$\frac{d(a/w)}{dN} = C\left[\frac{\sigma_{max}}{1-\frac{a}{w}} - \sigma_w\right]^2$	σ_{max}	C, σ_w
15	Liu	1963	$\frac{da}{dN} = C(\Delta\sigma)^2 a$	Δσ	C
16	McClintock	1963	$\frac{da}{dN} = C\frac{\sigma_{max}}{R_m}\sqrt{a}$	σ_{max}	C, R_m
17	Valluri, Glassco, Bockrath	1963	$\frac{da}{dN} = Ca(\sigma_p - \sigma_{min})^2$ or $\frac{da}{dN} = C^* \frac{w}{\pi\ \mathrm{tg}\frac{\pi a}{2w}}\left(\sigma_p - \sigma_{min}\right)^2$	σ_{min}	σ_p, C, C^*
18	Broek, Schijve	1963	$\frac{da}{dN} = C_1\left[\sigma_{max} a^{3/2}\sqrt{1+40a/w}\right]^3 \exp(-C_2 R)$	R, σ_{max}	C_1, C_2
19	Liu	1964	$\frac{da}{dN} = C(\Delta\sigma)^2 \frac{w}{\pi\ \mathrm{tg}\frac{\pi}{2w}}$	Δσ	C
20	Krafft	1964	$\frac{da}{dN} = d_f\left(\frac{K_t}{K_c}\right)^4$	K_t	K_c, d_f
21	Krafft	1964	$\frac{da}{dN} = C\frac{(1-\gamma)^4}{1-(1+\gamma)^2}\left(\frac{K_f}{K_c}\right)^2$	K_f, γ	C, K_c
22	Manson	1965	$\frac{da}{dN} = \left[C_1\Delta\varepsilon_p + C_2(\Delta\varepsilon_p)^2\right]a$	$\Delta\varepsilon_p$	C_1, C_2

23	Man-son	1966	$\frac{da}{dN} = C\left(\Delta\varepsilon_p \sqrt{a}\right)^n$	$\Delta\varepsilon_p$	C, n
24	Weert-man	1965	$\frac{da}{dN} = \frac{\sigma_{max}^4 a^2}{7EG_{1c}R_m^2}$	σ_{max}	R_m, G_{1c}, E
25	McEvily John-ston	1965	$\frac{da}{dN} = \frac{2\sigma_{max}^4 a^2}{E\left(R_{P0,2} + R_m\right)\varepsilon_2 R_{P0,2}^2}$	σ_{max}, ε_2	$R_{P0,2}$, R_m, E
26	Mark-ovec	1966	$\frac{da}{dN} = C_1 + \exp\left[C_2 K_{max}\right]$	K_{max}	C_1, C_2
27	Forman Kear-ney Engle	1967	$\frac{da}{dN} = C\frac{\Delta K^n}{(1-R)K_c - \Delta K}$	ΔK	C, n, K_c
28	Erdog-an	1967	$\frac{da}{dN} = CK_{max}^m \Delta K^n$	K_{max}, , ΔK	C, m, n
29	Forst, Dixon	1967	$\frac{da}{dN} = \frac{\sigma_{max}^2 a}{E^2}\left(\ln\frac{4E}{\sigma_{max}} - 1\right)$ or	σ_{max}	E
			$\frac{da}{dN} = 256\frac{\sigma_{max}^3 a}{E^2 R_{P0.2}}$	σ_{max}	$R_{P0,2}$, E
30	Csere-panov	1968	$\frac{da}{dN} = C\left[\frac{K_{max}^2 - K_{min}^2}{K_c^2} + \ln\frac{K_c^2 - K_{max}^2}{K_c^2 - K_{min}^2}\right]$	K_{max}, K_{min}	C, K_c, $R_{P0,2}$
31	Bilby, Heald	1968	$\frac{da}{dN} = \frac{5}{3}\left(\frac{\pi}{4}\right)^3 \frac{K_{max}^4}{f(a)G_{1c}R_{p0,2}^2}$	K_{max}	G_{IC}, $R_{P0,2}$
32	Lander	1968	$\frac{da}{dN} = \frac{1-\nu^2}{2ER_m}\Delta K^2$	ΔK	E, R_m, ν
33	Pavl-enko	1969	$\frac{da}{dN} = C(1-R)^2 K_{max}^n$	K_{max}, R	C, n
34	Yokob-ori, Aisawa	1970	$\frac{da}{dN} = \varepsilon, F\left(\omega, \beta, m, R^*, T\right)\Delta K^n$	ΔK, ω	n, β
35	Yokob-ori	1970	$\frac{da}{dN} = bn_g\alpha$	α, n_g	α, n_g

36	**Klesnil Lukas**	1970	$\frac{da}{dN} = C(K_a^4 - K_{th}^4)$	K_a	C, K_{th}
37	**Erdog-an Rat-wani**	1970	$\frac{da}{dN} = C\frac{\left(\frac{2}{1+R}\right)^n (K_c - K_{th})^m}{K_c - \left(\frac{2}{1+R}\right)\Delta K}$	ΔK	C, n, m, K_c,
38	**Taire, Tana-ka**	1971	$\frac{da}{dN} = C_1 r_P^{m_1} \frac{r_P}{a} = C_2\left[\sec\frac{\pi\sigma}{2C_3 R_{P0.2}} - 1\right]$	σ	C_1, C_2, C_3, M_1,$R_{p0.2}$
39	**Rüba-kina**	1971	$\frac{da}{dN} = \frac{\pi^2 K_{max}^4}{48 R_{p0.2}^2} K_c$	K_{max}	$R_{p0.2}$, K_c
40	**Elber**	1971	$\frac{da}{dN} = C\left[(0.5 + 0.4R)\Delta K\right]^n$	ΔK	C, n
41	**Gure-vics**	1971	$\frac{da}{dN} = C\left[\frac{(\Delta K)^2 \bar{\gamma}_n}{\pi(\lambda+1)\tau_0^2}\right]^n$	ΔK, $\bar{\gamma}_{nn}$	C, n, τ_0, λ, $\bar{\gamma}_{n0}$
42	**Flewitt Heald**	1971	$\frac{da}{dN} = C\frac{K_{max}^4}{G_{Ick}\gamma_f \sigma_1^2}$	K_{max}, σ_1	G_{IC}, γ_f
43	**Moro-zov**	1971	$\frac{da}{dN} = C(1-R)\frac{K_{max}^2\sqrt{a}}{R_m K_c}$	K_{max}, R	C, R_m, K_c
44	**Thom-kins**	1971	$\frac{da}{dN} = \frac{\delta}{2} = C\left[\frac{K_{max}^2}{E R_{P0.2}} + \frac{4\pi\Delta\sigma\Delta\varepsilon_P a}{R_{P0.2}(1+n)}\right]$	K_{max}, $\Delta\varepsilon_P$, Δσ	E, $R_{p0.2}$, n, C
45	**Gille-mot**	1971	$\frac{da}{dN} = \frac{C}{W_c}\Delta\varepsilon_P^m$	$\Delta\varepsilon_P$	W_c, C
46	**Dona-bue**	1972	$\frac{da}{dN} = \frac{4C}{\pi R_{P0.2} E}(K_a^2 - K_{th}^2)$	K_a	K_{th}, $R_{P0.2}$, C, E
47	**Smith**	1972	$\frac{da}{dN} = C\left[K_{max}(1-R)^{0.5}\right]^n$	K_{max}, R	C, n
48	**Lee**	1972	$\frac{da}{dN} = C\left[K_{max}(1-R)^m\right]^n$	K_{max}, R	C, n, m
49	**Nord-berg**	1972	$\frac{da}{dN} = \frac{(\Delta K)^n}{(1-R)^m\left[(1-R)K_c - \Delta K\right]}$	ΔK, R	n, m, K_c

50	**Weer-man**	1973	$\frac{da}{dN}=\frac{\pi}{96R_{p0.2}^2}\frac{(\Delta K)^4 K}{K_c^2-K_{max}^2}$	ΔK, K_{max}	$Rp_{0.2}$, K_c
51	**Pook, Forst**	1973	$\frac{da}{dN}=\frac{9}{\pi}\frac{(1-\nu^2)\ \Delta K^2}{E}$	ΔK	ν, E
52	**Gille-mot**	1973	$\frac{da}{dN}=C\frac{(\Delta K)^2\Delta\varepsilon}{R_{p0.2}W_c}$	ΔK, Δε	$R_{p0.2}$, W_c, Δε, C
53	**Schw-albe**	1973	$\frac{da}{dN}=\frac{(1-2\nu)^2}{\pi^2(1+n_1^2)R_{P0.2}}\left[\frac{2R_{P0.2}(1+\nu)}{E}\right]^{1+n_1}\Delta K^2$	ΔK	$R_{p0.2}$, n, ν
54	**Mc Evily**	1973	$\frac{da}{dN}=\frac{16C}{\pi\ R_{P0.2}E}\left[K_a-K_{th}\left(\frac{1-R}{1+R}\right)^{1/2}\right]^2\frac{1-R}{1+R}$	K_a, R	C, $R_{p0.2}$, K_{ath}
55	**Jere-ma**	1975	$\frac{da}{dN}=C\left(\frac{K_{max}-K_{th}}{K_c-K_{max}}\right)^n$	K_{max}	C, n, K_{th}, K_c
56	**Pridd-le**	1976	$\frac{da}{dN}=C\left(\frac{\Delta K-\Delta K_{th}}{K_c-K_{max}}\right)^n$	ΔK	ΔK_{th}, K_c, n, C
57	**Bowie, Hoep-pner**	1976	$\frac{da}{dN}=e+(\nu-e)\left[-\ln\left(1-\frac{K}{K_c}\right)\right]^{1/K_1}$	ΔK, e, ν, K_1	ΔK_c, e, ν, K_1
58	**Kudrj-asov, Szmol-encev**	1976	$\frac{da}{dN}=C\frac{(1-R)^{n-1}(K_{max}-K_{th})^n}{K_c-K_{max}}$	K_{max}, R	K_{th}, C, n, K_c
59	**Bran-co Radon Culver**	1976	$\frac{da}{dN}=C\left(\frac{K_{mean}.\Delta K(\Delta K-\Delta K_{th})}{K^2{}_c-K^2{}_{max}}\right)^m$	K_{mean}, K_{max}, ΔK	C, m, ΔK_{th}, K_c
60	**Hudak et al.**	1977	$\lg\frac{da}{dN}=C_1-C_2 Arth\left\{\frac{\lg\left[(1-R)K_c\frac{\Delta K_{th}}{\Delta K^{2}}\right]}{\lg\left[\frac{\Delta K_{th}}{(1-R)K_c}\right]}\right\}$	ΔK, R	K_c, K_{th}, C_1, C_2
61	**Mc Evily**	1977	$\frac{da}{dN}=\frac{C}{R_{P0.2}E}[\Delta K-\Delta K_{th}(R)]^2\left(1+\frac{K}{K_c-K_{max}}\right)$	ΔK, R	C, $R_{P0.2}$, K_c, ΔK_{th}

62	**Irving, Mc Crtney**	1977	$\frac{da}{dN}=\frac{\pi}{4R_{p0.2}^2}\frac{(\Delta K)^2}{\ln[K_c^2/K_{max}^2]}$	K_{max}, ΔK	$R_{P0.2}$, K_{Ic}
63	**Ruff et al.**		$\lg\frac{da}{dN}=C_1+C_2 Arth\left[\frac{\lg\frac{K_c K_{th}}{K_{max}^2(1-R)^{2m}}}{\lg\frac{K_{th}}{K_c}}\right]$	K_{max}, K_a	K_c, K_{th}, m, C_1, C_3
64	**Hom-ma, Naka-zawa**	1978	$\frac{da}{dN}=\left\{C/\left[R_{P0.2}A^{\alpha}E^{\alpha(1-n_1)}\right]\right\}K_{max}^2 K_a^{\alpha/(2-n_1)}$	K_{max}, K_a	C, Rp0.2, A, α, E, n_1
65	**Marci**		$\frac{da}{dN}=C\left[\left(K_I^G-K_{Imin}\right)+\Delta K^T\right]\left[\left(K_{Imax}-K_I^G\right)-\Delta K^T\right]^2$	K_I^G, K_{Imax}, K_{imin}, $K_I^G=\alpha K_{Imax}$	C, ΔK^T

From the Table 3. it can be seen that for description of fatigue crack propagation circumstances

- the loading conditions are characterised by stress, stress intensity factor or strain parameters;
- the material response, behaviour is characterised by a great number of material parameters (from the experimentally determined ones up to standardised strength properties through the fracture mechanics parameters) [77, 80, 84, 87, 103].

The proposed relations describe either whole range of the fatigue crack growth (see the Fig. 15.) interval or a part of it.

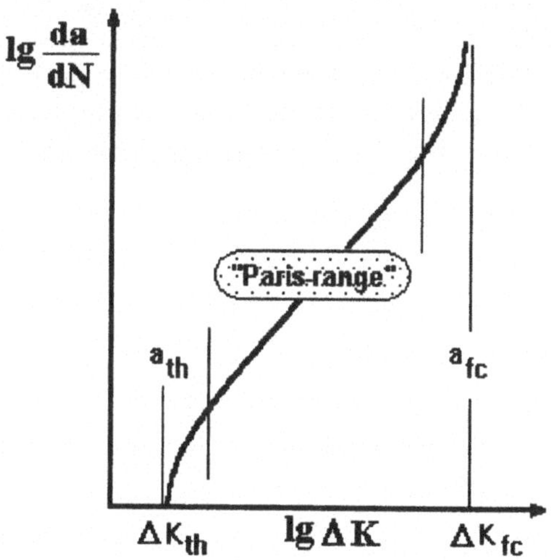

Figure 15. Different ranges of the fatigue crack growth from engineering application point of view

The short crack problem which may plays dominant role in the life time of cyclic loaded elements is not considered in the proposed models. The possible behaviour of short cracks are schematically summarised in Fig. 16 [48].

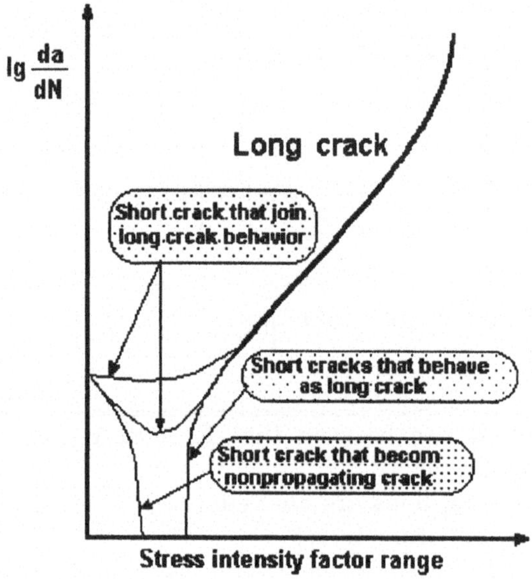

Figure 16. the possible behaviour of short cracks from engineering point of view.

In engineering applications the a_{th} - a_{fc} range has practical importance. The short crack problem cannot be considered in life time estimations. From this, it naturally follows that from engineering point of view either the "whole-range" models or the "part-range" models have practical interest. One of the most widely used "part range" model is proposed by Forman et. all (No 27 in Table 3) in 1967, or it modified version by Klesnil and Lukas in 1970 (No 36. in Table 3.). The "whole-range" models are summarised in Table 4.

TABLE 4. The "whole-range" models for description of fatigue crack growth

No	Author	Year	Relationship	Parameters	
				Loading	Material
1	Erdogan, Ratvani		$\upsilon = A(1-R)^q \dfrac{(K_{\mathrm{Im}ax} - K_{th})^m}{K_{fc} - K_{\mathrm{Im}ax}}$	K_{Imax}, R	A, q, m, K_{fc}, K_{th}
2	McEvily	1977	$\upsilon = A(1-R)^2 \dfrac{K^2{}_{\mathrm{Im}ax} - K_{th}}{K_{fc} - K_{\mathrm{Im}ax}}(K_{fc} - RK_{\mathrm{Im}ax})$	K_{Imax}, R	A, K_{fc}, K_{th}
3.	Wnuk			K_0, K_{Imax}, $\dot{K}_{\mathrm{Imax}}$, R	v_0, A, K_{fc},
			$\upsilon = \upsilon_0 \left\{ \ln \dfrac{K_{fc}^2 - K_{\mathrm{Imax}}^2}{K_{fc}^2 - K_0^2} + \dfrac{K_{\mathrm{Imax}}^2 - K_0^2}{K_{fc}^2} + \dfrac{A}{\dot{K}_{\mathrm{Imax}}}\left[\ln \dfrac{(K_{fc} + K_{\mathrm{Imax}})(K_{fc} - K_0)}{(K_{fc} - K_{\mathrm{Imax}})(K_{fc} + K_0)} - 2\dfrac{K_{\mathrm{Imax}} - K_0}{K_{fc}}\right]\right\}$, where $\dot{K}_{\mathrm{Imax}} = dK_{\mathrm{Imax}}/dt$; $K_0 = \inf(K_{th}, RK_{\mathrm{Imax}})$		
4	Freuden -thal		$\upsilon = \upsilon_0 \ln^m \left[\ln(K_{fc}/K_{th}) \lg^{-1}[K_{fc}/K_{\mathrm{Imax}}\right]$	K_{Imax}	v_0, m, K_{fc}, K_{th}
5	Kollipri-szt Davis, Feder-sen		$\upsilon = \upsilon_0 \lg^q \left[(1-R)K_{\mathrm{Im}ax}/K_{th}\right] K_{fc}/(1-R)K_{\mathrm{Im}ax}\,]$	K_{Imax} R	v_0, q, K_{fc}, K_{th}
6	Chu		$\upsilon = A(1-R)^{n-m} \dfrac{(K_{\mathrm{Im}ax} - K_{th})^n}{(K_{fc} - K_{\mathrm{Im}ax})^m}$	K_{Imax} R	A, n,m, K_{fc}, K_{th}
7	Chu		$\upsilon = A(1-R)^{n-m} \dfrac{K^n_{\mathrm{Im}ax} - K^n_{th}}{(K_{fc} - K_{\mathrm{Im}ax})^m}$	K_{Imax} R	A, n,m, K_{fc}, K_{th}
8	Jarema, Mikitisin	1975	$\upsilon = \upsilon_0 \left[\dfrac{K_{\mathrm{Im}ax} - K_{th}}{K_{fc} - K_{\mathrm{Im}ax}}\right]^q$	K_{Imax}	v_0, q, K_{fc}, K_{th}
9	Branco, Radon, Culver	1976	$\upsilon = \upsilon_0 (1-R^2)^q \left[\dfrac{K^2{}_{\mathrm{Im}ax} - K^2{}_{th}}{K^2{}_{fc} - K^2{}_{\mathrm{Im}ax}}\right]^q$	K_{Imax} R	v_0, q, K_{fc}, K_{th}
10	Andre-jkiv		$\upsilon = \upsilon_0 \left[\dfrac{(K_{\mathrm{Im}ax} - K_{th})^q}{(K_{fc} - K_{th})^2 - (K_{\mathrm{Im}ax} - K_{th})^q}\right]$	K_{Imax}	v_0, q, K_{fc}, K_{th}

11	Dower		$\upsilon = A(1-R)^{s+r-q} K_{\mathrm{Im}ax}^{s} \frac{(K_{\mathrm{Im}ax}-K_{th})^q}{(K_{fc}-K_{\mathrm{Im}ax})^r}$	K_{Imax}, R	A, r, s, q, K_{fc}, K_{th}
12	Bowie, Hoepner		$\upsilon=\upsilon_0\{\lg^q[K_{fc}/(K_{fc}-K_{Imax})]-\lg^q[(K_{fc}-K_{th})]\}$	K_{Imax}	v_0, q, K_{fc}, K_{th}
13	Anmsz, Vellejsz, Szansz		$\upsilon=\upsilon_0 \exp\left[(1-R)^m \frac{K_{\mathrm{Im}ax}^{2m}-K_m^{2m})}{(K_{\mathrm{Im}ax}^m}\right]$	K_{Imax}, R	v_0, m, K_m
14	Daggen		$\upsilon = A\left[K_{\mathrm{Im}ax}\frac{K_{\mathrm{Im}ax}-K_{th}}{K_{fc}-K_{\mathrm{Im}ax}}\right]^q$	K_{Imax}	A, q, K_{fc}, K_{th}
15	Saxena, Hudak		$\upsilon=\upsilon_0(1-R)^n\left[\frac{K_{fc}^n-K_{\mathrm{Im}ax}^n}{K_{\mathrm{Im}ax}^n}+\frac{B(1-R)^{n-m}}{K_{\mathrm{Im}ax}^n}\right]^{-1}$	K_{Imax}, R	v_0, B, m, n, K_{fc},
16	Tóth, Romvári	1980	$\upsilon=\upsilon_0\left\{-\lg\left[1-\left(\frac{K_{\mathrm{Im}ax}-K_{th}}{K_{fc}-K_{th}}\right)^m\right]\right\}^q$	K_{Imax}	v_0, m, q, K_{fc}, K_{th}

Analysing in Table 4. the collected relationships the following conclusion can be made:

- **the loading condition is characterised either by $K_{Imax,}$ only or both $K_{Imax,}$ R;**
- **the material response is characterised by 3-6 parameters from which two engineering meaning, i.e. the values of K_{fc} and K_{th}.**

From a comparison of the collected and Tables 3 and 4 which summarise relationships for fatigue crack growth (FCG) the following conclusions can be drawn:

- the empirical laws are valid only for a part of the FCG range;
- material response in the crack vicinity is characterised by different type and numbers of parameters;
- engineering damage process taking place during FCG is not included.

Only one of them is based on the description of the damage process which occurs during crack growth [71], and its parameters were determined for a number of materials [72, 98].

The aim of this model can briefly summarised as follows:

- The engineering damage process during fatigue crack growth can be described by failure of material bonds (i.e. atomic bonds or elementary specimens in the crack vicinity) or by decreasing the load capacity as a consequence of crack extension;
- The boundary conditions are expressed such that
    - ⇒ the value of damage of the structural element is equal to zero (**d =0**) for crack length $\mathbf{a_{th}}$ where the stress intensity factor range is $\Delta \mathbf{K_{th}}$, and the fatigue crack growth rate is equal to zero i.e. **da/dN=0**,
    - ⇒ the value of damage of the structural element is equal to unity (**d =1**) for crack length $\mathbf{a_{fc}}$ where the stress intensity factor range is $\Delta \mathbf{K_{fc}}$, and the fatigue crack growth rate is equal to infinity i.e. **da/dN≈∞** (this condition is be expected when considering the order of fatigue crack growth rate),
    - ⇒ At a given moment of crack propagation the value of damage is characterised on the one hand by a decrease of the ratio of the existing material bonds - which can be expressed by the fatigue crack growth rate - and on the other hand by the load capacity pertaining to this crack length.

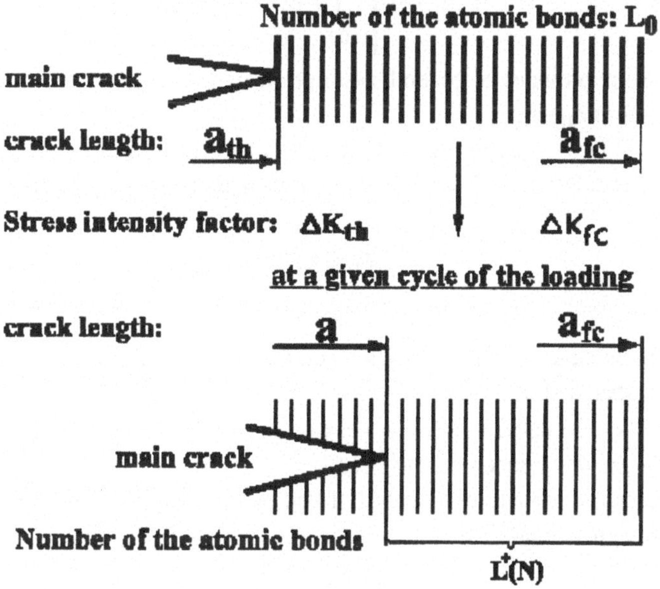

Figure 17. The fatigue crack propagation process based on crack growth characterised by engineering damage process

During the crack propagation process from a_{th} to a_{fc} a number of elementary bonds (atomic bonds, elementary tensile specimens) in the crack tip vicinity must fail. The number of the existing bonds within the range $a_{fc}-a_{th}$ is denoted by L_0. At a given number of load cycle (N), the crack length is equal to **a**, and the existing bonds ratio is assumed to be equal to

$$L(N) = L^*(N)/L_0 \tag{1}$$

$$-dL = L(N)\, f(N)\, dN \tag{2}$$

where the damage process is characterised by the function **f(N)**. If **f(N)** is equal to zero, it means that during this load cycle there is no failure of elementary bonds. The value of **f(N)**=1 means that all the existing elementary bonds fail in that dN cycle. It follows that **f(N)** has the range of $0 < f(N) \leq 1$. The number of elementary bonds after N load cycles can be calculated using the equation (2)

$$L(N) = \exp\left[-\int_{0,5}^{N} f(N)dN\right] \tag{3}$$

The damage value at this moment can be regarded as the unit minus the ratio of the existing bonds, i.e.

$$d = 1 - L(N) = 1 - \exp\left[-\int_{0,5}^{N} f(N)dN\right] \tag{4}$$

This relationship provides a general description of the damage process during the crack growth from $a_{th}-a_{fc}$.

The question is how to apply eqn. (4) for the characterisation of the fatigue crack propagation. Considering the boundary conditions detailed earlier, the function

$$g = \int_{0,5}^{N} f(N)dN \tag{5}$$

⇒ has to be equal to zero, i.e. $g \equiv 0$, when the damage value is equal to zero, i.e. d=0, since $d=1-1/e^0=0$

⇒ and has to be equal to infinity, i.e. $g \equiv \infty$, when the damage value is equal to unity, i.e. d=1, since $d=1-1/e^{\infty}=1$.

These circumstances are approximately fulfilled by the simple power type relationship of fatigue crack growth rate, i.e.

$$g = \left(\frac{da}{dN}\right)^b \tag{6}$$

The crack growth rate value is approximately zero at the crack length of a_{th} and infinity at the critical crack length of a_{fc}. Considering the fatigue crack propagation and the unstable crack growth rates, this supposition can be accepted. It means that the damage process during the fatigue crack propagation can be expressed by a Weibull function of the crack growth rate in the form of

$$d = 1 - \exp\left[-B\left(\frac{da}{dN} \right)^{b} \right] \tag{7}$$

The same approach can be applied for characterisation of interactions of dislocations and the Cottrell atmosphere in the dislocation theory. The same idea is formulated in the statistical theory of brittle fracture.

At a given moment, the damage value of a given element can be described by the loading capacity. Since the crack length of the structural element is equal to **a** at a certain moment, and the critical situation in the crack vicinity can be characterised by $K_{max.}$ in each cycle, the damage value can also be expressed by

$$d=[(K_{max}-K_{thmax})/(K_{fcmax}-K_{thmax})]^{n_2} \tag{8}$$

Besides the maximum values of the stress intensity factor, other fracture mechanics parameters in the crack tip vicinity can also be used, i.e. the stress intensity range, the effective value of the stress intensity range, the J-integral range, etc. It can be seen that this simple power type relationship is in good agreement with the boundary conditions, i.e. $d=0$ at $K_{max}=K_{thmax}$ and $d=1$ at $K_{max}=K_{fcmax}$. In the following discussion, the crack tip vicinity conditions will be described by the stress intensity factor range, i.e. by ΔK.

Combining equations (7) and (8) and applying other denotes an engineering damage process based fatigue crack propagation law can be proposed as in No 16 in Table 3. The parameters of this model can be calculated on the basis of experimental da/dN vs. ΔK curve [71, 72, 98]. One of the result is shown in Fig. 18. It can be seen that the proposed model closely describes the FCG diagram in its whole range.

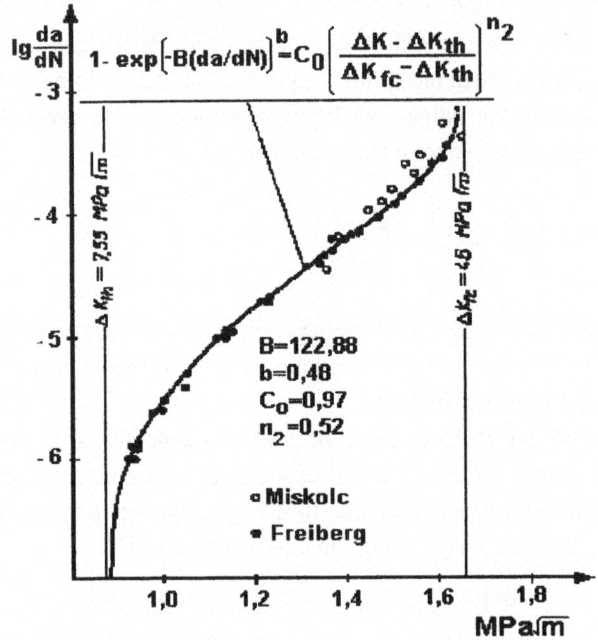

Figure 18. Fatigue crack growth diagram which was experimentally determined in different institutes and evaluated by the relationship No 16 in table 4.

4. Fatigue crack growth laws for variable amplitude loading

The damage process taking place at the crack vicinity depends on the local loading conditions and plastic behaviour of material. As the local condition is determined by external loading conditions, the crack propagation circumstance varies with the external loading parameters i.e. with overload, with variable load amplitude. For description of these situations the most widely used relationship was proposed by ELBER in 1971 (see No 40 in TABLE 3). This expression can be written in the following form:

$$da/dN=C(\Delta K_{eff})^{n} \tag{9}$$

where ΔK_{eff} is the effective value of the stress intensity factor which can be expressed

$$\Delta K_{eff} = U\,\Delta K \tag{10}$$

The value of can be calculated by the following way:

$$U = \frac{\sigma_{max} - \sigma_{op}(\sigma_{cl})}{\sigma_{max} - \sigma_{min}} \tag{11}$$

where σ_{max} and σ_{min} the maximum and minimum stress level in the given load cycle σ_{op} and σ_{cl} the crack opening and closing stress level respectively [89]. The value of **U** depends on the asymmetry factor of load. Some experimental results are summarised in TABLE 5 for aluminium alloys and steels [106].

TABLE 5. Different models for calculation of U value in Eg. (10)

	Material	Authors	U=f (R/K or K_{max})
1	RA, Ti-6 Al-4V	Katcher and Kaplan	U=0.73+0.82R 0.08<R<0.35
2	2219-T851	Katcher and Kaplan	U=0.68+0.91R 0.08<R<0.32
3	2024-T3 Al-alloy	Schijve	$U=0.55+0.35R+0.1R^2$ $\alpha=0.1$
4	2024-T3 Al-alloy	Elber	U=0.5+0.4R-0.1 -0.1<R<0.7
5	2219-T851 Al-alloy	Bell and Creagar	$U = \frac{1-C_f}{1-r}$, $C_f = \frac{\sigma_{op}}{\sigma_{max}}$
6	Steel A and C	Maddox	U=0.75+0.25R
7	Titanium alloy	Bachmann and Munz	$U=\frac{1}{1-R}\left[1-\frac{6{,}67R-4{,}27}{K_{max}}\right]$
8	6063-T6 Al-alloy	Srivastava and Garg	$U=\left(\frac{1{,}35R+5{,}925}{1000}\right)\Delta K+1{,}35R+0{,}223$
9	6063-T6 Al-alloy	Chand and Garg	$U=\frac{K_{max}}{1000}\frac{(880R+60)}{880}$

10	6063-T6 and 6061-T6 Al-alloy	Lal	$U=\frac{R}{60}+\frac{2}{3}\left(\frac{\Delta\sigma}{\sigma_y}\right)\left(\frac{1-n}{4}\right)$
11	6061-T6 Al-alloy	Kumar and Garg	$U=0.69+0.5R+0.12R^2 \qquad 0<R<0.3$
12	6063-T6 Al-alloy	Kumar and Garg	$U=0.55+0.6R+0.12R^2$

5. The fatigue crack growth resistance of materials

As it has been pointed out the responses of materials against crack growth can be characterised by a couple of material parameters. The selected ones depend on the type of crack growth law used. From an engineering point of view, the most widely applied is the Paris law in which the crack growth resistance is expressed by two experimentally determined parameters; **C** and **n**. A great number of tests were performed during the last three decades to determine the parameters of the Paris-law. In the second half of the seventies the first conclusions were published, that a close correlation among the parameters in the Paris-law exists and can be expressed by a straight line between the **LgC** and **n**. This statement is discussed in the literature. On the basis of the collected experimental results the correlation between the **LgC** and **n** for different types of materials is summarised in TABLE 6 where the parameters $\mathbf{A_1}$ and $\mathbf{B_1}$ of the $\mathbf{C=A_1/B_1{}^n}$ relationships are shown. The general loading condition (frequency, wave form and the environmental circumstances - i.e. room temperature, medium) for these data are the same. Because the effect of the factor of asymmetry on the second range of the fatigue crack propagation, described by the Paris-law, is not remarkable, the value of the asymmetry factor varies within the limits of $R\approx0.05\div0.7$. The correlation between A_1 and B_1 is of a rather stable nature as proven by an existing databank, which includes data for more than 500 steels.

Analyses of the power type relationship are used in evaluating the results of mechanical testing (i.e. in the field of low- and high cycle fatigue, creep). It can be proved that there is a correlation between the constants which appears as if the mechanism of damage process can be characterised by the parameters $\mathbf{A_1}$ and $\mathbf{B_1}$ for a group of materials. Relationships of this type have their advantages in assessment of the reliability of the cracked structural elements. For example, only one of the two parameters (i.e. the exponent of **n**) has to be considered in calculations of risk analysis.

TABLE 6. Correlation of the material parameters in Paris law

Type of material	No. of data	A_1	B_1	Correlation %	Range of validity
Steels	352	1.03E-4	27.20	98.9	$1.05 \le n \le 11.00$
Al-alloys non-tempered	47	5.15E-4	5.09	96.5	$2.00 \le n \le 5.69$
Al-alloys tempered	23	4.58E-5	39.79	99.7	$1.87 \le n \le 14.43$
Ti-alloys	43	2.25E-4	17.72	98.3	$2.04 \le n \le 6.21$
Cast-irons	45	4.62E-5	18.14	98.8	$3.00 \le n \le 8.25$

Where ΔK is measured in $MPa\sqrt{m}$, and da/dN is given in mm/cycle.

Another practical advantage of this close connection is that the material response can be characterised by only a single parameter. Considering that the material in the crack vicinity in the process zone is a self-organised system, this single material response cannot be a strength property or a deformation property, but both of them are included in an energy criterion. On the basis of this simple idea a connection was found between the specific absorbed fracture energy determined on smooth tensile specimen and the exponent of the Paris-law [78, 79]. The absorbed specific energy is the area under the true stress-true stain diagram. On the basis of the results of 18 different kinds of steels (mild-, rail-, case hardened-, Naxtra-, bainitic structure weldable-, St38, St52 and St45/3 type construction- and stainless steels) with the yield strength of R_Y=334-1074 MPa, and with an exponent of the Paris Law of n=1.91-5.11, a very good correlation has been found in the form

$$n = (5.0652 \pm 0.1168) - (0.00168 \pm 0.0001)\, W_0 \qquad (12)$$

with a correlation coefficient of 97.64%, and W_0 in units of MJ/m^3. This is represented in Fig. 18.

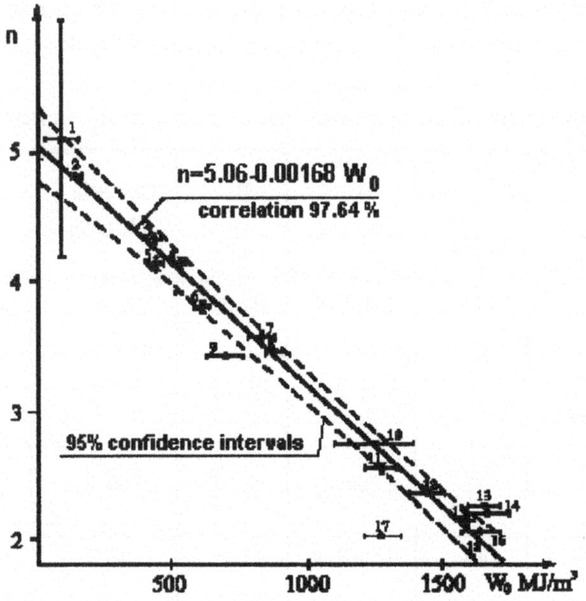

Figure 18. Correlation between the absorbed specific energy till fracture (W_0) and the exponent of Paris law (n).

This relationship shows that, by increasing the specific fracture energy of the material, the fatigue crack growth resistance increases as well. In principle it is obvious, because in the crack tip vicinity a given value of energy must be absorbed in the process zone in order to totally damage this zone.

6. Crack propagation index of structural elements

As mentioned before the dominance of crack propagation knowledge follows from steady development of non-destructive testing (NDT) methods applied during manufacturing and periodical control of structures [51, 52, 105]. Continuous use of up-to-date methods enables

- the observation and control of defects in structural elements of decreasing size with increasing reproducibility; and
- the detection of flaws in constructions which so far have been regarded as "defect-free".

If a crack like defect is detected, then many questions will arise and need to be answered, such as:

- Does the structure need immediate repair?

- Can it continue to operate?
- If it can continue to operate, what are the conditions?

During the supervision of a given construction element the most important questions are the following ones:

- Which of these possible flaws is the most dangerous one?
- How can the dangerousness of a flaw be unambiguously characterized?
- How does the dangerousness depend on the type of loading?

Many questions can be posed and have to be answered for the estimation of the risk of the final decision.

One of the basic problems in reliability assessment is: how the *crack propagation sensitivity of the structural element* having a crack like defect can be defined in terms of characteristic numbers? The definition is basically important because the NDT observations - loading conditions - crack growth resistance test results are connected [49, 65, 104] by means of a *crack propagation sensitivity index of the structural element* using fracture mechanics.

DEFINITIONS

Crack propagation sensitivity index of structural elements

Quasistatic loading

The crack propagation sensitivity is the derivative of the ***K*** vs. ***a*** function
(either the stress intensity factor or other fracture mechanics parameter vs. crack length).

Cyclic loading

The crack propagation sensitivity is the slope of the *logarithm of residual life time* vs. *crack length* function.

According to this definition the *crack propagation sensitivity index of the cyclically loaded structural element* is the slope of the curve relating the *logarithm of residual life time vs. crack length* which is illustrated in Fig. 19.

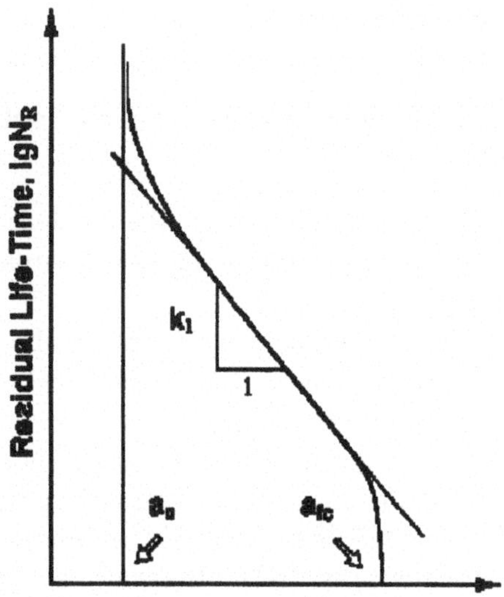

Figure 19. The definition of the *crack propagation sensitivity index* of the cyclically loaded structural element.

Considering the residual lifetime N_R to be given by

$$N_R = \int_a^{a_{crit}} \left(\frac{da}{dN}\right)^{-1} da \tag{13}$$

where the fatigue crack growth rate, da/dN is a function of geometry, loading condition, crack position, crack length and material properties, the *crack propagation sensitivity index* depends on

- type of construction element;
- crack position;
- crack length;
- loading condition (modes I., II. or III.);
- load level (non-linear); and
- local fatigue crack growth resistance of material.

Therefore, when the NDT requirement system is prescribed for different types of cyclically loaded construction elements made of different material, the above mentioned parameters have to be considered for reaching the same reliability of the safety assessment. This is illustrated in Fig. 20/a. and 20/b. for different structural elements and for the same element made of different materials.

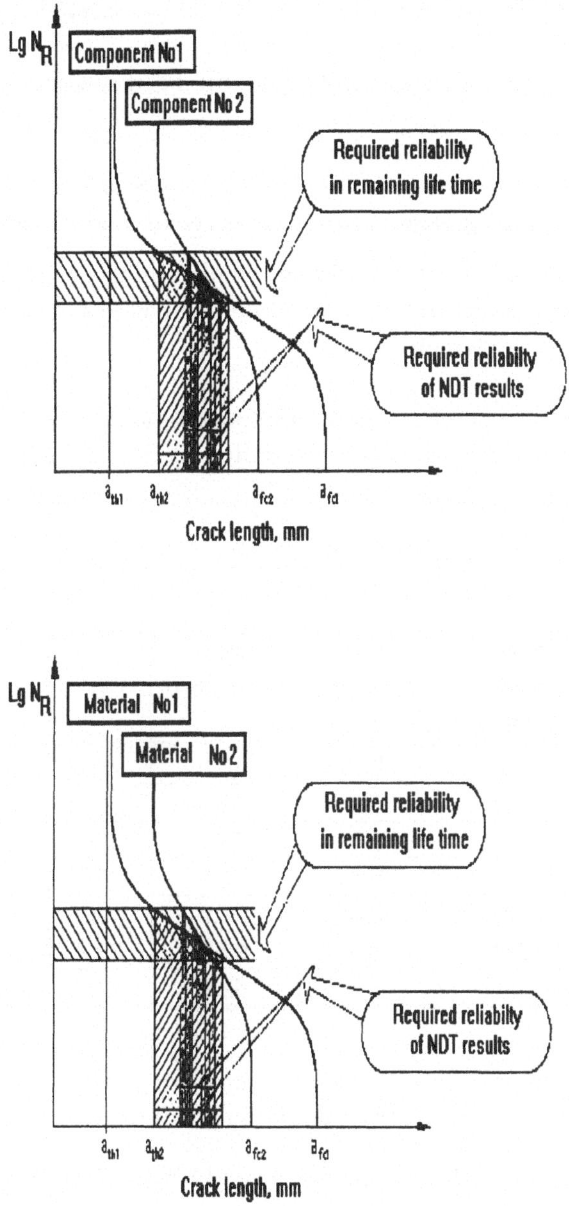

Figure 20. The effect of structural elements and materials on crack sensitivity index of cyclic loadaed components.

7. Practical application of crack propagation sensitivity index

In this section the following problems will be discussed by worked out examples:

- effect of yield stress on the dangerousness of the crack like defects with different position in the same type of construction element;
- effect of crack growth resistance distribution on the dangerousness of the crack like defects with different position in the same type of construction element.

Selecting different type of steels with different yield stress, the parameters of the fatigue crack growth law represented by TOTH model are shown in Table 7.

TABLE 7. The charactersistics of selected materials with different yield strength

Type of steel	Yield Strength [MPa]	ΔK_{th} [MPa√m]	ΔK_{fc} [MPa√m]	B	b	n_2
St 38	281	5.3	40.0	101.8	0.459	0.575
St 52	334	5.9	46.0	122.9	0.458	0.524
H 60	379	5.4	44.8	202.4	0.523	0.608
H 75/3	673	5.8	63.2	316.8	0.586	0.486
NAXTRA	801	5.4	71.0	144.6	0.506	0.523
NK	1150	4.5	97.9	73.4	0.502	0.612
K 13	1480	4.5	62.0	110.9	0.521	0.611

Selecting a plate type structural element with width **w=20 mm** and thickness **t=1** mm, the following calculations will be performed:

Determination of the a_{th} and a_{fc} values and the crack growth sensitivity index k_1 for:

- central cracked and edge cracked elements under tension and bending;
- cyclically loaded with stress amplitude $\Delta\sigma$=*75% of the actual yield stress.*

The main goals of these calculations are to show that

- the same value of the safety factor based on the yield stress cannot be used for reliability assessment of cyclically loaded structural components;
- the surface cracks are more dangerous than the embedded ones;
- the crack sensitivity index increases by increasing the strength of steels .

The a_{th} and a_{fc} values of elements cyclically loaded with stress amplitude equal to 75% of the actual yield strength are summarised in Fig 21a. and Fig.21b.

These figures are proof that:

- using the same value of the safety factor which is based on the yield stress, crack like defects become more and more dangerous with increasing strength;
- surface cracks are more dangerous than embedded ones which emphasizes both theoretical and practical importance of surface cracks [55, 86, 90-93,95-97].

The *crack growth sensitivity index against yield stress for cyclically loaded elements* is illustrated in Fig. 22. The loading stress amplitude was 75% of the actual yield strength. These curves strengthen the earlier conclusions which should be completed by the fact that, during supervisions, NDT methods have to be used to detect the crack lengths with better and better accuracy to reach the same risk in residual life time assessment. It has to be emphasised that the values of the *crack growth sensitivity index* rise by up to three times by increasing the yield strength. It can also be seen that a surface crack in a structural element under tension is the most dangerous case.

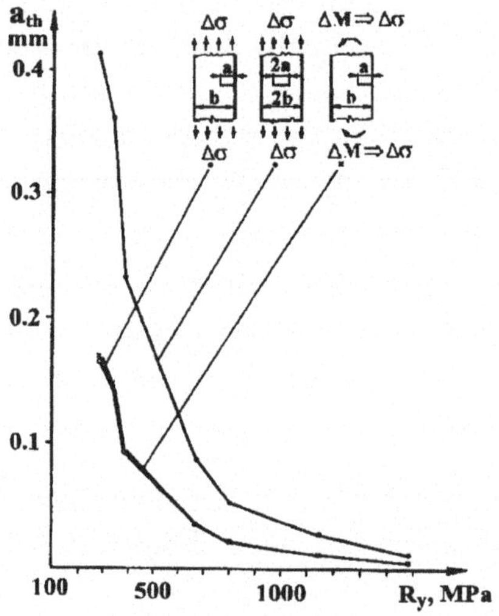

a,

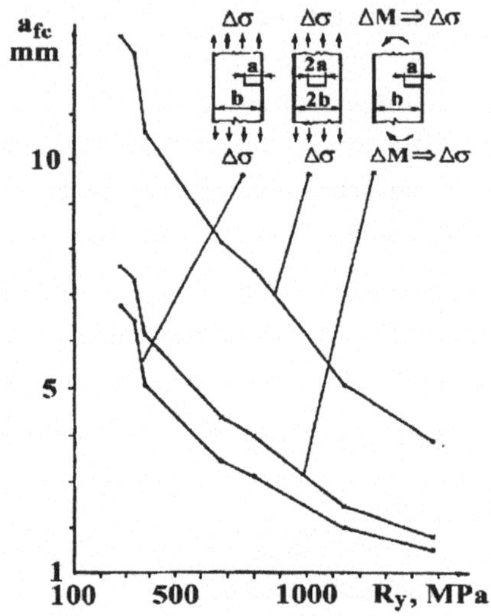

b,

Figure 21.The a_{th} and a_{fc} values of elements cyclically loaded with stress amplitude equal to 75% of the actual yield strength

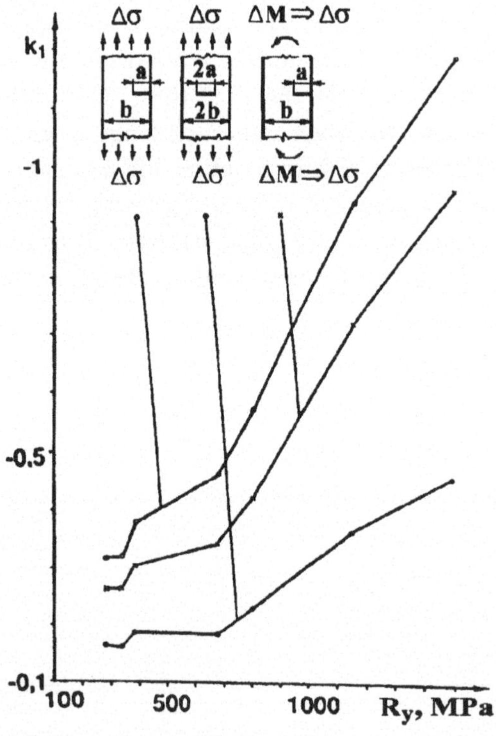

Figure 22. The crack growth sensitivity index of cyclically loaded elements

8. Conncluding remarks

Considering the aim of this paper and the presented results, the following conclusions can be drawn:

1. The development of fatigue experience is motivated by the most important inventions. On the basis of this statement four very important, but overlapped periods can be defined, i.e. the railway age, automobile period, aeroplane age and "living with defects" period.
2. Different models are proposed for description of fatigue crack propagation circumstances under constant and variable amplitude loading conditions.
3. The crack growth resistance of materials in these relationships are characterised by different parameters.
4. It was proposed that such a fatigue crack growth law be based on description of an engineering damage approach.
5. Close correlation exists between the parameters of the Paris law, which makes possible life time estimations using a single material parameter.
6. The exponent in the Paris law can be estimated by specific fracture energy which is the area under the true stress vs. true strain tensile diagram.
7. The reliability assessment of cracked structural components needs to be based on the co-operation of specialists working in the field of *NDT-Mechanical Testing-Fracture Mechanics*.
8. A system for characterisation of *crack propagation sensitivity index of the construction elements* for *quasistatic* and *cyclic* loading conditions has been proposed.
9. The application of the *crack propagation sensitivity index of the construction elements* provides the possibility to join reliability assessment calculations and the reproducibility of the NDT or crack growth resistance test results.
10. The effect of surface cracks on the reliability of structural components is more dangerous than that of other types of cracks, and therefore mechanical description and detection of surface flaws have to be central problems.

Acknowledgements: Part of this work has been initiated be the research supports of Hungarian National Scientific Foundation OTKA T-4408 and OTKA T-15601which is gratefully acknowledged.

References

1. Albert W.A. (1838) Driving ropes in the Harz (in German). *Arc. Minneralogie, Geognoise, Bergbau und Hüttenkunde* **10,** 215-234
2. *The History of the First Public Railway. Stockon and Darlington. The opening day and what followed* (1912) Edited by M. Havisides. Heavisides and Son.
3. Tóth L., Haase D.W., Sebek M (1994) *Short History of Material Testing.* (in Hungarian). MTS Training Centre. Miskolc. 50p.
4. Smith R. A.(1990) The Versailles Railway Accident of 1842 and the First Research into Metal Fatigue. *Proc. of 4th. International Conference on Fatigue and Fatigue Thresholds.* 15-20 July, Honolulu. Ed. by H. Kitagawa, T. Tanaka. Materials and Component Engineering Publications Ltd. Birmingham. Vol. IV. 2033-2041.
5. Rankie, W.J.M. (1843) On the causes of the unexpected breakage of the journals of railway axles; and on the means of preventing such accidents by obsering the law of continuity in their construction. *Min. Proc. Inst. Civ. Enggineers* **2**.105-108.
6. Braithwaite F. (1853) On the Fatigue and Consequent Fracture of Metals. *Min. Proc. Inst. Civ. Eng.,* **13** 463-475.
7. Wöhler A. (1858) Report on tests of the Königl. Niederschleesisch-Märkkischen Eisenbahn made with apparatus for the measurement of the bending and torsion of railway axles in service (in German). *Zeitsch. Bauwesen* **8** 642-651.
8. Mann J.Y. (1970) *Bibliography on the Fatigue of Materials, Components and Structures 1838-1950.* Prgamon Press, Oxford.
9. Schütz W. (1970) To the history of fatigue resistance (in German). *Mat.-wiss. u. Werkstofftechn.* **24** 203-232.
10. Messadié G. (1991) *Great Inventions through History*. Chambers. 237 p.
11. Rolt L. T. C. (1970) Victorian Engineering. Penguin Books. 300 p.
12. Bunch B., Hellemans A. (1993) *The Timetables of Technology*. Simon & Schuster. 490 p.
13. Paturi F. R. (1991) *Chronic of Tecnical Sciences* (in Hungarian) Officina Nova. Budapest.
14. Basquin O.H. (1910) The Exponential Law of Endurance Tests. *Proc. Am. Soc. for Testing Materials.* Vol. **10.**
15. *Fracture Mechanics Retrospective: Early Classic Papers.* (1913-1965). (1987) Edited by Barsom J. M. ASTM.

16. Weiss, V., Yukawa Y. (1965) *Critical Appraisal of Fracture Mechanics.* ASTM STP 381.
17. Rossmanith H.P.(1993) Biographical Sketch of Prof. Dr. G.R. Irwin. *ISTLI Founding Symposium. Vienna. 18-19 November.*
18. Panasjuk V. V. (1993) An Outline of the Development of Fracture Mechanics and Strength of Materials Investigations. ICF 8. Kiev. A View from the Eastern Europe. Lviv,. p.7-48.
19. Girvin H. F. (1948) *A Historical Appraisal of Mechanics.* International Textbook Company. Scranton. Pennsylvania.. 253 p.
20. Timoshenko S. P. (1953) *History of Strength of Materials.* Mc.Graw-Hill Book Company. Inc.
21. Timoshenko S. P.: Strength of Materials (1936). D. V. Nostrand Company, Inc. 1st edition in 1936, 2nd edition in 1941., 3rd edition in 1956.
22. *Fatigue of Aircraft Structures* (1963) Edited by Barros W., Ripley. Pergamon Press.. International Series of Monographs in Aeronautics and Astronautics.
23. *Mechanics during the last 50 years in Soviet Union* (1972) (in Russian) Edited by: Szedov, L. I., Lavrentev, M. A., Mikhajlov, G. K., Muszhelisvili H. I., Chernov, G. G. Nauka. Moszkva. Parton V. Z., Cherepanov G. P. Fracture Mechanics (in Russian) 365-468.
24. Barenblatt G. I. (1961) The Mathematical Theory of Equilibrium Cracks in Brittle Fracture. *Prikladnaja Matematika i Tekhnicheskaja Fizika.* **No4**, 55-129.
25. Wieghard K.(1907): Über das Spalten und Zerreißen elastischer Körper. *Zeitschrift für Mathematik und Physik* **55** No1-2. 60-103.
26. Inglis C. E. (1913) Stresses in a Plate due to the Presence of Cracks and Sharp Corners. *Transactions of Naval Architect.* **55** 219-230.
27. Griffith A. A. (1920) The Phenomena of Rupture and Flow in Solids. Philosophical Transactions of the Royal Society. Vol. 221.
28. *Fatigue Crack Growth. 30 Years progress* (1984) Edited by R.A. Smith. Pergamon Press.
29. Tóth L. (1995) Brief History of Fracture Mechanics. (in Hungarian). *Gép* 1995 **No.8**. 22-28.
30. Westergaard H. M.(1939) Bearing Pressures and Cracks. *Trans. ASME* **61**, A-49.
31. Sneddon I. N.(1946) The Distribution of Stress in the Neighbourhood of Crack in an Elastic Solid. *Proc. R. Soc. Ser.* **A 187**, 229.
32. Irwin G. R.(1948) *"Fracture Dinamycs". Fracturing of Metals.* American Society for Metals. Clevland, 147-166.
33. Irwin G. R. (1956) Onset of Fast Crack Propagation in High Strength Steel and Aluminium Alloys. *Sagamore Research Conference Proceedings.* Vol. 2. 289-305.
34. Irwin G. R. (1957) Relation of Stress Near a Crack to the Crack Extension Force. *Int. Congr. Appl. Mech. Brussels.* 8 245-251.

35. Cherepanov G. P. (1967) On Crack Propagation in Continuum. *Prikladnaja Matematika i Mekhanika* **No 3** 467-493.
36. Rice J. R. (1968) A Path Independent Integral and Approximate Analysis of Strain Concentration by Notches and Cracks. *Trans. ASME. J. Appl. Mech* 379-386.
37. Begley J. A., Landes J. D.(1972) The J-Integral as a Failure Criterion. *ASTM STP 514* 1-20.
38. Wells A. A.(1961) Critical Crack Opening Displacement as Fracture Criterion. *Proc. of the Crack Propagation Symp. Cranfield.* Vol. I. 210-221.
39. Dugdale D. S. (1960) Yielding of Steel Sheets Containing Slits. Int. Journ. of Mech. and Phys. of Solids. Vol. 8. 100-104.
40. Sih G. C.(1973) Some Basic Problems in Fracture Mechanics and New Concepts. *Engineering Fracture Mechanics* Vol.5. 229-234.
41. Paris P. C., Gomez R. E., Anderson W. E. (1961) A Rational theory of Fatigue. *The Trend in Engineering* **13** No.1. 9-14.
42. Paris P.C., Erdogan F. (1963) A critical analysis of crack propagation laws. *Trans. ASME, Series D.* **85** 528-534.
43. Paris P. C. (1982) Twenty Years of Reflection on Questions Involving Fatigue Crack Growth. in *Fatigue Thresholds - Fundamentals and Applications.* Edited by Bäcklund J., Blom A. F., Beevers C. J., EMAS, Warley, Birmingham. 3-10.
44. Grigorjan A. T., Pogrebiysskijj I. B. (1972) *History of Mechanics* (in Russian) Nauka. Moskva.
45. *Fatigue Crack Growth. 30 Years of Progress* (1984) Pergamon Press. Edited by R. A. Smith.
46. Derry T. K., Williams T. I. (1993) *A Short History of Technology. From the Earliest Times to A.D. 1900.* Dover Publications Inc. New York. 782 p.
47. Shin-ichi Nishida (1990) Failure *Analysis in Engineering Applications.* Butterworth, Heinemann.
48. *Fatigue Design Handbook.* Society of Automotive Engineers. Second Edition (1988).
49. Tóth L (1990) A computer aided assessment system of reliability cyclic loaded construction elements having flaws. *Proc.1st Int.Conf. on Computer-Aided Assessment and Control of Localized Damage.* Portsmouth, UK. 1990 (Edited by M.H.Aliabadi, C.A.Brebbia, D.J.Cartwright), Vol.1. pp.39-53, Springer-Verlag.
50. *Metals Handbook, Volume 8. Mechanical Testing.* American Society for Metals. (1985).
51. *Metals Handbook, Volume 17. Nondestructive Evaluation and Quality Control.* (1989).
52. *NDE Handbook* (1989) *Non-Destructive Examination Methods for Condition Monitoring.* (Edited by K.G. BØving), Butterworths.

53. Broek D. (1988) *Practical Use of Fracture Mechanics.* Kluwer Academic Publishers.
54. Blumenauer H., Pusch G. (1993) *Technical Fracture Mechanics* (in German). Deutscher Verlag für Grundstoffindustrie, Leipzig Stuttgart.
55. Carpinteri A. (1990) Surface flaw under cyclic bending loading. *Proc.1st Int.Conf. on Computer-Aided Assessment and Control of Localized Damage.* Portsmouth, UK. 1990 (Edited by M.H.Aliabadi, C.A.Brebbia, D.J.Carrtwright), Vol.1. pp.147-158, Springer-Verlag.
56. Tada H, Paris P.C. and Irwin G.R. (1973) *The Stress Analysis of Cracks Handbook.* Del Research Corp., Hellertown, Pa., U.S.A.
57. Sih, G.C. (1973) *Handbook of Stress-Intensity Factors for Researchers and Engineers.* Leigh University, Bethlehem Pa.
58. Rooke, D.P., Cartwright D.J. (1976) *Compendium of Stress Intensity Factors.* Her Majesty's Stationery Office, London.
59. *Stress Intensity Factors Handbook* (1987) Edited by Y. Murakami, Pergamon Press
60. Savruk M.P. (1988) *Stress Intensity Factors of Bodies Having Cracks* (in Russian). In Fracture Mechanics and Strength of Materials Vol.2. Edited by V.V. Panasjuk, Naukova Dumka, Kiev.
61. Pook L.P. (1986) Keyword scheme for proposed computer-based bibliography of stress intensity factor solution. *NEL Report N°704,* Department of Trade and Industry.
62. Pook L.P. (1989) Unacceptable differences in published stress intensity factor solutions. *Fat. Fract. Eng. Mat. Struct.* N°1, 67-69.
63. Benthem J.P., Koiter W.T. (1973) Asymptotic approximations to crack Problems. In *Mechanics of Fracture* Edited by G.C. Sih, 131-178. Nordhoff, Leyden.
64. Kinzler R. (1993) *Concept of Fracrure Mechanics* (in German). Vieweg, Braunschweig.
65. Tóth L., Lukács J. and L. Olajos (1984) A computer based database for calculation of the stress intensity factor of different structural elements (in Hungarian). *pp.235-239 GTE, Miskolc.*
66. Pook L.P. (1983) *The role of Crack Growth in Metal Fatigue.* The Metal Society, London.
67. Tóth, P. Romvári, Gy. Nagy (1980) Remarks on the fatigue crack propagation laws (in Hungarian), *Gép* **33** 325-333.
68. Tóth, P. Romvári, Gy. Nagy (1980) Remarks on the fatigue crack propagation laws (in Russian), *Problemy Prochnosty,* N°12 pp.18-28.
69. L.Tóth, P.Romvári, Gy. Nagy (1983) Remarks on the practical use of statistical methods in fracture mechanics considering the reproducibility of crack growth behaviour of materials (in Russian). *Problemy Prochnosty* N°11 54-59.
70. Romaniv O.N., Jarema A.Ja. et. al, (1990) *Fatigue and Fatigue Crack Growth Resistance of Construction Materials* (in Russian), in Fracture

Mechanics and Strength of Materials Vol.4. Edited by V.V. Panasjuk, Naukova Dumka, Kiev.

71. Tóth L. (1981) Describing the fatigue crack growth circumstances by damage process (in Hungarian). *Gép* **33**, 257-262.
72. Tóth, L., Romvári P. (1981) A damage accumulation model for description of fatigue crack propagation. In *Basic Mechanisms in Fatigue of Metals* Edited by P. Lukas and J. Polak, 297-304, Academia, Prague.
73. Troshhenko V.T., Sosnovvsky L.A.(1981) *Handbook of Fatigue Resistance of Metals and Alloys* (in Russian). Vol.1 and 2. Naukova Dumka, Kiev.
74. Troshhenko V.T., Pokrowsky V.V. and A.V. Prokopenko (1987) *Fatigue Crack Growth Resistance of Metals at Cyclic Loading* (in Russian). Naukova Dumka, Kiev.
75. Kogaev V.P., Makhutov N.A.,. Gusenkov A.P (1985) *Strength and Life Time Dimensioning of Constructions- and Maschine Elements* (in Russian). Mashinostroenie, Moskva.
76. Makhutov N.A (1981) Deformation Criteria of Fracture and Design Of Structural Components for Strengths. Moskva. Mashinostroenie.
77. Ivanova V.S., Terentew V.F. (1975) *Nature of Mertals Fatigue* (in Russian). Metallugija, Moskva.
78. Tóth L., Romvári P. (1982) A correlation of absorbed specific energy with the exponent in Paris equation of fatigue crack growth. *Proc. Symp. Absorbed Specific Energy and/or Strain Energy Density Criterion.* (Edited by F. Gillemot and G. Sih) 355-358, Akadémiai Kiadó, Budapest.
79. Tóth L., Romvári P. (1981) Fatigue crack growth resistance and absorbed specific energy of materials (in Hungarian). *Gép* **33**, 281-285.
80. Sosnowsky L.A. (1987) *Statistical Nature of the Fatigue Fracture* (in Russian). Nauka i Tekhnika, Minsk.
81. Trufjakov V.I. (1990) *Strength of Welded Joints at Cyclic Loading* (in Russian). Naukova Dumka, Kiev.
82. Panasjuk V.V. (1991) *Mechanics of Quasibrittle Fracture of Materials* (in Russian). Naukova Dumka, Kiev.
83. Sähn S., H. Göldner (1989) *Fracture and Assessment Criteria in Strength Konzept* (in German). VEB Fachbuchverlag, Leipzig.
84. *Cyclc Deformation and Fatigue of Metals* (1993) Edited by M. Bily. Elsevier, Amsterdam.
85. *Mechanics of Materials and Fracture*(1992) (in Italian) Edited by A. Carpinteri. Pitagora Editrice Bologna.
86. Carpinteri A., (1982) Fatigue growth of a surface crack in an elastic plate subjected to cyclic tensile loading. *Proc. 8th Cong. Mat. Test.* 327-331, Budapest.
87. *Cyclic Deformation and Fatigue of Metals* (1985) (in Russian) Edited by V.T.Troshhenko. Vol.II. Naukova Dumka, Kiev.

88. *Mechanics of Fatigue Crack Closure.* (1988) Edited by J.C. Newman and W. Elber, ASTM STP 982.
89. Schijve J.(1988).Fatigue crack closure: Observations and technical significance. In *ASTM STP 982,* 5-34.
90. Carpinteri A., (1992) Crack propagation under cyclic loading. *Fat. & Fract. of Eng.Mat. & Struct.* **15**, N°4, 265-376.
91. Carpinteri A. (1991) Stress-intensity factors for semi-elliptical surface cracks under tension and bending. *Eng Fract. Mech.* **38**, N°4/5, 324-334.
92. Carpinteri A. (1992) Elliptical-arc surface cracks in round bars. *Fat. & Fract. of Eng.Mat. & Struct.* **15**, N°11 1141-1153.
93. Carpinteri A. (1992) Stress intensity factors for straight-fronted edge cracks in round bars. *Eng Fract. Mech.* **42**, N°6 1035-1040.
94. Carpinteri A. (1992) An automated procedure for fatigue crack growth analysis. *Computers and Stuctures* **44**, N°6 1317-1338.
95. Carpinteri A. (1993) Surface flaw stress intensity factor computation with quarter-point elements. *Journ. Strain Anal.* **28**, N°2, 117-123.
96. Carpinteri A.(1993) Shape change of surface cracks in round bars under cyclic axial loading. *Int.J. Fatigue.* **15**, N°1, 21-26.
97. Carpinteri A. (1990) Combined tension and bending stress-intensity factor of surface crack modelled as spring elements. *Theo. Appl. Fract. Mech..* **14**, N°3, 243-251.
98. Tóth L. (1994) Reliability Assessment of Cracked Structural Elements under Cyclic Loading. in Handbook of Fatigue Crack Propagation in Metallic Structures. Edited By A. Carpinteri. Elsevier. 1643-1683.
99. Halmshaw R. (1995) The Discovery of X-Raxs and the Early History of Industrial Radiography. INSIGHT, *Non-Destructive Testing and Condition Monitoring.* **37** No.9. 669-671.
100. Knell M. (1995) A Brief History of X-ray Film- "A valuable Engineering Tool, using a new Kind of Light" INSIGHT, *Non-Destructive Testing and Condition Monitoring.* **37** No.9. 673-675.
101. Hoppemkamps U. (1996) The first DGS-Diagram of 1960 by Dr. J. Krautkrämer. Privat communication.
102. Allen R.J., Booth G.S., Jutla T. (1988) A Review of Fatigue Crack Growth Characteristisation by Linear Elastic Fracture Mechanics (LEFM). Part I. - Principles and Methods of Data Genaration. Fatigue and Fracture of Engoneering Materials and Structures **11** No 1. 45-69.
103. Allen R.J., Booth G.S., Jutla T. (1988) A Review of Fatigue Crack Growth Characteristisation by Linear Elastic Fracture Mechanics (LEFM). Part II. -Advisory Documents and Applications Within National Standerds. Fatigue and Fracture of Engoneering Materials and Structures **11** No 2. 71-108.
104. *Resistance of Materials against Deformation and Fracture* (1993) Vol. 1 and 2. Edited by V.T. Troshhenko, (in Russian) Kiev, Naukova Dumka.

105. *Equipments for NDT Control of Materials and Structures* (1976) Vol.1 and 2. Edited byV.V. Klujev, (in Russian), Moskva. Mashinostroenie.
106. Sigh S.B., Kumar R. (1993) Experimental Observations of Fatigue Crack Growth in IS-1020 Steel under Constant Amplitude Loading. *Int.J. Ves.& Piping* **53** 217-227.
107. Bily M. (1989) *Dependability of Mechanical systems* Elsevier 390 p.
108. Cacko J., Bily M., Bukoveczky J. (1988) Random Processes: Measurement, Analysis and Simulation. Elsevier 234 p.

NON-LINEAR DEFORMATION AND FATIGUE FRACTURE IN ENGINEERING DESIGN

NIKOLAY A. MAKHUTOV AND MICHAEL M. GADENIN

International Institute of Engineering Safety

4 Griboedov Str.

101 830 Moscow, Russia

1. Engineering Approaches in Structures Design

Futhure largescale technology projects will entail the risk of many hazards with failure having regional, national and global consequences.

Further development of complex technical systems within the lifetimes ranging from seconds (rocket-space vehicles) to 50-100 years (nuclear reactors, engineering facilities) without regard to new safety criteria which characterise these systems threatening people and the environment, should be considered unacceptable. Quantitative substantiation for conditions of emergency should be calculated not only for normal operation conditions, but also for external ones which are caused by, for example, fractures, explosions, fires, leakages of radioactive and toxic substances, earthquakes, hurricanes, tsunami, aircraft and space vehicle crushes and subversive actions. Safety assurance problem will be of vital importance for the immediate future in Russia due to the expiration of design life of a large number of power units (including atomic ones), chemical and transportation equipment, complete replacement or modernisation of which will require significant financial and intellectual expenditures.

The following classification of these objects can be offered; taking into account their design structural peculiarities, the level of potential hazard to people and environment in case of technical and natural catastrophy generation [1-13]:

R.A. Smith (ed.), Reliability Assessment of Cyclically Loaded Engineering Structures, 273–300.

- nuclear power engineering sites;
- chemical plants;
- special equipment - rockets, space vehicles, computer-aided systems;
- unique engineering structures;
- civil engineering sites;
- traditional and non-traditional power engineering sites;
- objects of machine building and metallurgy industries;
- transport systems;
- main pipelines;
- equipment for operation in low temperature conditions (Arctic equipment).

Owing to comparative similarity of the general mechanisms of emergency propagation in various types of facilities, it would be advisable, while forming general structural codes, to foresee two levels of life and safety norms and standards.

For the above mentioned structures with general service lives between 10^1 to 10^9 s it is possible to allocate the following ranges of loading cycles:

10^0 to 10^1 - extreme cycles (start-up, test, breakdowns);

10^2 to 10^3 - operational cycles (start-up, regulation of capacity, operation of protection systems);

10^4 to 10^5 - operational cycles (technological cycles, regulation);

10^6 to 10^8 - operational cycles (technological, transport cycles, changes of pressure);

10^9 to 10^{12} - operational cycles (vibration, changes of temperature and pressure).

The fatigue of metal constructional materials between 10^0 and 10^{12} cycles that are used in complex technical objects, has four characteristic regimes:

10^0 to 10^3 - low cycle quasistatic or fatigue fracture in the presence of large microplastic deformations in a zone of failure (when amplitude of stresses $\sigma_a >> \sigma_y$, where σ_y is the yield stress);

10^3 to 10^5 - low cycle fatigue fracture in the presence of small macroplastic deformations in a zone of failure (when $\sigma_e \leq \sigma_a \leq \sigma_y$, σ_e is the limit of elasticity);

10^5 to 10^8 - classical high cycle fatigue fracture in the presence of microplastic deformations in micro and macro volumes near a zone of failure (when $\sigma_a \leq \sigma_e$);

and 10^9 to 10^{12} - fatigue fracture at very high numbers of cycles in the presence of microplastic deformations in microvolumes near a zone of failure. ($\sigma_a << \sigma_e$)

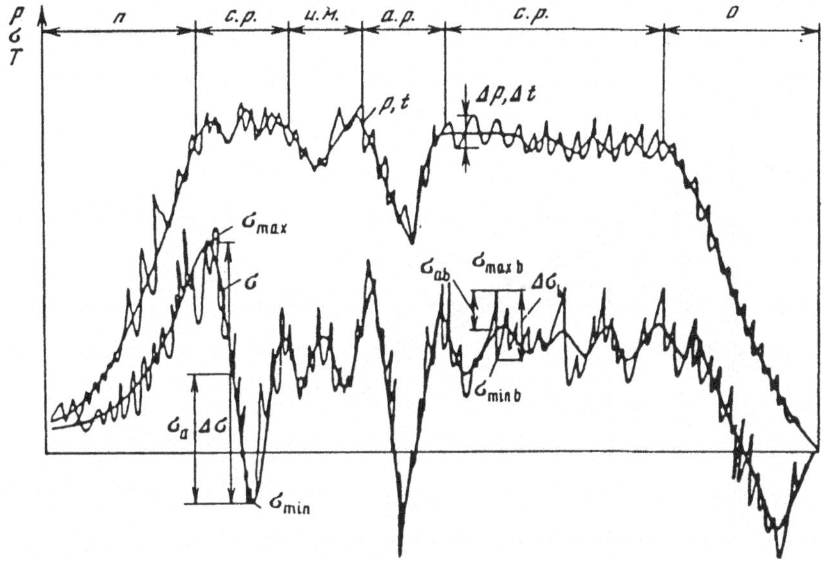

Fig. 1. The diagram of change for service loading parameters

On Figure 1 is shown a schema of change at of time τ of service loading parameters for the body of a PVR-reactor. Such parameters are maximal and amplitude values:

- pressure of the heat carrier $\eta(\eta_{max}, \eta_a, \Delta\eta)$;
- temperature $t(t_{max}, t_{min}, t_a, \Delta t)$;
- nominal and local stresses $\sigma(\sigma_{max}, \sigma_{min}, \sigma_a)$ or strain $e(e_{max}, e_{min}, e_a)$.

Main service modes are:

- starts-up (St) and hydrotests (Hd);
- stationary modes (Sr);
- change of capacity (Cp);
- emergency modes (Eg);
- stop-down (Sp).

The occurance of changes of pressure Δp, changes of temperatures Δt, and vibrations give high-frequency stresses amplitudes σ_{ah}. It creates two or multifrequent loading with the frequencies relations $f_a / f_{ah} = 10^1 \div 10^5$ and asymmetry factor $r_\sigma = \sigma_{min} / \sigma_{max} > 0$.

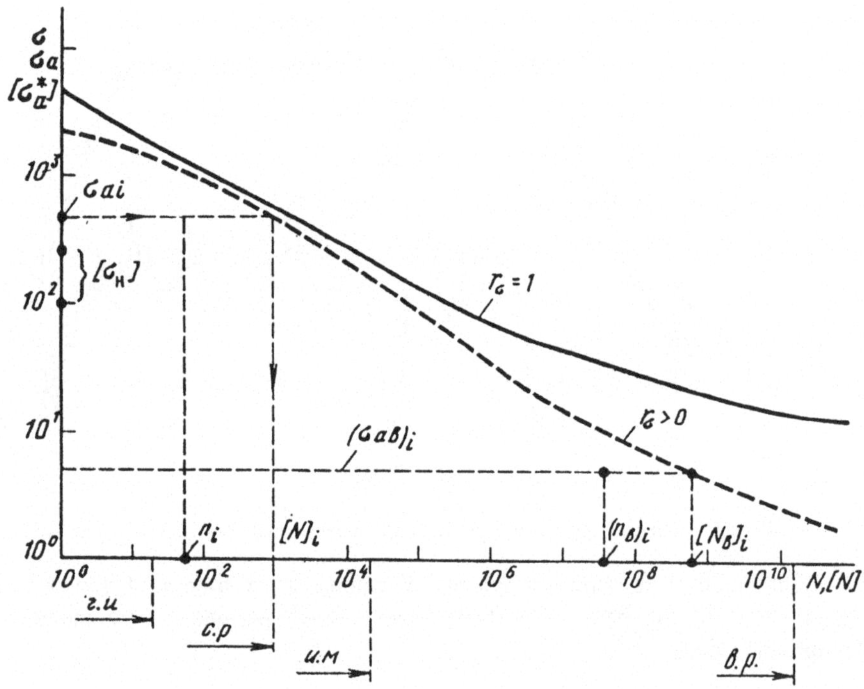

Fig. 2. The diagram of determination of the strength and resource characteristics

Engineering diagram for generalised fatigue curve for co-ordinates "σ, σ_a – N" is shown on Figure 2. On this figure:

σ_{ai}, $(\sigma_{ah})_i$ - an amplitude of main and vibrating stresses for an *i*-mode;

n_i, $(n_h)_i$ - a number of cycles for main and vibrating loadings;

$[\sigma_a^*]$, [N] - design stresses amplitude and lifetime (number of cycles);

$[\sigma_n]$ design nominal stresses.

From these stresses it is possible to determine from a traditional plot of static strength on ultimate σ_u and yield limits σ_u limits, the operating conditions of the body.

2. Deformation Fracture Criteria Under Static and Low Cyclic High Temperatures Loading

The low and high cycle loading of construction elements at high temperatures will usually cause an elasto-plastic deformation in the zones of stress concentration. The elasto-plastic deformation redistribution in the concentration zones for the zero (initial) and subsequent half-cycles of loading at given nominal stress amplitudes is facilitated as a result of resistance changes to elasto-plastic deformations under high temperature conditions. The number of cycles to fracture at room and high temperatures depends on the level of deformation redistribution in the zones of stress concentration, and on the fracture deformation criteria. This internal non-stationary character of the process of plastic deformation in the concentration zones at stationary external loadings leads to a deviation of the number of cycles to fracture at a given value of initial deformations or stresses in the concentration zones form that in the case of a stationary loading of smooth specimens at a given stress or strain of the same magnitude [5,7 - 12].

The stress-strain elasto-plastic relations under cyclic loading determined on smooth specimens [1 - 3] are used for the evaluation of the strain redistribution in concentration zones,. The analysis of this redistribution reveals that the amplitude will increase in the case of plastic deformations in the concentration zones of cyclic softening materials, and decrease if materials of cyclic hardening are tested,. The influence of increasing test temperature, as shown in [6, 14 - 16], depends on the type of material, test temperature, and strain rate. Decrease of the cyclic plastic deformation with temperature increase is typical for low carbon ageing steels, whereas an increase is characteristic of low alloy cyclic softening steels. The influence of strain rate on the amplitude of plastic deformation is connected to the rheological properties of the metal at the corresponding temperature. Extensive experimental investigations on the resistance to low cycle fracture at high temperatures [8 - 11, 14] showed that, when the creep effect is negligible, the fracture cycle number at strain-controlled loading is a power function of the plastic deformation amplitude. At temperatures where a creep effect exists, the number of cycles to fracture depends on the elasto-plastic deformation amplitude. In this case a combined cyclic and creep damage effect is observed [5, 7 - 10, 18]. Experimental data [5 - 8, 11, 13, 18] show the dependence of the endurance in the stress concentration zone, on the mechanical properties of the material, obtained by testing smooth specimens with a uniform state of stress at normal and high temperatures.

Experimental and analytical results for low cycle failures in the stress concentration zones, in connection with the deformation condition, were analysed for three types of steel:

I. low carbon, low-alloy steel, 0.28% C;

II. low-alloy stainless Cr-Mo-V steel;

III. stainless austenitic steel, Type 18-8

The test temperature for steel I was 20 to 350°C, for steel II - 20 to 450°C, and for steel III 20 to 550°C. The creep deformation and damage, under the above temperature and test frequencies of 1 - 2 cycles in minute, corresponding to a total test time of 10 - 40 hours, were not significant.

The mechanical properties of the steels were: elastic limit σ_T, yield point $\sigma_Y = \sigma_{0.2}$, ultimate strength σ_u, cross-section reduction ψ_k, strengthening index m_o, for a normal temperature $t = 20°C$, as given in the following table:

Table 1

Steel	σ_T MPa	$\sigma_{0.2}$ MPa	σ_u MPa	ψ_k %	m_0
I	262	320	650	61.2	0.195
II	485	520	693	70.3	0.180
III	233	282	681	73.2	0.188

The influence of test temperature on the basic properties in a static tensile test, expressed in relative values, is shown in Figure 3. The yield point $\sigma_{0.2}$, of all three types of steel decreases with an increase of temperature. The ultimate strength σ_u of steels II and III decreases with rising temperature, whereas the ultimate strength σ_u of steel I in the temperature range of strain ageing 150 to 300°C will increase. The plasticity of non-ageing steel II will rise and that of steel III decrease with temperature. A decrease of plasticity in the temperature range of strain ageing of steel I is observed. At deformation rates of between 1.6 and 3.10^{-3} s^{-1} the time of static loading to failure of the steels tested was equal to 2 - 4 min. Increase of the loading time to failure up to 20 hours at temperatures 350°C for steel I, 450°C for steel II, and 500°C for steel III hardly influenced the change of plasticity.

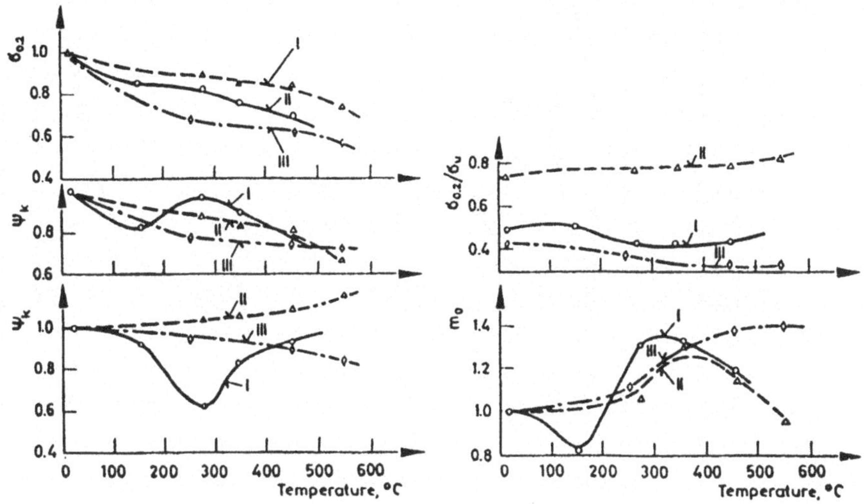

Fig. 3. Temperature dependence of mechanical properties

Under low-cycle loading conditions at a uniform stress state the resistance to plastic deformation is described by the stress-strain diagrams under loading, with a given stress or strain rate (Figure 4). These diagrams are plotted as stress values $\bar{\sigma}$; related to the elasticity limit $\bar{\sigma} = \dfrac{\sigma}{\sigma_T}$ and deformation values $\bar{e}$ related, in turn, to the deformation at the elasticity limit $\bar{e} = \dfrac{e}{e_T}$. An alteration of the loop width $\bar{\delta}^{(k)}$ and an accumulation of the one-side plastic deformation $\bar{e}^{(k)}$ are observed under stress-controlled loading, as a result of the difference $\bar{\Delta}^{(k)}$ of the loop with between even and uneven half-cycles. An alteration of the maximum stress of the cycle under strain-controlled loading also takes place.

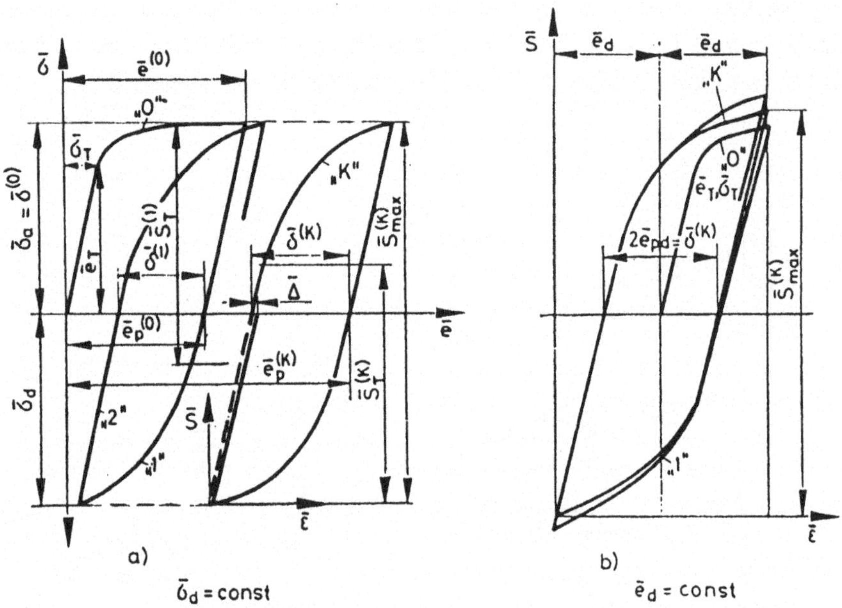

Fig. 4. Stress-strain cyclic diagrams

The loop width $\overline{\delta}^{(k)}$ at loading with a given stress amplitude, according to the results obtained above [5, 7, 8, 14 - 17] can be expressed by

$$\overline{\delta}^{(k)} \quad A\left(\overline{e}^{(o)}-1\right)F(k) \tag{1}$$

where:

A is an experimental defined material characteristic,

$\overline{e}^{(0)}$ is the elasto-plastic deformation in the initial zero half-cycle,

F(k) is a function of half cycle number k is determined experimentally at $\overline{\sigma}^{(a)}$ = const.

For cyclic softening materials

$$F(k) = \exp C\left(\overline{e}^{(0)} - 1\right)(k-1) \tag{2}$$

whilst for hardening materials.

$$F(k) = K^{-B(\overline{\sigma}^{(0)}-1)} \tag{3}$$

where: C, B are material characteristics ($C \geq 0$, $B \geq 0$).

For cyclic stable material $\delta = 0$ and

$$F(k) = 1 \tag{4}$$

Stresses $\overline{\sigma}^{(0)}$ at a power type approximation of the stress-strain diagram of zero half-cycle in the elasto-plastic range ($\overline{\sigma}^{(0)} \geq 1$) are expressed as depending on the initial deformation $\overline{e}^{(0)}$ by the relation

$$\overline{\sigma}^{(0)} = \overline{e}^{(0)m_0} \tag{5}$$

where m_0 is a characteristic of a power type hardening of the material $(0 \leq m_0 \leq 1)$. The values m_0 for normal temperature are given in Table 1., and their dependence on temperature is illustrated in Figure 3.

According to equations (1) - (5), the loop width in the k half cycle for a given material is determined by deformation $\overline{e}^{(0)}$ and the values A and C (or B). The values A and C are indirectly dependent on the testing temperature. Generalisation of the experimental results at normal and high temperatures (t = 20 to 550°C) for steels with an ultimate stress from 430 to 1300 MPa has shown (Figure 5) that the values A and C increase with increasing the ratio of yield point to the ultimate stress $\sigma_{0.2}/\sigma_u$. It may be supposed that the dependence of A and C on the ratio $\sigma_{0.2}/\sigma_u$ is not influenced by the test temperature, but the ratio $\sigma_{0.2}/\sigma_u$ depends on this temperature according to Figure 3. The experimental data shown in Figure 3 can be approximately expressed by

$$A = 0.16\left(1+\frac{1}{1-\sigma_{0.2}/\sigma_u}\right) \tag{6}$$

$$C = 1.5\text{x}10^{-3}\left(\frac{1}{1-\sigma_{0.2}/\sigma_u}-2\right) \tag{7}$$

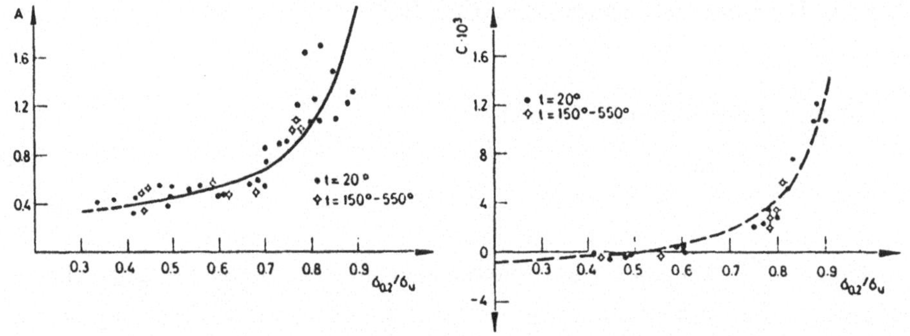

Fig. 5. Dependence of parameters A and C on the mechanical properties under static tension

According to relation (7), steels with $\sigma_{0.2} / \sigma_u$ >0.5 are cyclic softening (C>0), and those with $\sigma_{0.2} / \sigma_u$ >0.5 are hardening (C<0) types.

For cyclic hardening steels $(0.15 < m_0 \leq 0.3)$ at deformations $\bar{e}^{(0)} < 10$ as shown by the experimental data, between C and B a nearly linear relation exists:

$$C = -3{\cdot}10^{-3}\,B \tag{8}$$

Comparison of the values of loop width $\bar{\delta}^{(k)}$ according to equation (1), taking into account the values A and C conforming to (6) and (7) (smooth and dotted lines), and the same values of the experimental results (points) at normal and high temperatures for steels I, II, III are given in Figure 4. Low carbon steel I, cyclic stable at 20°C can be cyclic hardening at temperatures up to 350°C. At higher temperatures up to 450°C, the cyclic hardening of this steel will change to cyclic softening. The intensity of the cyclic softening of steel II increases with rising temperature.

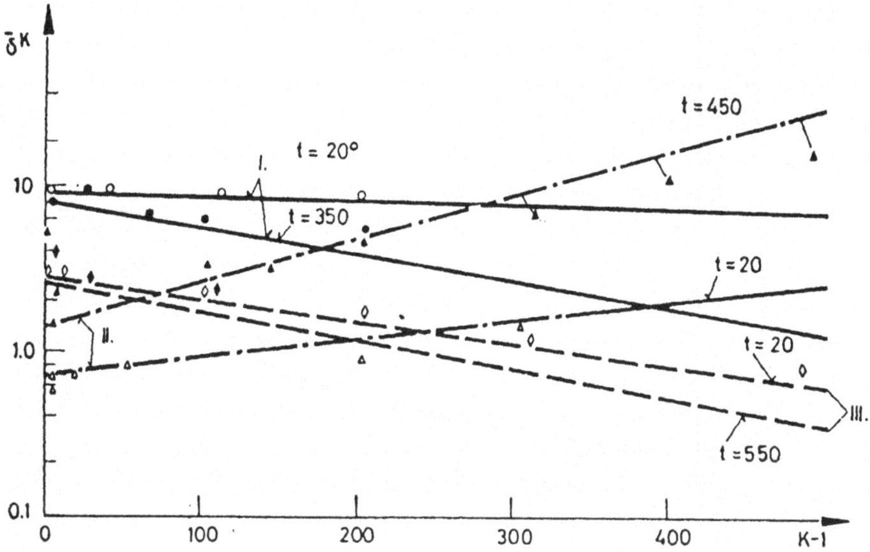

Fig. 6. Dependence of loop width on the number of cycles

The tendency to cyclic hardening of steel II with an increasing temperature increases.

The value of $\overline{\Delta}^{(k)}$ which defines the one-side accumulation of plastic deformation taking into account (1) is equal to

$$\overline{\Delta}^{(k)} = (A - A_*)\left(\overline{e}^{(0)} - 1\right)F(k) \tag{9}$$

where the value $(A - A_*)$ reflects the cyclic anisotropy of the steel.

There is a relation between the values A and $(A - A_*)$, as shown by the experimental data given in Figure 7. This relation can be approximately defined as:

$$(A - A_*) = 2\cdot 10^{-2}\left(\frac{1}{1-0.7A} - 1.3\right) \tag{10}$$

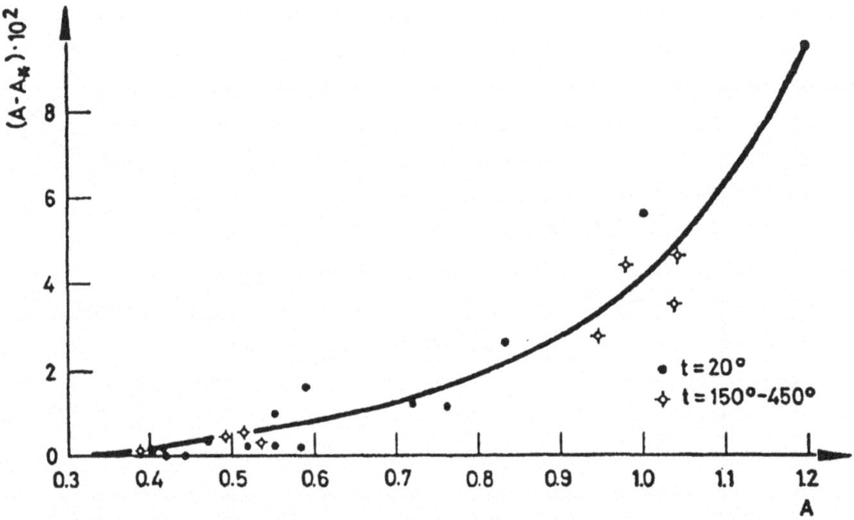

Fig. 7. Dependence between (A−A) and A at different temperatures*

Equation (10) reflects the tests results of low carbon and low alloy steels at temperatures up to 450°C, and this expression is independent of the temperatures.

Thus the resistance to cyclic elasto-plastic deformation at high temperatures can be determined by the dependence of the parameters A and C on $\sigma_{0.2}/\sigma_u$ as well as on the B and $(A - A_*)$ values and, therefore, also on the properties of the stress-strain diagram under static tension. The ratios $\sigma_{0.2}/\sigma_u$ depend on the temperature, hence the dependence of parameters A and C.

The equation of the stress-strain curve of cyclic deformation depends on the number of half-cycles k in the co-ordinate system S - ε, the zero point of the co-ordinates coincide with the beginning of load release (Figure 4), and can be written in the form (1-3)

$$\bar{\varepsilon}^{(k)} = \bar{S}^{(k)} + \bar{\delta}^{(k)} \tag{11}$$

where $\overline{\varepsilon}^{(k)}, \overline{S}^{(k)}$ are the current values of the elasto-plastic deformations and stresses; $\overline{\delta}^{(k)}$ is the loop width according to equation (1) at a stress amplitude of $\overline{\sigma}_a = \frac{1}{2}\overline{S}^{(k)}$.

According to (1), (5) and (11)

$$\overline{\varepsilon}^{(k)} = 2\overline{e}^{(0)m_0} + A\left(\overline{e}^{(0)} - 1\right)F(k) \tag{12}$$

which means that the form of the cyclic stress-strain curve depends on the form of the stress-strain curve under static tension at a corresponding temperature, and reflects the temperature dependence of the parameters A and C and, in this connection, also B and $(A - A_*)$.

Equation (11) of the cyclic stress-strain curve can be written in the analogue form (5):

$$\overline{S}^{(k)} = \overline{\varepsilon}^{(k)m(k)} \tag{13}$$

where m(k) is a characteristic of the strengthening of the current half-cycles. The stress and strain in the co-ordinate system $\overline{\sigma}^{(k)} - \overline{e}^{(k)}$ are equal to $\overline{\sigma}^{(k)} = \frac{\overline{S}^{(k)}}{2}$, $\overline{e}^{(k)} = \frac{\overline{\varepsilon}^{(k)}}{2}$.

Taking into account (1), (5), (11) and (13), we get

$$m(k) = \frac{\lg \overline{e}^{(0)1/m_0}}{\lg\left[\overline{e}^{(0)1/m_0} + \frac{A}{2}\left(\overline{e}^{(0)} - 1\right)F(k)\right]} \tag{14}$$

For cyclic hardening metals, the value of F(k) will decrease and the values m(k) increase with increasing cycles. At cyclic softening the inverse takes place. For a cyclic stable steel the value m(k) does not depend on the number of half-cycles.

Under loading, with a given amplitude of elasto-plastic deformation $\overline{e}_a = \frac{\overline{\varepsilon}^{(k)}}{2} = \text{const.}$, (Figure 2), the maximum stress values $\overline{S}^{(k)}_{max}$, change according to equation (11) as the loop width $\overline{\delta}^{(k)}$ for cyclic hardening steels and the stress $\overline{S}^{(k)}_{max}$ rise. For cyclic softening steels the value $\overline{S}^{(k)}_{max}$ decreases.

The fracture resistance under loading at controlled deformation amplitudes $\bar{e}_a = \text{const.}$ for a large group of metals can be described by the well known Manson-Coffin equation:

$$2\bar{e}_{ap} \cdot N_{ef}^{m} = \bar{C} \tag{15}$$

where $\bar{e}_{ap}$ - is the amplitude of plastic deformation;

N_{ef} - cycle number to fracture;

$m, \bar{C}$ - are steel characteristics.

The index ef of the cycle number N indicates fatigue fracture under loading, with a constant strain amplitude $\bar{e}_a$.

The index m for low carbon, low alloyed and stainless austenitic steels with an ultimate stress up to 700 MPa is equal to 0.5. Value $\bar{C}$ is defined by the contraction of the neck at a static fracture.

$$\bar{C} = \frac{1}{2e_T} \ln \frac{100}{100 - \Psi_k} = \frac{1}{2} \tag{16}$$

where $\bar{e}_k$ is the relative logarithmic deformation of the neck.

From (15) and (16) it follows that

$$4 \frac{\bar{e}_{ap}}{\bar{e}_k} N_{ef}^{m} = 1 \tag{17}$$

In Figure 8 a comparison of the values obtained by equation (17) (smooth line) with experimental data points for steels I and II at temperatures in the range from 20 to 450°C, and the total time to fracture up to 40 hours is shown. The data of Figure 8 confirm the possibility of using index m independent of temperature, and the critical value of plastic deformation $\bar{e}_k$ changing with the temperature increase (Figure 1). Fractures determined by equation (17) take place through crack formation, and are of the fatigue type.

The condition of fatigue failure can be expressed by using an experimentally confirmed [4, 5, 7] linear damage cumulative rule, in terms of relative fatigue life n/N_{ef}, and supposing

$$2\bar{e}_{ap} = \bar{\delta}^{(k)} \tag{18}$$

on the basis of the power type equation of the fatigue curve for constant strain loading $(\bar{e}_a = \text{const.})$, in the form (15), with index m independent of temperature, by the following formula:

$$\int_1^{k_{ef}} \left(\bar{\delta}^{(k)}\right)^{1/m} dk = 2\left(\bar{C}\right)^{1/m} \tag{19}$$

where k_{ef} is the critical half-cycle number to fracture.

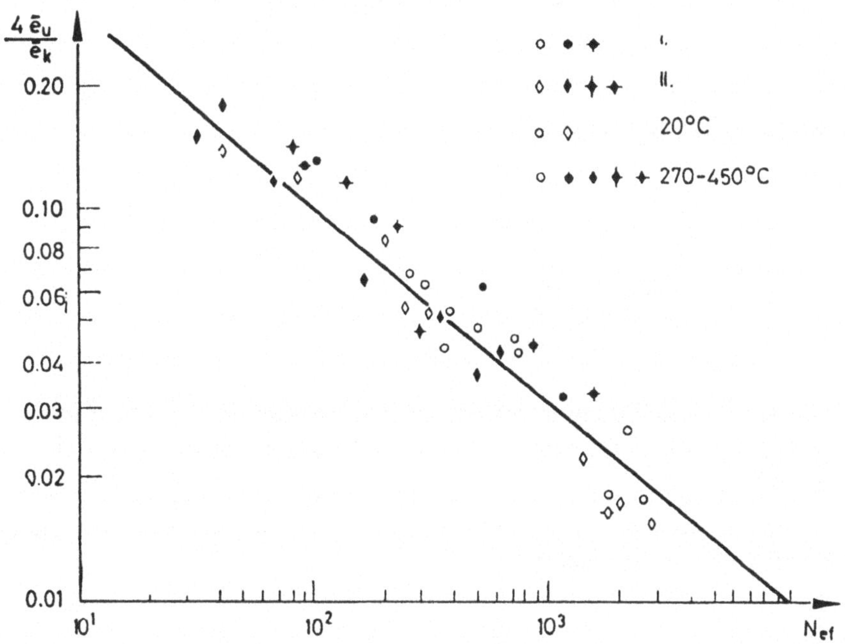

Fig. 8. Low cycle fatigue curve expressed in relative values of deformation amplitudes

Equation (19) is used [5, 8, 14 - 17] for the determination of the critical cycle number $N_{\sigma f}$ at loading with given stress amplitudes $\overline{\sigma}_a = \text{const.}$, when failure is preceded by the formation of fatigue microcracks. Index σf indicates a fatigue failure under loading, with a constant stress amplitude $\overline{\sigma}_a$. Integration of equation (19), taking into account (1) - (4), gives for a cyclic softening steel:

$$N_{\sigma f} = \frac{1.15}{C\left(\overline{e}^{(0)} - 1\right)} \lg\left\{ \frac{2\overline{C}}{m} \left[\frac{\overline{\overline{C}}}{A\left(\overline{e}^{(0)} - 1\right)^{1-m}} \right]^{1/m} + 1 \right\} + 0.5 \qquad (20)$$

for a cyclic hardening steel:

$$N_{\sigma f} = 0.5 \left\{ 2\left[1 - \frac{b}{m}\left(\overline{e}^{(0)m_0} - 1\right)\right] \left[\frac{\overline{\overline{C}}}{A\left(e^{(0)} - 1\right)} \right]^{1/m} \right\}^{m\left[m - B\left(\overline{e}^{(0)m_0} - 1\right)\right]} \qquad (21)$$

and for a cyclic stable steel:

$$N_{\sigma f} = \left[\frac{\overline{\overline{C}}}{A\left(e^{(0)} - 1\right)} \right]^{1/m} \qquad (22)$$

Value $\overline{\overline{C}}$ is defined by taking into account the decrease of plasticity as the result of a displacement of the loop in the zero half cycle at loading with a given stress amplitude.

This displacement is equal to the sum of elastic deformation at relief and half of the loop within the first half-cycle

$$\overline{\overline{C}} = \overline{C} - \left[\overline{e}^{(0)m_0} + \frac{A}{2}\left(\overline{e}^{(0)} - 1\right) \right] \qquad (23)$$

As it was mentioned above, the values A, B, C, m_0 are dependent on the test temperature. Equations (20, (21) and (22) allow prediction of the cycles before fatigue failure at a constant stress load ($\overline{\sigma}_a$ = const.) depending on the stress amplitude. The value $\overline{\sigma}_a$ is connected with the deformation $\overline{e}^{(0)}$ in the zero half-cycle according to relation (5).

For cyclic anisotropic, cyclic stable, and softening steels under load with a constant stress amplitude $\overline{\sigma}_a$ = const., the value of the one-side accumulated plastic deformation during k half cycles is determined by integrating with k the function $\overline{\Delta}^{(k)}$, according to equation (9)

$$\overline{e}_p^{(k)} = \int_1^k \overline{\Delta}^{(k)}\, dk \tag{24}$$

In the low cycle regime (below 10^3), the one-side accumulation of plastic deformation leads to a quasi-static failure with neck formation. A criterion for such type of failure is the value of one-side accumulated plastic deformation equal to the deformation $\overline{e}_k$ in the neck at static tension rupture. A quasi-static failure of the steels mentioned above in the field of low fatigue life $N_{\sigma s}$, will precede fatigue failure. Index σs at the cycle number N indicates a quasi-static failure under load with a constant stress amplitude.

The number of cycles needed for quasistatic failure N at loading at a given stress amplitude is determined from the condition that the plastic deformation $\overline{e}_k$ is equal to the sum of plastic deformation in the zero half-cycle $\overline{e}_p^{(0)}$, cyclic accumulated during k half-cycles, plastic deformation $\overline{e}_p^{(k-1)}$, and the plastic deformation of the last half cycle

$$\overline{e}_p^{(0)} + \overline{e}_p^{(k_{\sigma s}-1)} + \overline{\Delta}^{(k_{\sigma s})} = \overline{e}_k \tag{25}$$

Deformations $\overline{e}_p^{(0)}$ and $\overline{\Delta}^{(k_{\sigma s})}$ are determined by the curve of static deformation. Deformation $\overline{e}_p^{(0)}$ is equal to $\left(\overline{e}^{(0)} - 1\right)$. Denoting with $\overline{e}_{pk}$ the value of the deformation, we get

$$\overline{e}_k - \overline{\Delta}^{(k_{\sigma s})} + 1 = \overline{e}_{pk} \tag{26}$$

and, by integrating the function $\overline{\Delta}^{(k)}$ according to expression (24) with respect to (9), (2), (3), (4), the cycle number $N_{\sigma s}$, can be determined

For a cyclic softening steel:

$$N_{\sigma s} = \frac{1.15}{C\left(\overline{e}^{(0)} - 1\right)} \lg\left[\frac{C}{(A - A_*)}\left(\overline{e}_{pk} - \overline{e}_{(0)} + 1\right)\right] + 0.5 \tag{27}$$

for a cyclic stable steel:

$$N_{\sigma s} = 0.5\,\frac{\overline{e}_{pk} - \overline{e}^{(0)}}{(A - A_*)\left(\overline{e}^{(0)} - 1\right)} + 0.5 \tag{28}$$

and for cyclic hardening steels (at $B < 0.12$):

$$N_{\sigma s} = 0.5\left\{\frac{\overline{e}_{pk} - \overline{e}^{(0)}}{(A - A_*)\left(\overline{e}^{(0)} - 1\right)}\left[1 - B\left(\overline{e}^{(0)m_0} - 1\right)\right]\right\}^{1/\left[1 - B\left(\overline{e}^{(0)m_0} - 1\right)\right]} \tag{29}$$

Dependence of the critical cycle number $N_{\sigma s}$ on temperature can be determined by the corresponding temperature dependence A, B, C, $(A - A_*)$, $m_0, \overline{e}^{(k)}$. Stress amplitude $\overline{\sigma}_a = \overline{\sigma}^{(0)}$ is determined by the deformation $\overline{e}^{(0)}$ in the zero half-cycle of (5).

Comparison of the data calculated by equations (27), (28) and (29) (smooth line) with the experimental data (points) for steels II and I in the temperature range from 20 to 450°C is given in Figure 9. A quasistatic failure of steel II, characterised by the value $\psi = \psi_k$ at the temperatures mentioned takes place in the range of cycles from 1 to ~10^3. At cycles above 1.5×10^3 the failure under load with a given stress amplitude leads to a fatigue type failure. The fatigue life N_{ef} can be calculated by equation (20).

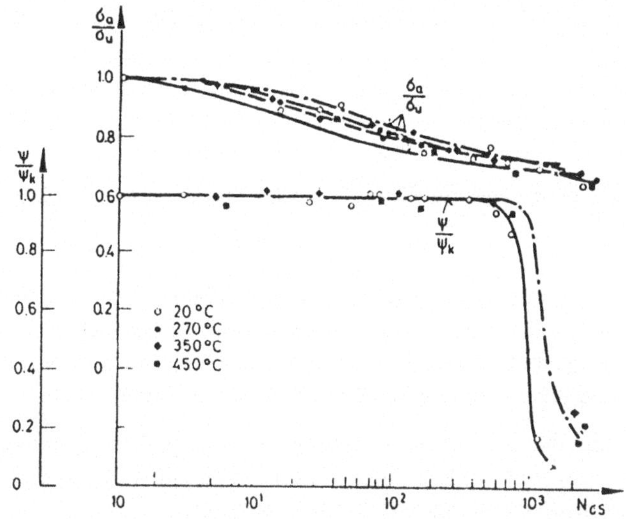

Fig. 9. Low cycle fatigue curves in relative values of stress amplitude, and the curves of the relative contraction coefficient values at cyclic rupture

With cyclic stable steels I between 20 and 150°C, a quasistatic failure takes place at cycles between 50 and 100. The range of cycle numbers at which a quasistatic failure takes place in the case of this steel may decrease to 10. At a temperature of deformation ageing, the cycle numbers at which a quasistatic failure will change into fatigue failure are denoted by crosses in Figure 9. Fatigue failure takes place after the decrease of plastic deformation by 60 - 90%.

As for the cycle number at which transition for quasi-static to fatigue failure occurs, this (Figure 9) is defined from the condition that the sum of accumulated quasi-static and fatigue damages is equal to a critical value, as shown by comparison of the calculated and experimental data. This value is the sum of the ratio of one side accumulated plastic deformations $\bar{e}_p^{(k)}$ to the local deformations at static tension rupture $\bar{e}_{pk}\left(d_s = \dfrac{\bar{e}_p^{(k)}}{\bar{e}_{pk}}\right)$ and of the ratio of the accumulated number of cycles N to the fracture number of cycles N_{ef}, according to equations (20) - (22) ($d_f = N/N_{ef}$)

$$d_s + d_f = 1 \tag{30}$$

Expression (13) of the cyclic stress-strain curve is used for the evaluation of the deformation state kinetic in the concentration zones, while equations (19), (24) and

(30) are employed for the evaluation of the fatigue life in these zones at high temperatures.

3. Stress-Strain Fracture Criteria Under High Cyclic Loading

Under real maintenance conditions, energy plants, both traditional thermal and nuclear, exist under complex thermomechanical conditions of loading that can be either low frequency and high strain amplitude ones with cycles upto $5x10^4$ or high frequency and low strain amplitude ones, at cycles from 10^5 to 10^{12}.

As a rule, those are cycles of setting the main parts of equipment in motion and their stopping, combined with the cycles of loading, that can be of high frequency due to vibration, hydrodynamics and thermal variations.

For cycles in the range 10^5 to 10^{12} it is possible to take following values of parameters:

- for curve of static loading by equation (5) - $m_0 = 1$;
- for curve of cyclic loading by equation (13) $m(k) = 1$.

By analogy to equation (15) for amplitudes of elastic strain it is possible to record

$$2\bar{e}_{ae} \cdot N_{ef}^{m_e} = \bar{C}_e \tag{31}$$

where m_e, $\bar{C}_e$ are the characteristics of steel.

With the account (16) and (17) it is possible to take

$$\bar{C}_e = \frac{S_k}{e_Y 4^{m_e} E} \tag{32}$$

where S_k is a resistance of rupture in a local zone at static failure.

For cycles with the range 10^0 to 10^{12}, it is possible to take

$$\bar{e}_a = \bar{e}_{ap} = \bar{e}_{ae} \tag{33}$$

Then on the basis of equations (15), (16), (31) and (32)

$$\bar{e}_a = \frac{1}{4e_Y N_{ef}^m} \ln \frac{100}{100 - \psi_k} + \frac{S_k}{2e_Y E(4N)^{m_e}} \tag{34}$$

The material characteristics of m and m_e are dependent on the ultimate limit, and from experimental data:

$$m = 0{,}5 \text{ for } \sigma_u < 700 \text{ MPa}$$

$$m = 0{,}5 + 0{,}0{,}002\,(\sigma_u - 700) \text{ for } \sigma_u \geq 700 \text{ MPa}$$

$$m_e = 0{,}135 \lg (S_k / \sigma_{-1})$$

For steels

$$S_k = \sigma_u (1 + 1{,}4 \cdot 10^{-2}\, \psi_k)$$

$$\sigma_{-1} = 0{,}4\, \sigma_u \text{ for } \sigma_u < 700 \text{ MPa}$$

$$\sigma_{-1} = (0{,}54 - 2 \cdot 10^{-4}\, \sigma_u)\, \sigma_u \text{ for } \sigma_u \geq 700 \text{ MPa}$$

4. Stress-strain Fracture Criteria Under Two-Frequency Cyclic Loading

Components used in the energy, transport, and aircraft industries, etc., are, during their service operation, frequently subjected to two or even multi-frequency cyclic loading. The shape of the complex stress cycles depends on the type of operating load and can be defined by their frequencies and amplitudes. Such modes may be, for example, due to the main operation mode of the structure being superimposed by high frequency stresses caused by hydro and aerodynamic vibrations, operation control, or other vibration mechanisms. In many cases of highly loaded structures the main loading occurs in the low-cycle fatigue range. Figure 10 shows a general case of such a loading mode and the experimental strain-stress diagram.

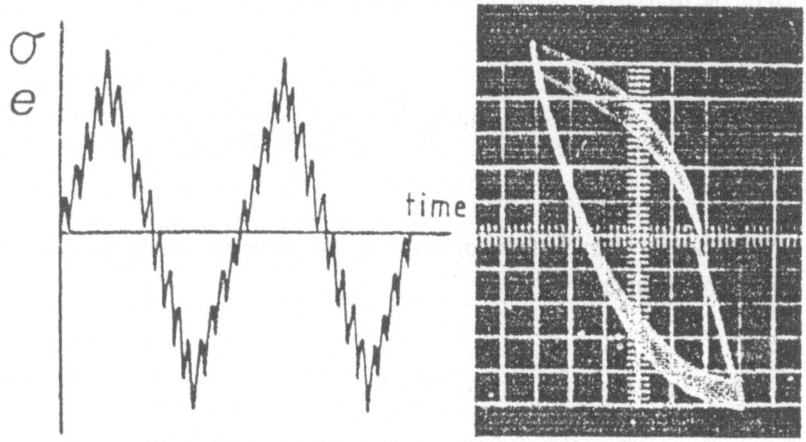

Fig. 10. Typical to-frequency loading and elasto-plastic diagram corresponding to this mode

Fatigue curves for the strain control mode of loading are presented in Figure 11. This figure shows that specimen life decreases with an increase in the superimposed deformation amplitude e_{a2} as compared with single-frequency loading.

For the general case of one-frequency low-cycle fatigue loading the total damage $\sum d$ consists of fatigue damage d_f and static damage d_s [6,7,15]

$$\sum d = d_f + d_s \tag{35}$$

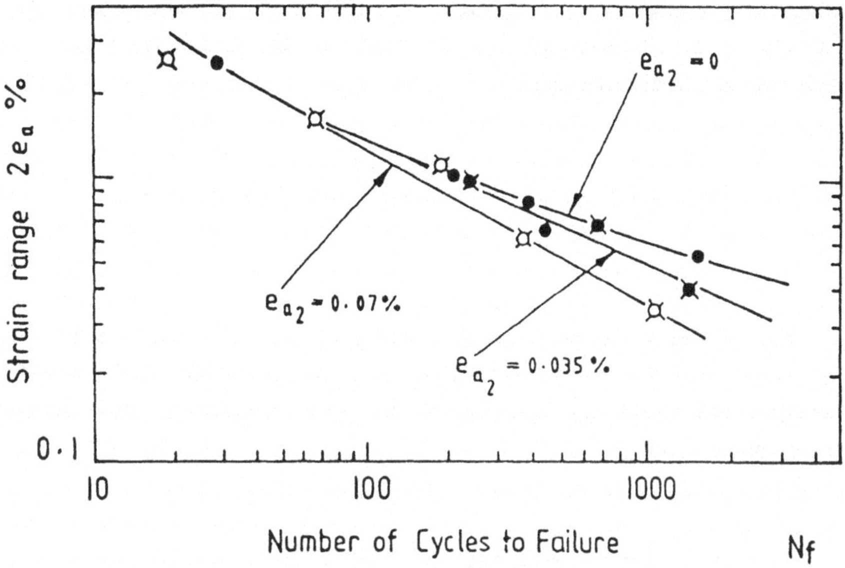

Fig. 11. Experimental fatigue curves for one-frequency and two-frequency low-cycle strain-control loading

The fatigue damage depends on the cycle strain amplitude e_a and may be evaluated from the equation of the fatigue curve as (34) (in which changes of material mechanical properties in the loading process are reflected) by presenting values of reduction in area ψ and ultimate limit σ_u as a function of time τ which in turn is related to the loading frequency f and cycle time τ_c as $\tau_c = 1/f$ and $\tau = \tau_c N$:

$$\bar{e}_a = \frac{1}{e_Y\left(4N^m + \frac{1+r}{1-r}\right)} \ln \frac{100}{100 - \psi_o\left(\frac{\tau_o}{N\tau_c}\right)^{m_\psi}} + \frac{K\sigma_u\left(\frac{\tau_o}{N\tau_c}\right)^{m_{\sigma u}}}{e_Y E\left(1 - K\frac{1+r}{1-r}\right)^{N^{m_e}}} \tag{36}$$

Then in each cycle fatigue damage d_f is

$$d_f = \frac{1}{N[e_a(n), t, \tau]} \tag{37}$$

where n is number of cycles applied, τ is time and t is temperature. The static damage d_s is characterised by the plastic strain accumulated during cyclic deformation e_n and is defined by

$$d_s = \frac{e_n}{\varepsilon_{t\tau}} \tag{38}$$

For one and two frequency, cyclic loading the material damage consists of fatigue d_{f1} derived from total strain amplitude $e_a = e_{a1} + e_{a2}$, and fatigue damage d_{f2} from the superimposed strain e_{a2} and finally the static damage d_s from accumulated deformation e_n i.e.

$$\Sigma d = d_{f1} + d_{f2} + d_s \tag{39}$$

Damage d_2 is evaluated with the help of equation (36) by applying it to the high-frequency loading, i.e. substituting e_{a2} for e_a and taking into consideration the frequency f_2, and the variable from cycle to cycle asymmetry coefficient r(n*). Then in each superimposed cycle d_{f2} is defined as:

$$d_{f2} = \frac{1}{N^*[e_{a2}, \tau, t, r(n^*)]} \tag{40}$$

Thus the total damage at the moment of fracture for two-frequency low-cycle loading, taking into consideration equations (36-40), will be defined as:

$$\Sigma d + \int_0^{N_f} \frac{dn}{N[e_a, n, t, \tau]} + \int_0^{N^*_f} \frac{dn^*}{N^*[e_{a2}, t, \tau, r(n^*)]} + \int_0^{N_f} \frac{e_n dn}{e_t, \tau} \tag{41}$$

where $N_f^* = N(f_2/f_1)$ and $n^* = n(f_2/f_1)$.

The results of the analysis for experimental data at test of steel X18H10T (t=650°C, f_1 = 1, f_2 = 30 c/s) according to equation (41) when d_s = 0 (strain control test, e_a = const) are given in Figure 12, which shows that under the investigated condition $\Sigma d \approx 1.0$.

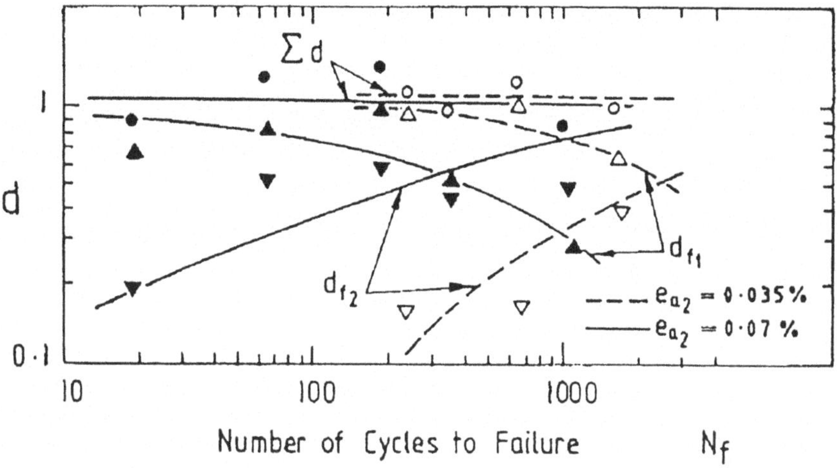

Fig. 12. Damage components of fatigue life under two-frequency loading

Similar results have been obtained for stress control test, two frequency loading of steel X18HIOT (t = 650°C) with f_2/f_1 = 80 when the processes of cyclic creep occur.

Hence the life estimate at two-frequencies low-cycle loading may be made on the basis of equation (41) with $\Sigma d = 1.0$.

The experimental data analysis showed that the superimposed strain modifies the character of deformation diagrams. For strain control loading this deformation increases the cyclic hardening of the material; for stress control loading with time dwells, the development of cyclic creep strain and irreversible accumulated strain increases.

Experimental studies of resistance to the strain of structural materials with the number of cycles at and to their destruction, have been carried out under the said conditions of two-frequency loading, the stages of cracks initiation in both austenitic stainless steel (of 18-8 type) and pearlitic steel (Cr-Mo-V) in the temperature range 20°C to 650°C where the frequency ratio is f_2/f_1 = 10 to 1000 being studied [1, 15-17].

The experimental data on the deformation diagrams and the fatigue curves at the stage of crack initiation have proved to be governed by their respective power equations.

5. Life Time Prediction in Cycle Loading Condition

At non-stationary modes of low cycle loading (Figure 1, 2) it is possible to use the rule of linear summation of damages. These damages are connected nonlinearly with stresses and strain.

With this purpose the coefficients of stocks on strain amplitudes n_e and on numbers of cycles n_N are entered

$$[\bar{e}_a] = \frac{\bar{e}_{af}}{n_e}, \qquad [N] = \frac{N_{af}}{n_N} \tag{42}$$

The values n_e and n_N are usually taken in range $1.25 \le n_e \le 2$ and $3 \le n_N \le 20$.

Equations (15), (31) and (34) permit to construct design curves of fatigue in co-ordinates "$[\bar{e}_a]$ - [N]" or "$[\sigma_a]$ - [N]", where $[\sigma_a] = [\bar{e}_a]\sigma_y$

On these curves at specific number of cycles $N_i = n_i$ (Figure 2) for i-mode and k-modes is determined permitted accumulated damage and condition of strength to be established under loading by main cycles of loads

$$\sum_{i=1}^{k} \frac{N_i}{[N_0]_i} = d \le [d_n] \tag{43}$$

Limiting value $[d_N] = 1$.

If on an i-mode takes place two-frequency loading, that

$$d_i = d_{oi} + d_{bi} \tag{44}$$

where d_{oi} - damage from main loads;

d_{oi} - damage from two-frequency loading.

Taking into account (42) - (44)

$$d_{bi} = \chi_i \frac{N_i}{[N_0]_i} \tag{45}$$

where χ_i - the factor of reduction of life, depends on the relation of frequencies f_b/f_o and stresses amplitudes σ_{ab}/σ_a for high-frequency and main loading.

Factor η grows from 1.3 to 2.0 with growth of ultimate limit from 500 to 1000 MPa.

The equations above enable fatigue curves in a range of cycles from 10^0 to 10^{12}, to be constructed; allowing the determination of local stresses and strains and hence the calculation of strength of designs at low cyclic loading.

References

1. N.A.Makhutov, M.M.Gadenin. *The Influence of High Temperature and Two-Frequency Loading on The Low Cycle Fatigue Resistance*, Third International Conference on Material Science Problems in NPP Equipment Production and Operation, St. Petersburg, 1994, p. 123.

2. N.A.Makhutov. *Basic Analysis of Technogenic Catastrophes*, Safety and Reliability of Complex Technical Systems, Norway, NTVA-rapport, №3 - 1994, p. 139-148.

3. M.M.Gadenin, V.P.Petrov. *The Development of Scientific Basis of Engineering Safety.* Safety and Reliability of Complex Technical Systems, Norway, NTVA-rapport, №3 - 1994, p. 93-100.

4. Н.А.Махутов, Е.В.Грацианский, М.М.Гаденин... *Природные и техногенные катастрофы: проблемы безопасности.* - М.: ВИНИТИ, 1993, т.1, 350 с, т.2, 479 с.

5. Н.А.Махутов. *Деформационные критерии разрушения и расчет элементов конструкций на прочность.* - М.: Машиностроение, 1981, 272 с.

6. Н.А.Махутов. *Сопротивление элементов конструкций хрупкому разрушению.* - М.: Машиностроение, 1973, 201 с.

7. С.В.Серенсен, Р.М.Шнейдерович, А.П.Гусенков, Н.А.Махутов... *Прочность при малоцикловом нагружении.* - М.: Наука, 1975.-285 с.

8. В.П.Когаев, Н.А.Махутов, А.П.Гусенков. *Расчеты деталей машин и конструкций на прочность и долговечность.* - М.: Машиностроение, 1985, 223 с.

9. *Нормы расчета на прочность оборудования и трубопроводов атомных энергетических установок* (коллектив авторов с участием Н.А.Махутова). - М.: Атомэнергоиздат, 1989, 524 с.

10. Н.А.Махутов, В.В.Стекольников, К.В.Фролов... Конструкции и методы расчета водо-водяных энергетических реакторов. - М.: Наука, 1987, 232 с.

11. Н.А.Махутов, К.В.Фролов, В.В.Стекольников, М.М.Гаденин... *Прочность и ресурс водо-водяных энергетических реакторов.* - М.: Наука, 1988, 312 с.

12. Н.А.Махутов, К.В.Фролов, В.В.Стекольников... *Экспериментальные исследования деформаций и напряжений в водо-водяных энергетических реакторах.* - М.: Наука, 1990, 296 с.

13. Н.А.Махутов, Е.М.Морозов. *Прочность и разрушение. Машиностроение*, Энциклопедия, т.1-3, книга 1. - М.: Машиностроение, - с. 129-173.

14. С.В.Серенсен, Р.М.Шнейдерович, Н.А.Махутов... *Поля деформаций при малоцикловом нагружении.* - М.: Наука, 1979, 277 с.

15. Н.А.Махутов, М.М.Гаденин, Д.А.Гохфельд... *Уравнения состояния при малоцикловом нагружении.* - М.: Наука, 981, 244 с.

16. Н.А.Махутов, В.В.Зацаринный, Ж.Л.Базарас, М.М.Гаденин... *Статистические закономерности малоциклового разрушения.* - М.: Наука, 1989, 252 с.

17. Н.А.Махутов, М.И.Бурак, М.М.Гаденин... *Механика малоциклового разрушения.* - М.: Наука, 1986, 264 с.

18. K.V.Frolov, YU.L.Izrailev, N.A.Makhutov, E.M.Morozov, V.Z.Parton. *Thermal stresses and strength of Turbines: Calculation and Design.* Hemisphere Publishing Corp. N.Y., W., Ph., K. 1988, 379 p.

GASEOUS ATMOSPHERE INFLUENCE ON FATIGUE CRACK PROPAGATION

J. PETIT, G. HENAFF, S. LESTERLIN and C. SARRAZIN-BAUDOUX
Laboratoire de Mécanique et de Physique des Matériaux, URA CNRS N° 863,
ENSMA, Téléport 2, BP. 109, 86960 FUTUROSCOPE Cedex, France.

1. Introduction

The effect of ambient air on the fatigue strength of metals was initially suspected by Haigh [1] in 1917. Since numerous investigations have clearly recognised the detrimental effect of normally non corrosive gaseous atmospheres on the fatigue behaviour of metals and alloys, as well on the initiation component as on the propagation component of the fatigue life [2, 3]. During the last decades, fatigue crack propagation (FCP) has been widely investigated, specially in the near-threshold area, where environment microstructure and closure have been shown to play a major role [3]. Such pronounced interest for the near-threshold crack propagation behaviour is two fold. From an engineering point of view there has been a need of reliable and reproducible data in order to assess the damage tolerance of components which have to withstand a high number of low-amplitude loading cycles. On the fundamental side, the basic mechanisms governing FCP in this region where ill-known in sharp contrast with the so-called Paris regime formally extensively investigated.

Following the initial work of Dahlberg [5], Hartman [6] and Bradshaw and Wheeler [7,8, figure 1], the deleterious effect of ambient air as compared to an inert environment like high vacuum, has been clearly related to the presence of moisture in the surrounding environment for most of the metallic materials fatigued at room temperature [3, 9,24]. At higher temperatures the respective role of water vapour and oxygen is still disputed [25].

The main difficulty encountered to understand the role of water vapour resides in the complex interactions of an active environment with other parameters which influence the propagation, including intrinsic parameters as alloy composition and microstructure or extrinsic parameters as loading conditions, specimen geometry, crack depth, crack closure and temperature.

This paper proposes a survey of the influence of gaseous moist environments on fatigue crack propagation at mid and low rates, on the basis of a framework describing the intrinsic fatigue crack propagation which is essential to uncouple the respective influence of microstructure, closure, environment, and to analyse their interactions.

R.A. Smith (ed.), Reliability Assessment of Cyclically Loaded Engineering Structures, 301–342.

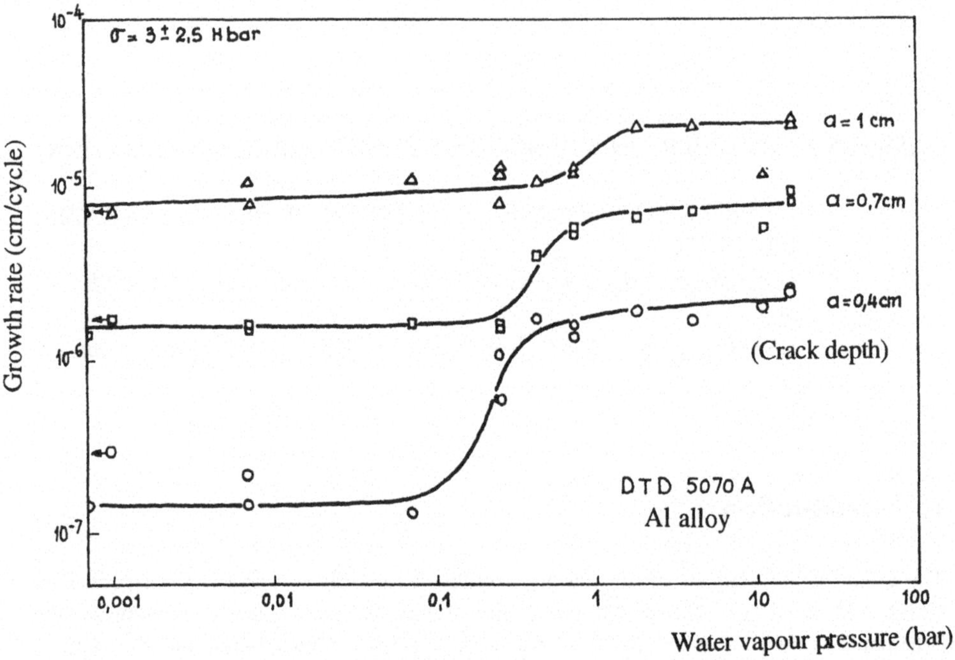

Figure 1. Influence of partial pressure of water vapour on the growth rate of a fatigue crack in a DTD 5070 A Aluminium alloy [7].

2. Intrinsic Fatigue Crack Propagation.

2.1. Propagation stages

The characterisation of the intrinsic behaviour of fatigue cracks means the elimination of any environmental and closure effect. High vacuum (~ 10^{-4} Pa) for tests run at frequencies of about 20 to 50 Hz, can be considered as an inert environment in most cases. Closure correction can be made using the compliance method as initially suggested by Elber [26]. Compliance variations are generally detected by mean of CTOD gauges mounted at the notch of the specimens or of back face strain gauges [27].

On the basis of numerous experimental data obtained on various Aluminium alloys (including technical alloys and high purity single crystals), steels and Titanium alloys, the intrinsic FCP has been analysed in accordance to three basis crack propagation regimes as illustrated in figures 2 to 10 .

The fastest intrinsic stage I, has been identified on single crystals of Al alloys. An illustration of a stage I crack grown in a peak aged Al-Zn-Mg high purity alloy is given in figure 2. Typically, the crack develops within a {111} planes pre-oriented for single slip [28,29]. This regime is also observed on various materials in the early growth of microstructural short cracks.

The intermediate intrinsic stage II is commonly observed on polycrystals and single crystals when crack propagation proceeds at macroscopic scale along planes normal to the loading direction [30]. Such propagation is favourised by microstructures which promote homogenous deformation and wavy slip as large or non coherent precipitates or small grains size (Figures 4,5). The figure 3 illustrates a typical change from a near threshold stage I to a mid ΔK stage II propagation in an Al-Zn-Mg single crystal.

The slowest regime, or intrinsic stage I-like propagation corresponds to a crystallographic crack growth observed near the threshold in polycrystals or in the early stage of growth of naturally initiated microcracks when ageing conditions or low stacking fault energy generate heterogeneous deformation along single slip systems, within individual grains [17, 31] (see example in figures 6 and 7). Crack branching and crack deviation mechanisms [32] and barrier effect of grain boundaries [33], are assumed to lower the stress intensity factor at the crack tip of the main crack.

The stage II regime is in accordance with a propagation law derived by Petit et al [34] from the models initially proposed by Rice [35] and Weertman [36] :

$$\frac{da}{dN} = \frac{A}{D_0^*}\left(\frac{\Delta K_{eff}}{E}\right)^4 \qquad (1)$$

where A is a dimensionless parameter, E the Young modulus and D* the critical cumulated displacement leading to rupture over a crack increment ahead of the crack tip.

Intrinsic data for well identified stage II propagation are plotted in figure 8 in a da/dN vs ΔK_{eff} diagram for a wide selection of Al alloys, and in figure 9 in a da/dN vs $\Delta K_{eff}/E$ diagram for a selection of steels and some data for a TA6V alloy in comparison to the mean curve for Al alloys. The diagrams constitute an excellent validation of the above relation and confirm that the LEFM concept is very well adapted to describe the intrinsic growth of a stage II crack which clearly appears to be nearly independent on the alloy composition, the microstructure (when it does not introduce a change in the deformation mechanism), the grain size, and hence the yield stress. The predominant factor is the Young modulus of the matrix, and the slight differences existing between the three base metals can be interpreted as some limited change in D_0^* according to the alloy ductility [17]. As a consequence, most of the changes observed between the nominal stage II propagation of differents alloys are inherent to the changes in the Young modulus and in the contribution of crack closure.

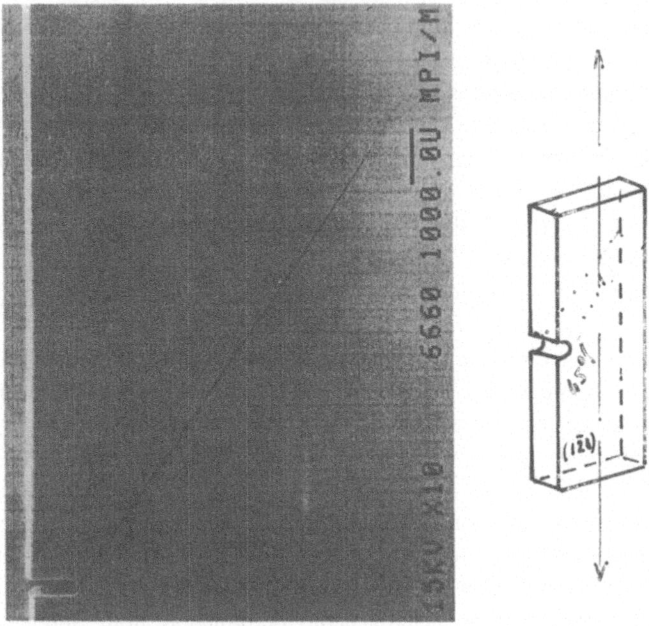

Figure 2. Stage I crack in Al-4.5 wt % Zn - 1.25 wt % Mg high purity single crystal in peakaged condition and oriented for single slip. Test run in high vacuum at 35 Hz and R = 0.1

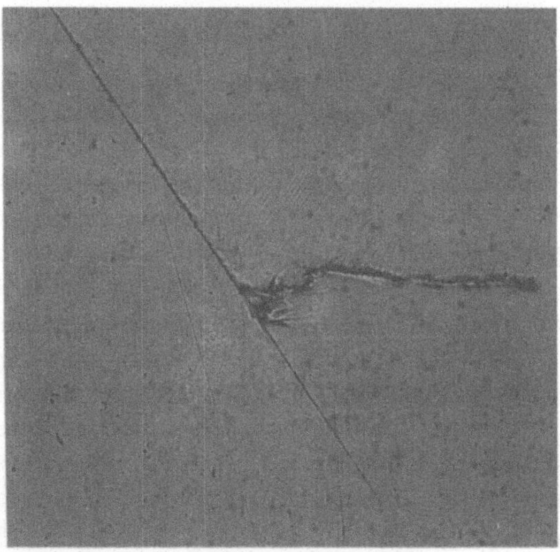

Figure 3. Stage I to stage II transition in high-purity Al - 4.5% Zn - 1.25 wt % Mg single crystal in overaged condition and oriented for single slip. High vacuum, 35 Hz, R = 0.1. The transition was obtained at a ΔK_{eff} range of 4 MPa

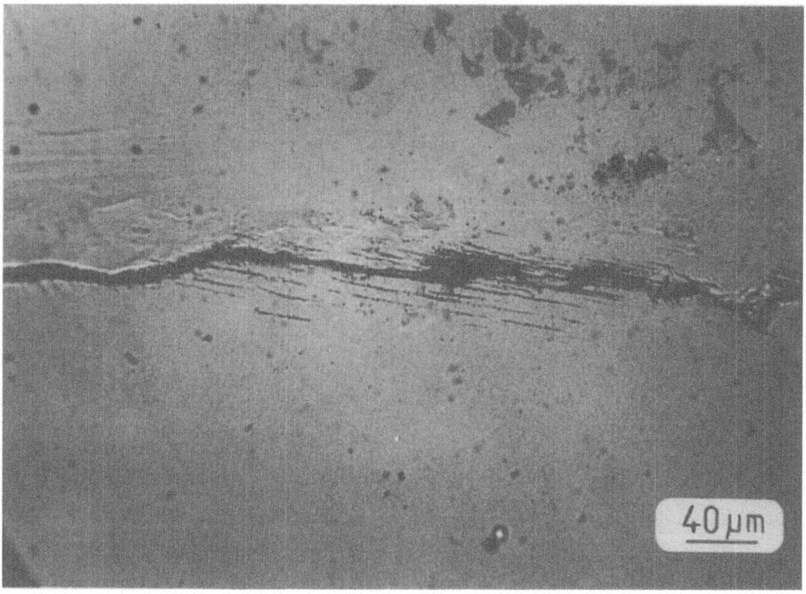

Figure 4. Stage I crack in 2024 T351 Al alloy tested in high vacuum. $da/dN = 2.10^{-10}$ m/cycle, 35 Hz and R = 0.1

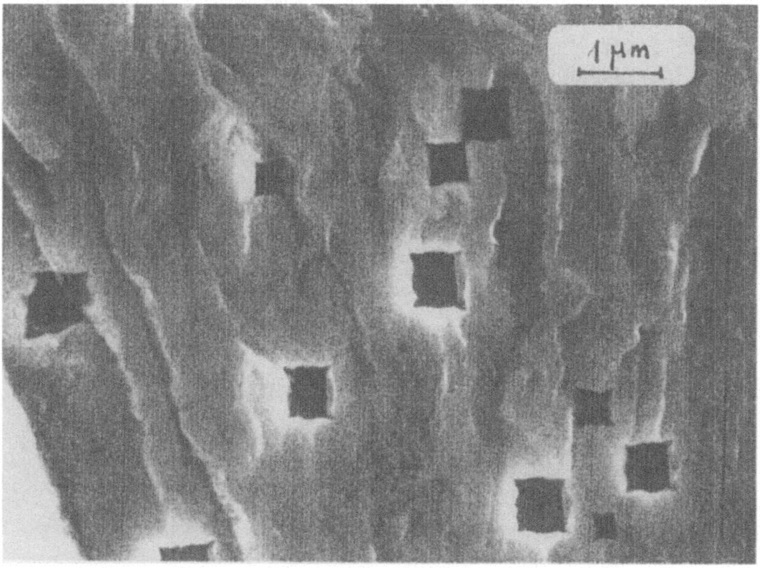

Figure 5. Stage II crack propagation in 2024 T351 Al alloy at low rates (~ 10^{-9} m/cycle) Etch pits are in accordance with a propagation along (100) planes in a <100>direction.

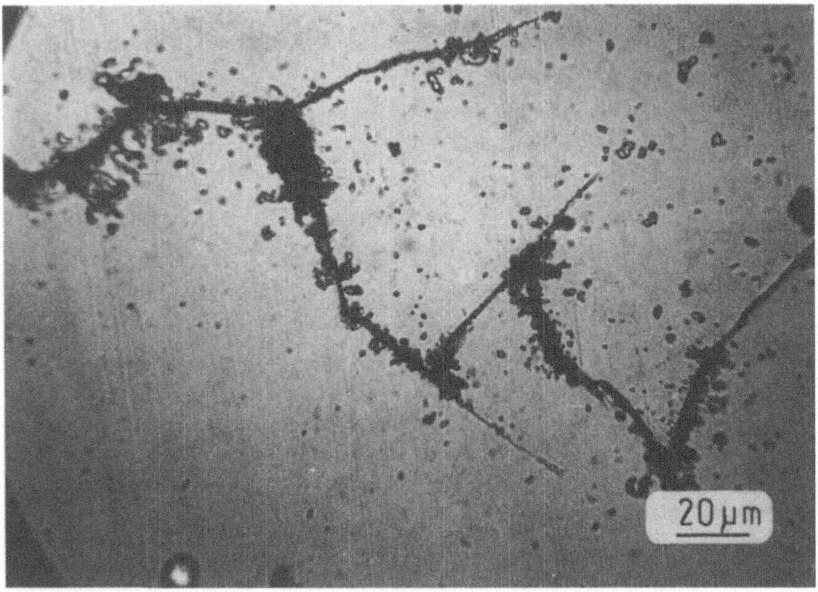

Figure 6. StageI-like crack in 2024 T351 alloy tested in high vacuum at 35 Hz and R = 0.5. da/dN = 5.10^{-11} m/cycle

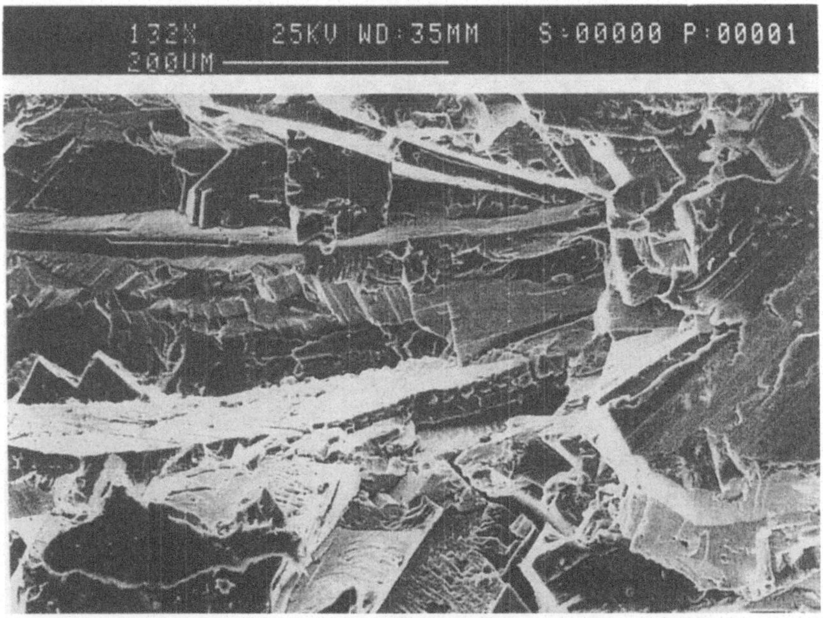

Figure 7. Microfractographic morphology of the stage I-like crack illustrated in figure 6. All facets correspond to (111) planes.

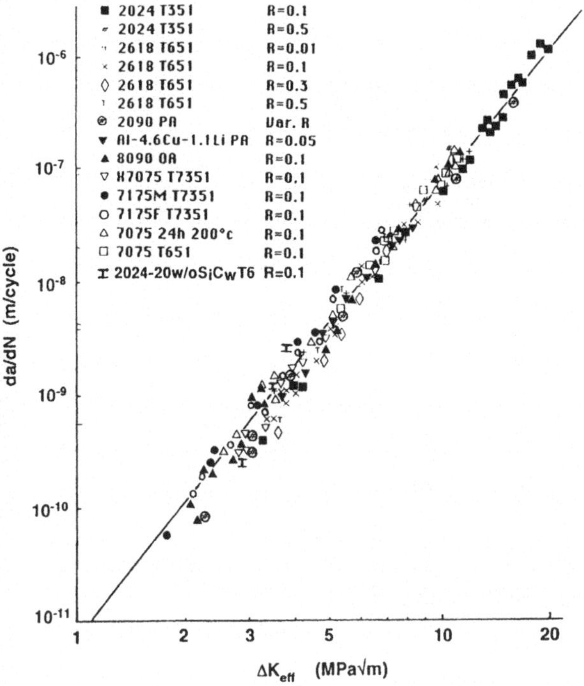

Figure 8
Intrinsic stage II propagation for a range of Aluminium based alloys. Tests performed in high vacuum at frequencies of 20 to 50 Hz.

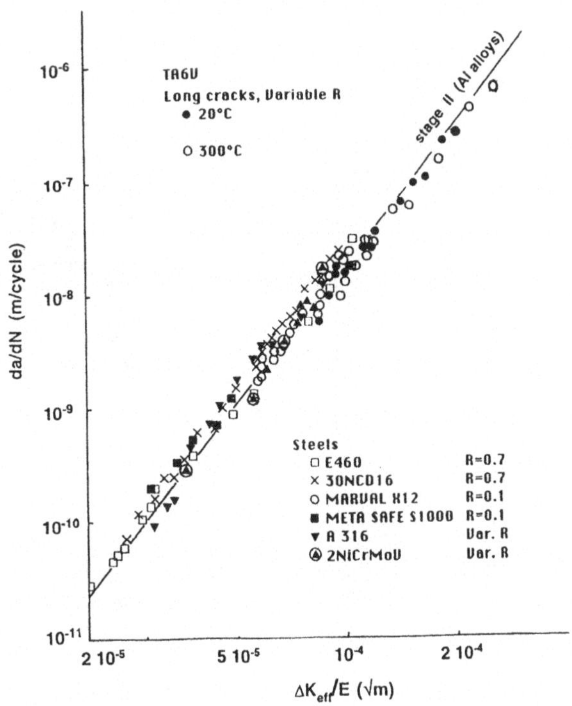

Figure 9
Intrinsic stage II propagation for a selection of steels and some data on TA6V alloys. The mean curves for Al alloys is reported from figure 8.

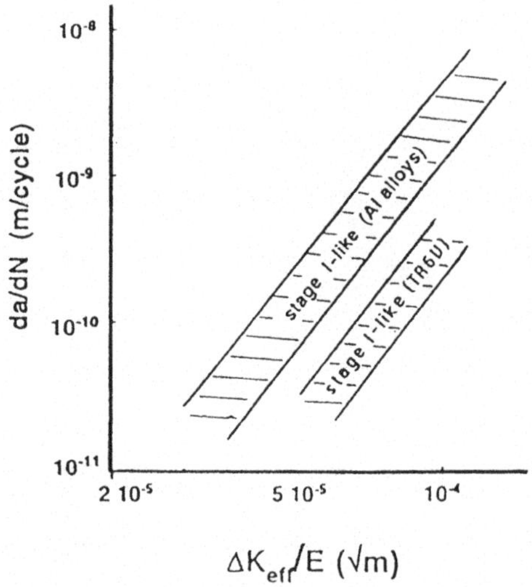

Figure 10a. Comparison of intrinsic stage I-like propagation for Al alloys and TA6V alloys in a da/dN vs ΔK_{eff} /E diagram.

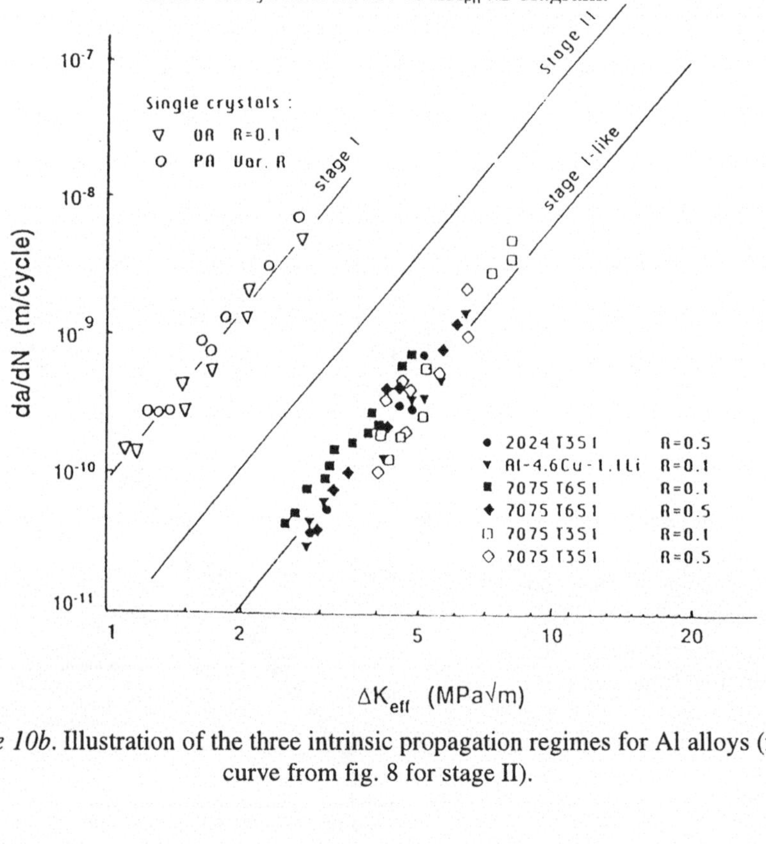

Figure 10b. Illustration of the three intrinsic propagation regimes for Al alloys (mean curve from fig. 8 for stage II).

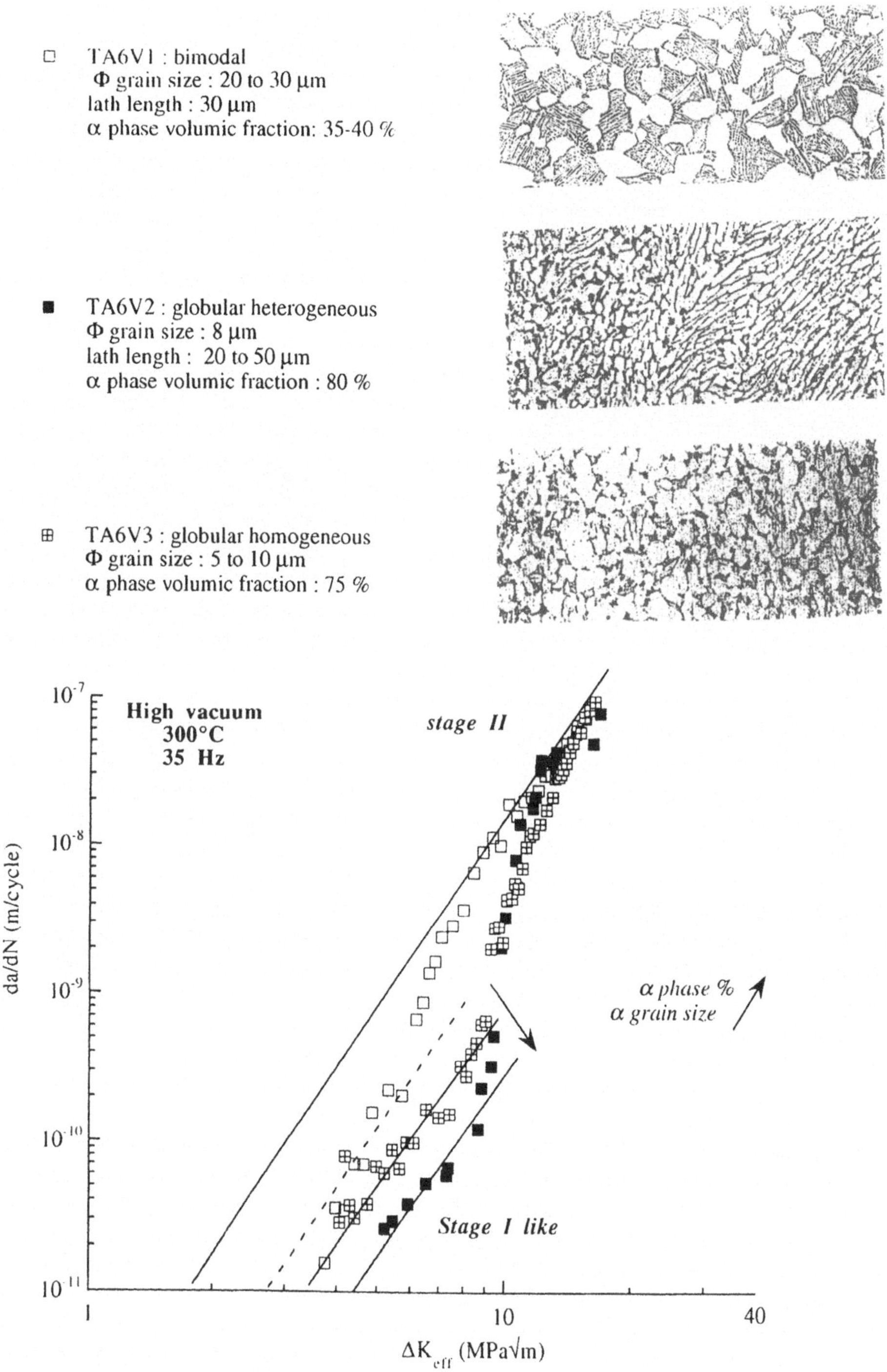

Figure 11. Influence of microstructure on the intrinsic near-threshold fatigue crack propagation of Ti-6Al-4V alloy. High vacuum, 35 Hz, R = 0.1.

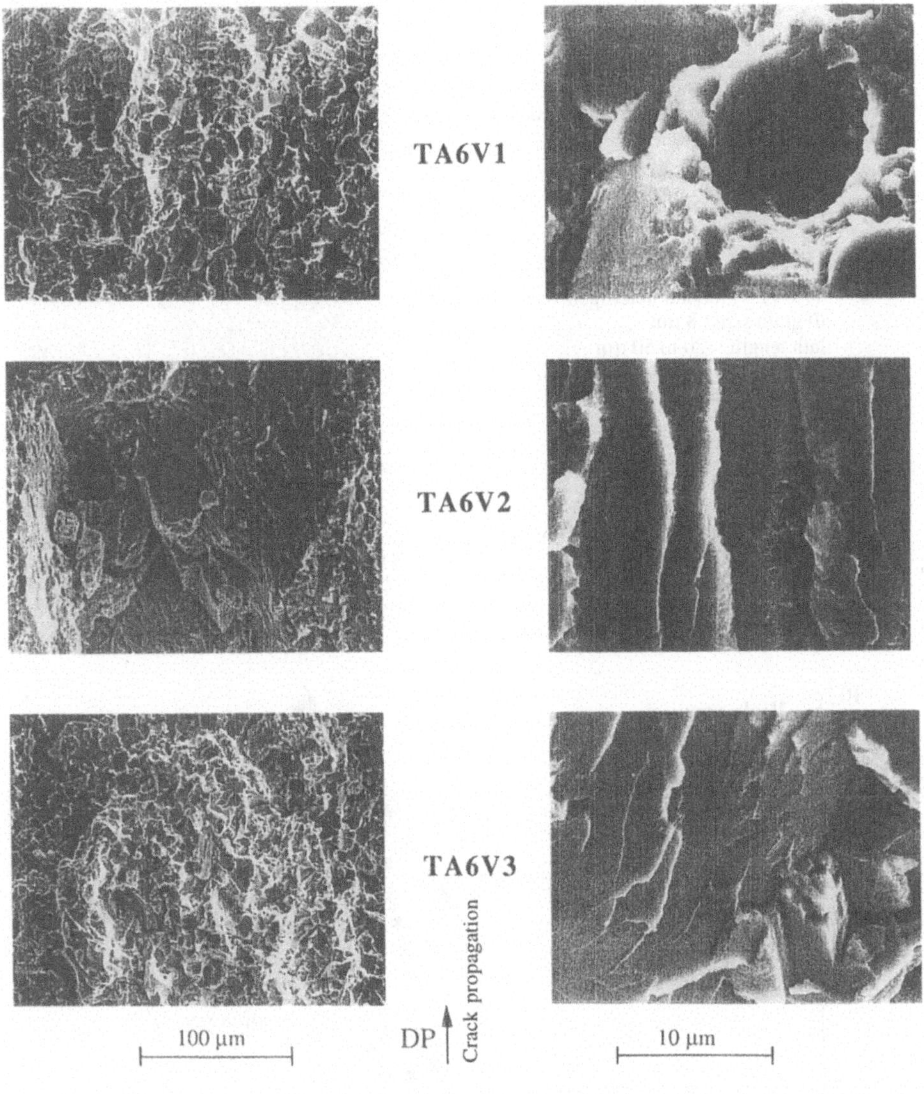

Figure 12. Fracture surface morphology of TA6V alloy in the near threshold stage I-like regime. Tests performed in high vacuum, at 300°C, 35 Hz and constant Kmax (no closure) : TA6V : binodal microstructure ; TA6V2 : heterogeneous globular microstructure ; TA6V3 : fine homogeneous globular microstructure (see fig. 11).

In sharp contrast with stage II, the stage I-like propagation cannot be rationalized using the above relation (Figure 10). This regime is highly sensitive to all factors which can favourise the strain localisation within a single slip system at the microstructural scale. Such localisation can occur at very different stress levels with respect to the grain size, the dimension and the shearability of the precipitates, the nature and the volumic fraction of the phase where such localization takes place. The induced retardation depends upon the grain size and upon the number of available slip systems in each grain and near the grain boundaries.

For example, the figures 11 and 12 give an illustration of the influence of the size and of the volumic fraction of primary α grains in a TA6V alloy treated in three different conditions. The higher the volumic fraction or/and the larger the grain size, the more accentuated the retardation. In addition the retardation is more well marked when the number of available slip systems is limited (Ti alloys) or is nearly absent when some secondary slip systems can be activated near the boundaries [37] and facilitate the crossing of slip barriers as observed in Al-Li alloys with Lithium addition higher than 2 wt %.

2.2. Intrinsic propagation of microcracks.

The intrinsic behaviour of naturally initiated microcracks can be analysed on the basis of the above framework consisting of three main crack growth regimes. An illustration is given in figure 13c for two Al alloys. As observed in illustration 13a, a microcrack initiated at the surface of a specimen of a 7075 type alloy in T651 peak aged condition, grows in the stage I regime in the first grain. Such propagation is favourised by GP and S' shearable precipitates which promote the localisation of the deformation within PSB's [31]. When the crack has crossed several grains, the stage I-like propagation regime is encountered giving highly retarded rates.

For larger crack extend and increased ΔK range, the propagation switches to the intermediate stage II regime. So, when the relation of the crack propagation with respect to the microstructure is well established, the LEFM concept, i.e. the ΔK concept, can be applied as well for short crack, which does not present closure, as for long crack after correction for closure.

3. Environmentally-assisted fatigue crack propagation.

3.1. Identification of the governing mechanisms.

Following the rationalization of the intrinsic stage II propagation as presented above, some similar rationalization of FCG in air could be expected after correction for crack closure and temperature effects ($\Delta K_{eff}/E$). Figure 14 presents a compilation of stage II propagation data obtained in ambient air for almost the same alloys as in vacuum (see, Figures 8 and 9). Obviously any rationalization does not exist in air. The sensitivity to air environment is shown strongly dependent as well on base metals, addition elements, and microstructures (see 7075 alloy in three different conditions) as on R ratio and growth rate. However a common critical rate range at about 10^{-8} m/cycle can be pointed out for all materials. This critical step is associated to stress intensity factor ranges at which the plastic zone size at the crack tip is of the same order as grain or sub-grain diameters. In addition there is a general agreement to consider that, for growth rates lower than this critical range, crack propagation results from a step-by-step advance mechanism instead of a cycle-by-cycle progression as generally observed in the Paris regime in air [38].

Historically, Snowden [39] has first suggested that the environmental effect on fatigue behaviour of metals must be described in terms of the number of gas molecules striking the crack tip surface and being adsorbed on fresh metal surface exposed to active species in the part of the loading cycle during which the crack is open.

According to gas kinetic theory, the number of gas molecules striking a unit area in a unit time is given by the relation :

$$n_o = \frac{N.P}{(2\pi MkT)^{1/2}} \qquad (2)$$

where N is the Avogadro number, P the partial gas pressure, M the gas molecular weight, k the gas constant. However, assuming that the upper critical pressure in a S-shaped pressure controlled phenomenon (figure 1) corresponds to the value of n_o at saturation of an adsorbed monolayer, the calculated critical pressures were lowered by a factor 100 to 1 compared with experimental data. Similar discripancies were observed by Bradshaw and Wheeler [7,8].

A first modification to this approach was proposed by Achter [40] who reconsidered coverage condition at the crack tip and the impedance factor related to the restricted gas flow in the fatigue crack ; but substantial discripancies were still observed between calculated and experimental critical pressure values.

These approaches based on correlation between rate variations and pressure of active gas at the crack tip did not settle the actual governing mechanism. In addition unity sticking coefficients were assumed, the influence of R ratio or/and of crack closure was not taken into account, and geometrical crack surfaces were considered instead of physical surfaces which could be much larger.

At the end of the sixties, Wei et al proposed a constitutive model to describe the influence of water vapour on fatigue crack propagation in the Paris regime on steels and Al alloys [11-15]. Crack propagation enhancement is attributed to Hydrogen embrittlement. Hydrogen production is assumed to result from dissociative chemical adsorption on freshly created surfaces at the crack tip.

On another side, Lynch [9-10] and Bouchet et al. [16] have suggested that the alteration of the fatigue resistance in active environment was the results of physical adsorption onto freshly created surfaces. Active species adsorption (or chemisorption) on a few atomic layers would be sufficient to enhance fatigue crack propagation by facilitating dislocation nucleation. Such approach, different from Wei's one, is closer to the description by Petch [41] who has proposed an expression describing the surface energy variation for a Langmuir isotherm[42] induced by adsorption of a diatonic molecule.

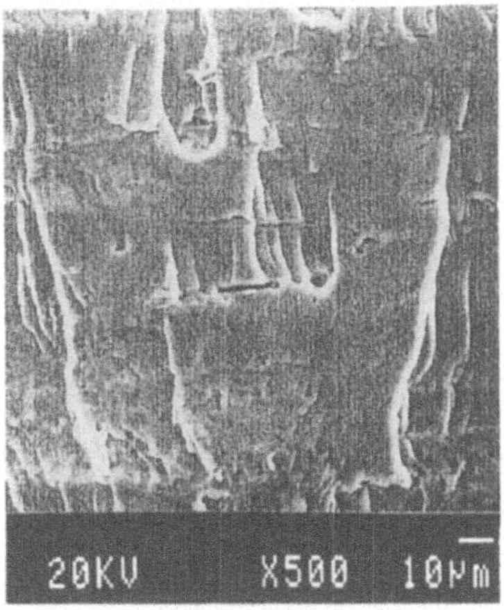

b)

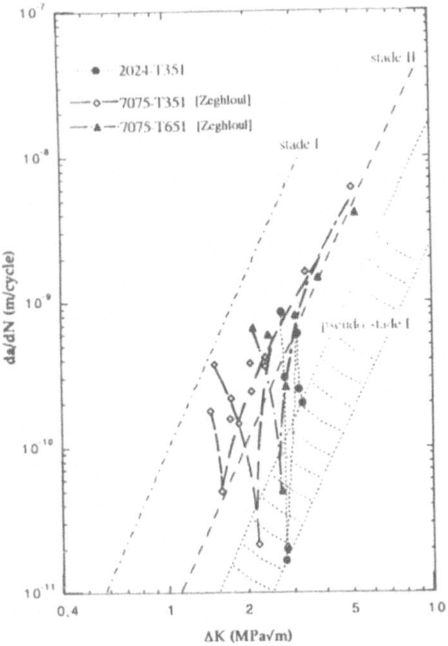

a)

c)

Figure 13. Intrinsic propagation of 3D microcracks naturally initiated in Al alloys. Tests run in high vacuum at 35 Hz and R = 0.1 :
a) Short crack data compared to the three intrinsic regimes.
b) Fracture surface of a stage II crack in 7075 T7351
c) Stage I facet at the initiation site in 7075 T651
d) Crystallographic surface morphology of stage I-like propagation in 2024 T351 in near-threshold conditions.

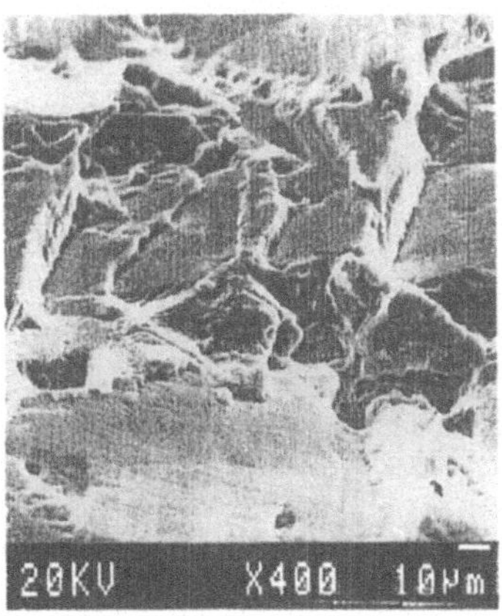

d)

a)

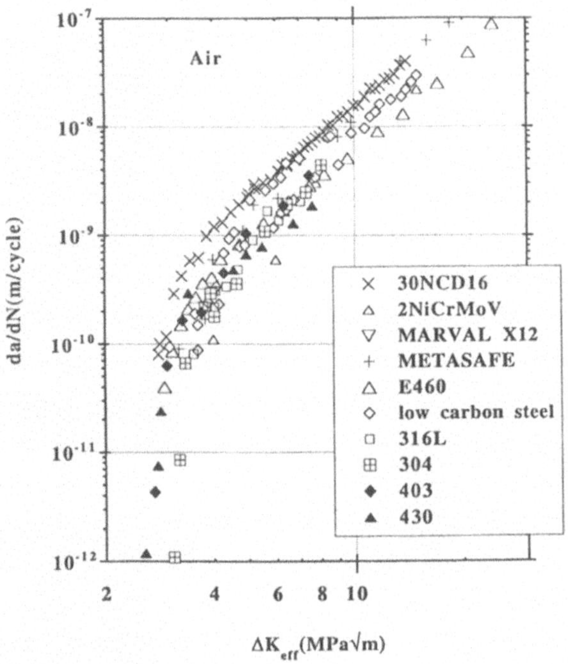

b)

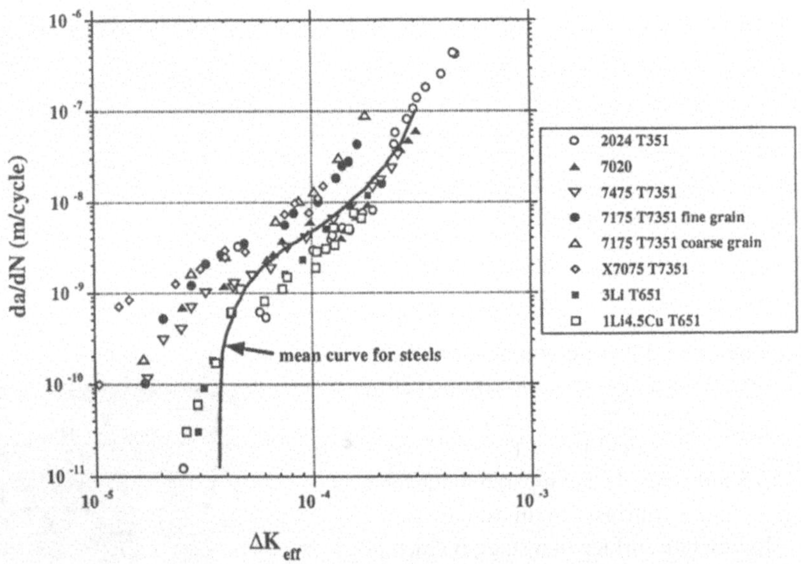

Figure 14. Effective crack propagation of steels (a) and Al alloys (b) tested in ambient air at frequencies of 20 to 50 Hz.

On the basis of experimental data obtained on Aluminium alloys and steels in air, high vacuum and purified nitrogen containing traces of water vapour, a comprehensive model has been established by J. Petit et al. including these two different mechanisms for environmentally assisted crack growth [3,17,21] as schematically illustrated in Figure15 :

- at growth rates higher than a critical rate $(da/dN)_{cr}$, which depends upon several factors as surrounding partial pressure of water vapour, load ratio, test frequency, chemical composition and microstructure, the crack growth mechanism is assisted by water vapour adsorption but it is still controlled by plasticity as in vacuum.

- at growth rates lower than $(da/dN)_{cr}$, the Hydrogen assisted crack growth mechanism becomes operative, Hydrogen being provided by adsorbed water vapour when some critical conditions are fulfilled. Both mechanisms will be detailed in the following.

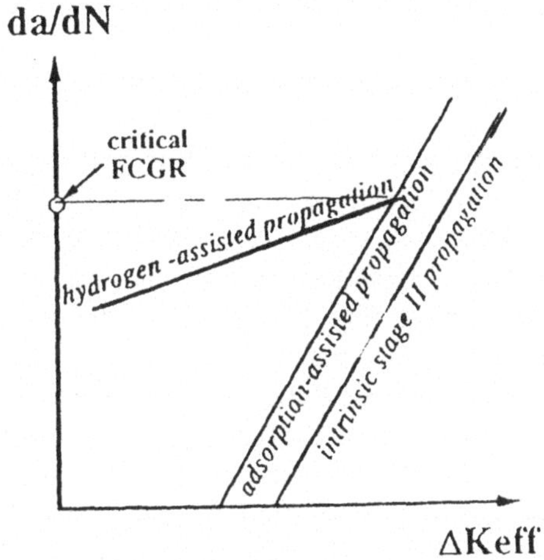

Figure 15. Schematic illustration of environmentally assisted stage II fatigue crack growth mechanisms.

3.2. Adsorption assisted propagation.

3.2.1 *Background*

In a study related to the fatigue of glass, Orowan originally suggested that adsorption of gas species on clean crack surfaces could reduce the surface energy of a Griffith crack [43]. Indeed gas species are submitted to an attractive force field perpendicular to the surface of a metal and therefore tends to accumulate on this surface. The subsequent decrease in surface energy is accounted by the Gibbs adsorption equation :

$$d\gamma = -\Gamma d\mu = -\Gamma RT \ln a \qquad (3)$$

where Γ denotes the surface energy, a is the activity of the adsorbed species, T the temperature, p the pressure and G the concentration of adsorbed molecules. This relationship was later reconsidered by Petch and Stables [44] in the case for iron-base metals. Oriani [45] demonstrated by thermodynamical considerations that adsorption induces a change in the stress experienced by the atoms located at a crack tip. One might expect an enhanced decohesion of the lattice or and an enhanced slip activation at the crack tip. Thus, according to Lynch [9, 10] active species adsorption or chemisorption on a few atomic layers would be sufficient to enhance fatigue crack propagation. This author suggests that in an inert environment, most of the dislocation sources participate in blunting the crack tip or accommodate plastic deformation ahead of the tip, so that only a few dislocations contributes to the crack advance process. In an active environment, adsorption would induce a localisation of the dislocation activity so that more dislocations are available for crack growth by facilitating dislocation nucleation. The role of adsorption on fatigue crack propagation was also reconsidered by Bouchet [16]. In particular he proved that adsorption of water vapour only plays a significant role above a critical threshold in tems of moisture content.

3.2.2. *The existence of a pure adsorption-assisted regime*

Results recently obtained under very low water vapour partial pressure (10^{-3} Pa) and under closure-free conditions in the case of a high-strength low-alloy steel [21] shed a new light on this point. They first clearly demonstrated that this residual moisture content is sufficient to induce a significant increase in crack growth rates especially when the load frequency is decreased (fig. 16). In addition these results provide insight about the kinetics of this adsorption process. The curve obtained at low frequency (0.2Hz) exhibits a linear relationship on the explored rate range with a Paris law exponent $m = 4$ like the intrinsic stage II regime [34], as shown on figure16. This regime is characterised by a fourfold increase in crack growth rates as observed in air in the upper range. At intermediate frequencies the behaviour first follows this regime and then, above a critical value of da/dN tends to recover the intrinsic behaviour.

A model has been proposed to provide a comprehensive picture of the processes involved in the behaviour observed on figure 17 [21]. The approach developed here is actually extremely close to the problem formerly addressed by Orowan [43]. The general framework is based on the concept of Langmuir isotherms [42].The pressure at the crack tip, which governs the adsorption kinetics, results from a competition between transport of active species, consumption by adsorption of water vapour molecules on fresh surfaces and crack growth rate. The transport of active species to the crack tip is calculated by taking into account the crack impedance effect. Indeed the crack behaves like a narrow channel connecting the bulk environment to a control volume located at the crack tip, inducing a pressure in this control volume lower than the nominal pressure in the bulk environment. The impedance is actually mostly governed by the distance between the crack walls, that means by the crack opening displacement and therefore the stress-intensity factor value. For the very low partial pressures used in these investigations, the flow within the crack is assumed of molecular type, that means

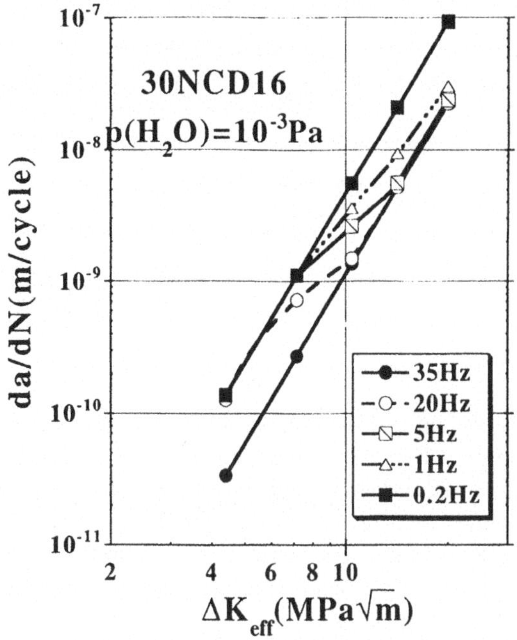

Figure 16.
Influence of test frequency on the propagation of a fatigue crack in high vacuum on a high strength steel. R = 0.7

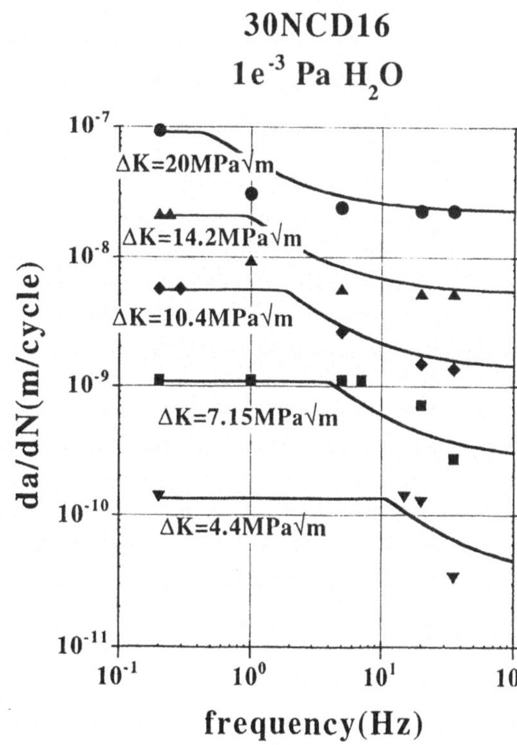

Figure 17.
Comparison of growth rates measurements with respect to test frequency (same experiments as in figure 16) to curves given by modelling of adsorption assisted propagation.

that the access of water vapour molecules to that crack tip is hampered by collisions between the molecules and the walls, and not by collisions with other molecules.
Besides the similarity in the slopes of the characteristic regimes suggests that the adsorption of water vapour molecules on freshly cracked surfaces by itself does not deeply affect the basic propagation mechanism which might still be described by a relationship similar to Eq. [1] :

$$\frac{da}{dN} = \frac{A}{D^*}\left(\frac{\Delta K_{eff}}{E}\right)^4 \tag{4}$$

Adsorption of water vapour molecules is assumed to lessen the value of the surface energy γ. However in the case of ductile alloys the energy required to create a unit area of cracked surface U might be several orders of magnitude higher than γ. Actually U is assumed to be derived from γ by a multipliying factor which might be viewed as a ductility factor. The same kind of dependence is assumed to prevail for the critical displacement D* which is thus closely related to the surface coverage rate θ by the following equation :

$$D^* = \frac{D_0^* D_1^*}{\left(D_1^* + \theta\left(D_0^* - D_1^*\right)\right)} \tag{5}$$

At the end of the day the model provides a relationship between the coverage rate θ (defined as the ratio between the number of sites occupied by an adsorbed molecule and the total number of available sites on the rupture surface) and the load frequency for a given value of the stress intensity factor.

The computations show a nice agreement with experimental data (solid lines on figure 17). This model has also been proved to succeed in explaining results obtained under gaseous atmospheres by Liaw et al.[46]. It also reveals successfull in accounting for the adsorption-assisted propagation at elevated temperature in the case of titanium alloy (figure 18) thoughout a broad spectrum of frequency and water vapour partial pressure [47]. However in this case it was assumed that equilibrium does no longer correspond with a saturating coverage rate, that means that θ is lower than 1 due to a promoted influence of desorption as temperature raises.

A compendium of data from ancillary testings and from literature proves that the adsorption-assisted regime might be encountered for a wide variety of materials and environmental conditions (figure 19).

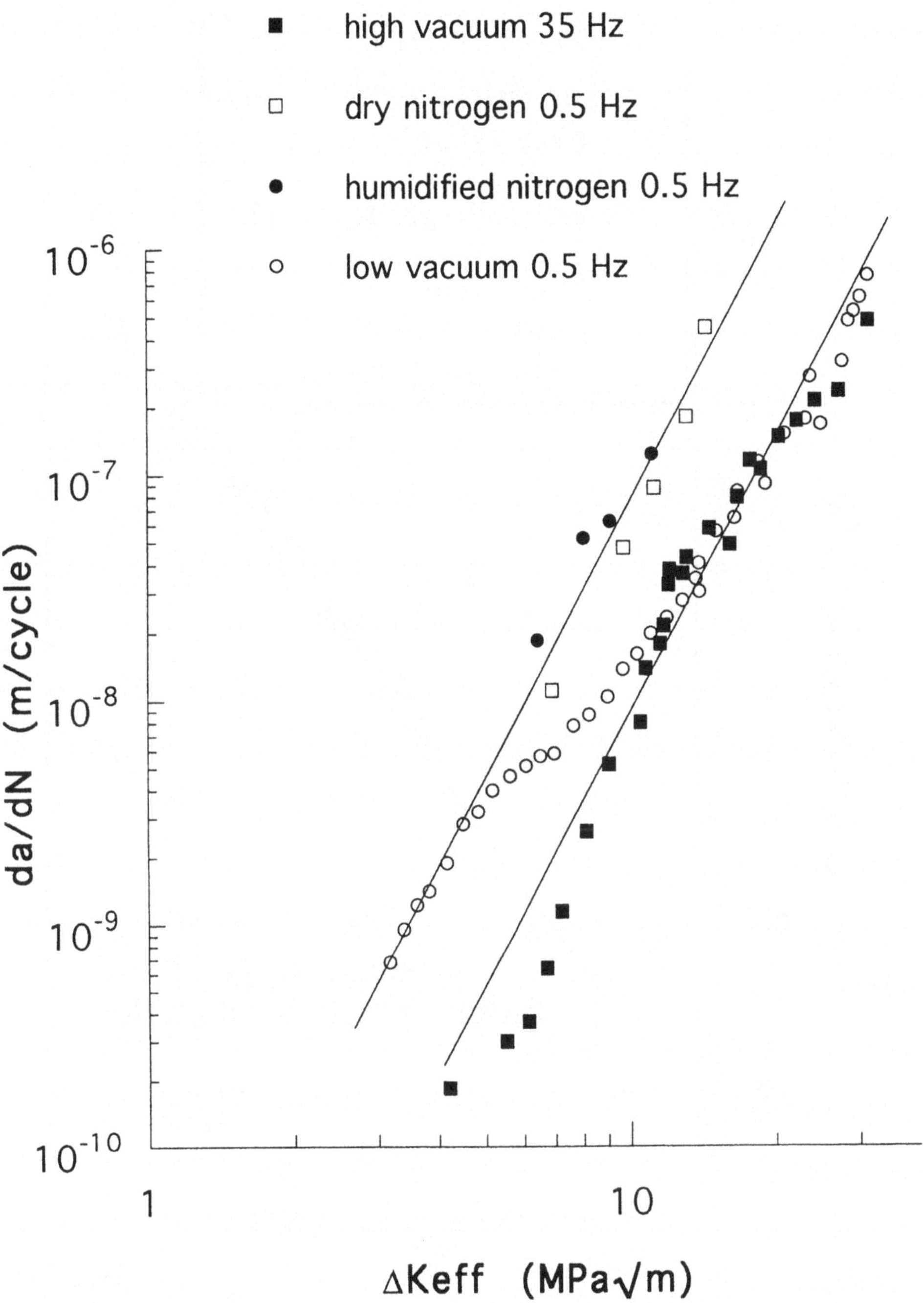

Figure 18. Influence of test frequency and water vapour partial pressure on the fatigue crack propagation behaviour in a TA6V alloys at 300°C. Characterization of an adsorption assisted regime induced by low pressure of water vapour.

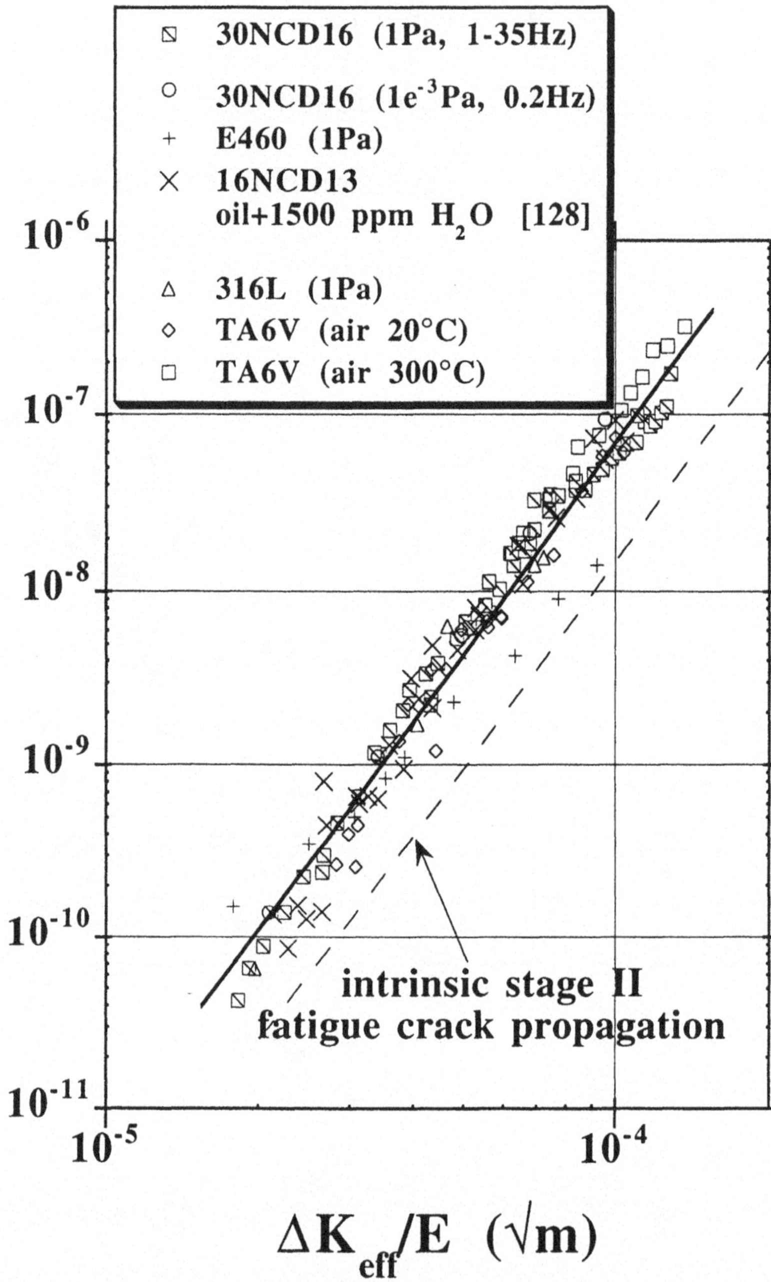

Figure 19. Compendium of data for various alloys illustrating and adsorption-assisted regime induced by low partial-pressure of water vapour.

3.3. Hydrogen-assisted fatigue crack propagation

Wei and co-workers [11-14], have proposed a sequential process for environmentally-assisted fatigue crack propagation in an hydrogenous atmosphere (figure 20). The sequential events would be :

- transport of the active species to the crack tip
- physical adsorption
- chemical adsorption and dissociation
- hydrogen penetration
- diffusion and embrittling reaction.

The same process is considered to describe environmentally assisted fatigue crack propagation in the Paris regime on steels. Hydrogen production is assumed to result from the dissociative chemical adsorption of molecules after physical adsorption on surfaces freshly created at the crack tip. Hydrogen then diffuses into the process zone where the embrittling reaction takes place at some specific but ill-identified sites. However most of the experimental results support the existence of such a regime in moist atmosphere only near threshold conditions and for growth rates lower than a critical rate as initially proposed by Achter [40]. In the Paris regime and at room temperature, in most cases, the adsorption assisted propagation is generally observed [3, 21].

Indeed the Hydrogen assisted regime becomes operative when several conditions favouring high Hydrogen concentration into the process zone at the crack tip are fulfilled :

- conditions of access to the crack tip for active species which lead to sufficient partial pressure of water vapour to create an instantaneous adsorbed monolayer : surrounding pressure, frequency, growth rate, R ratio ; in such conditions, the mechanism is reaction-controlled.

- sufficiently low stress intensity factor to reach a regime with a stationary crack and localized plastic deformation in a limited number of slip systems within a single grain at the crack front ;

- a long time enough to allow Hydrogen to diffuse by dislocation dragging so as to attain a critical Hydrogen concentration for metal embrittlement.

Such conditions are encountered in ambient air or in humidified inert gas for growth rates lower than the critial rate $(da/dN)_{cr}$ (Fig. 21).

Bowles [48] improved the expression initially derived by Acher [40] for the critical conditions (pressure, growth rate, frequency,...) required for an environmental effect in the case of a crack propagating along a (100) plane in a f. c. c. lattice :

$$\left(\frac{da}{dN}\right)_{crit} (mm/cycle) = 3.61 x 10^{-3} \frac{p}{f\sqrt{MT}} \qquad (6)$$

However this expression does not take into account the crack impedance effect highlighted by Bradshaw [49].

Hydrogen is assumed to evolve from the dissociation of adsorbed water vapour molecules. According to Wei et al. [11], the rate constant of this reaction would be 10^8 to 10^9 times higher for Aluminium alloys than for steels. However Hydrogen assisted-propagation in Aluminium alloys does not behave very differently from steels from a kinetical point of view [20]. Besides the threshold exposure (10^4 Pa x s) as measured by Simmons et al. [11] for this reaction is not reached for the experiments under vacuum conditions, which supports the apparent lack of hydrogen effect. Hydrogen assistance first requires the attainment of saturating adsorption at the crack tip but is triggered below a critical crack growth rate value depending on the loading conditions. In nitrogen the critical da/dN value for embrittlement is lower than in air [18, 50, 51] due to the lower water vapour content and the subsequent lower exposure. However, as soon as the hydrogen process is triggered, crack growth rates are readily enhanced and as a consequence the effective thresholds measured in air and in nitrogen are nearly equal. But moisture content does not seem to be the mere variable controlling the triggering of the hydrogen-assisted process. Indeed, results obtained under low vacuum conditions (1Pa), that means for a nearly similar moisture content, show that no hydrogen assistance takes place and saturating adsorption is the mere mechanism involved in environmental influence in this particular case (figure 19). Therefore the identification of the precise role of both total pressure and partial pressure of water vapour clearly requires further investigations.

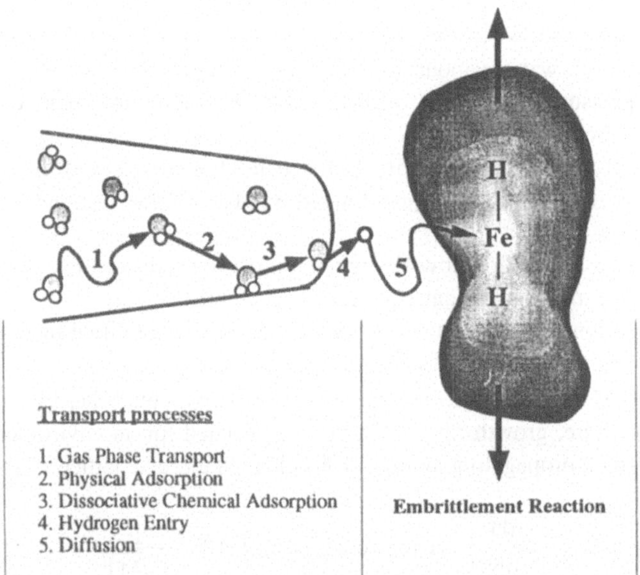

Figure 20. Sequencial process for environmentally-assisted fatigue crack propagation in an hydrogeneous atmosphere proposed by Wei and co-authors [11 - 14].

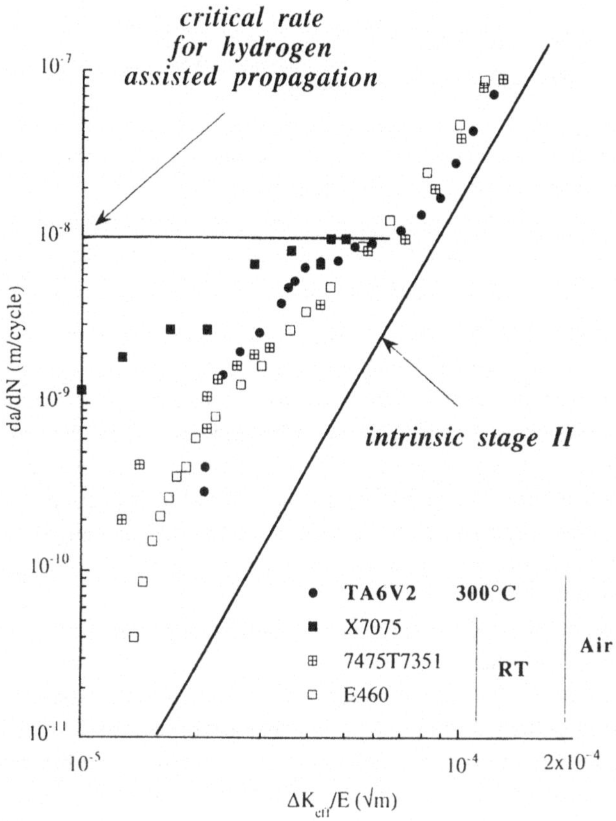

Figure 21. Critical rate for hydrogen assisted propagation in different materials tested in Laboratory air at 35 Hz.

4. Constitutive fatigue crack propagation law.

Up to now no fatigue crack growth law taking into account the different processes exposed above is available. Theoretical models such as Weertman's model do not account for environmental effects and therefore they are merely valid for fatigue crack propagation in inert atmospheres. Similarly fatigue crack growth laws considering the strained material at the crack tip as a low-cycle fatigue microsample [52, 53] indirectly integrates environmental effects mainly through the use of a Manson-Coffin law.

The occurrence of hydrogen-assisted crack growth is associated to a typical change in the slope of the propagation curves which becomes close to 2 to 1 at low rates, and the transition from one regime (adsorption assisted) to the other (hydrogen-assisted) often corresponds to a more or less well marked plateau range (Fig14, 21)

$$\frac{da}{dN}=\frac{A}{D_1^*}\left(\frac{\Delta K_{eff}}{\mu}\right)^4+\frac{B}{\sigma\mu}\left(\Delta K_{eff}^2-\Delta K_{eff,th}^2\right) \tag{7}$$

where $\Delta K_{eff,th}$ denotes the threshold obtained under closure-free conditions, B is dimensionless a constant and σ a strength parameter.

The first term accounts for the adsorption-saturating regime previously described. The second one was subsequently added in an attempt to describe the hydrogen-assisted propagation regime. The ΔK_{eff}^2 dependence might be viewed as a coarse description of the dislocation dragging via the CTOD. $\Delta K_{eff,th}$ would thus denote a threshold value of this sweep-in mechanism to enable the attainment of a critical hydrogen concentration at the crack tip. However, up to now, one needs to experimentally determine this value. The comparison between experimental data and computations in the case of a high-strength low-alloy steel presented on figure 22 brings out a nice agreement.

However such a formulation for hydrogen assisted propagation is still highly empirical. The problems raise from the lack of a sound understanding of what happens ahead of the crack tip. Some critical issues which require a detailed knowledge to achieve this goal are listed below:

- examination of the role of pressure on surface reaction kinetics and possible competitive adsorption of other constituent;
- identification of the dislocation sweep-in mechanisms in the strained lattice;
- determination of the net force acting on a dislocation within the plastic zone;
- identification of trapping sites, including possible diffusion redistribution;
- determination of the critical amount of hydrogen and its interactions with the failure process.

Although the above mechanisms might be expected to be strongly sensitive to metallurgical factors, the similarity in near-threshold behaviour exhibited by typically different microstructures (figure 14) suggests that the overall behaviour is mainly governed by the mechanical loading and its connection with dislocation dragging. Finally the temperature dependence of these phenomena obviously constitutes a prime issue to investigate.

5. Near-threshold closure effect

Since Elber's pioneer work [26], the role of crack closure in influencing fatigue crack growth has become a topic of considerable interest. The numerous experimental works dedicated to crack closure provide evidence that, if plasticity induced-closure plays a dominant role under plane stress, marked crack closure can occur even in the near-threshold regime where predominantly plane strain conditions exist.

As described by Suresh, Zaminski and Ritchie [55] crack closure can arise near-threshold due to the rough nature of the fracture surface when crack tip opening displacements (ΔCTOD) become comparable to the size-scale of the fracture surface asperities. Roughness induced closure has been shown to be promoted by coarse-grained materials and microstructures with coherent and shearable precipitates, which induce

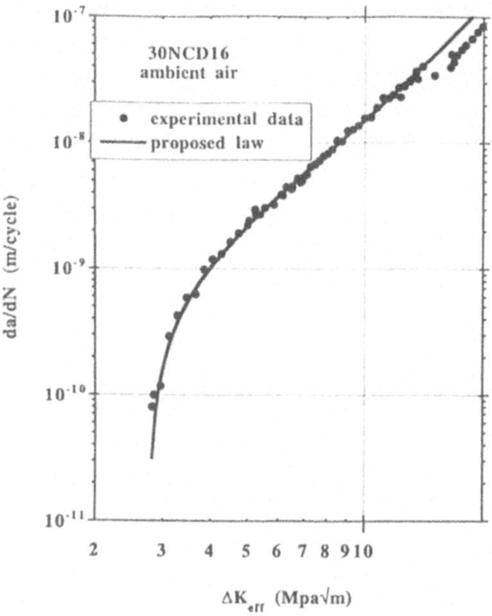

Figure 22.
Comparison of experimental data and computation of relation (7)

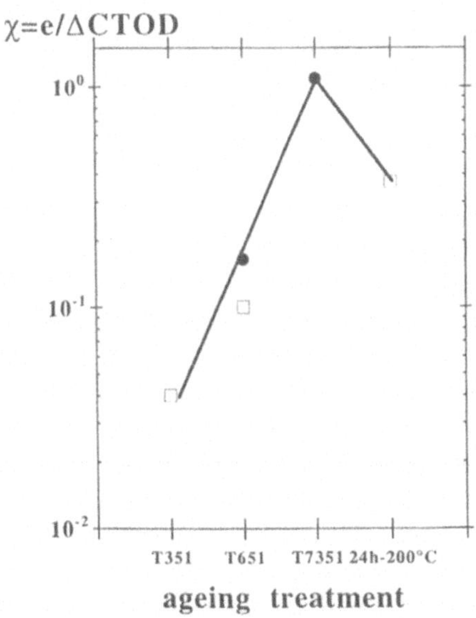

Figure 23.
Evolution of χ, ratio of the excess of oxyde thickning against the Crack Tip Opening Displacement range wikth respect to ageing treatment for a 7075 Al alloy tested near threshold in ambient air at 35 Hz and R = 0.1.

coarse planar slip or single slip processes (Stage I-like regime) and favour crack deflection and crack branching mechanisms. However, in active environment, crack wedging by crack surface oxidation (described as oxide induced-closure) can contribute to an enhanced closure effect. Since, the existence of crack closure due to corrosion debris influencing near-threshold propagation and threshold values has been proved by numerous authors for various metals. Moreover oxide built-up has been shown highly sensitive to microstructure [55]. It has been assumed that oxidation sharply modifies the access condition of gaseous active species up to the crack tip (p.e. water vapour trapping by oxide deposits). This effect explains in particular the well marked threshold generally obtained in oxidizing atmospheres compared to vacuum. Measurements of thickness and extent of oxide films on fracture surface of specimens tested in moist environment, performed on various metals by X-ray photo-electron spectroscopy, scanning Auger spectroscopy, and secondary ion mass spectroscopy, according to loading conditions and aging treatments, suggest that environmentally assisted FCG near the threshold in Aluminium alloys and steels is controlled by two competitive mechanism: crack closure due to oxidation and embrittling effect of moisture both depending on microstructure. An illustration is given in figure 23 where the oxide thickening as evaluated by SIMS measurements in the near-threshold region of the fracture surface of a 7075 aluminium alloy is shown maximum for the T7351 condition, i.e. when the excess oxide thickness is of the same order as the ΔCTOD at threshold [56]. Results from experiments on E460 structural steel in high vacuum, slightly wet nitrogen, and laboratory air [18] have shown a similar competitive effect between oxide induced closure and embrittling mechanism associated with water vapor on the FCG rate near the threshold. Operating at low R ratio, higher threshold levels in moist air than in vacuum can be even obtained in some cases. Similar results were reported by Smith [57] on 2NiCrMoV rotor steel, the threshold at $R = 0.6$ in moist air being slightly lower than in dry air, while at $R = 0.1$ the threshold in moist air is higher by a factor 1.6 to 1.

In conclusion, the understanding of environmental effect on nominal near-threshold FCG requires a detailed analysis of closure effects associated to roughness and oxidation according to loading conditions and microstructure.

The sensitivity of these factors to ambient humidity, temperature, test frequency, loading conditions, specimen type and related trend to favor mode II (p.e. four point bending or CT specimens) [58], explains substantial apparent scatter in nominal crack propagation behaviour. Existing normalisations do not account for the influence of these different parameters. On another hand, existing modellings of crack closure remain relatively limited. Numerical models for plasticity induced closure has been proposed by Newman et al. [59 and R.C. Mc Clung [60]] using a finite element technique, which has been successfully applied in the Paris regime by several authors. But near threshold i.e. in condition where oxide and roughness induced closure processes are dominant, no satisfactory modelling still exists.

	ENVIRONMENT	TOTAL PRESSURE	P_{H2O}	P_{O2}
△	Ambient air	10^2 kPa	1.3 kPa	$2x10^4$ Pa
▲	Humidified nitrogen	10^2 kPa	1.3 kPa	< 1 Pa
+	Low vacuum	1.33 Pa	1 Pa	< $4x10^{-4}$ Pa
●	Medium vacuum	10^{-2} Pa	$8x10^{-3}$ Pa	-
○	High vacuum	$3x10^{-4}$ Pa	$2x10^{-4}$ Pa	< 10^{-6} Pa

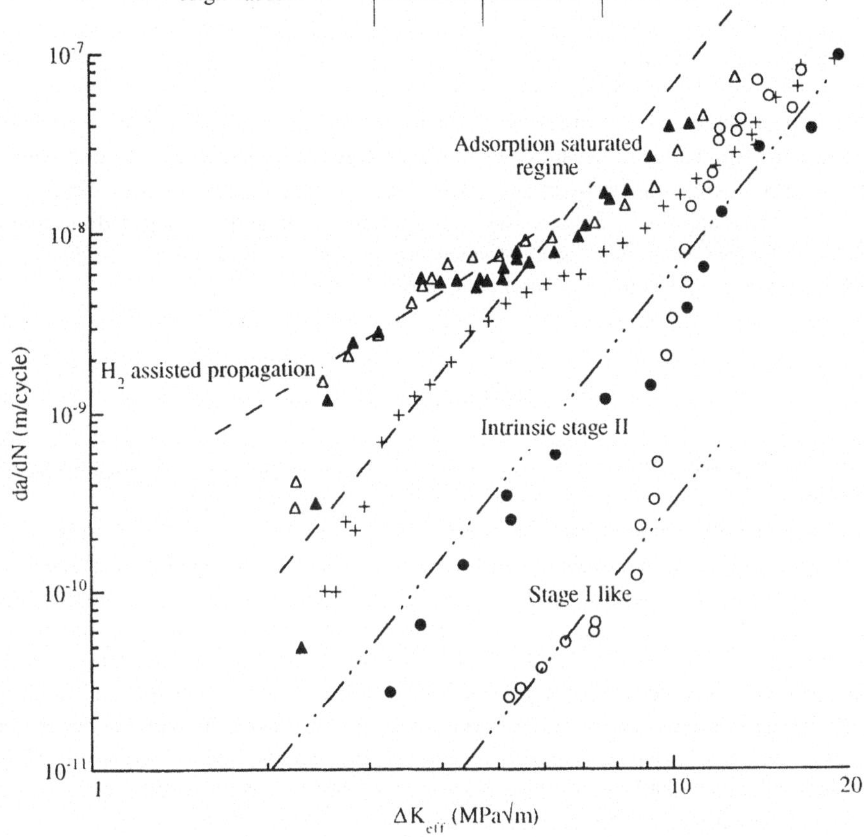

Figure 24. Influence of water vapour partial pressure on the fatigue crack growth behaviour of a TA6V at 300°C.
Comparison to the different propagation regimes.

6. Influence of high temperature

As described by Suresh [61], under cyclic loading conditions at elevated temperature, the conditions governing the crack growth can be estimated by comparing the cycle time t_c (i.e. the duration of the fatigue cycle) with the transition time t_T [62]. Small creep conditions occur when $t_c << t_T$ and extensive creep conditions prevail when $t_c >> t_T$. Based on this comparison, three distinct regions of fatigue crack growth can be identified. At lower temperatures and higher cycle frequencies (i.e. low t_c), fatigue crack growth is essentially cycle dependent and it is reasonably well characterised by ΔK. At very high temperatures and low frequencies (i.e. high t_c), crack growth is completely controlled by time dependent processes. For this purpose modelling introducing fracture mechanic parameter such as C_t proposed by Gieseke and Saxena [63] can be considered. In the region lying between these two extremes, fatigue crack growth is an outcome of the synergistic interactions between cycle-dependent and time-dependent processes.

In the first approach, Saxena [64] has proposed a linearly superposition of the mechanical fatigue and time-dependent creep component of crack growth to derive an owerall crack extension rate. The total growth rate is then given by :

$$\left(\frac{da}{dN}\right)=\left(\frac{da}{dN}\right)_{fatigue}+\left(\frac{da}{dN}\right)_{creep} \tag{8}$$

In the following, conditions for enhanced creep fatigue are not considered and conditions for mechanical fatigue or corrosion fatigue are emphasised.

It is generally admitted that the effect of air environment at elevated temperature results from oxidation. The role of oxides in fatigue crack propagation has been, for example, recently reviewed by J.E. King and P.J. Cotteril [65]. The role of oxide built up and subsequent oxide wedging of crack in the near threshold area has been widely demonstrated [46, 47, 55, 60]. But the relation between oxide formation observed on the crack surfaces of postmorten specimens and the crack growth rates in the mid-range, is mostly inferred, and the active role of oxygen at the very crack tip during the cycling process is not really demonstrated. Moreover recent experiments on Titanium alloys have clearly demonstrated a predominant detrimental effect of water vapour even at very low partial pressure and at temperatures ranging up to 500°C [25, 47, 66, 67].

Crack growth investigations on a Ti-6Al-4V alloy with a heterogeneous microstructure consisting of fine equiaxe α grains (8 μm of diameter) and elongated α platelets (50 μm of size), were conducted at 300°C in high vacuum, air, and nitrogen containing controlled amounts of water vapour and Oxygen (Figure 24). The FCG is shown strongly dependent on the partial pressure of water vapour while any influence of oxygen is not detected. It is of importance to notice that these tests were conducted at constant K_{max} i.e. in condition without closure in all the explored rate range and hence without interaction between environment and closure.

The fatigue crack growth behaviour of various TA6V and Ti alloys at 300°C [67] is shown to involve similar environmentally assisted mechanisms like steels and Al alloys at room temperature, i.e. and adsorption assisted-regime and an hydrogen assisted-regime which becomes operative for rates lower than a critical step as illustrated in figure 24 for Ti-6Al-4V alloy.

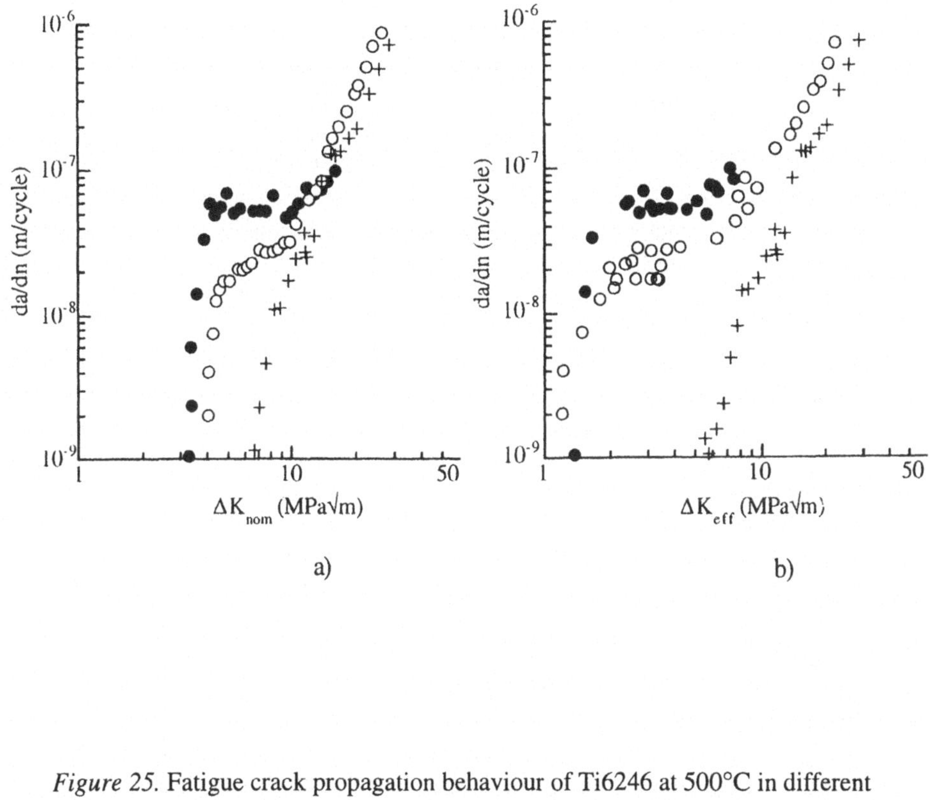

Figure 25. Fatigue crack propagation behaviour of Ti6246 at 500°C in different environmental conditions : a) nominal propagation curves (R = 0.1, 35 Hz). b) Effective data after closure correction or from tests at constant Kmax without closure.

Similar experiments were performed at 500°C on a Ti 6246 alloy with a fine Widmanstätten microstructure [47]. The crack propagation behaviour was determined in air, high vacuum, low vacuum (10^{-2} Pa) and in humidified Argon. In figure 25, the growth rates in moist argon are shown to be faster than in air with same partial pressure of water vapour (1,3 KPa). Indeed, the presence of oxygen appears to be more inhibiting than active. But oxide built-up explains for both environments the enhanced contribution same order as the ΔCTOD near the threshold. Lowering the frequency down to 0.1 Hz induces a sharp acceleration as well in air as in moist argon with still faster growth rates in the latter environment (Figure 26). These experiments conducted in the mid-rate range and in conditions without closure or with closure correction, support a great enhancement of the effect of water vapour at low frequency.

The morphology of the fracture surfaces as illustrated in figures 27 and 28 show different typical features related to oxidation. In humid argon (fig. 27a) the very flat transgranular crack path is covered by a thin and brittle rutile layer (about 100 nm as indicated by a blue coloration) exhibiting very thin microcracks which seem to correspond to the relaxation of some residual surface strains. The cross section in fig. 28a shows that these microcracks have a small extent and remain located within the oxidized surface. In air (Fig. 27b and 28b) more deeper microcracks are associated to interlamellar oxidation, the oxide layer been globally thinner (50 to 80 nm) and more heterogeneous than in argon.

In low vacuum, very little oxidation is observed (fig. 27c and 28c), the embrittling effect of environment inducing numerous microcracks which develop in the same direction as crack propagation. It is noticeable that so different apparent influence of oxygen has no substantial influence on the growth rates. Finally oxidation seems to occur on cracked surfaces but not at the very crack tip during the rupture process.

As a creep-fatigue process could be operative, cyclic loadings with long hold time were applied. These experiments clearly demonstrate the existence of a saturation of the influence of environment since increased holding times have no additional effect (figure 26). These observations substantiate the existence of a corrosion fatigue mechanism due to the presence of water vapour and are in contradiction with a time-dependent creep fatigue process. Careful observations and analysis of surface oxidation by mean of the RBS technique [47], support the absence of correlation between surface oxidation and crack growth rates. Very recent tests were conducted in low vacuum (10^{-2} Pa). Mass spectrometer analysis showed a predominantamount of water vapour (80 %) and a very small residual pressure of oxygen (< 10^{-4} Pa). Crack growth rates determined at 10^{-2} Hz in comparison to 35 Hz are shown to induced acceleration of more than one order of magnitude at a ΔK range of $6\text{MPa}\sqrt{\text{m}}$ leading to rates like as in air (figure 26). These last experiments demonstrate the major role of water vapour.

The final question is the identification of the detrimental mechanism controlling the propagation at such an elevated temperature. Hydrogen embrittlement can be still considered, but a specific action of OH^- ions resulting from water vapour dissociation cannot be a priori excluded [47]. Further investigations are required to answer this question.

The water vapour assisted propagation at 500 °C can be described using a relation based on the crack tip opening displacement $\Delta\delta_t$ which is related to the crack increment per cycle [68 - 70]. Since $\Delta\delta_t$ is related to the stress intensity factor range as :

$$\frac{da}{dN} \simeq \frac{1}{2}\Delta\delta_t = \frac{1}{2}\beta\frac{(\Delta K)^2}{\sigma_y' E} \tag{9}$$

where σ'_y is the cyclic yield stress, E the Young modulus and β a dimensionless parameter.

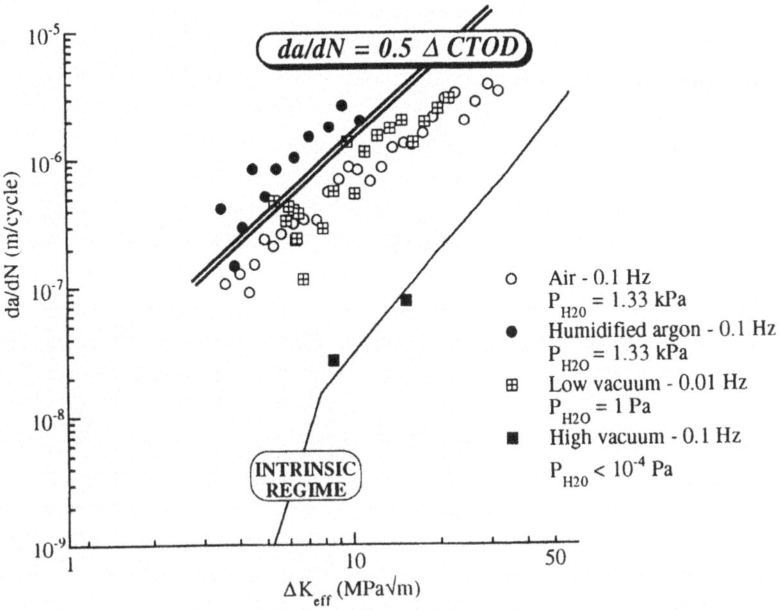

Figure 26. Fatigue crack propagation at low frequency in different environments. Comparison of water-vapour assisted corrosion fatigue propagation data with the conventional ΔCTOD model.

As plotted in figure 29, the experimental data provided by different testing conditions support the existence of a quite well defined corrosion assisted regime described by :

$$\frac{da}{dN} = \frac{1}{2\pi}\left(\frac{\Delta K^2}{\sigma'_y E}\right) \qquad (10)$$

In figure 29, experimental data obtained in air at 593°C on a Ti 1100 alloys are reported from Ghonem and co-authors [71, 72] in comparison with the above presented results. A good agreement is obtained supporting the existence of a similar FCG corrosion assisted mechanism in the range 500°C - 600°C. However these authors evoke mainly a creep-fatigue controlling mechanism. The exploration of the range 500°C to 600°C would give more precise information on the respective role of water vapour and oxygen. Similar water vapour assisted crack propagation has to be expected on all metallic alloys including steels and Ni-based alloys in some specific temperature ranges.

a)

Humid argon

0.1 Hz

b)

air

0.1 Hz

c)

Low vacuum

0,01 Hz

crack propagation

→

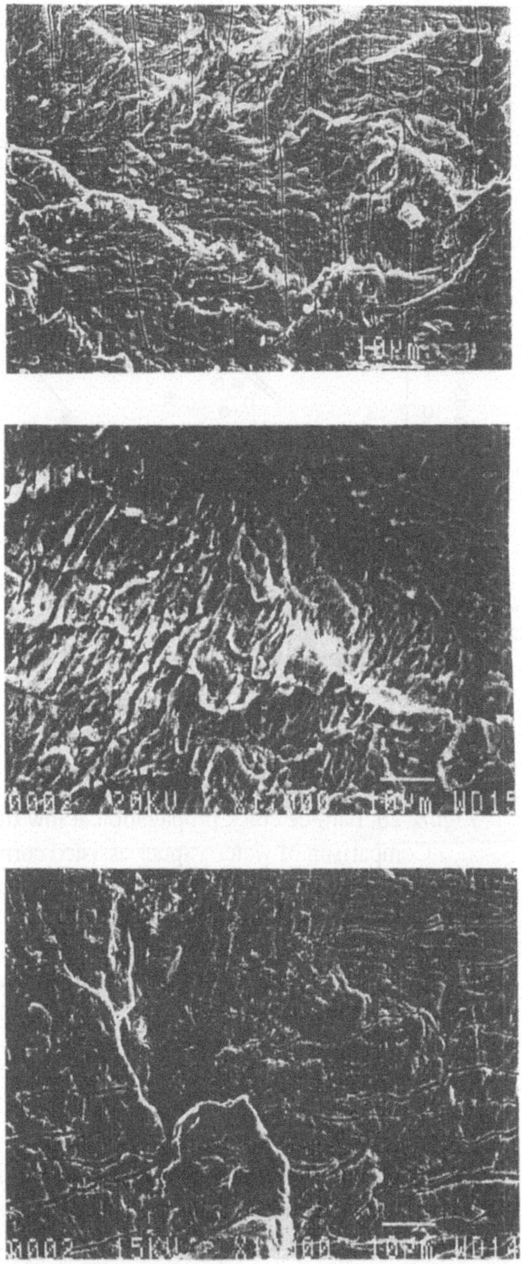

Figure 27. Fracture surface morphology of Ti 6246 tested at 500°C in different environments.

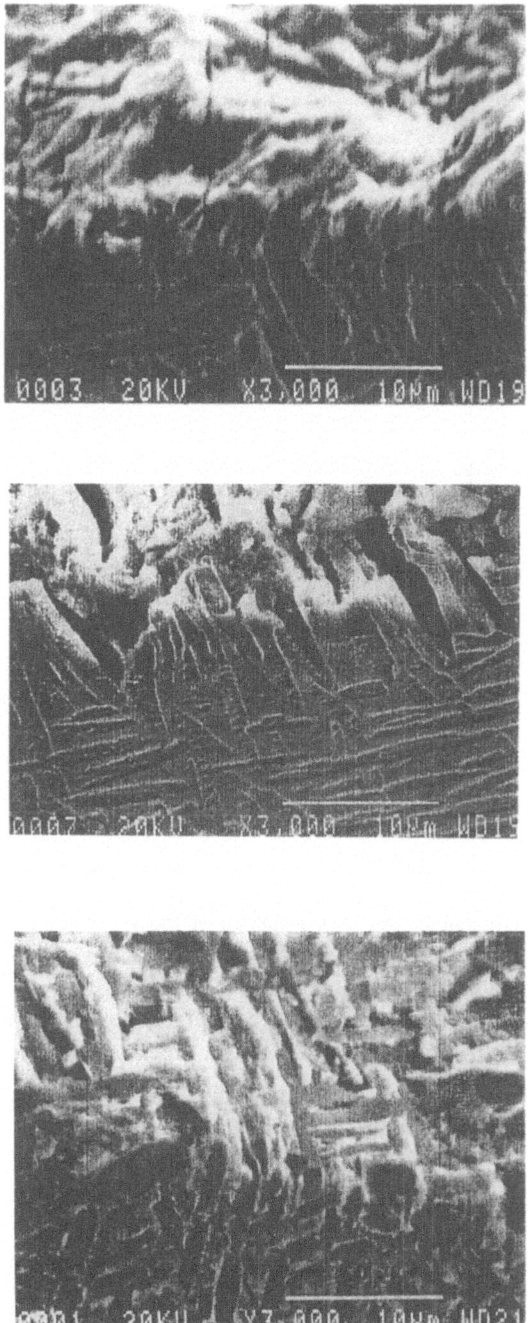

a)

Humid argon

0.1 Hz

b)

Lab. air

0.1 Hz

c)

Low wacuum

0,01 Hz

Figure 28. Cross section of Ti6246 tested at 500°C in three environmental conditions.

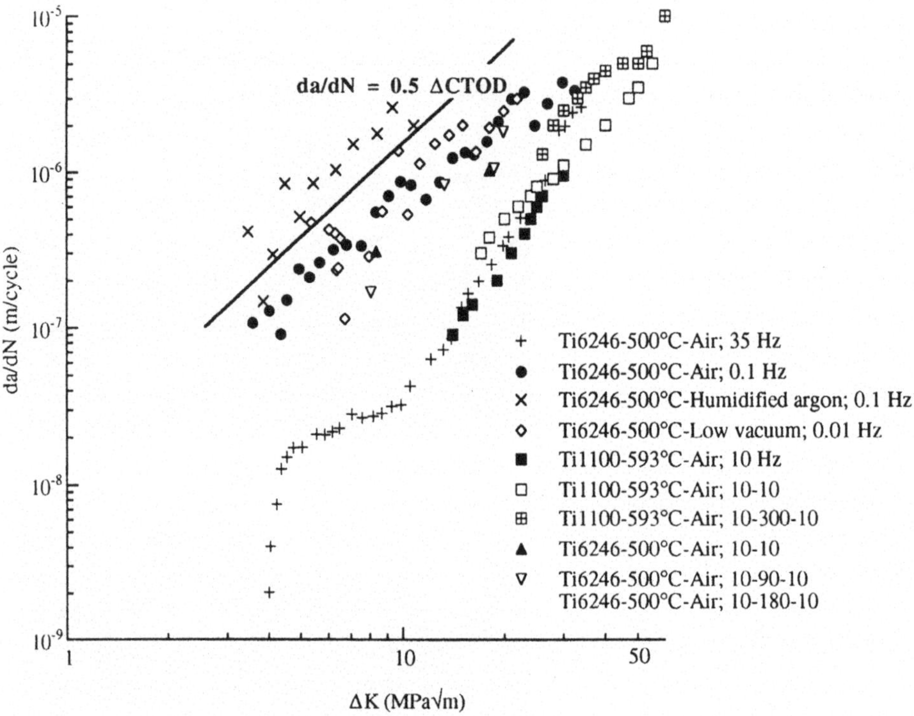

Figure29. Comparison of fatigue crack growth data at 500°C in Ti6246 and at 593°C on Ti 1100 (71,72).

7. Conclusions

This paper proposes a survey of the influence of gaseous atmosphere, and primarily of ambient air, on the propagation of fatigue crack. The first part is devoted to the characterisation of three intrinsic fatigue crack growth regimes on the basis of numerous experimental data obtained in high vacuum and with closure correction, on Al alloys with various ageing conditions, steels and Titanium alloys.

The faster stage I, identified on Al based single crystals, is also observed in the early growth of short cracks.

The intermediate stage II is commonly observed when the crack advance proceeds along planes normal to the load axis.

The slower regime called stage I-like corresponds to a crystallographic crack growth when the deformation is localised within single slip system in each individual grains. This regime prevails near threshold conditions.

The intrinsic stage II can be described by the law :

$$\frac{da}{dN}=\frac{A}{D_o^*}\left(\frac{\Delta K_{eff}}{E}\right)^4$$

where A is dimensionless, D_o^* the critical cumulated displacement leading to rupture, E the Young Modulus and Δk_{eff} the effective stress intensity factor range (i.e. corrected for closure).

The stage II crack growth rates plotted with respect to $\Delta K_{eff}/E$ appears to be nearly independent on the alloy composition, the microstructure (when it does not induce a localisation of deformation), the grain size and the yield stress. Limited change can be observed in D_o^* according to the alloy ductility.

A very large influence of microstructure for polycrystals is observed on the retardation induced on the stage I-like regime and on the stress range required for localised deformation.

The specific environmental crack growth enhancement in conditions excluding interaction with closure effects (oxidation, wedging, limited water vapour transport), has been analysed by comparing effective data in gaseous environment containing well-controlled amount of water vapour and oxygen, to intrinsic data obtained in comparable loading conditions :

i) The effective propagation in ambient air is characterised in most cases by a strong environmental enhancement of the crack growth, especially near threshold, and is much more accentuated for Al alloys than for steels and Ti alloys.

ii) Ambient air as well as humid environment favour stage II propagation in most cases and in a wide range of growth rates including the near-threshold area. In contrast to intrinsic stage II, environmentally-assisted effective stage II is highly sensitive to several factors including alloy composition, microstructure, grain size and yield strength.

The behaviour in moist environment of Al alloys and steels at room temperature and Ti alloys up to 300 °C, has been described by superimposing two distinct processes:

i) Adsorption of water vapour molecules which promotes the growth process without altering the basis intrinsic mechanism of damage accumulation. Adsorption of gaseous species onto fresh surface (Rhebinder effect) is analysed as a decrease in the critical cumulated displacement D* described in term of the surface coverage coefficient θ, as initially defined by Langmuir. This regime is generally operative in the mid-rate range at atmospheric pressure, and can be active near-threshold condition at sufficiently low pressure or/and by lowering the test frequency.

ii) Hydrogen-assisted propagation as initially described by Wei and co-authors ; hydrogen is provided by the dissociation of adsorbed water vapour molecules and is then dragged by mobile dislocations into the highly plastically strained material at the crack tip where the very embrittling reaction takes place. Critical conditions for such embrittling process would thus mainly depend on parameters as water vapour pressure, time (frequency) and temperature. This regime is generally observed in near-threshold conditions, at growth rate below a critical step about 10^{-8} m/cycle which corresponds to stress intensity factor ranges at which the plastic deformation becomes localised within each individual grain along the crack front.

iii) In case of enhanced oxide built-up conjugated with prominent closure by oxide wedging, a competitive influence of water vapour embrittlement and oxide protection can be observed. In such condition closure correction does not account for the influence of R ratio because of complex interactions between closure, oxidation and water vapour action.

iv) At 500 °C in Ti 6246 alloy, a predominant effect of water vapour has been demonstrated while the role of oxygen appears limited to oxidation of cracked surface which mainly favourises near-threshold closure. A water vapour assisted corrosion fatigue regime is identified and compared to a ΔCTOD controlled relationship of the form :

$$\frac{da}{dN} = 0.5\,\Delta CTOD = \frac{1}{2\pi}\frac{\Delta K_{eff}^2}{\sigma_y E}$$

8. References

1. Haigh, B.P. (1917) Journal Inst. of Metals, 18, 55-63.

2. Bennet, J. A. (1963) Effect of reactions with the atmosphere during fatigue of metals, *Fatigue: an interdisplinary approach, 10th Sagamore Army materials research conference* 209-227.

3. Petit, J., De Fouquet, J. and Hénaff, G. (1994) Influence of ambient atmosphere on fatigue crack growth behaviour of metals, in A. Carpinteri (eds), *Handbook of fatigue crack propagation in metallic structures*, Elsevier, pub., 1159-1204.

4. Davidson D. and Suresh S. (1984) *Fatigue crack growth threshold concept*, TMS AIME Pub., Warendale, Pensylvanie, USA.

5. Dahlberg, E. P. (1965) Fatigue crack propagation in high strength 4340 steel in humid air, *A.S.M. Transactions Quarterly* **58**, 46-53.

6. Hartman, A. (1965) *Int. J. Fract. Mech.*, 1, 167-187.

7. Bradshaw, F.J. and Wheeler, C. (1966) *Appl. Mat. Research*, 112-120.

8. Bradshaw, F.J. and Wheeler, C. (1969) *Int. J. Frac. mech.*, 5, 255-268.

9. Lynch, S. P. (1978) Mechanisms of Fatigue and Environmentally Assisted Fatigue, *Fatigue Mechanisms, ASTM STP 675*, ASTM pub., 174-213.

10. Lynch, S. P. (1988) Environmentally-assisted cracking: overview of evidence for an adsorption-induced localised-slip process, *Acta Metallurgica* **36**, 2639-2661.

11. Simmons, G. W., Pao, P. S. and Wei, R. P. (1978) Fracture Mechanics and Surface Chemistry Studies of Subcritical Crack Growth in AISI 4340 Steel, *Metallurgical Transactions* **94**, 1147-1158.

12. Wei, R. P. (1979) On understanding environment-enhanced fatigue crack growth: a fundamental approach, *ASTM STP 675*, ASTM pub., 816-840.

13. Wei, R. P., Pao, P. S., Hart, R. G., Weir, T. W. and Simmons, G. W. (1980) Fracture Mechanics and Surface Chemistry Studies of Fatigue Crack Growth in an Aluminium Alloy, *Metallurgical Transactions* **11A**, 151-158.

14. Wei, R. P. and Simmons, G. W. (1981) Recent progress in understanding environment-assisted fatigue crack growth, *International Journal of Fatigue* **17**, 235-247.

15. Pao, P.S., Gao, M. and Wei, R.P. (1988) Critical assessment of the model for environmentally-assisted fatigue crack growth. *ASTM-STP 924*, Ed. R.P. Wei and R.P. Gangloff, ASTM pub. 182-195.

16. Bouchet, B., de Fouquet, J., Aguillon, M.. Influence de l'environnement sur les faciès de rupture par fatigue d'éprouvettes monocristallines et polycristallines d'alliage Al-Cu 4 %, *Acta Metal.*, 23, 1325-1336.

17. Petit, J. (1983) Some aspects of near-threshold fatigue crack growth: microstructural and environmental effects, *Fatigue Crack growth Tresholds Concepts,* eds Davidson, D. and Suresh, *TMS AIME* pub., 3-25.

18. Bignonnet, A., Petit, J. and Zeghloul, A. (1990) The influence of environment on fatigue crack growth mechanisms, in P. Scott (eds), *Environment Assisted Fatigue, EGF7*, Mechanical Engineering Publications, pp. 205-222.

19. Gudladt, H. J. and Petit, J. (1991) Stage II crack propagation of Al-Zn-Mg single crystals in dry & wet atmospheres, *Scripta Metallurgica &Materialia* **25**, 2507-2512.

20. Petit, J. and Hénaff, G. (1993) A survey of near-threshold fatigue crack propagation : mechanisms and modelling, *Proceedings of Fatigue 93* eds. Baïlon, J.P. and Dickson, I.J., EMAS pub., Vol. 1, 503-512.

21. Hénaff, G., Marchal, K. and Petit, J. (1995) On Fatigue Crack Propagation Enhancement by a Gaseous Atmosphere: Experimental and Theoretical Aspects, *Acta Metallurgica et Materialia* **43**, 2931-2942.

22. Petit, J. and Mendez, J. (1996) Some aspects of the influence of microstructure on fatigue, *proc. sixth Int. Fatigue Congress, Fatigue'96*, G. Lütjering and H. Nowack eds., Pergamon Press, 1, 15-26.

23. Enochs, J.S. and Devereux, O.F. (1975) Fatigue crack grouwth in 5032-H34, Aluminium in vacuum and active gas environments, *Metal Trans.*, 6A, 391-397.

24. Piasiek, R.S. and Gangloff, R.P., Environmental fatigue of Al-Li alloy, *Metall. Trans.*, Part. I in 22A, 2415-2428 and Part. II in 24 A, 2751-2762.

25. Lesterlin, S., Sarrazin-Baudoux, C. and Petit, J. (1996) Effects of temperature and environment interactions on fatigue crack propagation in a Ti alloy, *Scripta Materialia* **34**, 651-657.

26. Elber, W. (1971) The Significance of Crack Closure, *Damage Tolerance in Aircraft Structures, ASTM STP 486* ASTM STP pub., 230-242.

27. Kikukawa, M., Jono, M. and Mikami, S. (1982) Fatigue Crack Propagation and crack closure behavior under stationary varying loading-test results of aluminum alooy, *Journal of the Society on Materials Science Japan* **31**, 438-487.

28. Petit, J. (1992) Modelling of Intrinsic Fatigue Crack Growth, in A. F. Blom and C. J. Beevers (eds), *Theoretical and numerical analysis of fatigue*, EMAS, pp. 131-152.

29. Petit, J., Kosche, K. and Gudladt, H. J. (1992) Intrinsic Stage I Crack Propagation in Al-Zn-Mg Single Crystals, *Scripta Metallurgica et Materialia **26**, 1049-1054.*

30. Starke, Jr., E.A., Lin, F.S., Chen, R.T. and Heikkenen, H.C., 1984, The use of the cyclic stress strain curve and a damage model for predicting crack growth thresholds, *Fatigue crack growth Threshold concepts*, eds. Davidson, D. and Suresh, S., TMS AIME pub., 43-62.

31. Selines, R.J., (1971) The fatigue behaviour of high strength Aluminium alloys, *Ph D Thesis*, MIT, USA.

32. Suresh, S. (1985) Fatigue crack deflection and fracture surface contact: micromechanical models, *Metall. Trans.*, **16A**, 249-260.

33. de Los Rios, E. R., Mohamed, H. J. and Miller, K. J. (1985) A micromechanical analysis for Short Fatigue Crack Growth, *Fatigue & fracture of Engineering Materials and Structures* **8**, 49-63.

34. Petit, J. and Hénaff, G. (1991) Stage II intrinsic fatigue crack propagation, *Scripta Metallurgica* **25**, 2683-2687.

35. Rice, J. R. (1965) Plastic Yielding at a Crack Tip,, *International Conference on Fractur,* Sendaï Japan, pp. 283-308.

36. Weertman, J. (1966) Rate of Growth of Fatigue Cracks Calculated from the Theory of Infinitesimal Dislocations Distributed on a Plane, *International Journal of Fracture Mechanics* , **2**, 460-467.

37. Xu, Y.B., Wang, L., Zhang, Y., Wang, Z.G. and Hu, Q.Z. (1991) Fatigue behavior of an Aluminum-Lithium Alloy 8090-T6 at Ambient and Cryogenic Temperature, *Metall. Trans.,* 22A, 723-729.

38. Davidson, D. L. and Lankford, J. (1981) The effect of water vapor on fatigue crack tip stress and strain range and the energy required for crack propagtion in low-carbon steel, *International Journal of Fracture* **17**, 257-275.

39. Snowden, K. U. (1964) The Effect of Atmosphere on the Fatigue of Lead, *Acta Metallurgica* **12**, 295-303.

40. Achter, M. R. (1968) The Adsorption Model for Environmental Effects in Fatigue Crack Propagation, *scripta Metallurgica* **2**, 525-528.

41. Petch, N. J. (1956) The lowering of fracture stress due to surface adsorption, *Philosophical Magazine* **I**, 331-337.

42. Langmuir, I. (1918) The adsorption of gases on plane surfaces of glass, mica and platinum, *Journal of the American Chemical Society* **40**, 1361-1403.

43. Orowan, E. (1944) The Fatigue of Glass under Stress, *Nature* **154**, 341-342.

44. Petch, N. J. and Stables, P. (1952) Delayed Fracture of Metals under Static Load, *Nature* **169**, 842-843.

45. Oriani, R. A. (1984) On the Possible Role of the Surface Stress in Environmentally Induced Embrittlement and Pitting, *Scripta Metall.* **18**, 265-268.

46. Liaw, P. K., Hudak Jr, S. J. and Donald, J. K. (1982) Influence of gaseous environments on rates of near-threshold fatigue crack propagation in NiCrMoV steel, *Metall. Trans.* **13A**, 1633-1645.

47. Lesterlin, S. (1996) Influence de l'environnement et de la température sur la fissuration par fatigue des alliages de titane. *Thèse de Doctorat*, Université de Poitiers, France.

48. Bowles, C. Q. (1978) The role of environment, frequency and wave shape during fatigue crack growth in aluminium alloys, *Report LR-270*, Delft University of Technology, Department of Aerospace Engineering.

49. Bradshaw, F. J. (1967) The effect of gaseous environment on fatigue crack propagation, *Scripta Metallurgica* **1**, 41-43.

50. Hénaff, G. and Petit, J.(1996) A logical framework for the analysis of fatigue crack propagation enhancement by ambient atmosphere, *Physicochemical Mechanics of Materials*, 32, **2**, 69-88.

51. Suyitno, B. M., Chalant, G. and Petit, J. (1990) Environment and Frequency Effects on Fatigue Crack Growth of Micro-Alloyed Steel in the Threshold Region, in H. Kitagawa and K. Tanaka (eds), *Fatigue'90*, MCEP, pp. 1381-1386.

52. Lanteigne, J. and Bailon, J. P. (1981) Theoretical Model for FCGR Near the Threshold, *Metall. Trans.* **12A**, 459-466.

53. Roven, H. J. and Nes, E. (1991) Cyclic Deformation of Ferritic Steel - II. Stage II Crack Propagation (Overview n°94), *Acta Metallurgica et Materialia* **39**, 1735-1754.

54. Hénaff, G. and Petit, J. (1993) Pure corrosion fatigue crack propagation, "Corrosion-deformation interaction", in T. Magnin and J. M. Gras (eds), *Proceedings CDI92*, Les Editions de Physique, pp. 599-618.

55. Suresh, S., Zamiski, Z. A. and Ritchie, R. O. (1981) Oxide-induced closure closure: an explanation for near-threshold fatigue crack growth behavior, *Metall. Trans.* **12A**, 1435-1443.

56. Kwon, J. H. (1985) Influence de l'environnement sur le comportement en fatigue d'un acier E460 et d'un alliage léger 7075 près du seuil de fissuration, *Thèse de Doctorat* , Université de Poitiers, France.

57. Smith, P. (1987) The effects of moisture on the fatigue crack growth behaviour of a low alloy steel near threshold, *Fatigue & Fracture of Engineering Materials & Structures* **10**, 291-304.

58. Zeghloul, A. and Petit, J. (1985) Environmental sensitivity of small crack growth in 7075 aluminium alloy, *Fatigue & Fracture of Engineering Materials and Structures* **8**, 341-348.

59. Fleck, N. A. and Newman, Jr. C. J. (1988) Analysis of crack closure under plane strain conditions, in J. C. Newman Jr, and W. Elber (eds), *Mechanics of fatigue crcak closure, ASTM STP 982*, American Society for Testing and Materials, pub., 319-341.

60. McClung, R. C. (1994) Finite Element Analysis of Specimen Geometry Effects on Fatigue Crack Closure, **17**, 861-872.

61. Suresh, S. (1991) *Fatigue of Materials* , Cambridge Solid State Science Series, Cambridge pub. U.K.

62. Riedel, H. (1983) Crack tip stress field and crack growth under creep-fatigue conditions. *In Elastic Plastic Fracture, Special Technical Publication 803*, **I**, pp. 505-520. Philadelphia, ASTM.

63. Gieseke, B. and Saxena, A. (1989) Correlation of creep-fatigue crack growth rates using crack-tip parameters. *Advances in Fracture Research* (eds. K. Salana, K. Pavi-Chandar, D.M.R., Taplin and P. Prama Rao) Pergamon Press, Oxford, **1**, pp. 189-196,.

64. Saxena, A. (1988) A model for predicting the effect of frequency on fatigue crack growth at elevated temperature, *Fatigue of Engng Mat. and Struct.*, **3**, 247-255.

65. King, J.E., Cotteril, P.J., (1990) Role of oxides in fatigue crack propagation, *Mat. Sc. and Tech.* **6**, 19-31.

66. Sarrazin-Baudoux, C., Lesterlin, S. and Petit, J (1996) Atmospheric influence on fatigue crack propagation in Titanium alloys at elevated temperature, *ASTM-STP 1297*, in press.

67. Sarrazin-Baudoux, C., Lesterlin, S. and Petit, J (1996) Fatigue behaviour of Titanium alloys at room temperature and 300°C in ambient air and high vacuum, *Proceedings of the 6th International Fatigue Congress Fatigue'96*, Eds. G. Lütjering and H. Nowack, Pergamon press, **II**, 783-788.

68. Laird, C. (1967) The influence of metallurgical structure on the mechanisms of fatigue crack propagation, *in Fatigue Crack Propagation ASTM STP* 415, 131-168.

69. Lardner, W. (1967) A dislocation model for fatigue crack growth in metals. *Philosophical Magazine* **17**, 71-82.

70. Pelloux, R.M.N. (1970) Crack extension by alternating shear. *Engng Fract. Mech.* **1**,.697-702.

71. Ghonem, H. and Foerch, R. (1991) Frequency effects on fatigue crack growth behavior in a near-α Titanium alloy, *Mat. Sci. and Engng.,* **A138**, 69-81.

72. Foerch, R., Madsen, A. and GhoneM, H. (1992) Environmental interactions in high temperature fatigue crack growth of Ti 1100, *Metall. Trans.,* **24A**, 1321-1332.

CRACK PROPAGATION OF SEMI-ELLIPTICAL SURFACE CRACKS : A LITERATURE REVIEW

T Boukharouba*, J. Gilgert§ and G. Pluvinage#

(*) Université des Sciences et de Technologie H. Boumediene (IGM USTHB), B.P. 32 El-Alia Alger - Algeria.

(#) Laboratoire de Fiabilité Mécanique de l'Université de Metz, Ile du Saulcy 57045 France.

(§) École Nationale d'Ingénieurs de Metz (ENIM), Ile du Saulcy 57045 France.

SUMMARY

A review of the different problems relative to the crack propagation of a semi- elliptical surface crack is presented.

The particular following points are examined :

- crack front evolution during propagation of a semi-elliptical surface crack;
- crack propagation laws for the same kind of crack;
- local value of the stress intensity factor in this case;
- influence of loading mode and SIF value on the parameters C and m of the crack propagation law.

Particular attention has been payed on semi-elliptical crack in a pipe submitted to internal pressure.

1-Introduction

Propagation of a semi-elliptical crack is often observed on the failure surface of engineering structures like pipes, bolts, plates, pressure vessels etc....

An example can be seen in Figure (1) which shows the fracture appearance of a crankshaft from a three-cylinder tractor diesel engine. Fatigue fracture occured at point A. The surface is smooth, both ends of the crack front show initially very pronounced growth lines.

The safety of a pressure vessel can be asserted using the fracture criterion "leak before break" which needs to know the geometrical evolution of a crack during crack propagation.

R.A. Smith (ed.), Reliability Assessment of Cyclically Loaded Engineering Structures, 343–375.

This evolution is relatively complex as it can be seen on Figure (2). At the beginning, crack grows faster in the direction of thickness. After this stage the propagation is more important in the transwere direction **[1]**.

Therefore, the prediction of life duration and fracture conditions needs to know the following points :

- How is the evolution of the crack geometry ?
- What is the crack growth governing law ?
- Is this governing law a function of the stress intensity factor and what is the nature of such a SIF (local, average, global) ?
- Are the coefficients of the law intrinsic to the materials ?

This paper will try to answer to these questions.

Figure 1 : Example of the failure surface of a crankshaft of a three-cylinder tractor diesel engine.

2- CRACK FRONT EVOLUTION DURING PROPAGATION OF A SEMI-ELLIPTICAL SURFACE CRACK

Crack front evolution during propagation can be observed on the fracture surface by the shape of the front lines. These lines show the different steps of crack evolution and result from overload during cyclic loading. A typical example can be seen on figure (2) where the evolution of a surface crack in a steel plate submitted to four-point bending is easily observed [1].

Figure 2 : Aspect of a semi-elliptical surface crack, different steps showing crack propagation [1].

2.1-Case of a plate submitted to tension or bending

Several investigators have pointed out that the fatigue propagation of surface flaws in metallic plates are significantly affected by the crack shape change. The semi-elliptical form is preserved during the whole growth but the aspect ratio of the part through defects changes (figures 3a and 3b).

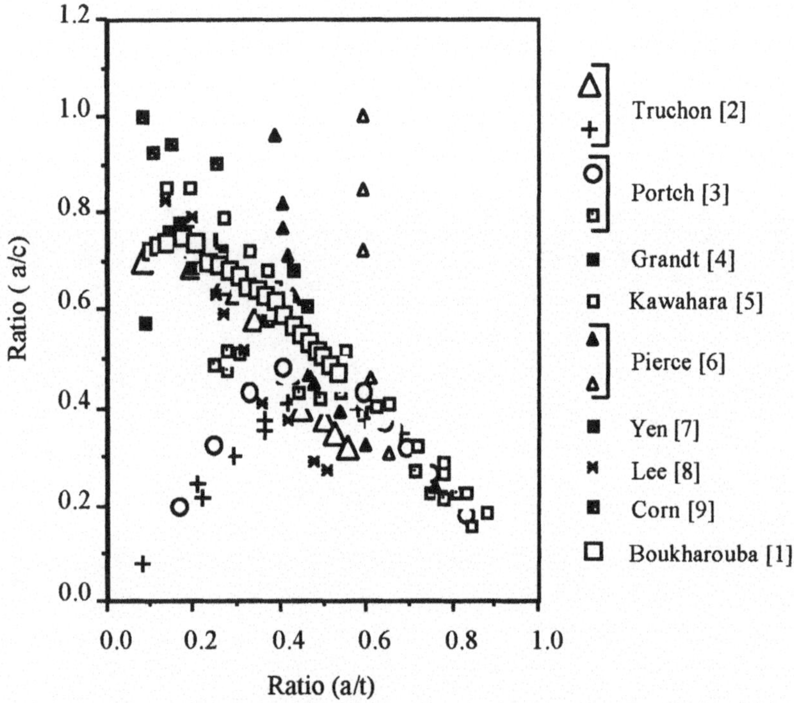

Figure 3a : Evolution of crack aspect a/c versus normalized crack depth a/t in bending.

The results from literature show that these flaws tend to follow a common propagation path in the diagram of the crack aspect ratio a/c against the relative crack depth a/t (t is the thickness)

As can be see on Figure (3) the ratio a/c first increases until a maximum and decreases quasi linearly and tends to an asymptote, independant of the initial value of a/c and a/t ratios. Several authors have proposed empirical relationships for crack aspect ratio versus relative crack depth (Table 1) of bending case.

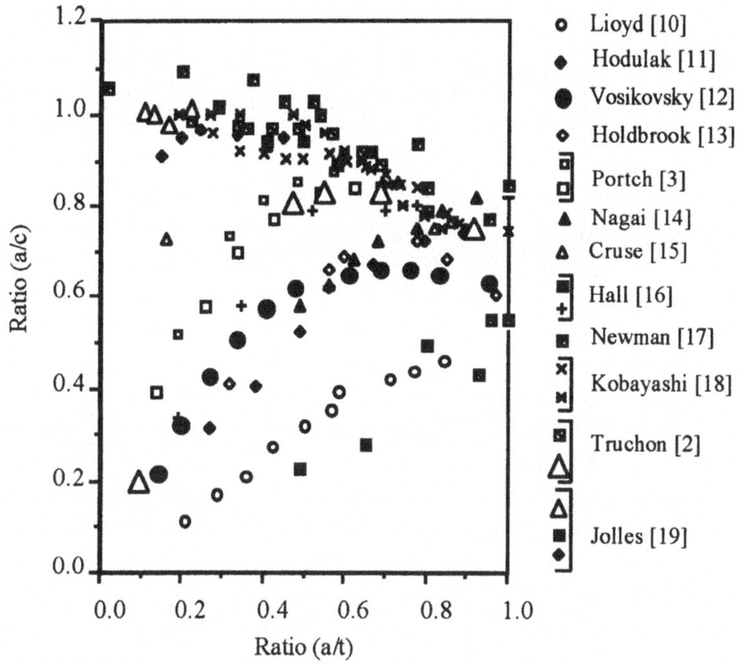

Figure 3b : Evolution of crack aspect a/c versus normalized crack depth a/t in tension.

Portch * [3]

$$\left(\frac{a}{c}\right)_i = \left[1.05 - \left(\frac{a}{t}\right)_i\right] - \left[1 - \left(\frac{a}{t}\right)_i\right]^4 . \left(1.05 - \lambda_p\right) \quad \text{for} \left(a/t\right)_i > \quad (1a)$$

$$\left(\frac{a}{c}\right)_i = I\left(\frac{a}{t}\right)_i \quad \text{for} \quad \left(a/t\right)_i < D \quad (1b)$$

$$I = \frac{\left(a/c\right)_0}{\left(a/t\right)_0} \quad \text{and} \quad D = \frac{3.2 - I}{3\left(I+1\right)} \quad (1c)$$

(*) : If D is calculated from equation (1c) to be > 1.0, relationship (1b) is used for the full range of $0 < (a/t)_i < 1.0$. If $D < 0.0$, relationship (1a) is used for the full range $0 < (a/t)_i < 1.0$. In this case λp is calculated from a using the initial (a/c) and (a/t) values. If $0 < D < 1.0$, then equation (1b) is used for the range $0 < (a/t)_i < D$. In this case, (a/c) is the aspect ratio corresponding to D and is calculeted from (1b). λ is then calculated from the equation (1a) using (a/c) and (a/t).

Table 1a : Empirical equations for the prediction of the evolution of the ratio a/c given in the literature. Bending case

Gorner** [20]

$$\left(\frac{a}{c}\right)_i = \left[1.05\left(\frac{a}{t}\right)_i - \left(\frac{a}{t}\right)_i^2\right]\left\{\left(\frac{a}{t}\right)_i^2 + \lambda_G\left[1.05 - \left(\frac{a}{t}\right)_i^2\right]\right\}^{-0.5} \qquad (2)$$

Iida *** [21]

$$\left(\frac{a}{c}\right)_i = \left[0.85 - 0.75\left(\frac{a}{t}\right)_i\right] \pm 0.0063\left[\left(\frac{a}{t}\right)_i - \lambda_I\right]^{-3.8} \qquad (3)$$

Mahmoud & al [22]

$$\frac{da}{dN} = \left(\frac{\overline{\Delta K_{ave,A}}}{0.9\,\overline{\Delta K_{ave,C}}}\right)^m \text{ with } \overline{C_{ave,C}} = (0.9)^m \cdot \overline{C_{ave,A}} \qquad (4)$$

Carpinteri**** [23]

$$\left(\frac{a}{c}\right) = a_0 + a_1\left(\frac{a}{t}\right) + a_2\left(\frac{a}{t}\right)^2 + a_3\left(\frac{a}{t}\right)^3 + a_4\left(\frac{a}{t}\right)^4 \qquad (5)$$

(**): λ_G is calculated from the initial values of a and c.

(***): λ_I is a constant calculated using the initial values of a/c and a/t. The plus sign is for a/c > 0.85 - 0.75 a/t and the minus sign for a/c < 0.85 - 0.75 a/t.

(****) :The parameter a_i depends on crack shape and m values (see table 2).

Table 1b : Empirical equations for the prediction of the evolution of the ratio a/cgiven in the literature. Bending case

	m	a_0	a_1	a_2	a_3	a_4
straight front	2	-0.520	7.341	-15.751	13.661	-4.240
	3	-0.709	10.127	-25.926	28.314	-12.035
	4	-0.884	12.882	-38.024	50.398	-26.992
circular front	2	1.011	-0.007	-0.862	0	0
	3	1.002	0.098	-1.335	0	0
	4	1.006	0.079	-1.503	0	0

Table 2 : a_i values given by Carpinteri [23], used in equation (5).

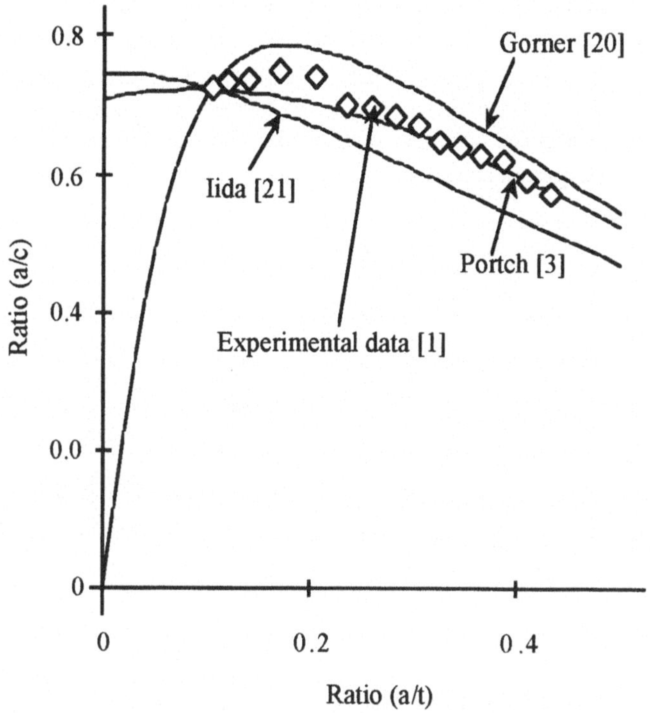

Figure 4 : Evolution of the aspect of a semi-elliptical surface crack versus normalized crack depth in the case of a plate submitted to bending.Comparison with empirical formula of Portch **[3]**, Iida **[21]** and Gorner **[20]**.

The abscissa of the maximum value a/c depends on the stress ratio. The asymptotic line is given by the following equation :

$$\left(\frac{a}{c}\right) = 1 - \left(\frac{a}{t}\right) \tag{6}$$

This asymptote is horizontal in the case of a cyclic axial loading (Figure 4) and the a/c ratio is close to 1 which corresponds to a semi-circular crack. The accuracy of the empirical formula of Portch **[3]**, can be seen on Figure (4) experimental dates are compared with the formulation of Iida **[21]**, Gorner **[20]** and Portch **[3]**.

2.2-Case of a pipe submitted to internal pressure

In case of a pipe submitted to an internal pressure, the evolution of the crack shape is different from the case of pure tension or bending **[24]**. After a transient stage, the crack shape becomes quasi semi-circular and keeps this geometry during all the crack propagation (figure 5).

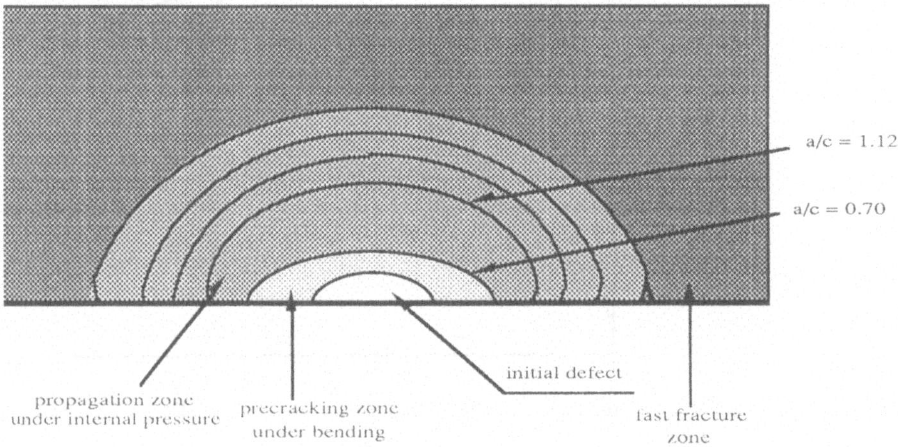

Figure 5 : Evolution of crack aspect in the case of a pipe submitted to an internal pressure **[24]**.

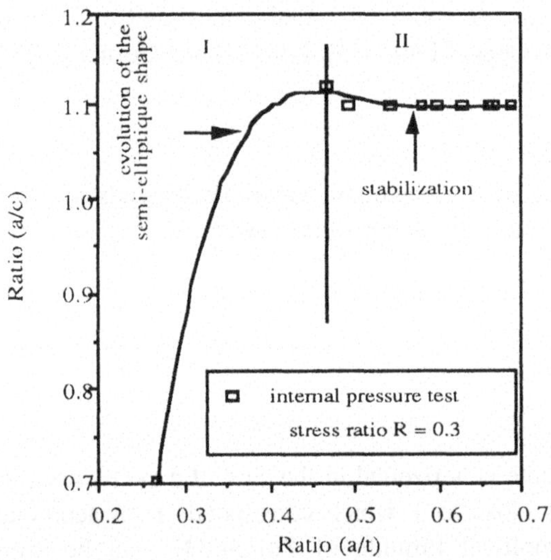

Figure 6 : Diagram of the crack aspect ratio a/c against the relative crack depth a/t, for a pipe submitted to internal pressure (R = 0.3) **[25]**.

3-CRACK PROPAGATION LAWS FOR SEMI-ELLIPTICAL SURFACE CRACK

3.1-General fatigue crack growth laws

Fatigue crack growth is governed by the amplitude of the stress gradient at the crack tip. This will be described by the stress intensity factor range ΔK. In 1963, Paris & Erdogan **[25]** showed that the crack growth rate is a power function of the stress intensity factor range :

$$\frac{da}{dN} = C(\Delta K)^m \qquad (7)$$

where :
- da/dN : crack growth rate [mm/cycle]
- ΔK : stress intensity factor range [MPa $m^{1/2}$]
- C and m : material constants

The stress ratio has some influence on fatigue crack growth as can be seen on figure (7). Tests have been made on a 35NCDV12 plate (dimensions 500∗ 100∗50 mm).

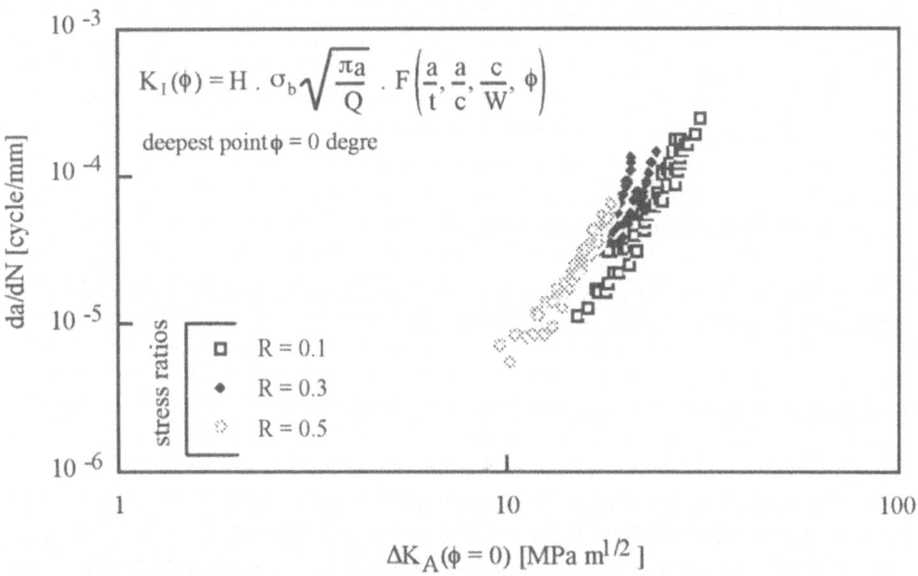

Figure 7 : Crack growth rate in the thickness direction against the local SIF range.

- Forman & al **[26]** have proposed to take into account the influence of the stress ratio R and the fracture toughness K_c.

$$\frac{da}{dN} = \frac{C \cdot (\Delta K)^m}{(1 - R) \cdot (K_c - \Delta K)} \tag{8}$$

- Pearson **[27]** have taken into account the stress ratio R, the fracture toughness K_c and also the fatigue threshold ΔK_{th} :

$$\frac{da}{dN} = \frac{C(1 - \beta)^{\alpha} \cdot (\Delta K - \Delta K_{th})^m}{K_c - (1 - \beta) \cdot \Delta K} \quad \text{with} \quad \beta = \frac{1 + R}{1 - R} \tag{9}$$

Where : ΔK_{th} : fatigue threshold [MPa $m^{1/2}$]
α : experimental constant

- Richards & Lindley **[28]** have proposed the following relationship :

$$\frac{da}{dN} = A\left[\frac{(\Delta K - \Delta K_{th})^4}{R_m^2 (K_C^2 - K_{max}^2)}\right]^m \tag{10}$$

Where : K_{max} : maximum strain intensity factor [MPa $m^{1/2}$]
R_m : ultimate strength
A : experimental constant

3.2-Local fatigue crack growth laws

For the particular points A and C (the deepest point and the surface point) the local crack growth law can be written as follows :

$$\frac{da}{dN} = C_{I,A} \Delta K_I^{m_{I,A}} \quad \text{and} \quad \frac{dc}{dN} = C_{I,C} \Delta K_I^{m_{I,C}} \tag{11}$$

a and c are respectively the depth and the half length of the crack. In this case, we assume that the knowledge of the local value of the stress intensity factor on point A, $\Delta K_{I,A}$ and at the deepest point $\Delta K_{I,C}$ is enough to predict the crack growth rate. In addition we assume that the parameters of the local Paris law are identical on point A and C.

$$C_{I,A} = C_{I,C} \quad \text{and} \quad m_{I,A} = m_{I,C} \tag{12}$$

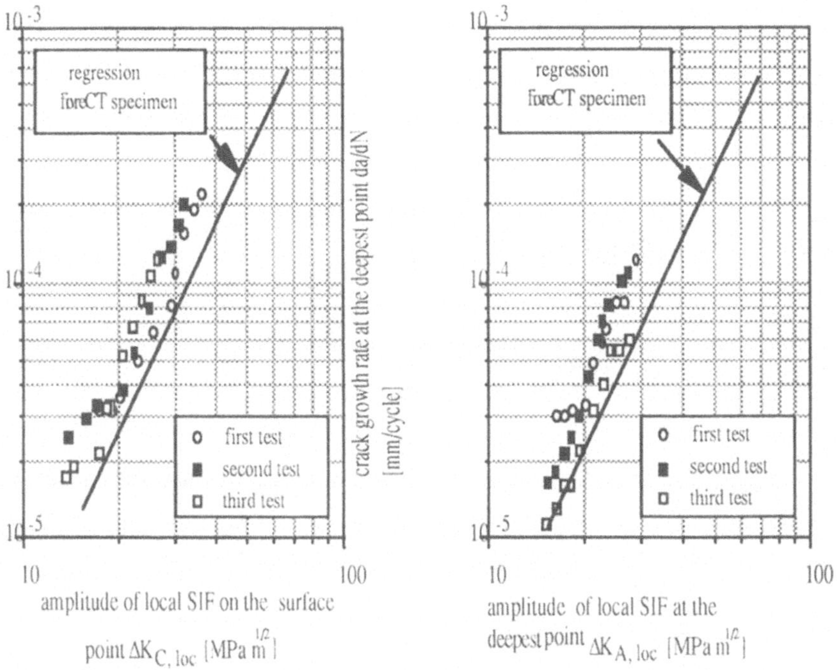

Figure 8 : Crack growth rate at the deepest point da/dN and on surface dc/dN against the local value of the SIF amplitude. Comparison with the material reference tests case of a steel plate submitted to two point bending.

The crack gowth rate at the surface point C, dc/dN and at the deepest point A, da/dN versus the local value of the stress intensity factor are presented in Figure (8), for a semi-elliptical surface crack in a steel plate. We can notice that the experimental curve does not fit with the reference crack growth curve got from a test performed on a CT specimen. According to the Newman & Raju's [17] expression of local value of ΔK the following differential equation is obtained :

$$\frac{da}{dc} = \left[\sqrt{\frac{a}{c}} \cdot \left(1.1 + 0.35 \left(\frac{a}{t} \right)^2 \right) \cdot \frac{\left(1 - R_b + M_1 R_b \right)}{1 - R_b + M_2 R_b} \right]^{-m} \tag{13}$$

with :

$$M_1 = 1 - 0.34\left(\frac{a}{t}\right) - 0.11\left(\frac{a}{c}\right)\left(\frac{a}{t}\right) \tag{14a}$$

$$M_2 = 1 + G_1\left(\frac{a}{t}\right) + G_2\left(\frac{a}{t}\right)^2 \tag{14b}$$

$$G_1 = -1.22 - 0.12\left(\frac{a}{c}\right) \tag{14c}$$

$$G_2 = 0.55 - 1.05\left(\frac{a}{c}\right)^{0.75} + 0.47\left(\frac{a}{c}\right)^{1.5} \tag{14d}$$

The bending ratio is $$R_b = \frac{\Delta\sigma_b}{\Delta\sigma_b + \Delta\sigma_t} \tag{14e}$$

Where $\Delta\sigma_b$ and $\Delta\sigma_t$ are the bending and tension components respectively. The assumptions (12) can be modified as follows $C_{I,A} \neq C_{I,C}$ and $m_{I,A} \neq m_{I,C}$. For instance Newman & Raju **[17]** have used the following assumption.

$$C_{1,C} = C_{1,A} \cdot (0.9)^m \quad \text{and} \quad m_{1,A} = m_{1,C} \tag{15}$$

Expression (13) is modified as follows :

$$\frac{da}{dc} = \left[\left(0.9\left(\frac{a}{c}\right)^{0.5}\right) \cdot \left(1.1 + 0.35\left(\frac{a}{t}\right)^2\right) \cdot \frac{(1 - R_b + M_1 R_b)}{1 - R_b + M_2 R_b}\right]^{-m} \tag{16}$$

3.3-Average crack growth law

This approach was proposed by Cruse & Besuner **[29]**. The variation of the average stress intensity factor ΔK along the crack front is used rather than local value. It is conjectured that the growth at a point on the crack front is influenced by the K values at adjacent points. Average ΔK value on $\overline{\Delta K_{ave\,C}}$ surface and at the deepest point $\overline{\Delta K_{ave\,A}}$ are calculated from the equations :

$$\left(\overline{\Delta K_{ave,A}}\right)^2 = \frac{1}{\Delta S_A} \int_{\Delta S_A} \Delta K_{local}^2(\phi)\, dS_A \qquad (17)$$

$$\left(\overline{\Delta K_{ave,C}}\right)^2 = \frac{1}{\Delta S_C} \int_{\Delta S_C} \Delta K_{local}^2(\phi)\, dS_C \qquad (18)$$

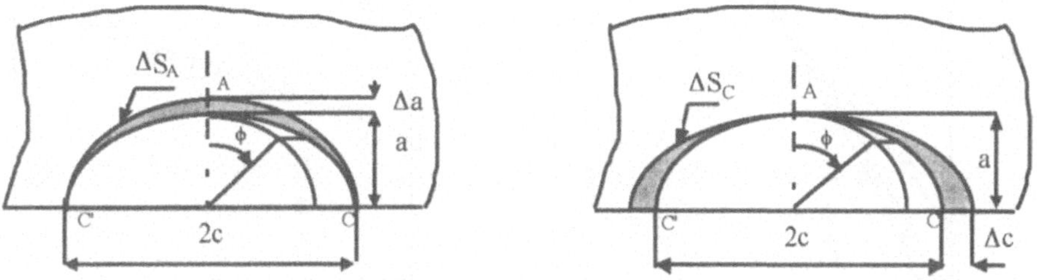

Figure 9 : Definition of the incremental surface crack growth ΔS_A and ΔS_C according to Cruse & Besuner [29].

Where $\Delta K_{local}(\theta)$ is the stress intensity factor at any angle θ, ΔS_A and ΔS_C are respectively the incremental surface crack growth in the A and C directions.

An example of the difference beetween the local stress intensity factor calculated by the Newman & Raju's [17], Cruse & Besuner [29] solutions and the finite elements method [1] for the points A and C is given on Figure (10).

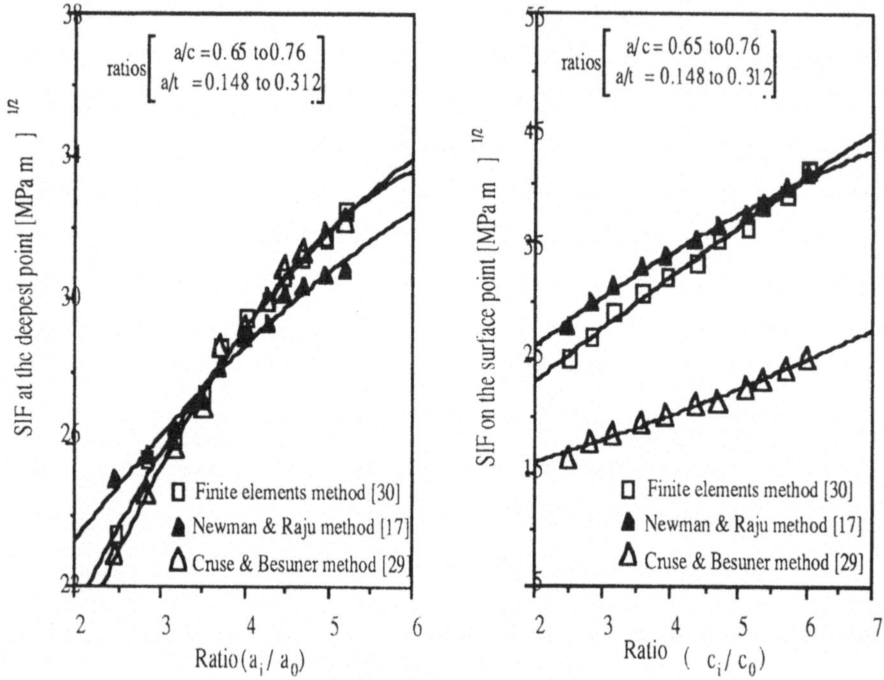

Figure 10 : Comparison of the local and average values of the stress intensity factors at points A and C obtained by finite elements [30], Newman & Raju's [17] and Cruse & Besuner's [29] methods.

This leads to the following relationship :

$$\frac{da}{dc} = \left(\frac{\overline{\Delta K_{ave,\,A}}}{\overline{\Delta K_{ave,\,C}}} \right)^m \tag{19}$$

Application of such a average law to crack propagation at point A and C of a semi-elliptical crack can be seen on Figure (11). The local stress intensity factor value was computed by a finite elements method [30]. A non agreement with the Newman & Raju's [17] solution was found near the surface.

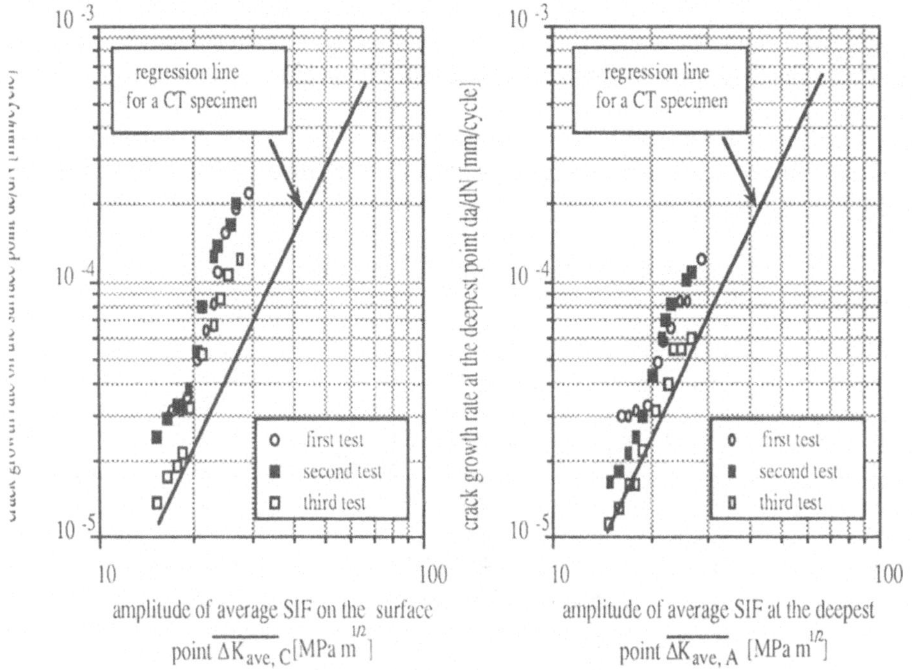

Figure 11 : Crack growth rate at the deepest point da/dN and on surface dc/dN versus the average value of the stress intensity factor. Comparison with the reference tests **[30]**.

4 - LOCAL VALUE OF THE STRESS INTENSITY FACTOR FOR A SEMI-ELLIPTICAL CRACK

4.1 - Case of a plate under tension or bending loading

Numerical methods

The local stress intensity factor for an embedded semi-elliptical crack in a plate in tension or under bending has been computed by several authors between 1962 and 1994 (table 3).

These authors **[1, 17 and 31 to 46]** have used different numerical methods mostly finite elements method (F.E.M) but also engineering estimation (EE), iterative alternating method (AM) or line spring model method (LSM).

Results are generally presented on graphs or as analytical equations. These analytical solutions are presented in Appendix (1) including the most popular given by Raju & Newman **[17]**.

All these analytical solutions can be presented according to the general following formula :

$$K_I(\phi) = \left(\sigma_t + H.\ \sigma_b\right) . \sqrt{\frac{\pi a}{Q}} . F\left(\frac{a}{t}, \frac{a}{c}, \frac{c}{W}, \phi\right) \quad (20)$$

With :

	σ_t	: applied tension stress
	σ_b	: applied bending stress
	a	: depth of the elliptical crack
	c	: half-length of the elliptical crack
	t	: specimen thickness
	H	: geometric correction for bending
	F	: geometric correction for tension
	Q	: shape factor for an elliptical crack
	ϕ	: angular coordinate of the ellipse
	W	: half-width of the component

The geometrical correction factor "F" in tension is plotted versus the ratio crack path over thickness a/t for the following conditions ($a/c = 0.2$ at the deepest point $\theta = 0$). A large difference exists between all these formulas which are presented in Figure (12).

The difference is very low for $a/t < 0.2$ but increases up to 80 % for $a/t > 0.6$. This local stress intensity factor varies along the crack front. At the present time, there is some controversy for the location of the maximum value, at the deepest point, at surface or beetween these two points.

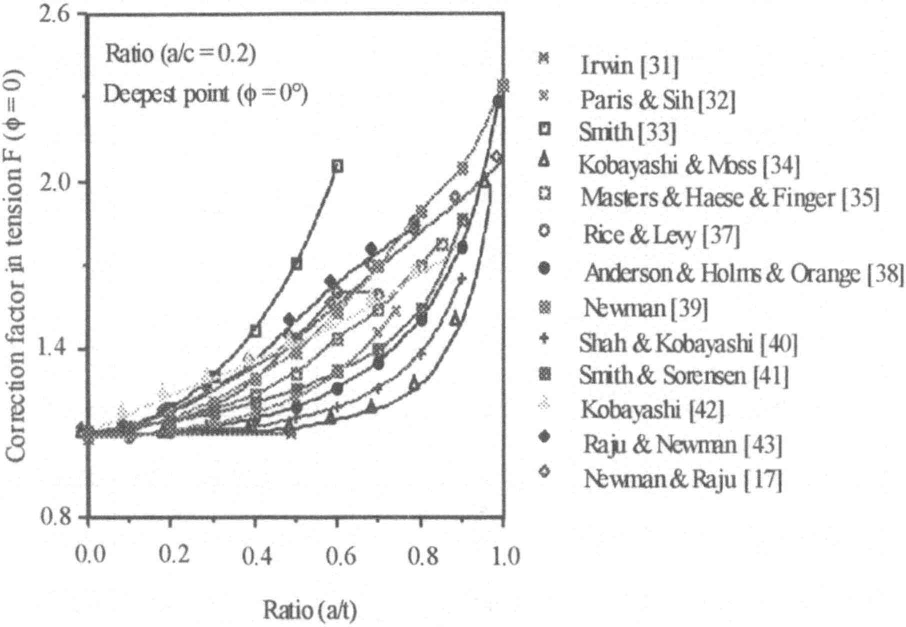

Figure 12 : Comparison of the evolution of the correction factor $F(\phi = 0) = f(a/t)$ given by different authors at the deepest point $\phi = 0°$ and for $a/c = 0.2$ [39].

Authors	Date	Ref.	Method	Range of validity			
				a/t	a/c	c/W	results
Irwin	1962	[31]	EE	^{2} 0.50	0 to 1	b	equation
Paris & Sih	1965	[32]	EE	^{2} 0.50	0 to 1	b	equation
Smith & al	1966	[33]	AM and EE	^{2} 0.75	0.2 to 1	b	graph
Kobayashi & al	1969	[34]	AM and EE	f (a/c)	0 to 1	b	graph
Masters & al	1969	[35]	EM	^{2} 0.98	0.1 to 0.8	b	graph

Table 3a

Smith & Alavi	1969	[36]	AM and EE	^{2} 0.85	0.4 to 1	b	graph
Rice & Levy	1970	[37]	LSM	^{2} 0.70	0 to 1	b	graph
Anderson & al	1970	[38]	EE	^{2} 1.0	0 to 1	b	equation
Newman	1972	[39]	EE	^{2} 1.0	0,02 to ∞	< 1	equation
Shah & al Kobayashi	1972	[40]	AM	^{2} 0.9	0.1 to 1	b	graph
Smith & al	1974	[41]	AM	^{2} 0.9	0.1 to 1	b	graph
Kobayashi	1976	[42]	AM and EE	^{2} 0.9	0.2 to 1	b	graph
Raju & al	1977	[43]	FEM	^{2} 0.8	0.2 to 2	b	graph
Newman & al	1978	[44]	FEM and EE	^{2} 1.0	0.03 to ∞	< 0.5	equation
Newman & al	1981	[17]	MEF	0.0 to 1.0	0.2 to 1.0	< 0.5	equation
Carpinteri	1990	[45]	LSM	0.1 to 0.4	0.2 to 1.0	b	equation
Carpinteri	1993	[46]	FEM	0.1 to 0.4	0.2 to 1.0	b	graph
Boukharouba & al	1994	[1]	FEM	0.07 to 0.3	0.6 to 0.85	< 0.5	graph

(EE) : Engineering Estimate, (AM) : Alternating Method, (LSM) : Line-Spring Model and (FEM) : Finite-Element Method. (b) : Effect of finite width was not considered.

Table 3b : Comparison of calculation from different authors for the SIF K_I (ϕ).

Experimental methods

In order to check the most accurate formula and to determine the point of maximum value, several experimental determinations of the local stress intensity factor have been done **[4, 47 and 48]** and are listed in table (4).

The experimental method used is generally photo-elasticity. An example of such an experimental determination is presented in Figure (13) from the work of Ruiz & Epstein **[47]**.

Authors	Date	Ref.	Method	Range of validity			
				a/t	a/c	c/W	results
Mars & Smith	1972	[48]	bending**	.027 to 0.46	0.46 to 1	b	table
Grandt & Sinclair	1972	[4]	bending*	0.15 to 0.35	0.7 to 0.9	b	graph
Ruiz & Epstein	1985	[47]	bending***	0.12 to 0.37	.34	b	graph
Ruiz & Epstein	1985	[47]	tension***	0.12	0.34	b	graph

(*) : deepest point, (**) : surface, (***) : along the crack front and (b) : effect of finite width not considered.

Table 4 : Experimental determination of the SIF by photoelasticimetry.

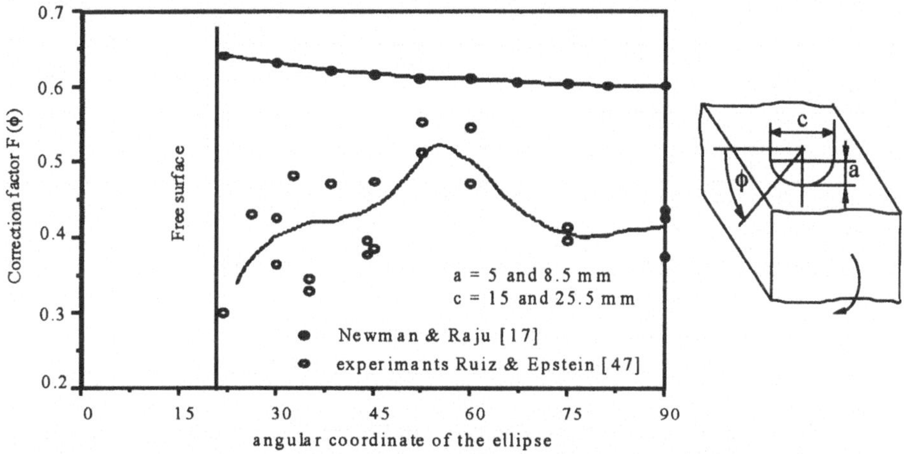

Figure 13 : Comparison of the variation of F(ϕ) correction factor. Numerical method from Newman & Raju **[17]** and experimental method by Ruiz & Epstein **[47]**.

It can be noticed in figure (13) a large scattering of the results and a large difference with Newman & Raju's solution. This large scattering given by the experimental methods makes it impossible to check the most valid formula and to know the point of the maximum value of the stress intensity factor.

4.2 - Case of a pipe submitted to internal pressure

Few works have been done to compute the value of the local stress intensity factor in a pipe submitted to an internal pressure. A listing of authors is given in Table (5). In this case a finite elements method has been generally used but also the weight functions and the boundary integral method.

The results are presented using the following general relationship :

$$K_I(\phi) = \frac{pR_{int}}{t} \cdot \sqrt{\frac{\pi a}{Q}} \cdot M\left(\frac{a}{t}, \frac{a}{c}, \frac{R_{int}}{t}, \phi\right) \tag{21}$$

M	:	geometrical correction factor for a crack in a pressurised pipe
p	:	internal pressure
Q	:	shape factor for an elliptical crack
ϕ	:	excentricity angle of the ellipse
R_{int}	:	internal radius of the pipe
a and c	:	semi-axes of the elliptical crack
t	:	thickness of the pipe wall

Authors	Date	Ref.	Method	Range of validity			
				a/t	a/c	t/R_{int}	results
Kobayashi & al	1977	[49-50]	FDP	0.25 to 0.8	.34	0.1	graph
Hilliot & Labbens	1979	[51]	EIB	0.25 to 0.8	.34	1.1	graph
McGowan & al	1979	[52]	FEM	0.25 to 0.8	.34	1.1	graph
Newman & Raju	1980	[53, 54]	FEM	0.2 to 0.8	0.2 to 1	0.1 and 0.25	equation
Boukharouba & al	1995	[24, 30]	FEM	0.46 to 0.7	0.7 to 1.12	0.25	graph

(WF) : Weight function, (BIE) : Boundary integral equation and (FEM) : finite elements method.

Table 5 : Works on pressure tubes for SIF calculations.

Results of 4 authors can be compared in Figure (14) for identical geometrical conditions (a/t = 0.8, a/c = 1.3 and t/Rint = 0.1). Except for the result given by Kobayashi **[49, 50]** the difference beetween solutions is small and less than 5 %. We can notice that the point of maximum value of the stress intensity factor is the surface point (θ = 90°). No experimental determination of such a stress intensity factor has been made.

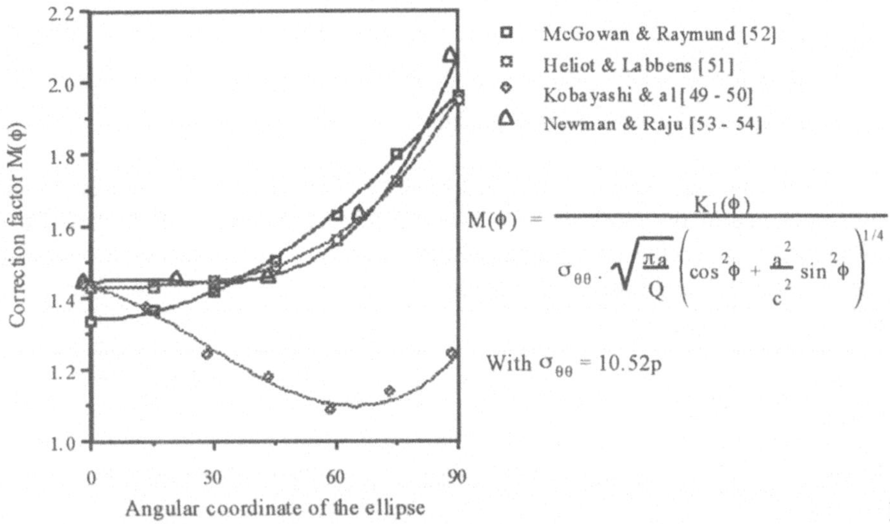

F

Figure 14 : Evolution of the shape factor M(ϕ) from different authors for the folowing conditions $a/t = 0.8$, $a/c = 1/3$ and $t/R_{int} = 0.1$.

5 - INFLUENCE OF LOADING MODE AND TYPE OF SIF ON PARAMETERS C AND m

5.1 - Case of a plate

Few works have been done to compare the various crack growth laws for a semi-elliptical surface crack (Mahmoud **[22, 56]** , Varfolomeyev & al **[55]** and Boukharouba & al **[30]**). The laws used are different and the parametrers C and m are different from parameters C* and m* measured on a reference test (using a CT specimen for instance).

5.1.1 - Analysis of Mahmoud

Mahmoud **[22, 56]** has studied data from literature for the crack propagation of a semi elliptical surface crack in bending (15 references) and tension (20 references). The experimental data used in our analysis are taken from references **[2-9, 21, 57-62]** in bending and **[2-3, 5, 9-16, 18-19, 62-65]** in tension. Results have been analysed using the local and the average crack propagation laws and particular local values of the parameters $C_{j,i}$ and $m_{j,i}$ given in Table (6). Experimental and predictive results have been compared using a parameter R_i called "residual" defined as :

$$R_i = (a/c)_{i,\, prédiction} - (a/c)_{i,experiment} \quad (22)$$

The standard deviation "d_s" is given by the following relationship :

$$d_s = \frac{\bullet\ R_i^2}{(n-1)^{0.5}} \qquad (23)$$

n is the number of experimental datas.

All results have been divided in 4 groups and are presented in figures (15 and 16). The choice of the most accurate law and the determination of the best fitted coefficient are given by a statistical criterion based on the combination of lowest value of residual R_i and standard deviation d_s.

Criterion	Point	$C_{J,i}$	$m_{J,i}$
$\Delta K_{loc,A}$ $\Delta K_{loc,C}$	A C	$C_{loc,A} = C_{loc,C}$	$m_{loc,A} = m_{loc,C} = m$
$\Delta K_{loc,A}$ $\Delta K_{loc,C}$	A C	$C_{loc,C} = (0.9)^m \,.\, C_{loc,A}$	$m_{loc,A} = m_{loc,C} = m$
$\overline{\Delta K_{ave\,A}}$ $\overline{\Delta K_{ave\,C}}$	A C	$C_{ave,A} = C_{ave,C}$	$m_{ave,A} = m_{ave,C} = m$
$\overline{\Delta K_{ave\,A}}$ $\overline{\Delta K_{ave\,C}}$	A C	$C_{ave,C} = (0.9)^m \,.\, C_{ave,A}$	$m_{ave,A} = m_{ave,C} = m$

A : deepest point, C : surface point, loc : local and ave : average.

Table 6 : Particular values of $C_{j,i}$ et $m_{j,i}$ given by Mahmoud [22, 56].

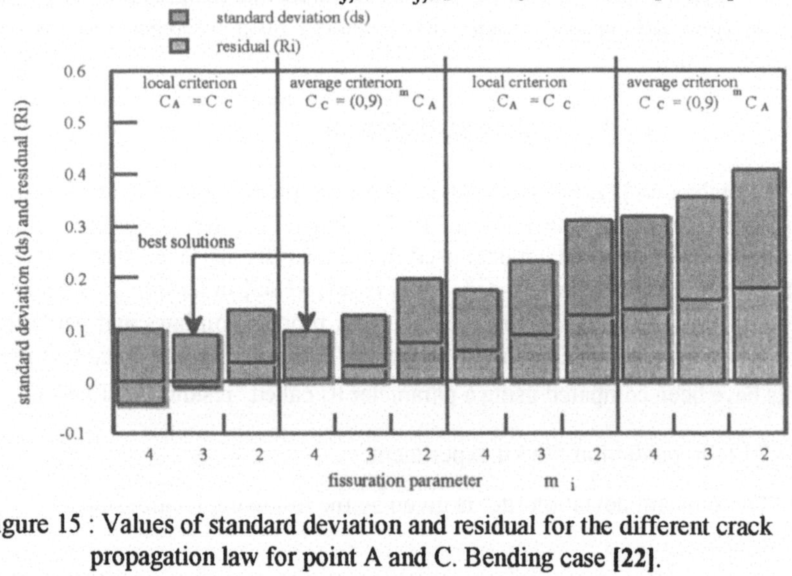

Figure 15 : Values of standard deviation and residual for the different crack propagation law for point A and C. Bending case [22].

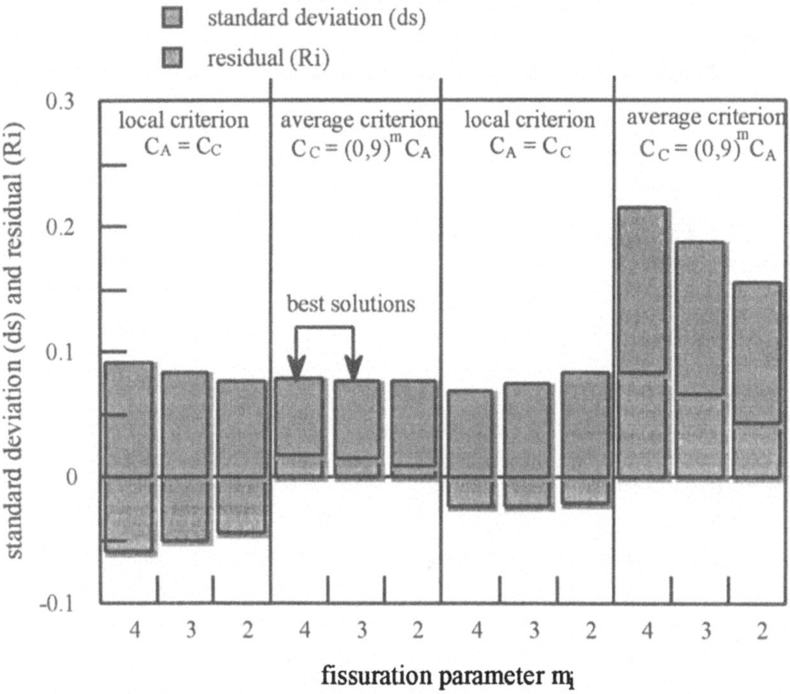

Figure 16 : Values of standard deviation and residual for the different crack propagation law for point A and C. Tension case [56].

These authors [22, 56] have found that the best solution for the crack propagation law is obained by using the local stress intensity factor and the particular following values.

$m_A = m_C = 3$ and $C_A = C_C$ in bending

$m_A = m_C = 4$ and $CA \neq C_C$ in tension

5.1.2 - Analysis of Varfolomeyev

Varfolomeyev [55] has analysed 5 tests he has performed on a steel plate (steel15kh2MFA, yield stress 584 MPa, ultimate stress 700 MPa) in four point bending with stress ratio R= 0.32. Results have been examined with the following fatigue crack growth laws.

- local stress intensity factor range ΔK_{loc} given by the formula of Newman & Raju [17].

- semi local strain intensity factor range, $\overline{\Delta K_{ave}}$ given by the formula of Cruse & Besuner [29]

- effective stress intensity factor range ΔK_{Ie}.

This effective stress intensity factor is defined by the following formula :

$$\Delta K_{I,e} = \gamma \cdot \Delta K_I \tag{24}$$

The γ coefficient is determined from experimental results. They used the following assumptions :

$$\Delta K_{loc,A} = \Delta K_{loc,C} = \Delta K_I$$

$$C_{loc,A} \circ C_{loc,C} \circ C^* \Delta K_{loc,A} = \Delta K_{loc,C} = \Delta K_I$$

$$m_{loc,A} = m_{loc,C} = m^*$$

which leads to :

$$\frac{da}{dN} = C_{loc,A}\left(\Delta K_{loc,A}\right)^m = C^*\left(\gamma_A \cdot \Delta K_I\right)^{m^*} \tag{25a}$$

$$\frac{dc}{dN} = C_{loc,C}\left(\Delta K_{loc,C}\right)^m = C^*\left(\gamma_C \cdot \Delta K_I\right)^{m^*} \tag{25b}$$

We can write :

$$\frac{C_{loc,A}}{C_{loc,C}} = \left(\frac{\gamma_A}{\gamma_C}\right)^m \quad \text{or} \quad \mu = \frac{\gamma_A}{\gamma_C} = \left(\frac{C_{loc,A}}{C_{loc,C}}\right)^{\frac{1}{m}} \tag{26}$$

μ characterise the difference between the da/dN and dc/dN predicted values from equation (25).

Experimental results are presented as the local crack growth da/dN and dc/dN rate versus ΔK, $\overline{\Delta K_{ave}}$ and ΔK_{Ie} (figures 16a to 16c).

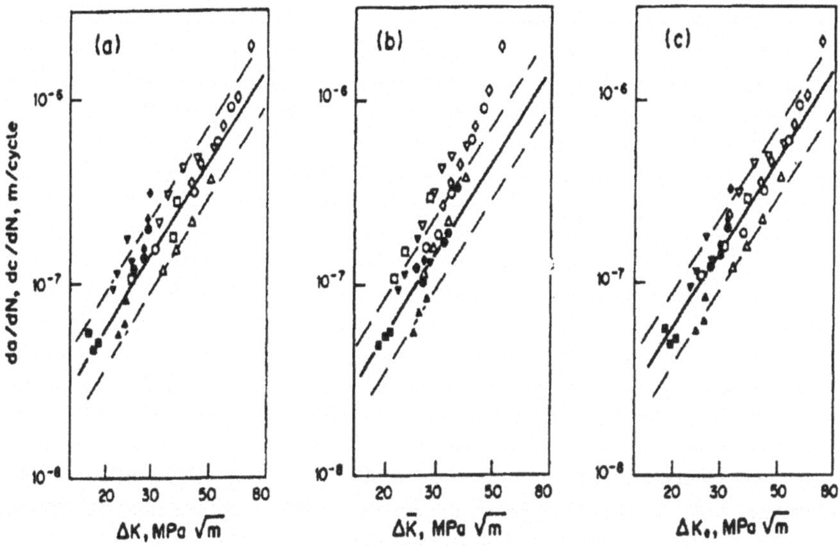

Figure 16 : Crack growth rate on the surface point dc/dN and at the deepest point da/dN against local, average and effective stress intensity factor range **[56]**.

They found that the effectif stress intensity factor range ΔK_{Ie} obtained from the value γ_A & γ_C and with $C_{e,A} \neq C_{e,C}$ and $m_{e,A} = m_{e,C}$ give the lowest discrepancy.

5.1.3 - Analysis of Boukharouba

Boukharouba & al **[1]** have analysed 12 fatigue tests on a steel plate steel 35 NCDV12, (yield stress 1282 MPa, ultimate stress 1433 MPa), submitted to three points bending with three stress ratios R = 0.1 & 0.3 et 0.5 at room temperature. From the experimental shape of the crack front they have computed the local stress intensity factor range by a finite elements method and the average one with the equations of Cruse & Besuner.

They have determined the best fitted values of the parameter $C_{j,i}$ along the crack front and particularly at the deepest and surface points. These values are compared to reference test values C* and m* obtained on a CT specimen. Results are presented on Figures (17a and 17b) where it can be noticed that the use of the local stress intensity factor range leads to values of $C_{j,i}$ and $m_{j,i}$ close to the references values C*" and m*.

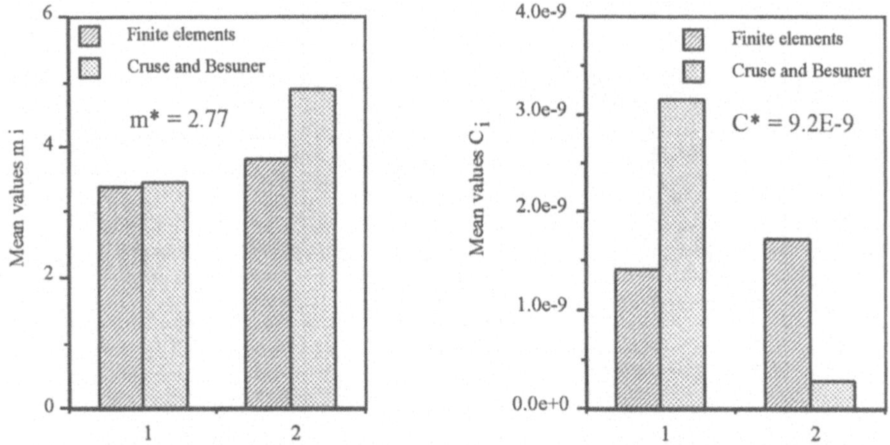

(1) : Deepest Point and (2) : Surface point.

Figure 17a : Mean values of $C_{j,i}$ and $m_{j,i}$ for Paris's law. Experimental determination on plates for stress ration R = 0.1 **[30]**.

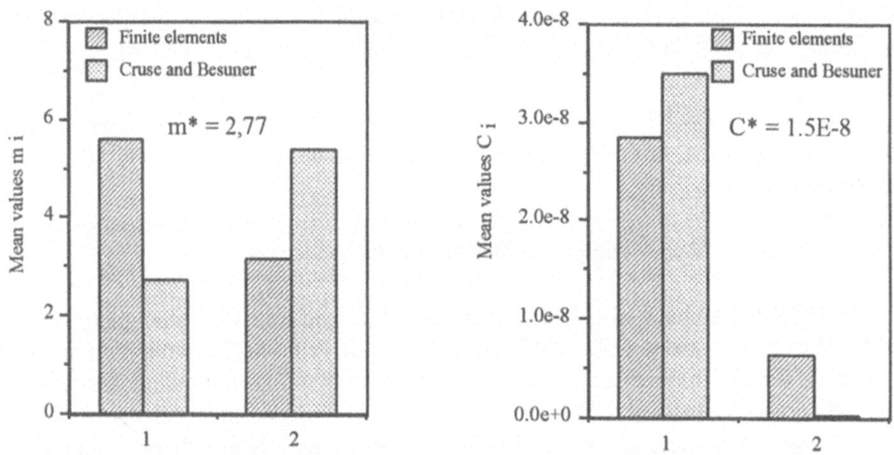

(1) : Deepest Point and (2) : Surface point.

Figure 17b : Mean values of $C_{j,i}$ and $m_{j,i}$ for Paris's law. Experimental determination on plates for stress ration R = 0.3 **[30]**.

5.2 - Pipe under internal pressure

Boukharouba and al **[24]** have analysed results of the fatigue crack growth of a surface semi-elliptical crack in a steel pipe. They computed the local stress intensity factor range by a finite elements method or using the Newman & Raju's solution. They compute also the average stress intensity factor range using the Cruse & Besuner **[29]** solution. Values of local crack growth rate da/dN and dc/dN versus the different stress intensity factor range are plotted in Figure (18).

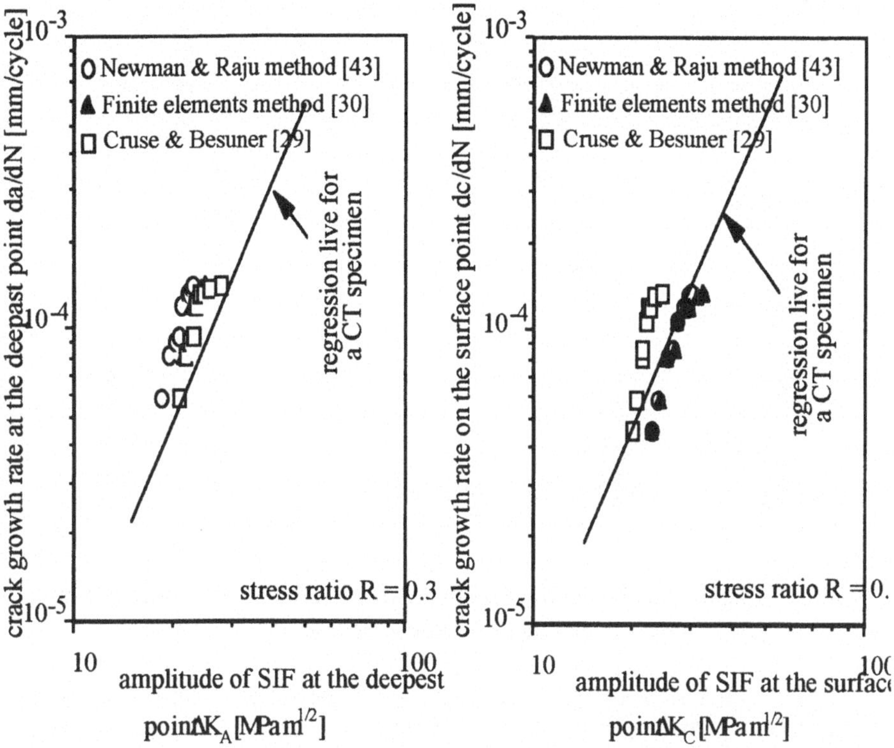

Figure 18 : Crack growth rate at the deepest point da/dN and on surface dc/dN against the local or the average value of the SIF range. Comparison with the material reference test. Case of a pipe submitted to internal pressure **[30]**.

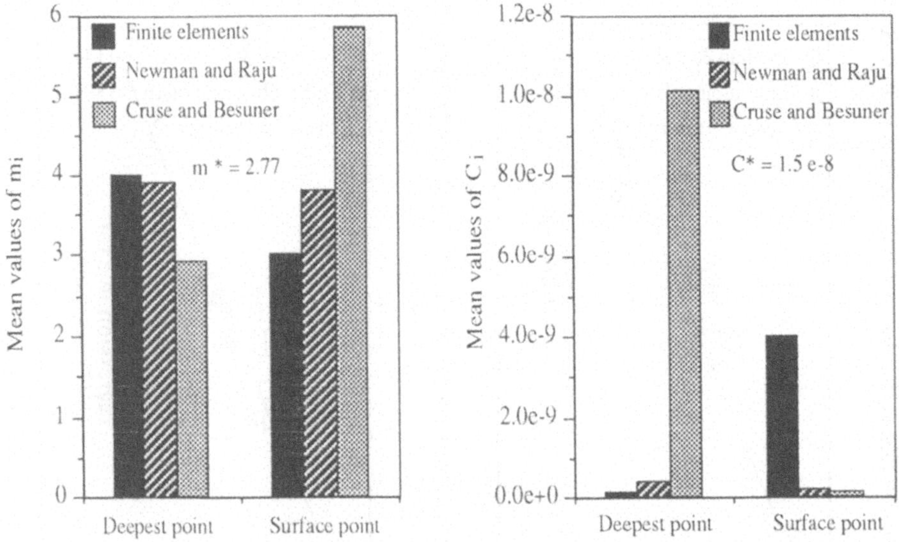

Figure 19 : Experimental determination of the mean value of crack propagarion $C_{j,i}$ and $m_{j,i}$ for Paris's law **[30]**.

The conclusions are different from those obtained on a plate. The values of the parameter Cj,i and mj,i from the average stress intensity factor range are closen to the references values C^* and m^*.

6 - CONCLUSION

The propagation of a semi-elliptical surface crack in a plate in tension or bending in a pipe submitted to an internal pressure is always an open question in spite of a number of experimental and numerical works.

Crack front evolution is generally well predicted by empirical relationships. This evolution is influenced by the loading mode and the fundamental mechanisms is not explained and is probably an example of energy minimisation.

The role of the local or average stress intensity factor is also not very clear. It has also been pointed out that the parameters C and m of the crack propagation law seem not intrinsic to the material but depend on the loading mode, the geometry and the local value of the stress intensity factor.

REFERENCES

1. Boukharouba, T.,. Chehimi, C., .Gilgert.,J . and Pluvinage G, (1994) "Behaviour of semi-elliptical cracks in finite plates subjected to cyclic bending", *Handbook of fatigue crack Propagation in Metallic Structures,* (Edited by Andrea Carpinteri), Vol. I, pp. 707-731.
2. Truchon, M. and Lieurade, H, (1981) "Experimental study of surface crack propagation" in. *Advances in Fracture Mechanics, Proc. 5th. Int. Conf. Fracture,* Cannes (Edited by D. Francois), Vol. 1, pp. 33-39. Pergamon Press, Oxford, NY.
3. Portch, DJ., (1979) "An investigation into the change of shape of fatigue cracks initiated at surface flaws". Central Electricity Generating Board, Report RD/B/N4645, Great Britain .
4. Grandt, A., and Sinclair, G., (1972) "Stress intensity factors for surface cracks in bending". *Stress Analysis ans Growth of Cracks,* ASTM STP 513, pp. 37-58.
5. Kawahara, M,. and Kurihara, M,. (1977) "Fatigue crack growth from a surface flaw". in *Proc. of the 4th Int. Conf. Fracture* (Edited by D. Taplin), Vol. 2, pp. 1361-1373, University of Waterloo, Canada.
6. Pierce, W. and Shannon, J., (1978) "Surface crack shape change in bending fatigue using an inexpensive resonant fatiguing apparatus". *J. Testing and Evaluation,* Vol. 6, N°. 3, pp. 183-188.
7. Yen, C., and Pendleberry, S., (Nov. 1962) "Technique for making shallow cracks in sheet matals". *Materials Research Standards,* 2, pp. 913-916
8. Lee, L., et al., (1982) "Experimental research on surface crack propagation laws for alloy steel". *Engng Fracture mech.* Vol. 16, N° 1, pp. 105-113 .
9. Corn, D., (1971) "A study of cracking techniques for obtaining partial thickness of pre-selected depths and shapes". *Engng Fracture Mech.* Vol. 3, pp. 45-52.
10. Lloyd, G., and Walls, J., (1980) "Propagation of fatigue cracks surface flaws in austenitic type 316 butt welds", *Engng. Fracture Mechanics,* Vol. 13, pp. 879-911.
11. Hodulak, L., Kordisch, H., Kunzelmann, S., and Sommer, E., (1979) "Growth of part through cracks", in *Fracture Mechanics, ASTM STP 677,* (Edited by C.W. Smith), pp. 399-410.
12. Vosikovsky, O., and A. Rivard, (1981) "Growth of surface fatigue cracks in a steel plate", *Int. J. Fatigue,* **3**, pp. 111-115.
13. Holdbrook S., and Dover, W., (1979) "The stress intensity factor of a deep surface crack". *Engng. Fracture Mechanics,* Vol. 12, pp. 347-364 .
14. Nagai, A., Toyosada, M., and Okamoto, T., (1975) "A study on the fatigue crack growth in 9% Ni steel plate", *Engng. Fracture Mechanics,* Vol. 7, pp. 481-490.
15. Cruse, T., Meyers, G., and Wilson, R., (1977) "Fatigue growth of cracks", in *Flaw Growth ans Fracture, ASTM STP 631,* pp. 174-189.

16. Hall, L., Shah, R., and Engstrom, W., (May 1974) "Fracture and fatigue crack growth behaviour of surface flaws and flaws originating at fastner holes", *Boeing Aerospace Co., AD/A-001 597*,.
17. Newman, N., and Raju, I., (1981) "An empirical stress intensity factor equation for the surface crack". *Engng. Fracture Mechanics,* Vol. 15, pp. 185-192.
18. Kobayashi, K., Narumoto, A., and Tanaka, M., (April 1977) "Prediction of crack propagation life in axial loading fatigue of structural steels", *Proc. oj Third Int. Conf. of Pressure Vessel Technology, ASME 11,* pp. 807-813, Tokyo.
19. Jolles, M., and Tortoriello, V., (1983) "Geometry variation during fatigue growth of surface flaws", in *Fracture Mechanics, ASTM STP 791,* (Edited by J. Lewis and G. Sines), Vol. I, pp. I297-I307.
20. Gorner, F., Mattheck, C., and Munz, D., (1983) "Change in geometry of surface cracks during alternating tension and bending", *Z. Werkstofftech,* Vol. 14, pp. 11-18.
21. Iida, K., (1980) "Aspect ratio expressions for part-through fatigue crack". IIW Doc. XIII-967-80.
22. Mahmoud, M.A., (1988) "Growth patterns of surface fatigue cracks under cyclic bending - A quantitative analysis", *Engng. Fracture Mechanics,* Vol. 31, N°. 2, pp. 357-369.
23. Carpinteri, A., (1994) "Propagation of surface cracks under cyclic loading", *Handbook of fatigue crack Propagaton in Metallic Structures,* (Edited by Andrea Carpinteri), Vol 1, pp. 653-705.
24. Boukharouba, T., Chehimi, C., Gilgert, J., and Pluvinage, G,, (1995) "Comportement de fissures semi-elliptiques contenues dans un tube sous pression interne". *Journée d'Etude de AFIAP, Revue Chaudronnerie, Tôlerie et Tuyauterie Industrielle du SNCT,* 17-20 Oct. Paris.
25. Paris, P., & Erdogan, F., (Dec. 1963)*Trans. ASME,* pp. 528-534.
26. Forman, R.G., & Kearny, V.E., & Engle, R.M., (1967)*Transactions of the ASME D89,* pp. 459.
27. Pearson, S., (1972) *Engng. Fract. Mechanics,* Vol. 4, pp.9.
28. Richards, C.E., & Lindley, T.C., (1972) *Engng. Fract. Mechanics,* Vol. 4, pp. 951.
29. Cruse, T.A., & Besuner, P.M., (1975) "Residual life prediction for surface cracks in complex structural details", *J. Air Craft,* Vol. 12, pp. 369-375.
30. Boukharouba, T., (1995) "Étude du comportement de fissures semi-elliptiques, application aux plaques en flexion trois points et aux tubes sous pression interne", *Thèse de Doctorat, en Sciences de l'Ingénieur N° 7642 de l'Université de Metz*,.
31. Irwin, G.R., *J. (1962) of Applied Mechanics, American Society of Mechanical Engineers,* Vol. 29, N°. 4, pp. 651-654 .
32. Paris, P.C., & Sih, G.C., (1965) in *Fracture Toughness Testing and its Application, ASTM STP 381, American Society for Testing and Materials,* pp. 30-83.

33. Smith, F.W., & Emery, A.F., & Kobayashi, A.S., (Dec. 1967)*J. of applied Mechanics,* Vol. 34, N°. 4, Transactions, *American Society of Mechanical Engineers,* Serie E, pp. 953-959.
34. Kobayashi, A.S., & Moss, W.L., (1969) in *Fracture,* Chapman and Hall, New York and London, pp. 31-45.
35. Mesters, J.N., & Haese, W.P., & Finger, R.W., (déc 1969) "Investigation of deep flaws in thin walled tanks", *NASA CR-72606, National Aeronautics and Space Administration.*
36. Smith, F.W., & M.J. Alavi, (1969) in *Proceedings, 1st International Conf. oj Pressure Vessel Tech.,* Delft, The Netherlands, American Society of Mechanical Engineers, pp. 783-800.
37. J.R. Rice & N. Levy, (1970) "The part-through surface crack in an elastic plate", *Technical Report NASA NGL 40-002-080/3, Division of Engineering,* Brown University , Transaction, American Society of Mechanical Engineers, *J. of Applied Mechanics,* Paper N°. 71-APM-20.
38. Anderson, R.B., & Holms, A.G., & Orange, T.W., (Oct. 1970) "Stress intensity magnification for deep surface cracks in sheets and plates", *NASA TN D-6054, National Aeronautics and Space Administration.*
39. Newman, J.C., (1973) "Fracture analysis of surface and through-cracked sheets and plates", *Symposium on Fatigue and Fracture,* George Washington University, 1972, (also, *Eng. Fract. Mechanics,* Vol. 5, pp. 667-689.
40. Shah, R.C., & Kobayashi, A.S., (1972) "Stress intensity factor for an elliptical crack approaching the surface of plate in bending", *Stress Analysis and Growth of Cracks, Proceedings of the 1971 National Symposium on Fracture Mechanics, Part I, ASTM STP 513,* pp. 3-21.
41. Smith, F.W., & Sorensen, D.R., (1974) "Mixed mode stress intensity factors for semi-elliptical surface cracks", *NASA CR-134684, National Aeronautics and Space Administration.*
42. Kobayashi, A.S., (1976) in *Proceedings, 2nd International Conference on Mechanical Behavior of Materials, American Society of Mechanical Engineers,* pp. 1073.
43. Raju, I.S., & Newman, J.C., (1979) "Improved stress intensity Factors for wide range of semi-elliptical surface cracks in finite-thickness plates", *Engng. Frac. Mechanics,* Vol. 11, N°. 4, pp. 817-829.
44. Newman, J.C., & Raju, I.S., (1983) "Stress intensity factor equations for cracks in three-dimensional finite bodies", *Fracture Mechanics, Fourteenth Symposium,* Vol. I, Theory and Analysis, *ASTM STP 791,* pp. I308-I326.
45. Carpinteri, A., (1990) "Combined tension and bending stress intensity fctor of surface crack modelled as spring elements", *Theoretical and Applied Fracture Mechanics,* Vol. 14, pp. 243-251.
46. Carpinteri, A., (1993) "Surface flaw stress intensity factor computation with quarter-point element", *J. of Strain Analysis,* Vol. 28, pp. 117-123.
47. Ruiz, C., & Epstein, J., (1995) "On the variation of the stress intensity factor along the front of a surface flaws", *Int. J. of Fracture,* Vol. 28, pp. 231-238.

48. Mars, G.R., & Smith, C.W., (1972) "A study of local stress near surface flaws in bending fields", *Stress Analysis and Grawth of Cracks, Proceedings of the 1971 National Symposium on Fracture Mechanics, ASTM STP 513,* pp. I-22 _I-36.
49. Kobayashi, A.S., & Emery, A.F., & Polvanich, N., & Love, W.J., (1977) *Int. J. of Pressure Vessels and Piping,* pp. 103-133.
50. Kobayashi, A.S., & Polvanich, N., & Emery, A.F., & Love, W.J., *J. (1977) oj pressure Vessel Technology,* Vol. 9, pp. 83-89.
51. Heliot, H., & Labbens, R.C., & Pellissier-Tanon, A., (1979) "Semi-elliptical cracks in a cylinder subjected to stress gradients", *Fracture Mechanics ASTM STP 677,* (Edited by C.W. Smith) pp. 341-364.
52. McGowan, J.J., & Raymund, M., (1979) "Stress intensity factor solutions for internal longitudinal semi-elliptical surface flaws in a cylinder under arbitrary loadings", *Fracture Mechanics Astm STP 677,* pp. 365-380.
53. Newman, J.C., & Raju, I.S., (Nov. 1980) "Stress intensity factors for internal and external surface cracks in cylindrical pressure vessels", *J. of Pressure Vessel Technology,* Vol. 10.
54. Newman, J.C., & Raju, I.S., (Nov 1982) "Stress intensity factors for internal surface cracks in cylindrical pressure vessels", *J. of Pressure Vessel Technology,* Vol. 104.
55. Varfolomeyev, V., & Vainshtok, V.A., & Krasowsky, A.A.Y., (1991) "Prediction of part-through crack growth under cyclic loading", *Eng. Frac. Mechanics,* Vol. 40, N°. 6, pp. 1007-1022.
56. Mahmoud, M.A., (1988) "Quantitative prediction of growth patterns of surface fatigue cracks in tension plates", *Engng. Fracture Mechanics,* Vol. 30, N°. 6, pp. 735-746.
57. Kawahara, M., and Kurihara, M., (June 1975) "A preliminary study on surface crack growth in a combined tensile and bending fatigue process". *J.S.N.A. Japan*, Vol. 137, pp. 86-92.
58. Iida, K., and Kawahara, M., (1975) University of Tokyo, Naut. Report 9011.
59. Newman, J., and Raju, I.S., (1979) "Analyses of surface cracks in finite plates under tension or bending loads". *NASA Technical Paper* 1579.
60. Scott, P., and Thorpe, T., (1981) "A critical rewiew of crack tip stress intensity factors for semi-elliptic cracks", *Fatigue Engng. Mater. Struc.,* Vol. 4, pp. 291-309.
61. Yu, R., and al, "A study of surface crack propagation on wide plates under bending loads". *Second Int. Conf. on Fatigue and Fatigue Threshold,* Vol. III, pp. 1705-1718, University of Brimingham (Fatigue 84).
62. Muller, H., Mullet, S., Munz, D., and Newman, J., (1986) "Extension of surface cracks during cyclic loading", in *Fracture Mechanics, ASTM STP 905,* (Edited by J. Lewis and G. Sine), Vol. I, pp. 625-64.
63. Nishioka, K., Hirakawa, K., and Kitaura, I., (1977) "Fatigue crack propagation behaviour of various steels", *The Sumitarno Search,* N°. 17, pp. 39-55.

64. Hoeppner, D., Pettit, D., Frayherson, C., and Hyler, W., (1968) "Determination of flaw growth characteristics of Ti-6Al-4v sheet in the solution treated and aged condition", *NASA CR-65811*.
65. Sommer, E., Hodulak, F., and Kordisch, H., (1977) "Growth characteristics of part-through cracks in thick-walled plates and tubes", *J. Pressure Vessel Technology, ASME,* pp. 106-111.
66. Dufresne, J., (1981) in *Advance in Fracture Research,* Vol. 2 ICF5-Conference, pp. 517-531.

FATIGUE OF CERAMICS AND INTERMETALLICS: APPLICATION TO DAMAGE TOLERANCE AND LIFE PREDICTION IN CYCLICALLY-LOADED BRITTLE MATERIALS

R. O. RITCHIE and K. T. VENKATESWARA RAO

Materials Sciences Division, Lawrence Berkeley National Laboratory
and Department of Materials Science and Mineral Engineering
University of California at Berkeley, Berkeley, CA 94720-1760, USA

1. Introduction

Over the past ten years, there has been an increasing interest in the use of high-strength, brittle materials, such as ceramics, intermetallics and their respective composites, for structural applications. This has been particularly focused at elevated temperature applications, but in the case of ceramics for biomedical implant devices, at lower temperatures too. Examples of such "advanced materials" are the use of silicon nitride ceramics for automobile turbocharger wheels and engine valves and pyrolytic carbon for prosthetic cardiac devices, and the contemplated use of composite ceramics for gas turbine blades. Similarly, intermetallic alloys, such as the γ-based titanium aluminides, have been considered for applications such as automobile engine valves and compressor blades in gas turbines. Whereas these materials offer vastly improved specific strength at high temperatures compared to conventional metallic alloys, they suffer in general from a pronounced lack of damage tolerance in the form of an extreme sensitivity to pre-existing flaws. Moreover, recently it has become apparent that even low ductility materials such as ceramics show a pronounced susceptibility to premature failure under cyclic fatigue loading [e.g., 1].

The majority of intermetallic and ceramic materials must be toughened, either through composite reinforcement or through the development of novel microstructures, to overcome their limitations of low toughness and ductility at ambient temperature [e.g., 2,3]. With advanced materials, such toughening invariably is *extrinsic*, i.e., based on the concept of crack-tip shielding [1,2]. This involves mechanisms that act primarily

R.A. Smith (ed.), Reliability Assessment of Cyclically Loaded Engineering Structures, 377–403.

in the crack wake to reduce the *local* driving force actually experienced at the crack tip through closure tractions imposed in the wake. The approach is rather distinct from toughening in metallic materials. Here, the approach generally involves *intrinsic* mechanisms that enhance the inherent microstructural resistance of the material by phenomena that are generally active in the process zone ahead of the crack tip. Thus, fracture may be envisaged as a mutual competition between intrinsic mechanisms ahead of the crack tip, which act to promote cracking, opposed by extrinsic mechanisms in the wake, which act to inhibit it (Fig. 1). Thermomechanical treatments aimed at variations in grain size, precipitate size and distribution in microstructures of metallic materials like steels, aluminum alloys and superalloys are good examples of intrinsic toughening. Extrinsic mechanisms, on the other hand, are associated with the tractions that may be induced by *in situ* phase transformations, interlocking grains, or by crack bridging resulting from the incorporation of fibers, whiskers or ductile reinforcements in the microstructure. Under cyclic loading, shielding can additionally result from wedging mechanisms in the form of crack closure [4,5]. A schematic illustration of various extrinsic mechanisms is given in Fig. 2 [1].

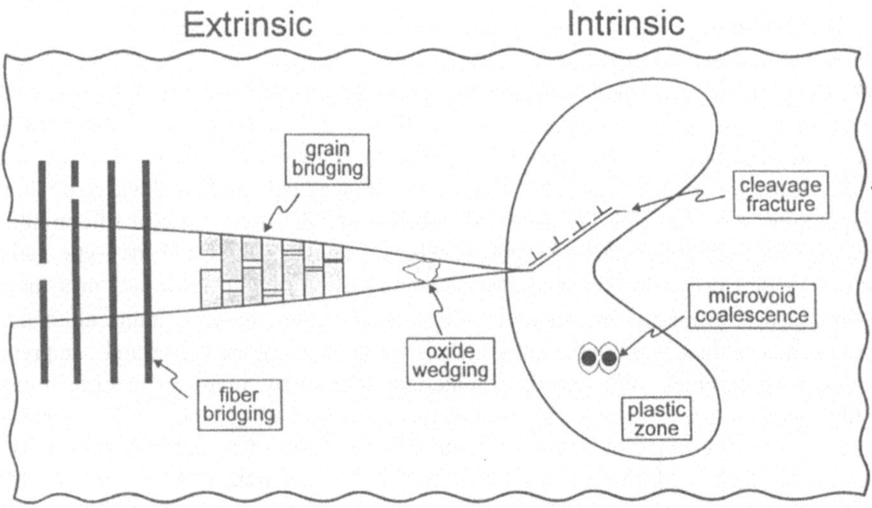

Figure 1. Schematic illustration of the mechanisms involved in crack growth, specifically, *intrinsic* mechanisms, such as transgranular cleavage and microvoid coalescence, are active in the process zone *ahead* of the crack tip and act to promote cracking; *extrinsic* mechanisms, such as fiber bridging, grain bridging and oxide wedging, are active in the wake *behind* the crack tip and act to inhibit cracking.

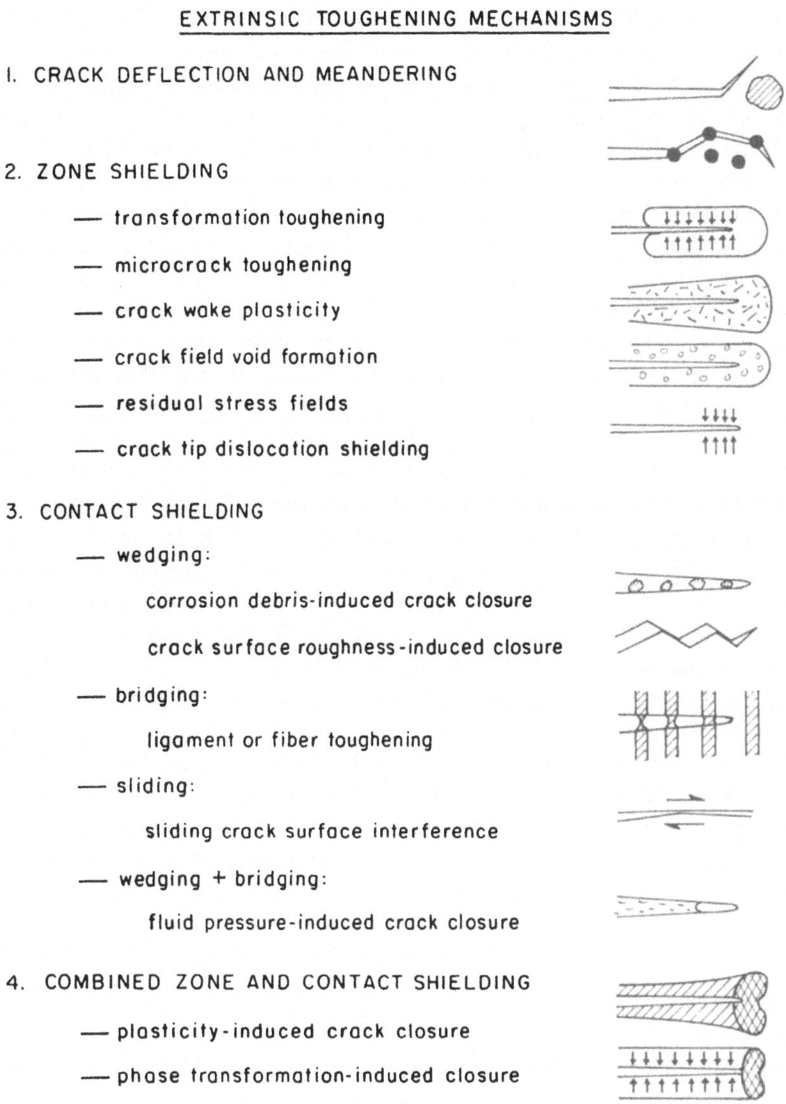

Figure 2. Schematic representation of the classes and micromechanisms of extrinsic toughening by crack-tip shielding [1].

Although crack-tip shielding in brittle materials such as ceramics and ceramic-matrix composites is critical in inducing some degree of toughness, such mechanisms can concurrently render these materials to be susceptible to cyclic fatigue failure [e.g., 6-9]. In most ceramics and their respective composites, there appears to be no intrinsic mechanism for cyclic crack advance, at least at low homologous temperatures; the advance of the crack tip proceeds by mechanisms identical to those occurring under monotonic loading (i.e., by static fatigue). However, under cyclic loads, a progressive degradation in the toughening (or shielding) mechanisms can occur behind the crack tip which locally elevates the near-tip driving force. It is this cyclic suppression of shielding that is considered to be the principal source for the susceptibility of brittle materials to cyclic fatigue failure. By contrast, the propagation of fatigue cracks in metallic materials is primarily caused by *intrinsic* damage processes occurring *ahead* of the crack tip; this typically involves crack advance by progressive blunting and resharpening of the crack tip [10], i.e., along alternating or simultaneous slip planes [11], clearly mechanisms distinct from fracture under monotonic loads. Shielding, in the form of crack closure (wedging) mechanisms, can act in the crack wake. However, since the physical mechanisms of crack advance and crack-tip shielding are quite different in metals and ceramics, the dependencies on the alternating and mean loads, specifically on ΔK and K_{max}, the alternating and maximum stress intensities, respectively, are also quite different. A schematic illustration illustrating these differences is shown in Fig. 3. Indeed, the two material classes of metals and ceramics represent the extremes of behavior with respect to the role of intrinsic vs. extrinsic effects on fracture and fatigue; crack growth in intermetallic materials generally lies somewhere in between. In this paper, examples of fracture and fatigue behavior in selected ceramic and intermetallic systems are presented with specific emphasis on the relative role of intrinsic and extrinsic mechanisms influencing crack growth under monotonic and cyclic loading. The influence of temperature on the fracture and fatigue properties and the resultant implications to structural design and life prediction are also discussed.

2. Ceramic Materials

Most ceramics cannot be toughened intrinsically; invariably toughness at ambient temperature is achieved using extrinsic mechanisms [2]. Notable among these are transformation toughening in *in situ* phase-transforming ceramics such as partially-stabilized zirconias, fiber bridging in ceramic-matrix composites such as Nicalon-fiber reinforced LAS or CAS glasses, ductile-phase toughening in ceramics reinforced with a metallic phase, such as Al-reinforced Al_2O_3, and most commonly, interlocking grain bridging in monolithic ceramics such as coarse-grained Al_2O_3 and grain-elongated

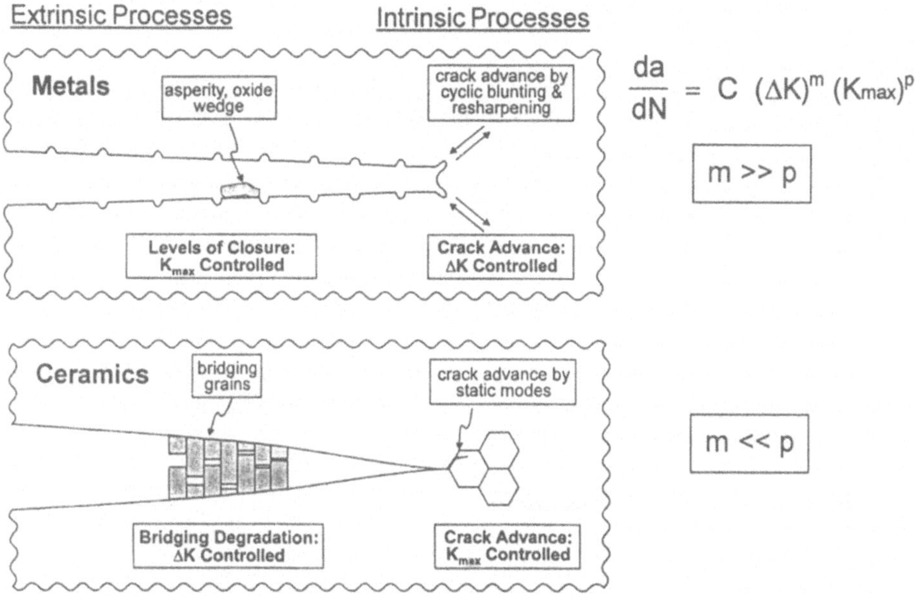

Figure 3. Schematic illustrations of the intrinsic and extrinsic mechanisms involved in cyclic fatigue-crack growth in (a) metals and (b) ceramics, showing the relative dependencies of growth rates, da/dN, on the alternating, ΔK, and maximum, K_{max}, stress intensities.

Si_3N_4. All these materials develop their toughness through a rising resistance curve (or R-curve), where the driving force for crack growth increases with crack extension as the tractions that are progressively developed in the crack wake locally shield the crack tip. Such wake toughening mechanisms, however, may degrade under cyclic loading, and it is believed that this is the source of cyclic fatigue failure in most ceramics [6-9]. Indeed, fatigue crack advance occurs by a mechanism essentially identical to that seen during monotonic loading but, due to the degraded shielding in the crack wake, at significantly reduced driving forces. A good example of such behavior, for an *in situ* toughened SiC ceramic [12,13], is described below.

2.1 HIGH-TOUGHNESS SILICON CARBIDE

SiC ceramics generally exhibit excellent creep and oxidation resistance, high thermal conductivity and low thermal expansion properties compared to most other ceramic materials at high temperatures [e.g., 14]. The widespread use of monolithic SiC has been limited, however, by its low toughness (K_c ~ 2-3 MPa√m) at ambient temperature. Recent developments in hot-pressed SiC using Al, B, and C as sintering additives (referred to as ABC-SiC) have shown that microstructures consisting of elongated SiC platelets, following the β(3C) to α(4H) phase transformation, exhibit toughnesses approaching nearly 10 MPa√m [12,13], while retaining a high modulus of rupture (~ 660 MPa). Such improvements in the room temperature toughness and strength, however, lead to a susceptibility to fatigue failure and can have a detrimental influence on mechanical properties at high temperatures. Results in ABC-SiC and underlying crack-growth mechanisms are discussed below.

2.1.1 *Behavior at Ambient Temperatures*

Fracture Toughness Properties. Unlike commercial SiC, ABC-SiC displays rising R-curve behavior under monotonic loading, with crack initiation occurring at K_i ~ 5.5 MPa√m (for small cracks ~ 100 μm or less, K_i is reduced to ~3.5 MPa√m) and the crack-growth resistance increases over several hundred microns of crack extension to a maximum value of K_c ~ 9 MPa√m (Fig. 4a). This represents more than a threefold improvement in the toughness of commercial SiC (Hexoloy SA), which shows no R-curve and fails catastrophically at K_c ~2.5 MPa√m. Analogous to behavior in Si_3N_4, the marked increase in fracture resistance of ABC-SiC can be ascribed to the heterogeneous, elongated plate-like grain structure. Crack paths and fracture-surfaces show clear evidence of extensive crack deflection along the α-grain boundaries, which in turn leads to intergranular cracking (Fig. 4c) and frictional grain bridging [13]. In contrast, Hexoloy SA exhibits transgranular cleavage cracking with no significant deflection at grain boundaries and hence no grain bridging, consistent with the absence of an R-curve.

Cyclic Fatigue Properties. Under cyclic loading, crack-growth rates in ABC-SiC are significantly faster than corresponding (static fatigue) growth rates under monotonic loads at equivalent stress-intensity levels; in both cases, growth rates show a marked dependence on the applied K. Although the underlying mechanism of crack advance in static and cyclic fatigue appears to be identical, the resulting cyclic fatigue fracture surfaces display evidence of profuse damage and debris (Fig. 4d) which is not observed on static fatigue surfaces (Fig. 4c). The debris results from repetitive contact between the crack faces during cyclic loading, and is a characteristic of the cycle-dependent decay in the interlocking grain bridges [e.g., 7,8]. Since Hexoloy SA is not toughened extrinsically, it displays no evidence of cyclic crack growth, as represented by the vertical line in Fig. 4b.

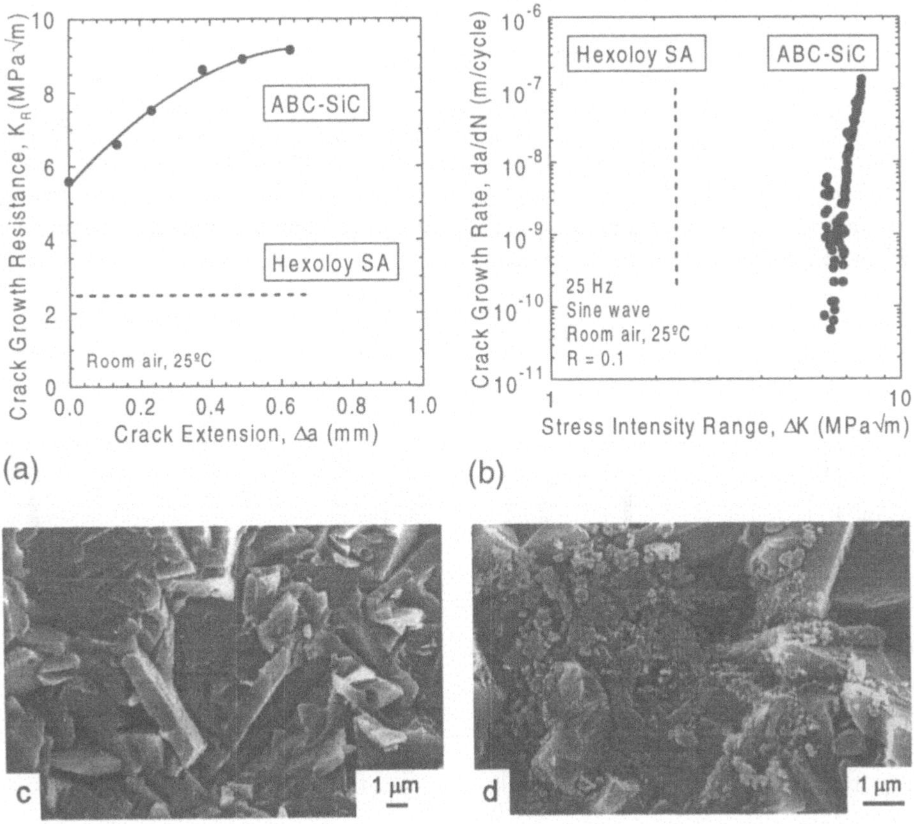

Figure 4. (a) Resistance curve and (b) fatigue-crack growth behavior in an *in situ* toughened ABC-SiC in room temperature air (R = 0.1, 25 Hz), compared to corresponding behavior in a commercial SiC (Hexoloy SA) that shows no apparent R-curve or susceptibility to fatigue. (c) Clean and faceted intergranular failures observed during monotonic loading, and (d) wear debris on the fracture surface from repetitive grain sliding under cyclic loading [13].

At higher (positive) load ratios ($R = K_{min}/K_{max}$), fatigue-crack growth rates are faster (at a given ΔK), with the value of the fatigue thresholds (ΔK_{TH}), below which "long-crack" growth is presumed dormant, being reduced (Fig. 5a). Such load ratio effects, however, can be normalized by characterizing in terms of the maximum stress intensity, K_{max} (Fig. 5b), demonstrating that fatigue-crack propagation in ceramics, unlike in metals [15], is mostly controlled by K_{max} rather than ΔK. The relative dependencies can be quantified by expressing the data in terms of a modified Paris power-law relation to includes the effect of both ΔK and K_{max}:

$$da/dN = C\,(\Delta K)^m\,(K_{max})^p, \qquad (1)$$

where C is a scaling constant (generally relatively independent of K_{max}, ΔK and R), and m and p are experimentally determined exponents. A regression fit to data in Fig. 5 yields $m \sim 1.9$, and $p \sim 36$, highlighting the marked dependence on K_{max}. For results obtained at a given load ratio, Eq. (1) can be rewritten, using $K_{max} = \Delta K/(1 - R)$, to give the more familiar form of the Paris equation:

$$da/dN = C'(\Delta K)^n, \qquad (2)$$

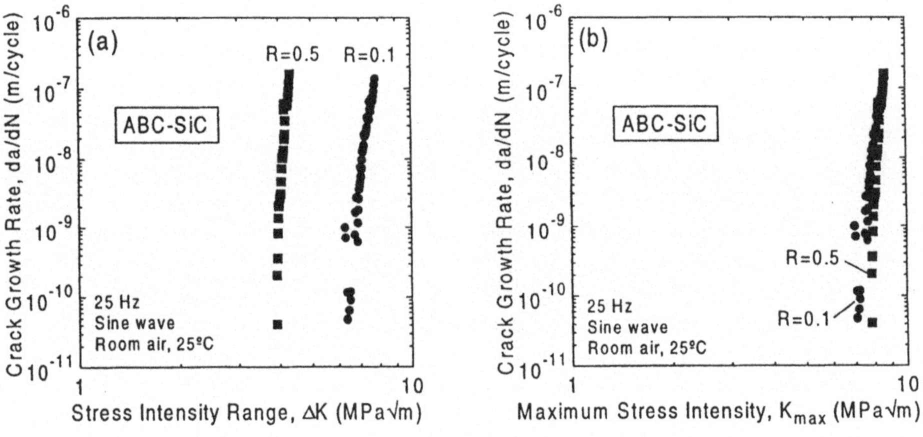

Figure 5. Variation in fatigue-crack propagation rates in ABC-SiC at different load ratios, plotted as a function of (a) the applied stress-intensity range, ΔK, and (b) the maximum stress intensity, K_{max}. Note that the data are normalized when characterized in terms of K_{max}, suggesting that the cyclic crack-advance mechanism is similar to that for static fatigue [13].

where $n = m + p$ and $C' = C/(1 - R)^m$. Values of the exponents n, m and p are listed in Table 1 for a range of material systems. It is apparent that the exponent n in ceramics is characteristically much higher than the values of n ~2-4 typically reported for metals (in the mid-range of growth rates) [15]. However, in ceramics this results from a marked dependence on K_{max}, i.e., $p \gg m$ whereas in metals, $p \ll m$.

The high growth-rate exponents, and in particular the marked K_{max} dependence of growth rates, in ceramics results from a crack-advance mechanism identical to that under static loading; the much smaller dependence on ΔK follows from the cyclic suppression of crack-tip shielding in the wake [6-9,16,17]. Specifically, in grain-bridging ceramics such as Al_2O_3, Si_3N_4 and the present ABC-SiC, the degradation in shielding is dominated by progressive wear of the frictional grain bridges, i.e., of material in the grain boundaries between the sliding grains, during the opening and closing of the crack. This rapidly reduces the bridging tractions between the interlocking grains, thereby increasing the *near-tip* driving force for crack extension. Similar mechanisms have been reported in ceramic composites toughened by the bridging of partially bonded whiskers or fibers [9]. Conversely, untoughened ceramics such as the commercial Hexoloy SA material are essentially immune to cyclic fatigue because transgranular crack morphology precludes the formation of any such shielding zones in the crack wake; these materials merely fail catastrophically when $K_{max} \rightarrow K_c$.

Finally, since the mechanisms of static and cyclic fatigue in most ceramics at low homologous temperatures appear to be identical, to the first approximation, the maximum stress intensity at the fatigue threshold, $K_{max,TH}$, should be approximately the same as the crack-initiation toughness, K_i, on the R-curve; experimental data for many ceramic systems [6,13,16-20] indeed confirm this notion.

TABLE 1. Power-law exponents in fatigue-crack growth relationships at 25°C

Material	Total Exponent n	ΔK Exponent m	K_{max} Exponent p	*Ref.*
Metals:				
Ni-base superalloy	3.4	3.0	0.4	[15]
Intermetallics				
$(\gamma+\alpha_2)$-TiAl	15.9	10.3	5.6	[32]
TiNb/γ-TiAl	12.4	6.3	6.1	[28]
$Nb_f/MoSi_2$	20.7	7.5	13.2	[38]
Ceramics				
Si_3N_4	30.3	1.3	29.0	[20]
ABC-SiC	37.9	1.9	36.0	[13]
Al_2O_3	31.6	9.8	21.8	[19]
SiC_w/Al_2O_3	15.0	4.8	10.2	[18]

2.1.2 *Behavior at Elevated Temperatures*

Fracture Toughness Properties. R-curve toughening is also evident in ABC-SiC at elevated temperatures (Fig. 6a). Both crack-initiation and steady-state toughness values are similar at ambient and 1200°C; the maximum toughness at 1200°C is some 11% higher, although the crack extends over larger dimensions (~2 mm) before reaching steady state. Such behavior can be attributed to softening of the largely amphorous grain-boundary phase, which reduces the magnitude of the bridging tractions but extends the length of the bridging zone. In fact, crack bridging by the glassy-phase ligaments can be observed to span the crack faces several millimeters behind the crack tip at the higher temperature (Fig. 6c), in contrast to the frictional interlocking of grains at ambient temperatures (Fig. 6d).

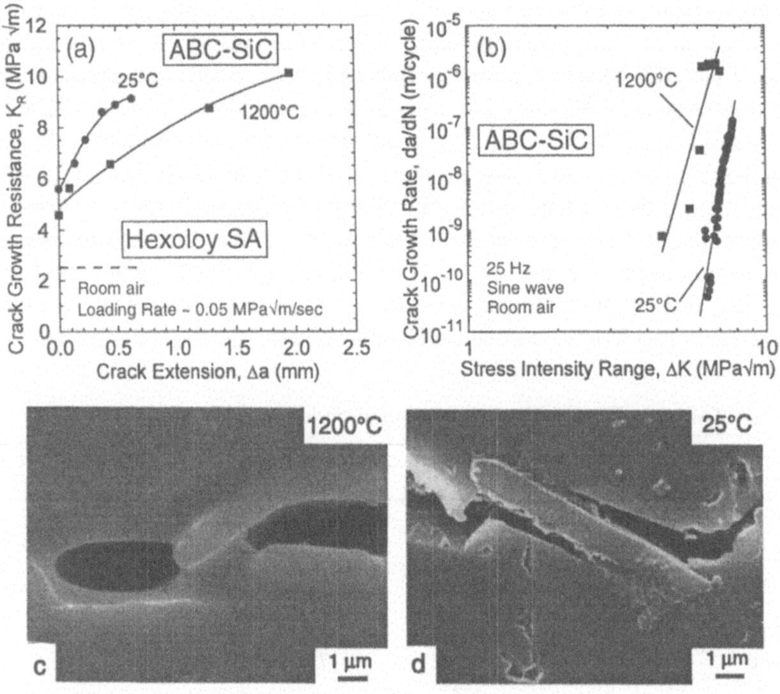

Figure 6. (a) R curves and (b) fatigue-crack propagation behavior in ABC SiC at 25° and 1200°C in air (at $R = 0.1$ with 25 Hz frequency). Note that R-curve toughening at 1200°C is associated with (c) bridging from viscous films present at grain boundaries, whereas at ambient temperatures, it involves (d) frictional grain bridging.

Cyclic Fatigue Properties. Mechanisms of fatigue-crack growth in ceramics at elevated temperatures are considerably more complex. Cyclic growth rates in ABC-SiC are nearly an order of magnitude faster at 1200°C than at 25°C, with the threshold ΔK_{TH} value reduced from ~ 6.5 to ~ 4 MPa$\sqrt{m}$ (Fig. 6b); in addition, growth rates exhibit a lower dependence on ΔK at the high temperature, similar to recent results on hot-pressed Si_3N_4 [21]. The acceleration of growth rates at elevated temperatures is likely due to enhanced creep deformation associated with the softening and cavitation of the glassy phase at grain boundaries, and to the reduced potency of the grain and viscous-phase bridging mechanisms under cyclic loading conditions at high temperatures.

Because of predominance of several competing mechanisms, namely intrinsic creep and perhaps fatigue damage, which promote crack growth, and grain and viscous-phase bridging, which act to oppose it, it is difficult to ascertain *a priori* whether cyclic fatigue or static creep crack growth is the more damaging at high homologous temperatures. It is clear that at ambient temperatures, cracks in toughened ceramics propagate much faster (at a given applied driving force) under cyclic compared to monotonic loads due to a cyclic suppression of the toughening; however, there is no consensus in the limited results that exist for high temperatures. For example, two separate studies on hot-pressed Si_3N_4 report that cyclic crack-growth rates are faster [21] or slower [22] than corresponding growth rates measured under static loading. Such results are most likely the result of the fact that the prominent mechanisms, namely *intrinsic* creep/fatigue damage ahead of the crack tip via cavitation and grain-boundary sliding and *extrinsic* shielding from grain and viscous-film bridging, all depend upon the deformation and fracture properties of the amphorous grain-boundary phase; since this phase will be viscous at the higher temperatures, the relative potency of the various mechanisms will depend critically on the temperature and strain rate, as well as composition and environment. Thus, small differences in temperature, cyclic frequency, composition and environment may be expected to lead to large differences in crack-growth behavior at elevated temperatures.

3. Intermetallic Materials

In general terms, fatigue-crack growth in metallic materials occurs by a largely ΔK-controlled intrinsic damage mechanism (unique to cyclic loading) which advances the crack tip and is opposed by K_{max}-limited crack closure (wedge shielding) mechanisms in the wake. Conversely, crack growth in ceramic materials occurs by a K_{max}-controlled crack-advance mechanism (identical to static loading) and is opposed by ΔK-limited wake shielding mechanisms that degrade under cyclic loads. Most intermetallic materials lie between these two extremes. For example, molybdenum disilicide behaves essentially as a ceramic, whereas titanium aluminide alloys display part metallic, part ceramic characteristics. Below we examine such behavior in monolithic and composite γ–TiAl alloys and in ductile-phase toughened $MoSi_2$ composites.

3.1 TITANIUM ALUMINIDE COMPOSITES

3.1.1 *Behavior at Ambient Temperatures*

γ-TiAl based intermetallic alloys are currently of interest as low-density alternatives to conventional titanium alloys for use in gas-turbine engines; however, due to their brittle nature at ambient temperatures, much effort has been aimed at improving their tensile ductility and fracture toughness. Both composite reinforcement and alloying techniques have been used with considerable success to toughen γ-TiAl, which has a intrinsic toughness of less than 10 MPa√m [23-28]. One composite approach involves ductile-phase toughening [29] through the addition of a small amounts of ductile metallic phases, e.g., Nb, TiNb or Ti-6Al-4V, into the microstructure. Depending upon how the crack interacts with this phase, which is a strong function of its morphology and interfacial bonding with the matrix, a variety of toughening mechanisms can ensue. Prominent among these are mechanisms which can increase the crack-initiation toughness, such as crack arrest, blunting and delamination at the interface, crack trapping by a particulate reinforcement, or crack renucleation across the ductile phase by a laminate reinforcement, and mechanisms of crack-growth (R-curve) toughening, such as extrinsic crack bridging by the uncracked ductile phase (Fig. 7).

Fracture Toughness Properties. A substantial improvement in fracture resistance of γ-TiAl can be achieved through reinforcement with ductile particles at room temperature [23,28]. As shown in Fig. 8a, adding ~20 vol.% of β-TiNb particles to TiAl increases crack-initiation toughness to ~18 MPa√m, more than twice that of pure γ-TiAl (~8 MPa√m); this can be attributed to crack trapping by TiNb particles and crack renucleation effects. The fracture resistance further increases with crack extension to above 30 MPa√m due to crack bridging by unbroken ductile TiNb ligaments in the crack wake (Fig. 8c). Indeed, these bridging zones under monotonic loading can be as large as ~3-6 mm [28], such that the results in Fig. 8a are undoubtedly influenced by large-scale bridging effects. The degree of toughening increases with volume fraction of reinforcements, but is independent of particle orientation. Similar toughening can be achieved through the addition of Nb particles; however, the TiNb reinforcements result in a larger toughness due to their much higher strength (yield stress for TiNb ~ 430 MPa compared to ~ 140 MPa for Nb).

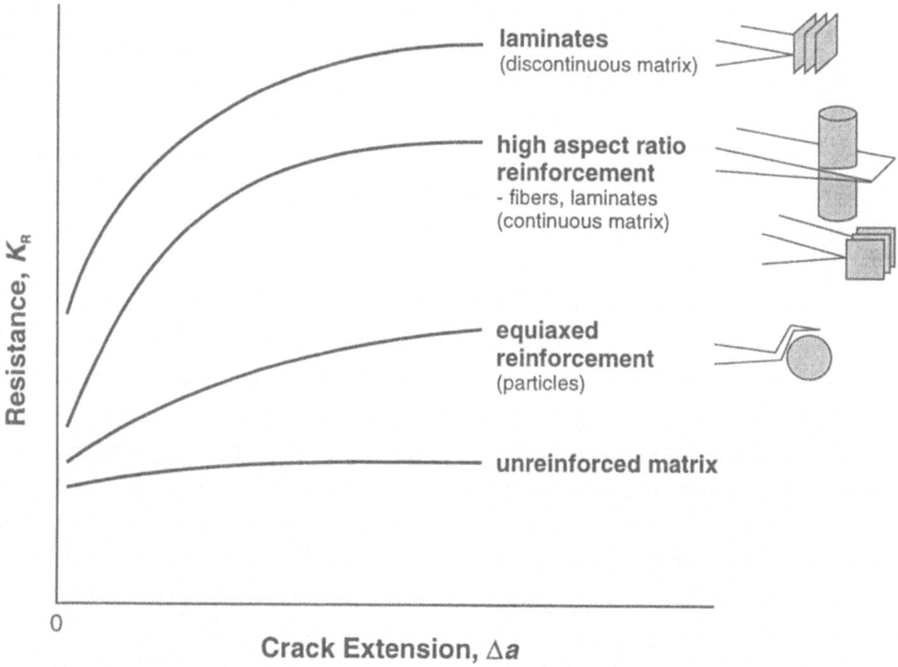

Figure 7. Schematic illustration of R-curve behavior for brittle-matrix composites reinforced with particulate, fibers or laminates as a ductile phase. Mechanisms that increase the crack-initiation toughness (where $\Delta a \rightarrow 0$) include crack trapping by particles and crack renucleation by laminates; corresponding mechanisms that increase the crack-growth toughness include crack bridging by intact ductile phases or uncracked ligaments.

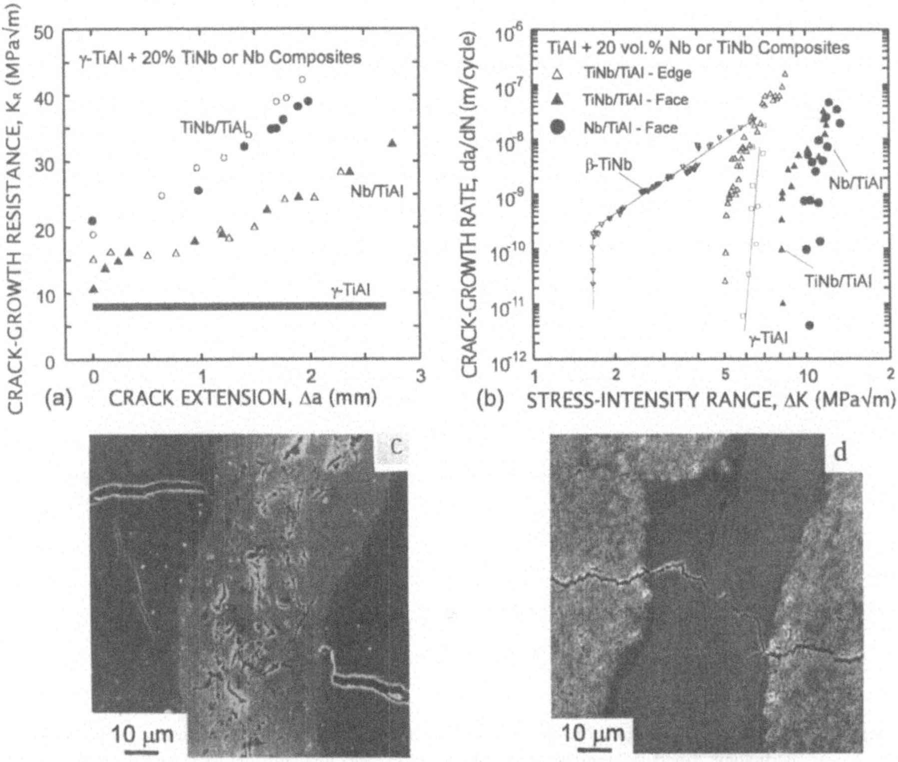

Figure 8. (a) Fracture toughness and (b) fatigue-crack growth behavior in a 20 vol.% TiNb-reinforced γ-TiAl composite at room temperature (at $R = 0.1$ with 25 Hz frequency), in the edge (C-R) and face (C-L) orientations; results are compared to data in pure γ-TiAl, β-TiNb and a Nb-particulate reinforced γ-TiAl composite. (c) Extensive crack bridging by the uncracked ductile phase under monotonic loading is severely degraded under cyclic loading due to (d) premature fatigue failure of the ductile ligaments [28].

Cyclic Fatigue Properties. Despite their effective role under monotonic loading, under cyclic loads, TiNb and Nb reinforcements result in only modest increases in the fatigue-crack growth resistance of γ-TiAl (Fig. 8b); moreover, this is achieved *only* for the C-L (face) orientation where faces of pan-cake shaped particles are oriented normal to the crack plane [28]. In fact, for the C-R (edge) orientation where particle edges are oriented perpendicular to the crack plane, the incorporation of the ductile particles actually *degrades* the fatigue-crack growth properties relative to monolithic γ-TiAl.

Such behavior in TiNb/TiAl is primarily due to the premature fatigue failure of ductile ligaments which restricts the development of a bridging zone and resultant shielding at the crack tip (Fig. 8d). In the face orientation, small shielding contributions from crack branching, matrix cracking and coplanar bridging within ~100-200 μm behind the crack tip do provide a moderate increase in fatigue resistance, although these dimensions are far below the bridging-length scales observed during monotonic loading. The Nb-particulate reinforced TiAl alloys does show somewhat better fatigue-crack growth resistance (Fig. 8b). This is because the Nb particles are more weakly bonded to the matrix (the reaction layer consists of the brittle σ phase) compared to the TiNb particles (which has a strong reaction layer consisting of ductile α_2+B_2 phases). The resulting delamination at the Nb particle interface acts to delay the fatigue failure of the Nb ligaments such that they are more effective in preserving the shielding from crack bridging under cyclic loads.

Thus, similar to ceramic materials, cyclic loading acts progressively to degrade the toughening associated with wake shielding, in this case through the premature fatigue failure of the ductile bridging ligaments. However, unlike most ceramics, there may be an intrinsic mechanism of fatigue-crack advance in many intermetallics. Some insight can be achieved by examining the separate dependencies of the growth rates on ΔK and K_{max}. In Fig. 9, fatigue-crack growth rates in the 20 vol.% TiNb/γ-TiAl alloy are shown at load ratios of $R = 0.1$, 0.5 and 0.7 as a function of ΔK and K_{max}. As noted previously, increasing R accelerates crack-growth rates for a given ΔK and lowers the fatigue

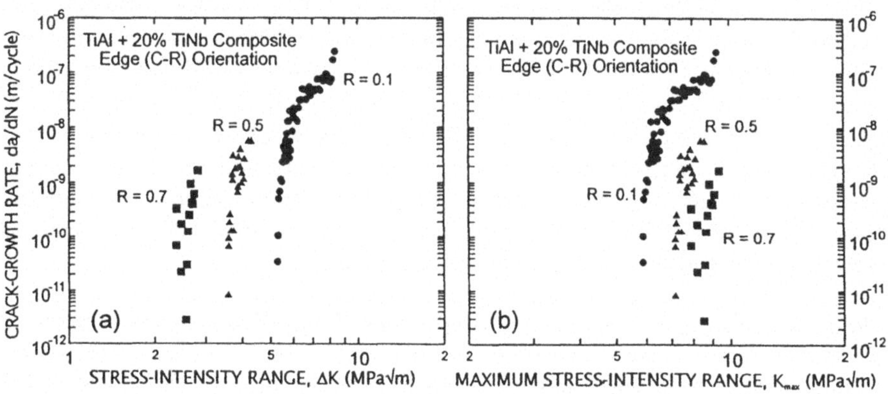

Figure 9. Fatigue-crack propagation rates in the TiNb/γ-TiAl composite at various load ratios, characterized in terms of (a) the applied stress-intensity range, ΔK, and (b) the maximum stress intensity, K_{max}. Note that, unlike the SiC ceramic in Fig. 5, K_{max} does not normalize the growth-rate data.

threshold. Measured crack-closure levels are quite substantial at $R < 0.4$ (closure stress intensities, K_{cl}/K_{min}, are ~0.4), suggesting that shielding is important at the lower load ratios [28]; however, marked differences in behavior between $R = 0.5$ and 0.7 imply that other factors are relevant. What is apparent is that the variation in growth rates with load ratio is not normalized by K_{max}, as was the case for ceramics (Fig. 5b), implying that cyclic crack growth in TiNb/TiAl composites is not governed solely by shielding mechanisms. This is also clear from examination of the respective dependencies on ΔK and K_{max} (Table 1); regression fits to the crack-growth data in Fig. 9 yield similar values for the ΔK (m ~ 6.3) and K_{max} (p ~ 6.1) exponents, implying a role of intrinsic fatigue damage in the composite. Such mechanisms of microstructural fatigue damage and cyclic crack advance are not understood in titanium aluminide intermetallics, but presumably are related to repeated blunting and resharpening of the crack in the TiAl and TiNb constituent phases, akin to behavior in ductile metals. In addition, static cleavage fracture modes are prevalent in the TiAl regions at higher growth rates, particularly at K_{max} levels approaching K_c of the γ-TiAl matrix.

3.1.2 *Behavior at Elevated Temperatures*

Fracture Toughness Properties. Ductile TiNb particles also enhance the fracture resistance of γ-TiAl composites at elevated temperatures, specifically at 650 and 800°C [30], although the measured toughness is somewhat lower at 650°C (Fig. 10a). Crack bridging is still prominent during monotonic loading at high temperatures, although the tractions imposed in the crack wake are lower due to a reduced particle strength [30]. However, this decrease in the efficiency of bridging is offset by an increase in intrinsic ductility and toughness of the TiAl matrix, especially above the ductile/brittle transition temperature for the γ-phase (~700°C) [24]; such competing effects account for the lower toughness of the composite at 650°C. However, ductile-phase toughening with a particulate phase such as TiNb or Nb may be severely compromised after prolonged exposure to elevated temperatures due to the poor oxidation resistance of the Nb-bearing phases. Protective coatings are thus essential to minimize environmental degradation if these materials are ever to find structural use.

Cyclic Fatigue Properties. At elevated temperatures, fatigue-crack growth is observed in the composite at ΔK levels between ~4-10 MPa√m (Fig. 10b), well below that needed to initiate crack growth under monotonic loading (where K_i ~ 12-18 MPa√m) [30]. Although growth rates are slightly faster at 650°C than at 25 and 800°C, the dependency on applied K remains high at all temperatures (n ~ 7-10). Similar to behavior at 25°C, the TiNb particles fail prematurely by transgranular shear in fatigue to leave virtually no bridging zone (Fig. 10d); such behavior is quite distinct from the creation of an extensive zone of uncracked TiNb ligaments under monotonic loading, which ultimately fail by microvoid coalescence (Fig. 10c).

As with behavior at ambient temperatures, a principal factor influencing fatigue-crack growth is again the cyclic-loading induced suppression of ductile-ligament bridging. However, because of the reduced efficiency of such bridging at high temperatures [30], the premature failure of the bridges in fatigue does not significantly accelerate crack-growth rates in the composite relative to unreinforced TiAl. Indeed, at 650°C, cyclic crack-growth rates are essentially identical in the monolithic and composite alloys.

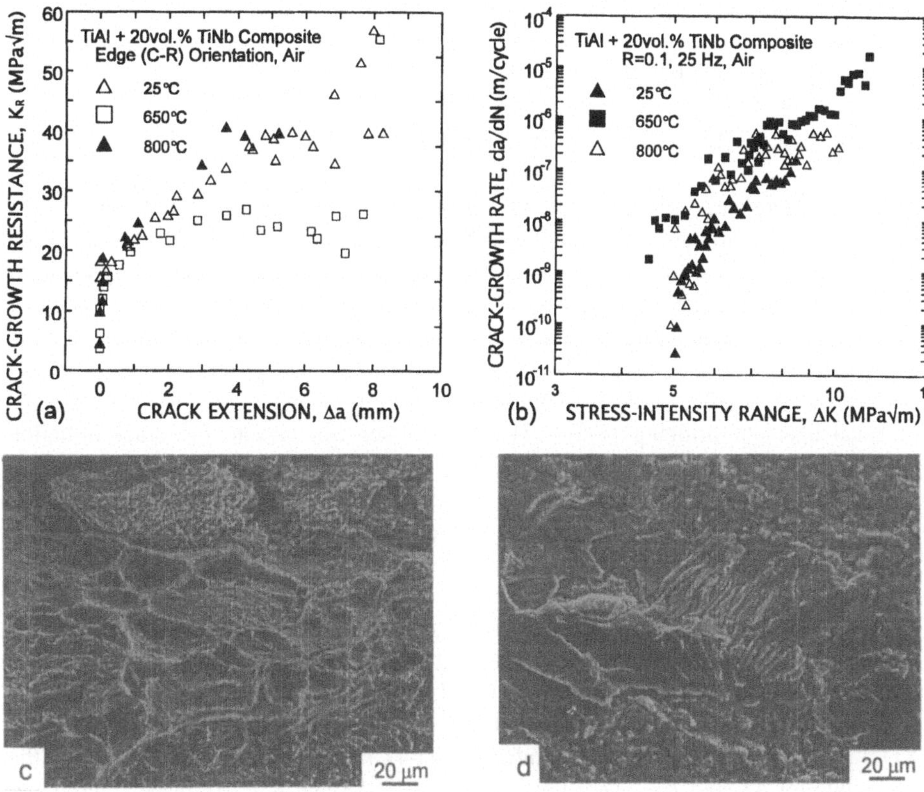

Figure 10. (a) Fracture toughness and (b) fatigue-crack growth behavior (at $R = 0.1$ with 25 Hz frequency), in the TiNb/γ-TiAl composite (edge (C-R) orientation) at various temperatures; corresponding fracture surfaces at 800°C, under (c) monotonic and (d) cyclic loading, are also shown. Note the clearly distinct (microvoid coalescence vs. transgranular shear) failure modes in the TiNb under monotonic and cyclic loading [30].

Thus in both ceramics and intermetallics, the extension of fatigue cracks is enhanced by the cycle-dependent degradation of bridging in the crack wake. There are significant differences though in the behavior of intermetallics. In TiNb/TiAl, fatigue-crack growth is seen at K levels far below the crack-initiation toughness, K_i, on the R-curve, specifically at a fatigue threshold of $K_{max,TH} \sim 0.25\text{-}0.4\ K_i$. This implies that in addition to the *extrinsic* effects of limited shielding under cyclic loading, there are *intrinsic* damage mechanisms uniquely associated with fatigue failure in intermetallics (Figs. 10c,d), as in metals. This is also evident in the differing dependencies of growth rates on ΔK and K_{max} in these material systems (Table 1). This is in sharp contrast to ceramics, where there are no intrinsic fatigue damage mechanisms at low homologous temperatures and crack advance occurs by identical mechanisms under monotonic and cyclic loads, such that $K_{max,TH} \sim K_i$.

3.2 MONOLITHIC TITANIUM ALUMINIDE ALLOYS

Fracture Toughness Properties. An alternative approach to toughening TiAl intermetallics is through microalloying, with small amounts of Cr, V, Mo and Nb, and thermomechanical treatment; this yields two-phase microstructures comprising lamellar colonies of alternating of γ and α_2 (Ti_3Al) layers [25-27]. Improved toughness, especially in coarse-lamellar structures, results from crack deflection and microcracking along interlamellar boundaries ahead of the crack tip; this generates a crack wake bridged by intact lath colonies undergoing shear deformation (referred to as uncracked- or shear-ligament bridging [27]). These processes can elevate the crack-initiation toughness of TiAl alloys to ~15-20 MPa√m; the associated wake shielding can result in a rising R-curve with maximum toughness values as well over 30 MPa√m (Fig. 11a), depending on the colony and lath size, and the specimen dimensions (where colony sizes are of the order of several mm) [31-33].

Cyclic Fatigue Properties. Fatigue-crack growth characteristics in lamellar TiAl intermetallic alloys at ambient and elevated temperatures show similar features to those noted in the TiAl composites [31-33]. Crack-propagation rates in ($\gamma+\alpha_2$) TiAl alloys show a high dependency on the applied K ($n \sim 16$); specifically, in a nearly-lamellar XD^{TM} processed 1 vol.% TiB_2-reinforced Ti-47.7Al-2Nb-0.8Mn (at.%) alloy, the exponents for ΔK and K_{max} were found to be $m = 10.3$ and $p = 5.6$, respectively [32]; i.e., large compared to metals yet with a higher ΔK dependence than in the TiAl composites. As with the toughness properties, the coarser-grained lamellar microstructures in general display the best crack-growth properties (Fig. 11b) [31-35]. Cyclic degradation of shear-ligament bridges is the primary extrinsic mechanism influencing crack propagation, yet intrinsic damage and crack-advance mechanisms are clearly active.

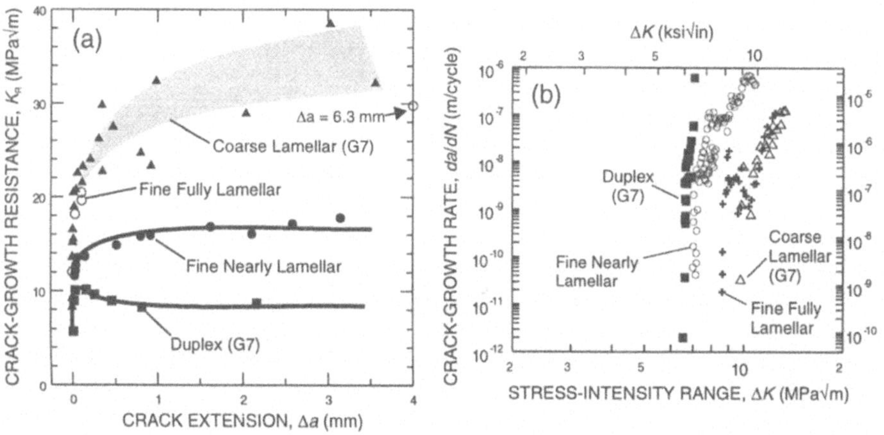

Figure 11. (a) Fracture toughness and (b) fatigue-crack growth behavior (at $R = 0.1$ with 25 Hz frequency), in monolithic ($\gamma+\alpha_2$) TiAl alloys at ambient temperatures with fine nearly lamellar, fine fully lamellar, coarse lamellar and duplex microstructures, showing the superior crack-growth resistance of the coarse lamellar structure under both monotonic and cyclic loading [32].

Fatigue behavior at elevated temperatures is considerably more complex, as growth rates are faster at ~650°C than at 25° or 800°C [33,34]. Moreover, at the higher temperature, extensive oxide formation, comparable in thickness to the crack-tip opening displacements points to significant shielding at 800°C from oxide-induced crack closure [35].

3.3 MOLYBDENUM DISILICIDE COMPOSITES

Fracture Toughness Properties. Materials based on the $MoSi_2$ intermetallic compound have attractive properties for structural use above 1200°C, yet like other ceramics and intermetallics, they suffer from low ductility and toughness at ambient temperatures [e.g., 3,36],. Ductile-phase toughening using high-aspect ratio reinforcements [37,38] has been a successful approach to increase the intrinsic toughness (K_c ~ 4 MPa√m) of $MoSi_2$. For example, using 20 vol.% Nb short fiber (wire) reinforcements, $MoSi_2$ displays R-curve behavior with the crack-growth resistance rising from ~5 MPa√m at the onset of cracking to approximately 14 MPa√m after ~1.5 mm of stable crack extension (Fig. 12a) [38]. Crack deflection, renucleation and crack bridging by ductile Nb phase are the principal mechanisms responsible for the higher toughness.

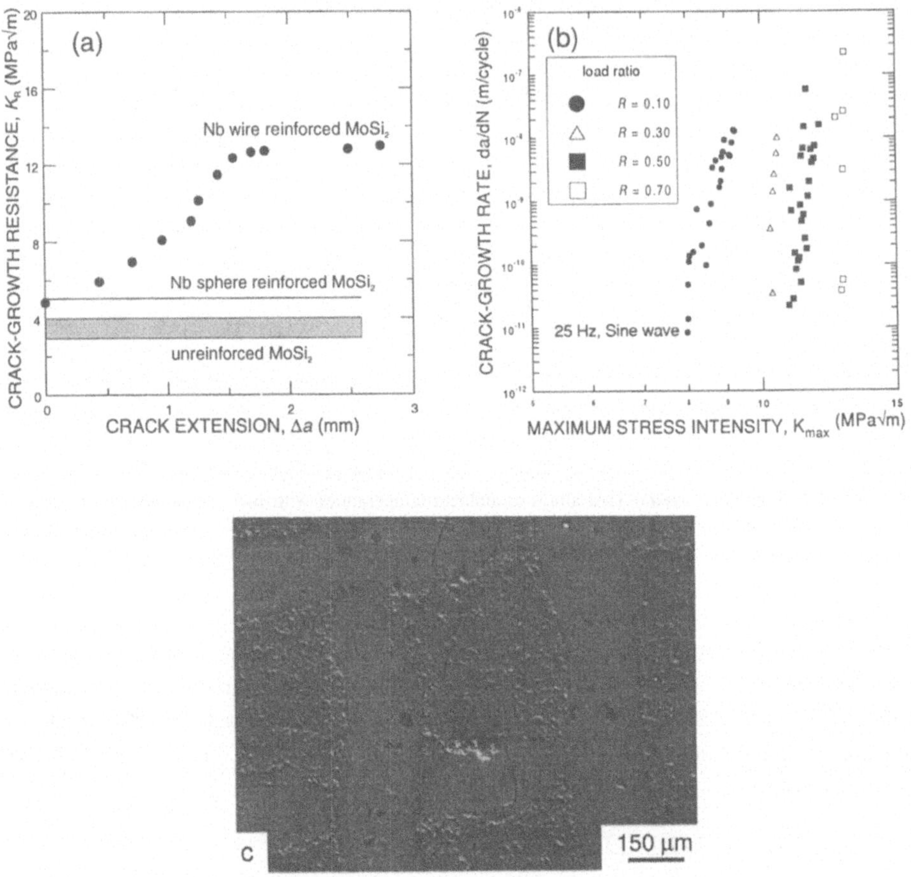

Figure 12. (a) Fracture toughness and (b) fatigue-crack growth behavior (at $R = 0.1$, 0.3, 0.5 and 0.7 with 25 Hz frequency) in a Nb-wire reinforced $MoSi_2$ composite at room temperature. Whereas high-aspect ratio Nb particles induce greater toughening than Nb particulate due to extensive crack bridging, such bridging is progressively degraded in fatigue. However, the loss of bridging in fatigue is compensated by marked contributions from crack deflection and closure, as shown in (c) by the meandering fatigue crack path along the reaction-layer interface. (The crack-growth direction is from top to bottom). Note that growth rates at varying R are not normalized by K_{max} [38].

Cyclic Fatigue Properties. While generally similar to behavior in TiAl alloys, the features of fatigue-crack propagation in the $Nb_f/MoSi_2$ composites are more "ceramic-like", primarily because no intrinsic fatigue damage is apparent in pure $MoSi_2$ [38]. Correspondingly, growth-rate exponents are high (n ~20), with the K_{max} dependence ($p = 13.2$) being considerably larger than the ΔK dependence ($m = 7.5$), although unlike ceramics, load-ratio effects are not normalized by K_{max} (Fig. 12b). Of interest, however, is that although bridging by the uncracked Nb phase is progressively degraded under cyclic loading, the fatigue-crack growth properties of the composite are not severely affected as other mechanisms of shielding are developed [38]. These arise principally from extensive crack deflection (at angles of up to 70°) along the weak $Nb/MoSi_2$ reaction-layer interface (Fig. 12c); this in turn promotes shielding in fatigue from premature contact and wedging of fracture surface asperities (roughness-induced crack closure [5]), akin to behavior in many structural materials [1,5].

4. Design and Life Prediction

The marked sensitivity of fatigue-crack growth rates to the applied stress intensity in intermetallics and ceramics, both at elevated and especially ambient temperatures, presents unique challenges to damage-tolerant design and life-prediction methods for structural components fabricated from these materials. For safety-critical applications involving most metallic structures, such procedures generally rely on the integration of data relating crack-growth rates (*da/dN* or *da/dt*) to the applied stress intensity (ΔK or K_{max}) in order to estimate the time or number of cycles, N_f, to grow the largest undetectable initial flaw, a_i, to critical size a_c, viz:

$$N_f = \frac{2}{(n-2)C'\,(Q\,\Delta\sigma)^n\,\pi^{n/2}}\left[a_o^{(2-n)/2} - a_c^{(2-n)/2}\right] \tag{3}$$

where $\Delta\sigma$ is the applied stress range, Q is the geometric factor; C' and n ($\neq 2$) are the scaling constant and exponent in the crack-growth relationship in Eq. 2. For advanced materials such as intermetallics and ceramics, this approach may be difficult to implement in practice due to the large values of the exponent n; since the projected life is proportional to the reciprocal of the applied stress raised to the power of n, a factor of two change in applied stress can lead to projections of the life of a ceramic component (where n can be as high as 20 or more) to vary by more than six orders of magnitude. Essentially, because of the high exponents, the life spent in crack propagation in advanced materials will be extremely limited or infinitely large, depending upon whether the initial stress intensity is above or below the fatigue threshold.

Accordingly, a more appropriate approach may be to design on the basis of a threshold below which fatigue failure cannot occur; this may involve a conventional fatigue-crack initiation threshold or fatigue limit determined from stress-life (S/N) data or, more conservatively, the ΔK_{TH} or $K_{max,TH}$ thresholds for no crack growth. However, even these approaches may not be conservative due to uncertainties in the definition of such thresholds. This is particularly important in many advanced materials as their growth-rate and S/N curves are not sigmoidal and correspondingly show no evidence of a threshold (Fig. 13). Moreover, as with metallic materials [e.g., 39,40], there is now increasing evidence for small-crack effects in both ceramics and intermetallics, where cracks of a size comparable with the scale of microstructure, the extent of local inelasticity ahead of the crack tip, or with the extent of shielding in the wake of the tip, can propagate at applied stress intensities *below* the "long-crack" ΔK_{TH} or $K_{max,TH}$ thresholds [e.g., 18]. Clearly for flaw-sensitive materials such as ceramics and intermetallics to be safely used in fatigue-critical applications, the critical levels of damage necessary for the *onset* of fatigue failure must be defined; whether this involves an S/N fatigue limit, long- or small-crack thresholds, or more refined statistical analyses is as yet still unclear.

5. Concluding Remarks

Major progress has been made over the last ten to fifteen years in significantly improving the fracture resistance of low-ductility materials such as ceramics and intermetallics using the extrinsic shielding approach to toughening. This has included the design of microstructures which develop zones of inelasticity, microcracking or bridging (by grains, particulate or fibers) that surround the crack. However, it is now evident that in most instances, cyclic loading can severely degrade such shielding; in fact, this provides the critical mechanism promoting fatigue-crack growth, even in materials such as ceramics which display no intrinsic mechanisms of cyclic crack advance.

Between the various classes of materials, fatigue-crack growth rate data suggest that the general dependency of growth rates on stress intensity (i.e., *n*) increases, and the fatigue threshold decreases, with decreasing ductility (Fig. 13). Indeed, whereas ΔK_{TH} thresholds in metals can be as small as $0.1K_c$, in the most brittle ceramics, $\Delta K_{TH} \sim K_c$. However, there is a certain commonality that can be deduced through a general consideration of which intrinsic and extrinsic mechanisms are active (e.g., Fig. 1) and their relative dependence on mean or alternating loads:

- In ductile metals (Fig. 3a), crack advance is promoted by an intrinsic mechanism ahead of the tip (unique to cyclic fatigue), for example, by alternating crack-tip

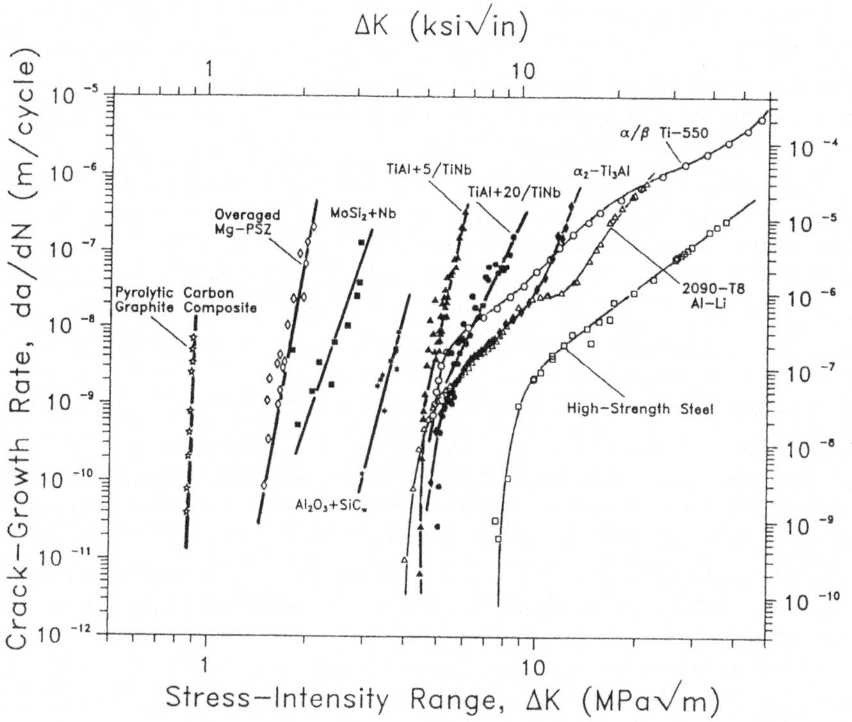

Figure 13. A compendium of cyclic fatigue-crack propagation data, *da/dN*, as a function of the applied stress-intensity range, ΔK, in metallic alloys, monolithic and composite intermetallics, and monolithic and composite ceramics.

blunting and resharpening. Since this is primarily controlled by the alternating plastic strain, the primary dependence is on ΔK, although a smaller intrinsic K_{max} dependence may arise due to the occurrence of static fracture modes, particularly as K_{max} approaches K_c. Crack advance is impeded by extrinsic shielding behind the crack tip, which in metals is primarily associated with crack wedging mechanisms (crack closure). Since the degree of wedging depends critically on the relative size of the maximum crack-tip opening displacements, this leads to a much smaller dependency on K_{max}, i.e., $p \ll m$.

- In brittle ceramics (Fig. 3b), crack advance is promoted by a mechanism ahead of the tip that is identical to that occurring under static loading, e.g., intergranular cracking. Since this is controlled by the maximum load, the primary dependence is on K_{max}. Crack advance is once again impeded by extrinsic shielding behind the crack tip, which in most ceramics and their composites is associated with crack bridging; however, as cyclic loading acts progressively to diminish the potency of such wake shielding, this leads to a much smaller dependency on ΔK, i.e.,. $p \gg m$.

- In intermetallics, fatigue properties are intermediate between the extremes of "metal-" and "ceramic-like" behavior, i.e., $p \sim m$. In brittle intermetallics such as $MoSi_2$, where there is no intrinsic cycle-dependent crack advance mechanism, $p > m$. In more ductile materials, such as the $(\gamma+\alpha_2)$ TiAl alloys, intrinsic fatigue damage mechanisms, similar to those in metals, clearly exist, i.e., $p < m$. However, in most intermetallics, extrinsic ("ceramic-like") crack-bridging mechanisms in the wake are generally degraded under cyclic loads, although the fatigue resistance is often not compromised as they can be replaced by the development of ("metal-like") crack-closure mechanisms.

It should be noted that the existence of active wake-shielding mechanisms such as crack bridging or crack closure has other significant implications for fatigue and fracture behavior. Clearly, there will be a loss of similitude between cracks of varying size if considerations are based on the applied (or nominal) stress intensities. Accordingly, since small cracks will have more limited wakes, they are likely to experience a higher *local* driving force due to reduced shielding. This results in R-curve behavior under monotonic loading and the well known sub-threshold growth of small cracks in cyclic fatigue. In addition, since shielding can have little influence on crack initiation, it is to be expected that microstructures designed for optimal crack-growth resistance may be quite different to those designed for optimal resistance to crack initiation or small-crack growth (see, for example, behavior in dual-phase steels [41]).

Finally, the marked sensitivity of growth rates to the applied stress intensity in ceramics and intermetallics implies that projected lifetimes will be a very strong function of stress and crack size; this makes design and life prediction using damage-tolerant methodologies extremely difficult. Accordingly, design approaches based on the definition of a critical level of damage for the *onset* of fatigue cracking must be contemplated, involving either an S-N fatigue limit or crack-propagation threshold, although even these approaches may not be conservative due to the sub-threshold extension of small cracks. Clearly, this is an area that mandates increased attention in the future if these materials are to find widespread structural use.

6. Acknowledgments

This work was supported by the U.S. National Science Foundation under Grant No. DMR-9522134 (for high temperature studies on ceramics), the Office of Energy Research, Office Basic Energy Sciences, Materials Sciences Division of the U.S.

Department of Energy under Contract No. DE-AC03-76SF00098 (for research on SiC), and the U.S Air Force Office of Scientific Research under Grant Nos. F49620-93-1-0107 and 0289 (for studies on intermetallics). Our particular thanks are due to C. J. Gilbert for his work on SiC, to K. Badrinarayanan, J. P. Campbell and A. L. McKelvey for their work on intermetallics, and to Professors G. R. Odette (University of California, Santa Barbara) and L. C. DeJonghe (University of California, Berkeley) for their collaboration and helpful discussion.

7. References

1. Ritchie, R. O. (1988) Mechanisms of fatigue crack propagation in metals, ceramics and composites: role of crack-tip shielding, *Mater. Sci. Eng. A* **A103**, 15-28.
2. Evans, A. G. (1990) Perspective on the development of high-toughness ceramics, *J. Am. Ceram. Soc.* **73**, 187-206.
3. Liu, C. T., J. O. Stiegler, and F. H. Froes (1991) Ordered intermetallics, *Metals Handbook, vol. 2*, ASM Intl., Materials Park, OH, p. 913.
4. Elber, W. (1970) Fatigue crack closure under cyclic tension, *Eng. Fract. Mech.* **2**, 37-45.
5. Suresh, S. and R. O. Ritchie (1984) Near-threshold fatigue crack propagation: a perspective on the role of crack closure, in D. L. Davidson and S. Suresh (eds.), *Fatigue Crack Growth Threshold Concepts*, TMS-AIME, Warrendale, PA, pp. 227-261.
6 Ritchie, R. O. and R. H. Dauskardt (1991) Cyclic fatigue of ceramics: a fracture mechanics approach to subcritical crack growth and life prediction, *J. Ceram. Soc. Japan* **99**, 1047-1062.
7. Lathabai, S., J. Rödel, and B. Lawn (1991) Cyclic fatigue from frictional degradation at bridging grains in alumina, *J. Am. Ceram. Soc.* **74**, 1340-1348.
8. Dauskardt, R. H. (1993) A frictional-wear mechanism for fatigue-crack growth in grain-bridging ceramics, *Acta Metall. Mater.* **41**, 2765-2781.
9. Rouby, D. and P. Reynaud (1993) Fatigue behaviour related to interface modification during load cycling in ceramic-matrix fibre composites, *Comps. Sci. Technol.* **48**, 109-118.
10. Laird, C. and G. C. Smith (1962) Crack propagation in high stress fatigue, *Philos. Mag.* **7**, 847-857.
11. Neumann, P. (1969) Coarse slip model of fatigue, *Acta Metall.* **17**, 1219-1225.
12. Cao, J. J., W. J. MoberlyChan, L. C. DeJonghe, C. J. Gilbert, and R. O. Ritchie (1996) *In situ* toughened silicon carbide with Al-B-C additions, *J. Am. Ceram. Soc.* **79**, 461-469.
13. Gilbert, C. J., J. J. Cao, W. J. MoberlyChan, L. C. DeJonghe and R. O. Ritchie (1996) Cyclic fatigue and resistance-curve behavior in an *in situ* toughened silicon carbide with Al-B-C additions, *Acta Mater.* **44**, in press.
14. Srinvasan, M. (1989) The silicon carbide family of structural ceramics, in J. B. Wachtman, Jr. (eds.), *Structural Ceramics*, Treatise on Materials Science and Technology, vol. 29, Academic Press, New York, NY, pp. 99-159.
15. VanStone, R. H. (1988) Residual life prediction methods for gas turbine components, *Mater. Sci. Eng. A* **A103**, 49-61.
16. Kishimoto, H. (1991) Cyclic fatigue in ceramics, *JSME Intl. J.* **34**, 393-403.
17. Jacobs, D. S. and I.-W. Chen (1995) Cyclic fatigue in ceramics: a balance between shielding accumulation and degradation, *J. Am. Ceram. Soc.* **78**, 513-520.

18. Dauskardt, R. H., M. R. James, J. R. Porter, and R. O. Ritchie (1992) Cyclic fatigue-crack growth in SiC-whisker-reinforced alumina ceramic composite: long and small-crack behavior, *J. Am. Ceram. Soc.* **75**, 759-771.
19. Gilbert, C. J., R. H. Dauskardt, R. W. Steinbrech, R. N. Petrany, and R. O. Ritchie (1995) Cyclic fatigue in monolithic alumina: mechanisms for crack advance promoted by frictional wear of grain bridges," *J. Mater. Sci.* **30**, 643-654.
20. Gilbert, C. J., R. H. Dauskardt, and R. O. Ritchie (1995) Behavior of cyclic fatigue cracks in monolithic silicon nitride, *J. Am. Ceram. Soc.* **78**, 2291-2300.
21. Ramamurty, U., T. Hansson, and S. Suresh (1994) High-temperature crack growth in monolithic and SiC_w-reinforced silicon nitride under static and cyclic loads, *J. Am. Ceram. Soc.* **77**, 2985-2999.
22. Liu, S-Y., I-W. Chen, and T-Y. Tien (1994) Fatigue crack growth of silicon nitride at 1400°C: a novel fatigue-induced crack-tip bridging phenomenon, *J. Am. Ceram. Soc.* **77**, 137-142.
23. Elliott, C. K., G. R. Odette, G. E. Lucas, and J. W. Sheckherd (1988) Toughening mechanisms in intermetallic γ-TiAl alloys containing ductile phases, in F. D Lemkey, A. G. Evans, S. G. Fishman, and J. R. Strife (eds.), *High-Temperature/High-Performance Composites*, Materials Research Society, Pittsburgh, PA, vol. 120, pp. 95-101.
24. Lipsitt, H. A., D. Schectman and R. E. Schafrik (1975) The deformation and fracture of TiAl at elevated temperatures, *Metall. Trans. A* **6A**, 1991-1996.
25. Kim, Y. W. and D. M. Dimiduk (1991) Progress in understanding gamma titanium aluminides, *J. Metals* **43**:8, 40-47.
26. Dève, H. E., A. G. Evans, and D. S. Shih (1992) A high-toughness γ-titanium aluminide, *Acta Metall. Mater.* **40**, 1259-1265.
27. Chan, K. S. and Y.-W. Kim (1992) Influence of microstructure on crack-tip micromechanics and fracture behaviors of a two-phase TiAl alloy, *Metall. Trans. A* **23A**, 1663-1677.
28. Venkateswara Rao, K. T., G. R. Odette, and R. O. Ritchie (1994) Ductile-reinforcement toughening in γ-TiAl intermetallic-matrix composites under monotonic and cyclic loading: effect on fracture and fatigue-crack propagation resistance, *Acta Metall. Mater.* **42**, 893-911.
29. Ashby, M. F., F. J. Blunt, and M. Bannister (1989) Flow characteristics of highly constrained metal wire, *Acta Metall. Mater.* **37**, 1847-1857.
30. Venkateswara Rao, K. T. and R. O. Ritchie (1995) Toughness and fatigue-crack growth in γ-TiAl intermetallic composites at ambient and high temperatures, in W. O. Soboyejo, T. S. Srivatsan, and R. O. Ritchie (eds.), *Fatigue and Fracture of Ordered Intermetallics and Compounds II*, TMS, Warrendale, PA, pp. 327-338.
31. Venkateswara Rao, K. T., Y. W. Kim, C. L. Muhlstein, and R. O. Ritchie (1995) Fatigue-crack growth and fracture resistance of duplex and lamellar microstructures in a two-phase ($\gamma+\alpha_2$) TiAl intermetallic alloy, *Mater. Sci. Eng. A* **A192/193**, 474-482.
32. Campbell, J. P., K. T. Venkateswara Rao, and R. O. Ritchie (1996) Fatigue-crack growth and fracture behavior in a XD™ γ-TiAl alloy with a fine lamellar microstructure, in G. Lütjering and H. Nowack (eds.), *Fatigue '96*, Proc. Sixth Intl. Congress on Fatigue, ed., Pergamon Press, New York, NY, vol. III, pp. 1779-1784.
33. Balsone, S. J., J. M. Larsen, D. C. Maxwell, and J. Wayne Jones (1995) Effects of microstructure and temperature on fatigue crack growth in a TiAl alloy Ti-46.5Al-3Nb-2Cr-0.2W, *Mater. Sci. Eng. A* **A192/193**, 457-464.
34. Venkateswara Rao, K. T., Y. W. Kim, and R. O. Ritchie (1995) High-temperature fatigue-crack growth behavior in a two-phase ($\gamma+\alpha_2$) TiAl intermetallic alloy, *Scripta Metall. Mater.* **33**, 459-465.

35. McKelvey, A. L., J. P. Campbell, K. T. Venkateswara Rao, and R. O. Ritchie (1966) High temperature fatigue-crack growth behavior in an XD™ γ-TiAl intermetallic alloy, in G. Lütjering and H. Nowack (eds.), *Fatigue '96*, Proc. Sixth Intl. Congress on Fatigue, ed., Pergamon Press, New York, NY, vol. III, pp. 1743-1748.

36. Petrovic, J. J. and A. K. Vasudévan (1992) A perspective on $MoSi_2$ based composites, in D. B. Miracle, D. L. Anton, and J. A. Graves (eds.), *Intermetallic Matrix Composites II*, Materials Research Society, Pittsburgh, PA, vol. 273, pp. 229-239.

37. Pickard, S. M. and A. K. Ghosh (1995) Toughened microstructures for ductile phase reinforced molybdenum disilicide, in J. Horton, I. Baker, S. Hanada, R. D. Noebe, and D. S. Schwartz (eds.), *High-Temperature Ordered Intermetallic Alloys VI*, Materials Research Society, Pittsburgh, PA, vol. 364, pp. 905-910.

38. Badrinarayanan, K., A. L. McKelvey, K. T. Venkateswara Rao, and R. O. Ritchie (1996) Fracture and fatigue-crack growth behavior in ductile-phase toughened molybdenum disilicide: effects of niobium wire vs. particulate reinforcements, *Metall. Mater. Trans. A* **27A**, in press.

39. Ritchie, R. O. and J. Lankford, eds. (1986) *Small Fatigue Cracks*, TMS-AIME, Warrendale, PA.

40. Miller, K. J. and E. R. de los Rios, eds. (1992) *Short Fatigue Cracks*, Mech. Eng. Publ. Ltd., London, U.K.

41. Shang, J.-K., J.-L. Tzou, and R. O. Ritchie (1987) Role of crack-tip shielding in the initiation and growth of long and small fatigue cracks in composite microstructures, *Metall. Trans. A* **18A**, 1613-1627.

FATIGUE PROBLEMS IN TRANSPORT APPLICATIONS

RODERICK A. SMITH
Royal Academy of Engineering British Rail Research Professor
Department of Mechanical and Process Engineering
The University of Sheffield
Mappin Street
Sheffield S1 3JD
UK

1. Introduction

In 1981, the author prepared a brief review of the history of the development of fatigue and fracture problems. That review is summarised as Appendix 1. The conclusions were:

- many practical problems arise in transport applications
- stress concentrating features (notches and joints) are important in real problems
- there is a return from small specimen testing to full-scale structural testing, an increasing interest in low cycle fatigue, corrosion effects and realistic loading patterns
- crack propagation can be reasonably well quantified, but crack initiation is still difficult to deal with quantitatively
- there is a wide gulf between microscopic, mechanistic understanding and practical macroscopic engineering design
- weight saving will result in more fatigue/fracture problems in the future

With the benefit of now more than fifteen years hindsight, I would not significantly change those overall conclusions. Perhaps the most significant developments in this period have been first the advent of cheap, local computing capabilities of enormous power and the advent of useful software, particularly accessible finite element stress analysis programs with powerful graphical interfaces. Secondly, microelectronics have been introduced into many engineering applications, replacing mechanical parts, but introducing new kinds of maintenance problems.

With regard to the understanding of fatigue and fracture, my conclusions have hardly changed despite the large increase in the number of publications on these topics. The key problem seems to be the dissemination of existing knowledge from research to practical engineering design.

R.A. Smith (ed.), Reliability Assessment of Cyclically Loaded Engineering Structures, 405–418.

The purpose of this paper is to review fatigue and fracture problems in passenger and freight transport, excluding the bulk transport of material in pipelines. In all transport cyclic loadings occur in a variety of ways: from rotating parts e.g. wheels, axles, pistons and crankshafts, from contact with the environment e.g. bumpy roads, dips in rails, gusts in the air and from starting/shutdown operations e.g. pressurisation once per flight, thermal loads in engines. Additionally ancillary equipment operating on vehicles is subject to vibration and its own internal operation cycles. The increase in the application of electronic components has led to a generic class of a new kind of failure with attendant reliability problems.

Land based transport is, of course, particularly susceptible to loadings from the wheel/running surface interface and arguably, the biggest problems occur with deterioration of the infrastructure. Bridges and viaducts deteriorate with use and the prediction of remaining safe lives is an acute problem. Costs of road repairs include large sums attributable to delays caused to traffic. Broadly speaking, fatigue of the major structural components of automobiles is not a major problem because lives are generally short, about 200,000 km or 10 years and have been limited by corrosion until recent improvements. Particular components in braking or steering gear do occasionally fail and sometimes prompt expensive recalls. Trains have much longer lifetimes, but are generally over-designed. Axles and rails occasionally cause difficulties despite over 150 years of effort.

Ships are very large welded flexible structures subjected to the joint attack of corrosion and fatigue. The poor safety record of bulk carriers gave rise to international concern in the 1980's and led to a tightening of inspection standards. Over 118 large ships have been lost since 1990, with 587 crew deaths. The design of large cargo ships remains a problem.

Superficially, aircraft are 'fatigue machines' and the consequences of failure of critical components are severe. Nevertheless, much attention has been given to the problem and many advances in our understanding of fatigue design and maintenance have come from the aircraft industry, often from the study of failures. Structural fatigue is a much smaller cause of accidents than human error, but several well publicised accidents have occurred through the failure of minor components such as bolts, thus emphasising the local nature of fatigue.

The paper will conclude by summarising the issues raised by efforts to maximise the life of ageing equipment. Economic forces have emphasised the need for such activities and have raised many challenges to our knowledge of the effects of cyclic service loads.

2. Fatigue Knowledge - Current Status

It is not the purpose of this paper to review the current status of fatigue knowledge. Several authoritative reviews have been published in recent years: Miller [1] discusses the central role of cracks, Nelson and Sheppard [2] review life estimation methods, the proceedings of a recent conference, "Fatigue '96", [3], contain many reports of current research problems.

It is sufficient for present purposes to understand that the fatigue life of a structure or component can be, somewhat arbitrarily, divided into various stages: the crack initiation period and the macro crack growth period, with the former divided into a nucleation and a micro growth phase. The size scale over which these events occur is a central issue. Figure 1 identifies some key orders of magnitude: the range of size scales involved is high and shown in a logarithmic scale. A bridge might be several kilometres long, a large ship 1/3km long, an aeroplane say 50m and a car 3m. Vital components in these structures may only be millimetres long, but the initiation and growth phases of fatigue might occupy the majority of the fatigue life before any cracks are detectable. Defects introduced during manufacture can erode the nucleation and macro crack stages, such that the fatigue life of a large welded structure may be determined by macro crack growth from an initial defect of several mm length, but at the other end of the spectrum the life of a 'clean' steel ball bearing will be dominated by nucleation and micro crack growth.

The reality is that this range of sizes causes many practical problems and in any particular case, the identification of the dominant size of the problem should be carefully identified, so that the appropriate theoretical or experimental prediction technique can be identified.

It is worth recalling that in general, we make tests on laboratory samples of typically mm size scale and use data so generated to make predictions of internal events in our structures, perhaps of a metre size scale, which occur at a micron size scale or even less. It should come as no surprise to learn that such extrapolations are fraught with difficulties!

2.1. OVERVIEW OF DESIGN METHODOLOGIES

The design techniques for transport equipment with regard to fatigue can be divided into several categories, following the suggestion of Schijive [4]:

Cracks Should Not Occur

1a Fatigue Crack Nucleation Not Allowed. Infinite life required, so load cycles in service should not exceed the fatigue limit. Many parts of engines such as crankshafts fall into this category.

1b Defects Should Not Grow. Defects are often unavoidable, such as scratches from machining, nicks from impact damage, weld defects and corrosion pits. In many components there is a requirement that such defects should not grow into the macro-size range, e.g. railway axles.

Cracks Will Occur

2a Incidental Fatigue Failures Are Acceptable. A finite life of such parts can occur in service, economy and safety are not directly involved, but a sufficient economical life must be guaranteed. Many small parts fall into this group, e.g. small return springs on ashtrays, knobs, often made of plastics and switches in automobiles. Considerable customer irritation impacting on overall quality perception is generated by this type of failure.

2b Sufficient Fatigue Life Without Inspections Required. Inspections are sometimes undesirable for economic reasons, being time consuming, interrupt operation or impracticable by lack of accessibility. A *safe-life* philosophy is adopted and large safety factors are usually applied, e.g. the structures of cars or train bodies.

2c Sufficient Crack Growth Life Required for Crack Detection by Inspection. Inspections for fatigue cracks are generally acceptable if safety or large economic consequences of a fatigue fracture are involved, e.g. aircraft hulls.

3. Land Based Transport

3.1. THE AUTOMOBILE - A TARNISHED GLORY?

That the automobile has liberated and enriched the opportunities which stem from the ability of the ordinary man to travel easily is a glorious achievement of this century's engineering and production technology. However, its very success in multiplying in an exponential fashion and concerns at the level of pollution produced by the internal combustion engine, have led many people to realise that the unbridled expansion of road transport cannot continue.

The fact that cars are produced in such large numbers, has led to several important characteristics of their fatigue design. Large amounts of money are available for research, development and testing programmes: the car is a mature product which has developed steadily in an evolutionary way: new modifications can be tested in prototypes and operated well ahead of actual service. As a result, one might suggest that fatigue is not a major constraint to the automobile industry - major components in engines and suspensions rarely fail. Nevertheless, considerable attention is paid to fatigue design and many practical advances in general areas such as cumulative damage due to random loadings, have stemmed from the work of automobile manufacturers, [5]. Pressures to reduce the weight of cars have been a recent challenge - partly met by the introduction of lighter materials (high strength low alloy steels, plastics and composites), but also by the use of increasingly sophisticated finite element stress analysis which has allowed excess lightly stressed mass to be shaved from components made from established materials. Although the lifetime of cars is relatively short, corrosion of structural parts has been a limiting factor, but recent improvements have helped to overcome this limitation.

The levels of maintenance required for modern cars are astonishingly low compared with even 25 years ago: a lesson which could be learned and applied to other areas of the transport industry. An element of concern is however the increasing electronic sophistication of modern vehicles which in many cases defeats simple diagnosis and repair when faults occur. Although solid state and 'chip' based electronics are generally reliable, a major source of failure is the thermal fatigue of soldered connections; an area which is now the subject of active research worldwide and a generic problem to equipment used in fields of transport.

3.2. INFRASTRUCTURE DAMAGE CAUSED BY HEAVIER ROAD VEHICLES

Vehicles cause damage to the surface of the road on which they pass. Continual passage of traffic causes cyclic loading, the damage accumulates and eventually the road surface breaks up and needs to be repaired. The repair costs, and the costs associated with delays to traffic whilst repairs are made are huge. The UK spent 1.17% of its GNP on roads in 1990, of which 46% was spent on maintenance. Despite the obvious economic consequences, the fatigue relationships between traffic quantity, axle loads and road surface lifetime are little understood. Cebon [6] in an extensive review, but concentrating on heavy vehicles, suggests a fourth power law between axle load and damage is essentially a rule of thumb, the relationship may be even higher, but in any case illustrates the high sensitivity of damage to the axle load. Now in recent years the overall weights and axle loads of heavy lorries have increased considerably. Forty four tonne lorries are common place through Europe, as are the sections of highway under repair caused by their passage, and the trend is to even higher weights, Sweden for example allows 60 tonne vehicles.

Bridges too are subject to cyclic loadings from traffic - on longer bridges the total weight of the vehicle is clearly more important than the axle load. The severity of this problem can be seen for a simple calculation. If a bridge was designed for lorries of ten tonnes and a fifty year life, then for the same number of lorries of forty four tonnes, the life will be reduced to 1.6 months if the fourth power law holds! Since in most countries the bridge stock is old, the question of remaining safe fatigue life is extremely important.

3.3. TRAINS - DUE FOR REVIVAL?

The development of railway systems since 1830, brought about an awareness of fatigue as a failure mechanism. The first railway accident involving major loss of life, at Versailles in 1842, Smith [7], involved the failure of a locomotive axle by fatigue and instigated many studies into fatigue failures, including the classic work of Wöhler, and the establishing of the 'fatigue limit' concept.

Trains have traditionally been over designed and have proved to be capable of operating for many years (30 years is a nominal design life, but 50 or even 60 years service has been common). These reserves of strength have meant that fatigue problems have been rare and hardly ever generic. Axles have, and still, break, bearings fail but generally railway vehicles have been robust. Modern design trends towards lightweighting and changes from traditional steel to aluminium and even composite materials, have led to some cracking problems. An interesting example, is the airtight cars of the Japanese Shinkansen. The cars are pressure-tight to avoid uncomfortable pressure pulses as trains pass at high speed in tunnels: the major cause of retirement of the vehicles is loss of air-tightness due to fatigue cracking from fastener holes.

In parallel with road traffic, fatigue problems arise with the associated infrastructure, rails, bridges and overhead electrical equipment.

3.4. BROKEN RAILS: A REMAINING PROBLEM FOR THE RAILWAY INDUSTRY

The passage of trains both wears the track and causes fatigue failure. Corrugations have been extensively studied for decades, but the exact mechanism of their formation remains a mystery. Heavy freight traffic causes rapid rail wear; a particular problem in the USA where freight movement is more important than passenger traffic. Table 1 illustrates that rail fractures still remain a major problem. The data is from the UK and is subject to some causes of uncertainty, particularly definition and reporting standards over the hundred plus years of the time span. However, it is clear that whilst failures of wheels and axles have reduced by a factor of 20 over the last century, failures of rails per train kilometre, have actually increased by a factor of more than 2. Heavier axle loads, more axles/train and the move to all-welded track have all contributed to these figures. Because the consequences of failure can be severe, much effort and cost is expended in detecting, monitoring and repairing cracks in railway lines. A quantitative understanding of the mechanisms and mechanics of fatigue failure of cracks in regions away from joints in rails is still lacking. The problem is important and the subject of active research in many countries. Another paper in this conference address this problem in detail [8].

4. Ships: The Unsolved Fatigue Problem

Escalating losses of bulk carriers during the 1980's caused considerable concern in the shipping community and led to a tightening of inspection standards. A total number of 118 of large vessels has been lost since 1990. Thirty four of these accidents involved loss of life and a total of 587 crew members have lost their lives over this six year period. The costs involved, apart from loss of ships, cargo and lives, are huge. The Exxon Waldez accident cost $1b in clean-up charges, a large routine repair costs about 100,000 ECU and delays cost up to 30,000 ECU/day. The costs of frequent inspections are high, and often the inspections are by necessity superficial. The world's trade depends on cheap transport by sea: the vessels used are often old and suffer from corrosion and fatigue cracking: the value of the ship is often less than the value of the cargo and, as always, time is money so that loading and unloading operations are sometimes conducted with undue haste and can induce damage into the ships hulls.

Why are losses of large ships so prevalent? Basically, ships are huge but fragile welded structures. A typical Very Large Crude Carrier (VLCC) is 330m long, and 56m wide. The side plates are 18 to 20mm thick. An egg is about 48mm wide and its shell some 0.35mm thick i.e. the thickness to width is of the order 1/140, relatively some 20 times thicker than the corresponding VLCC ratio of 1/2800. Given the huge mass of a loaded VLCC, it can easily be imagined that a 'gentle' impact with a dock side can cause severe indentation damage. A ship is often divided into six huge holds: if the cargo is loaded unevenly into the holds, large bending stresses can result. In recent years some ships have been instrumented to assist the Master to control the stresses induced by loading. However, many owners cut costs to a minimum: a driver or walker can now be equipped with a Global Positioning Satellite navigation system for a few hundred pounds: many ships do not carry such equipment! Modern ships are constructed by welding large plates and girders into a structure: the total length of weld in a VLCC runs into several hundred kilometres, often originally made in the shipyard under difficult conditions leading to misalignment and poor quality control. Inspection in service is difficult because of the size of the structures involved and hampered by poor accessibility. In addition, and vitally important, ships' hulls have traditionally and still are designed to codes based on static loads and without consideration of fatigue loadings of either the high cycle type caused by wave action or the low cycle type caused by uneven loadings. When the hostile corrosion conditions in which ships work are then taken into account, it should come as no surprise to learn that ships suffer structural deterioration with age: deterioration which often goes unnoticed until a large scale failure occurs.

The largest British ship lost at sea was the MV Derbyshire, a bulk carrier 294m long 44m wide, with a displacement of 192,000 tonnes. In 1980 the Derbyshire disappeared without sending a distress call in a storm in the Philippine sea. Many theories have been put forward to account for the sudden loss, but several centre round sudden catastrophic failure caused by fractures running from fatigue cracks in the area just ahead of the stern superstructure. A paper by Bishop and co-workers [9], discussed the unpredictability of fatigue cracking in large ships.

"We despair of ever estimating the fatigue life of a ship with any accuracy. Our reasons for pessimism include:

1. Uncertainty about material.
2. Impossible to predict fatigue properties.
3. Corrosion.
4. Welding not perfect.
5. Residual stress in hull.
6. Stress concentrations.
7. Mean stress effects not fully understood.
8. Stress states at all points impossible to predict.
9. Impossible to determine entire stress history at every point.
10. Crack detection very difficult.
11. Brittle fracture during lifetime.

The truth is that ships do crack and cracks grow".

This comprehensive list serves to indicate the large number of 'unknowns' still associated with fatigue in ships structures. We could add to the list the trend of replacing mild steel with thinner sections of high tensile steels. Corrosion occurs at the same rate, but because of the reduced thicknesses used, the higher grade steels must be replaced earlier if not properly coated and maintained. In addition, the stresses in the thinner plates are higher, leading to much greater rates of fatigue cracking.

This review would be incomplete without a mention of losses of passenger ships. In general, the problems here arise not from structural failures: the ships are smaller, better maintained and generally newer. At the time this paper was written news broke of a ferry loss on Lake Victoria, Tanzania with 500 feared drowned, probably caused, as many other accidents have been, by overcrowding. Even in heavily regulated European waters, tragedies have occurred: in 1987 the *Herald of Free Enterprise* capsized in Zeebrugge harbour, most probably because of sloppy operating practices induced by cost cutting management policies: but the loss of the *Estonia* in the Baltic sea in 1994, may have been due to excessive sea loadings on fatigue cracked bolts securing the bow door, used to allow vehicles to enter the loading decks. Both cases have prompted discussions of the poor stability of roll-on/roll-off ferries when water enters the vehicle decks.

Of all the transport industries, shipping is probably the "lowest tech", but it is interesting to recall that two of the most significant theoretical advances in fatigue and fracture arose out of studies of ships. Inglis' [10] famous paper on stress concentrations was published in the Transactions of Naval Architects, because he was studying the stress round cut-outs and portholes in ships hulls: this work was used by Griffith [11] to formulate his energy balance approach to fracture instability. The numerous brittle fractures of Liberty ships during World War II led to the intensive study of the effect of welding and low temperatures on the fracture of mild steel plates and arguably to increased interest in fracture research and the development of sharp crack fracture mechanics.

However, improvements are underway. In 1989, Lloyd's Register initiated a long term research and development programme with the aim of generating fatigue assessment procedures for the structural design of ships, Violette [12]. This work involved the examination of a large database of damage reports, and an extensive testing programme of fatigue testing performed at the Krylov Shipbuilding Research Institute, employing large scale testing machines capable of delivery forces up to 30 MN, on 96 separate time-varying controlled channels. Specimens of up to half-size modelling various constructional features of tankers were employed.

The result has been the publication of a software based design tool, Ship Right, for fatigue assessments, which has been applied to tankers from 1994 and bulk carriers from 1996. The software includes details of the likely sea states encountered on the trading routes of particular ships and the calculation of local cumulative fatigue damage at weld details subjected to the spectrum of stresses resulting from wave loading. The programme will be further developed as experience in its use and feedback from service observations become available. This work is a significant and welcome development.

5. Aircraft: Keeping Fatigue at Bay by Inspection

Aircraft are designed in such a way as to minimise their structural mass. The consequences of failure are usually severe. The failures of the pressure hulls of the early *Comet* aircraft prompted much research which contributed to the development of quantitative understanding of fatigue crack growth [13], particularly in aluminium alloys. A modern aircraft is a complex system and failure of an apparently insignificant part can lead to catastrophic loss: a classic case was the crash of DC-10 on take off from

Chicago on 25 May 1979. A bolt in the engine pylon/wing attachment bracket failed by fatigue. During the subsequent investigation the world's fleet of DC-10 aircraft was grounded, to await changes in maintenance techniques which overcame the problem, which was caused by loads generated by incorrect fitting of the engine pods to the wings. At the time of this accident about one fifth of the world's jet passengers was being carried by aircraft of this type.

It would be wrong, however, to suggest that fatigue is a major limitation on the safe operation of aircraft. A recent survey indicated that:

- only 2% of plane crashes are caused by structural failure
- 7% by maintenance defects
- 11% by terrorism or military action
- 12% by weather, thunder and lightning, ice, fog, wind shear, etc.

and a vast proportion, 67% are caused by human error, by pilots, aircrew and air traffic controllers. Only on rare occasions are the causes of an accident unknown. In recent years several accidents have been caused by unattributable defects in complex control systems and in 1996, the disintegration of a TWA 747 off New York has so far defied satisfactory explanation.

Peel [14] discussed the relative frequency of occurrence of different types of failure mechanisms in structural failures in aircraft. Fatigue, (47%) predominated, but stress corrosion (16%), corrosion (27%) and corrosion fatigue (10%) failures often occurred. It was suggested that the rectification of corrosion damage in military and civil aircraft consumed more effort than the repair of fatigue cracking. The statistics for helicopters give a rather different picture. Failures in the highly stressed mechanical transmission system lead to sudden loss of airworthiness and accident rates are greater on a passenger kilometre basis.

Why then, despite the obvious difficulties, do fatigue failures contribute so little to the loss of aircraft? The industry is subjected to very tight regulation which is, in the main, strictly enforced. Aircraft are subject to a range of inspections designed to detect cracks before they grow to dangerous sizes. Repair and replacement schedules have been defined and in critical parts in both hulls and engines, strip-down inspection and replacement is performed which has reduced failures to low levels. The situation is different for military aircraft where operational lives are much shorter than for civil aircraft, flight loadings more severe and performance criteria are more stringent. Expensive research programmes to reduce fatigue damage in military aircraft have had obvious spin-offs in the civil field. Until recently many national governments funded civil research in support of their national aircraft industries. The trend has now turned to international collaboration: the European Airbus is a good example.

Overall therefore, resources have been available in the aircraft industry to allow it to lead other branches of transport in fatigue and fracture design. Economic pressures are beginning to force extensions to design lives for existing aircraft and the weight problems associated with long-haul fuel loads in large capacity aircraft force the margins of structural design to be lowered. Only by strict compliance to high standards of safety related regulation will the aircraft industries impressive record in suppressing structural failures be kept or even enhanced in the future.

6. Concluding Remarks

This brief review has described how fatigue related problems in the major branches of the transport industry have been addressed to allow 'safe' economic operation. Of the major branches discussed, only large ships have generic fatigue problems. In land transport more needs to be known about the interaction of vehicles with their infrastructure - roads and bridges for automobiles and lorries, rails in the case of trains. The public have an expectation of a continuance of enhanced safety standards.

It is worth quoting some figures for the risks associated with various modes of travel [15].

Table 2 - Deaths per 10^9 km travelled, UK.

	1967-71	1972-76	1986-90
Railway Passengers	0.65	0.45	1.1
Passengers in schedules air services on UK airlines	2.3	1.4	0.23
Bus or coach drivers and passengers	1.2	1.2	0.45
Car or taxi drivers and passengers	9.0	7.5	4.4
Two-wheeled motor vehicle driver	163.0	165.0 }	
Two-wheeled motor vehicle passengers	375.0	359.0 }	104.0
Pedal cyclists	88.0	85.0	50.0
Pedestrians *	110.0	105.0	70.0

* Based on a National Travel Survey (1985/86) figure of 8.7 km / person / week
Source: Department of Transport

The risk of travel has commonly been assessed in terms of accidental deaths per 10^9 km travelled, Table 2. For rail travel it will be noted that there is an increase in the most recent figures, which reflects two major accidents in that period: the King's Cross Underground fire and the Clapham Junction railway accident. For road travel the risk will vary with many conditions (e.g. class of road, experience of driver, weather, lighting, wearing seat belt or crash helmet, as well as type of vehicle). The reduction of risk for car travel noted in Table 2 reflects the introduction of compulsory wearing of seat belts, greater enforce.nent of drink driving laws and the public attitudes to drink-driving, improvements in car design, a greater mileage on motorways, which have lower accident and casualty rates, and slower traffic in towns due to increasing congestion. For air travel the risk per flight (or per sector of a flight) is arguably more significant than per 10^9 km travelled, as a substantial proportion of all fatal accidents occur during take-off or landing. The reduction in risk of air travel noted in Table 2 reflect the improvement in reliability of aircraft and the extensive use of automatic landing of aircraft with a reduction of accidents due to pilot error.

Much work has been performed on aspects of the quantification of risk in recent years. In the field of transport, risks on public modes (rail, coach and aircraft) are significantly lower than in private modes (walking, cycling and the automobile). The public expects to be safer "when in someone-else's care". Nevertheless, there remains much to be done in educating the public of the link between safety and cost: further improvements from current acceptable levels of safety are generally costly (the gradient of the safety level/cost curve is very shallow at the top end). A major challenge of the public transport industry, particularly acute in aircraft, is the need to at least sustain, or, better, to improve safety and risk levels whilst at the same time operating under more stringent economic conditions.

7. References

1. Miller, K. J., (1991) Metal fatigue - past, current and future, *Proc. Instn. Mech. Engrs.* **205**, 291-304.
2. Nelson, D. V., and Sheppard, S. D., (1995) Fatigue and fracture estimation for metallic components: some current methods and future developments, *Trans. ASME.* **117 (B)**, 121-127.
3. Lütjering, G. And Nowack, H. (Eds) (1996) *Fatigue '96. Proc. Sixth Int. Fatigue Congress*, Pergamon, Oxford **(3 vols)**.
4. Op cit, Schijive, J., Prediction on fatigue life and crack growth as an engineering problem: A state of the art survey, **2**, 1149-1164.
5. Op cit, Bignonnet, A., Fatigue design in the automotive industry, **3**, 1825-1836.
6. Cebon, D., (1993) Interaction between heavy vehicles and roads, *Soc. Automative engineers, The 39th Ray Buckendale Lecture*, SP-951.
7. Smith, R. A., (1990) The Versailles railway accident of 1842 and the beginning of the metal fatigue problem, in H. Kitagawa and T. Tanaka (eds.), *Proc. Fourth Int. Conference on Fatigue and Fatigue Thresholds*, Materials and Component Publications Ltd., Birmingham **4**, 2033-2041.
8. Beynon, J. H. and Kapoor A., (1996) The interaction of wear and rolling contact fatigue, *NATO Advanced Research Workshop, Varna, Bulgaria.*
9. Bishop, R. E. D., Price, W. G., and Temavel, P., (1990) A theory on the loss of the MV Derbyshire, *Trans Inst. Navel Arch* **133**, 389-453.
10. Inglis, C. E., (1913) Stresses in plates due to the presences of cracks and sharp corners, *Trans. Inst. Naval Arch* **55**, 219-230 (and discussion 231-241).
11. Griffith, A. A., (1921) The phenomena of rupture and flow in solids, *Proc. Roy. Soc.* **A221**, 163-198.
12. Violette, F., (1995) Lloyd's Register's Integrated Fatigue Design Assessment System, in *Proceedings of Lyoyd's Register Ship Division Seminar, Detail Design - The Key to Success or Failure of Ships,* Lloyd's Register, London, 1-19.
13. Smith, R. A., (1986) Thirty years of fatigue and crack growth - An historical review, in R. A. Smith (eds.), *Fatigue Crack Growth - 30 Years of Progress*, Pergamon, Oxford, 1-16.
14. Peel, C. J., and Johnes, A., (1990) Analysis of failures in aircraft structures, *Metals and Materials* **6**, 496-502.
15. Anon (1992) Risk: Analysis, Perception and Management, Society, London, 78-79.

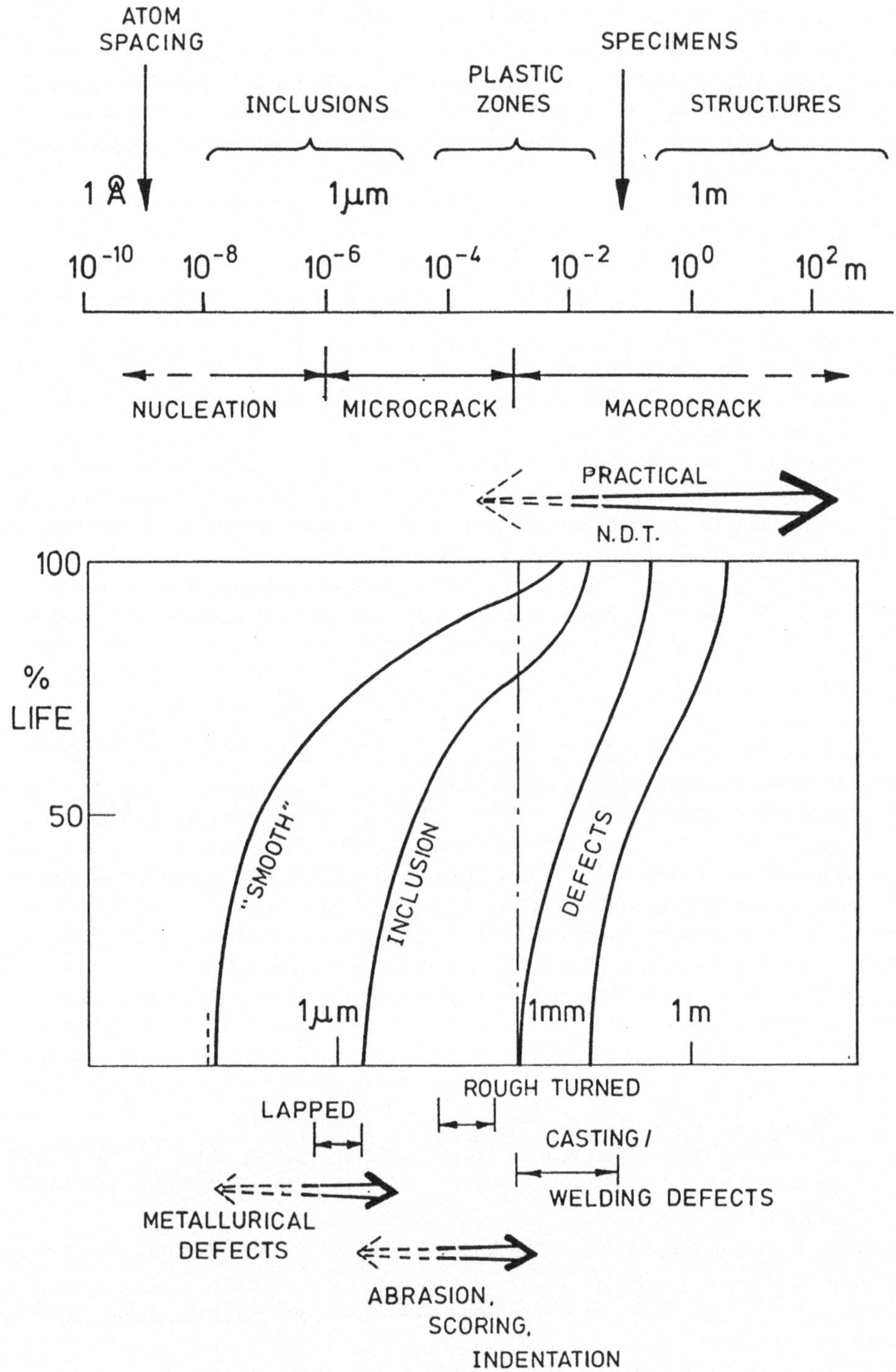

Figure 1. Size Scales of Fatigue and Fracture

Appendix 1. A Brief Outline of Fatigue and History

Date	Experimental Techniques, Materials	Ideas	Theory	Important Names
1825				Trevithick
Railways Axles	Full scale tests 3 pt bending (wrought iron)		Empirical	Albert 1830 Rankin 1840 Hodgkinson and Fairbairn 1840 - 65
1850		Crystalline metal? Importance of K_T		I. Mech. Eng. discussions 1849 - 52 R Stephenson, President
1850 Britannia Bridge				
1856	Steel			Bessemer
1879 Tay Bridge	Laboratory specimens, rotating bending	Importance of stress range	σ/N data Infinite life design, fatigue limit	Wöhler 1867
1900	Optical microscope	Slip bands and microcracks		Ewing 1902
		Mean stress effect	K_T for ellipse	Inglis 1913 Goodman 1914
1920	Bending and torsion	Multiaxial fatigue		Gough 1920
Axles, Springs in Automobiles		Importance of cracks in brittle fracture, energy balance	$\sigma\sqrt{a}$ = const.	Griffith 1921
1930		Residual stresses	Notch strain analysis	Haigh 1930
		Dislocations Elementary block concept	Notch strains	Taylor, Orowan 1934 Neuber 1937

Date	Experimental Techniques, Materials	Ideas	Theory	Important Names
1940 Brittle fracture in ships, welds				Tipper
		Cumulative damage	$\sum \frac{N_i}{N} = 1$	Miner 1945
			Stress analysis of crack, singularity	Sneddon 1946
1950 Comet	Closed-loop servohydraulic machines, Electron microscope, beginnings of computers, Strain gauges	Effect of plasticity on energy balance		Orowan 1955
			K-concept for cracks	Irwin 1957
1960			Finite element stress analysis	Zienkiewicz 1960
QEII turbine blades	Improved data collection and analysis	Low cycle fatigue, thermal strains	$\varepsilon_p N^\alpha$ = const	Manson/Coffin 1962
		Crack propagation	da/dN = $C(\Delta K)^n$	Paris 1963
		Elastic/plastic fracture	C.O.D.	Wells 1963
1970 North Sea Oil Concord	Micro-chip technology	" "	J-integral	Rice 1968
Energy crisis, conservation 1980		Fail safe, safe life and damage tolerant design. Only history will teach us the really important contributions of this recent era.		

HIGH NITROGEN STEELS BEHAVIOUR UNDER CYCLIC LOADING

STEFAN VODENICHAROV
Institute of Metal Science
67 Shipchensky Prohod str.
1574 Sofia, Bulgaria

1. Introduction

Metallurgical society is recently paying more attention to new type of materials, the high nitrogen steels and to the metallurgy under pressure, considered as a new international trend. During the period of 25 years extensive work has been conducted in this field in the Institute of Metal Science at the Bulgarian Academy of Sciences.

The term "high nitrogen steels" means steels, produced under gas pressure in which the nitrogen concentration is higher than its normal solubility. At this stage of development it can be taken for proven that high nitrogen steels have properties exceeding by 30-150% as compared to those of the nitrogen-free analogues in almost all steel classes, [1].

The interest shown to the nitrogen as an alloying element has several decades attracted the attention of scientists and steel producers from number of industrial countries. A special subject of this attention is it's austenizing ability which is 20 times higher than the nickel's one. This ability is a prerequisite for new type stainless and corrosion-resistant steels development - nickel-free high nitrogen steels. Their corrosion resistance is the same as this one for steels containing 10% nickel.

As an element forming interstitial solid solution, nitrogen causes deformation of the crystal lattice and thus considerably increasing the steel strength characteristics. Another way for obtaining higher mechanical properties for this type of steels is precipitation hardening - nitride phases precipitate in the temperature interval 650-900 ^{0}C. For this purpose elements of a strong nitride forming ability can be added, as V, Nb etc., and they will contribute for steel strength increase at high temperatures. Hence such steels can be used as refractory materials. The thrird way for increasing nitrogen steel strength characteristics is using strain hardening. Here also considerably higher parameters can be reached compared to Cr-Ni steels strength.

Increasing the nitrogen content, the degree of hardening for these steels also tends to increase, due to lower value of the stacking fault energy - the nitrogen and manganese strongly reduce this value.

The strong austenizing ability of the nitrogen leads to one of the most interesting properties of high nitrogen steels - they are entirely non-magnetic materials and in combination with their high strength such steels are very successful subsitutes for 24Ni-3W

R.A. Smith (ed.), Reliability Assessment of Cyclically Loaded Engineering Structures, 419–434.

steels, containing scarce and expensive alloying elements.

Increasing the nitrogen content in high nitrogen steels leads to about 20 times increase of their cavitational resistance, due to the favourable influence of the nitrogen on increasing steel resistance against deformation.

A possibility exists for high nitrogen steels austenitic sructure to be formed even when alloying them with high concentrations of ferrite forming elements - Si, Mo, etc. Such elements rapidly increase the corrosion resistance of such steels in high agressive oxide and chlorine environments.

2. Experimental Results and Discussion

Considerable attention has been paid in Material Testing Laboratory at the Institute of Metal Science to some high nitrogen steels behaviour subjected to cyclic loading and the results of such investigations will be presented hereafter.

The material investigated was high nitrogen steel developed in the Institute of Metal Science, namely Cr18Mn12N steel. This material is stainless high nitrogen nickel-free steel. It may find application as high strength non-magnetic material in bulding, toolmaking, nuclear power plant, chemical industry, refrigerator and food industry units, etc.

2.1. HIGH-CYCLE FATIGUE INVESTIGATIONS

Four modifications of this steel with different nirtogen quantities have been tested in order their behaviour under high-cycle loading conditions to be established, [2]. The chemical contents of these materials are given in Table 1. The steel ingots were produced using the counter-pressure casting method. They were subjected to homogenizing tempering at 1150 ^{0}C for 10 hours and followed by furnace-cooling. After cold rolling the steel ingots were heat treated again at temperature 1150 ^{0}C followed by water cooling. Thus an austenitic structure for the modifications with 0.50, 0.88 and 1.13% and ferritic-pearlitic one for the modification with 0.36% nitrogen content were obtained. The test specimens were cut along the rolling direction. The tensile test mechanical properties thus obtained are presented in Table 2.

TABLE 1: Chemical contents of the materials investigated, in volume percent

Steel type	C	N	Si	Cr	Mn	S
Type A	0.03	0.36	0.77	18,73	12.46	-
Type B	0.03	0.50	0.45	18.97	12.04	-
Type C	0.03	0.88	0.50	17.42	12.08	0.007
Type D	0.04	1.13	0.32	17.92	12.36	-

TABLE 2: Some mechanical propeties of the materials investigated

Material	Type A	Type B	Type C	Type D
R_e, [MPa]	590	640	860	960
R_m, [MPa]	870	950	1110	1180
A, [%]	23.0	16.8	13.2	1.5

The Dixon-Mood method was used for fatigue limit determination under high-cycle loading conditions, [3]. The stress ratio was R= -1 and the loading frequency was 166 Hz. The results for mean values and meansquare deviation of the fatigue limits are given in Table 3. The fatigue curves are given in Fig.1.

TABLE 3: Fatigue characteristics of the materials investigated

Material	Type A	Type B	Type C	Type D
σ_{-1}, [MPa]	332.2	340.0	381.0	387.1
S_σ, [MPa]	31.0	23.4	21.5	30.8
ε'_f, [%]	-	5.41	5.21	1.09
σ'_f/E, [%]	-	1.23	1.58	0.91
b	-	-0.044	-0.072	-0.028
c	-	-0.196	-0.194	-0.049
C	-	6.03×10^{-18}	9.93×10^{-17}	5.46×10^{-13}
n	-	5.97	4.88	2.22

A tendency can be observed that increasing the nitrogen content in this steel from 0.36 to 1.13% leads to fatigue limit increase. The worst behaviour is demonstrated by the type A modification containing 0.36% nitrogen due to the specific two-phase ferritic-pearlitic structure. Best behaviour in the endurance life zone of the curve possesses the type B modification with 0.50% nitrogen due to the fact that the nitrogen increase in this steel leads to material plasticity decrease. Using this fact the austenitic modifications fatigue behaviour at the higher stress amplitude levels can be explained.

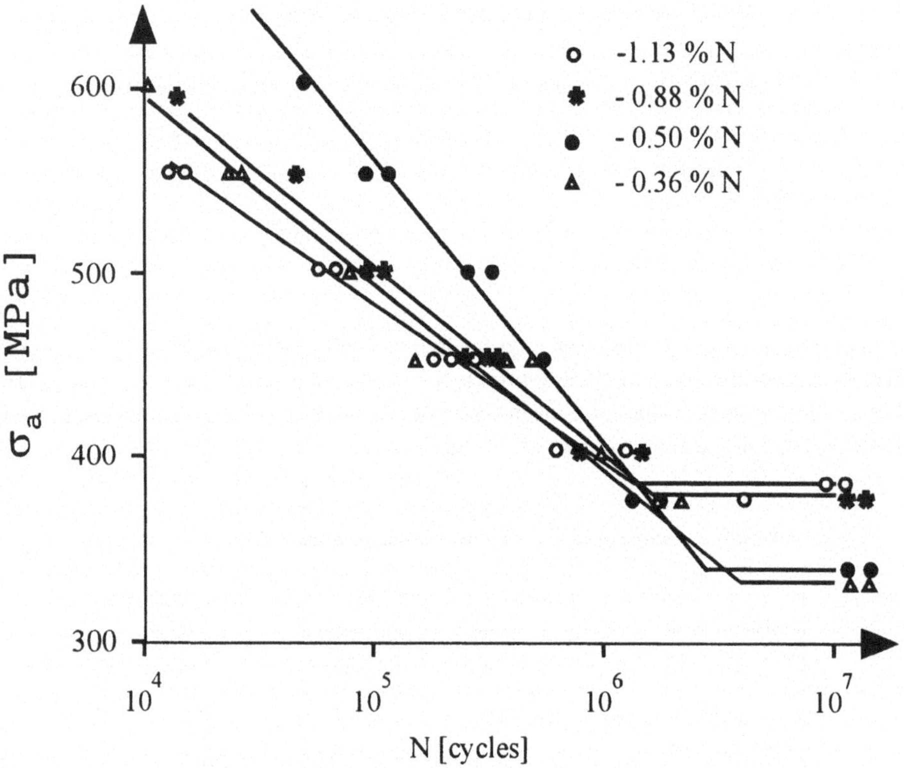

Figure1. Fatigue endurance curves for the materials investigated

2.2. LOW-CYCLE FATIGUE INVESTIGATIONS

Low-cycle fatigue tests for the modifications with austenitic structure were conducted on MTS-810.12 servohydraulic closed loop testing machine, [4]. The loading frequency was 1 Hz. The experiments were carried out under constant amplitude of the total deformation ($\Delta\varepsilon_t$), [5]. The stress ratio was R= -1. The total deformation amplitude $\Delta\varepsilon_t/2$ versus number of cycles curves for every modification were ploted and its elastic and plastic components as well, Fig.2. The empirical coefficients in Menson-Coffin and Basquin equations, [5-7] were determined, Table 3. The stress range ($\Delta\varepsilon$) values variations versus number of cycles curves for some constant deformation amplitude values were ploted, Fig.3. The cyclic stress-strain curves obtained were compared to the static tensile ones, Fig.4.

The intersection point between elastic and plastic deformation curves, Fig.2 is considered as transition region from low-cycle fatigue to high-cycle one, [5]. Therefore for all steel modifications investigated the low-cycle fatigue region is placed up to about $1.0\text{x}10^4$ cycles. In this region the fatigue behaviour of the materials depends mainly on their cyclic ductility, [8]. The data in Table 3 show that the nitrogen content increase in the steel from 0.50%, (type B) to 0.88 %, (type C) leads to decrease of the cyclic ductility coefficient, (ε'_f) and at the same time increasing the cyclic ductility exponent, (c). Further decrease of

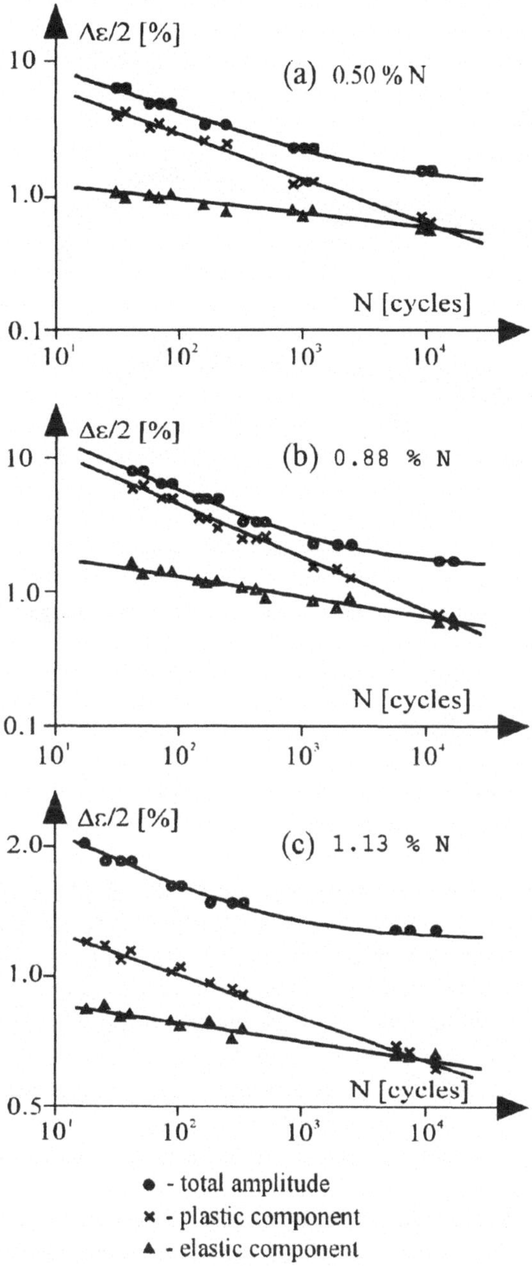

Figure 2. Log$\Delta\varepsilon_t$/2-logN curves and their elastic and plastic components

ε'_f and increase of c is observed for the modification containing 1.13 % N, (type D) but these tendencies are much stronger due to the worse plastic properties of the type C modification. Probably this behaviour is result of the increased number and dimensions of the nitride type precipitates in the modification type D compared to type C, [9]. These precipitates are localized both along the grain boundaries and in groups inside the grain bodies. Such precipiates are missing for type B, [9]. Therefore the nitrogen content increase in this steel leads to cyclic ductulity decrease. Best behaviour in low-cycle fatigue region demonstrates the modification with 0.88 % nitrogen content. Fig.3 shows that for modifications type B and C cyclic hardening occurs unlike of type D where cyclic softening can be observed. More intensive cycle hardening occurs for type C modification compared to type B. The comparison between cyclic, (1) and tensile static, (2) stress-strain curves, Fig.4. gives a quantitative presentation for the influence of the loading type on the material mechanical properties. Modification type D (1.13% N) is cyclic softening material, types B and C are cyclic hardening materials in all amplitude range investigated and this process is most intensive for modification type C (0.88% N).

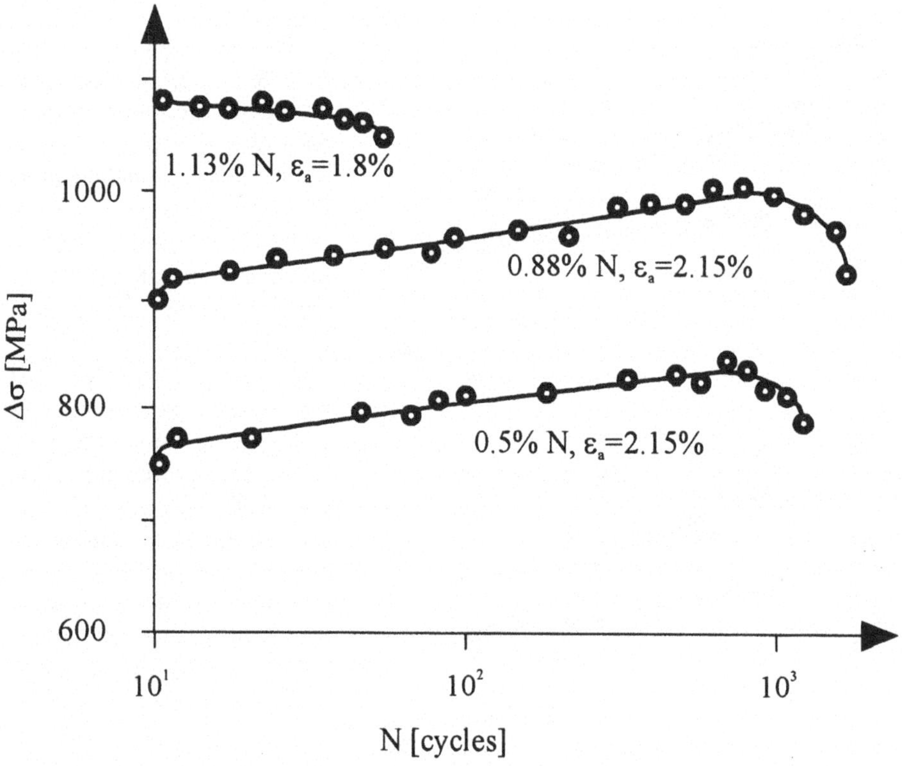

Figure 3. Some $\Delta\sigma$ - logN curves for the austenitic materials investigated

2.3. FATIGUE CRACK GROWTH RATE INVESTIGATIONS

Fatigue crack growth rate (FCGR) investigations were carried out for the modifications with austenitic structure, types B, C and D, [10,11], as well as for Cr18Ni9Ti stainless steel. Potential drop technique was used for crack growth monitoring. The results of these investigations are presented in Fig. 5 and the values of the empirical coeficients C and n for high nitrogen modifications are given in Table 3. Increasing the nirtogen content leads to decrease of n and increase of C, and the FCGR also decreases. Therefore the nitrogen has favourable effect regarding the fatigue crack growth resistance of this steel. This is also confirmed by the fact that the moment of macrocrack formation is delayed with the increase of nitrogen amount, i.e. the number of cycles to its appearance increases.

It is obvious that all high nitrogen steel modifications investigated have lower FCGR at equal ΔK values compared to Cr18Ni9Ti steel ones, Fig.5 [12]. Some Transmission Electron Microscope (TEM) investigations have been carried out in order attempt to be made for explanation of this phenomenon. Under cyclic loading a typical ladder structure with walls and channels is formed in Cr18Ni9Ti stainless steel, while no changes in the initial structure is observed in the high nitrogen modifications. The dislocations keep their plane distribution and low density typical for the materials before cyclic loading. Both steels are considered as materials with low stacking fault energy but it is consider-

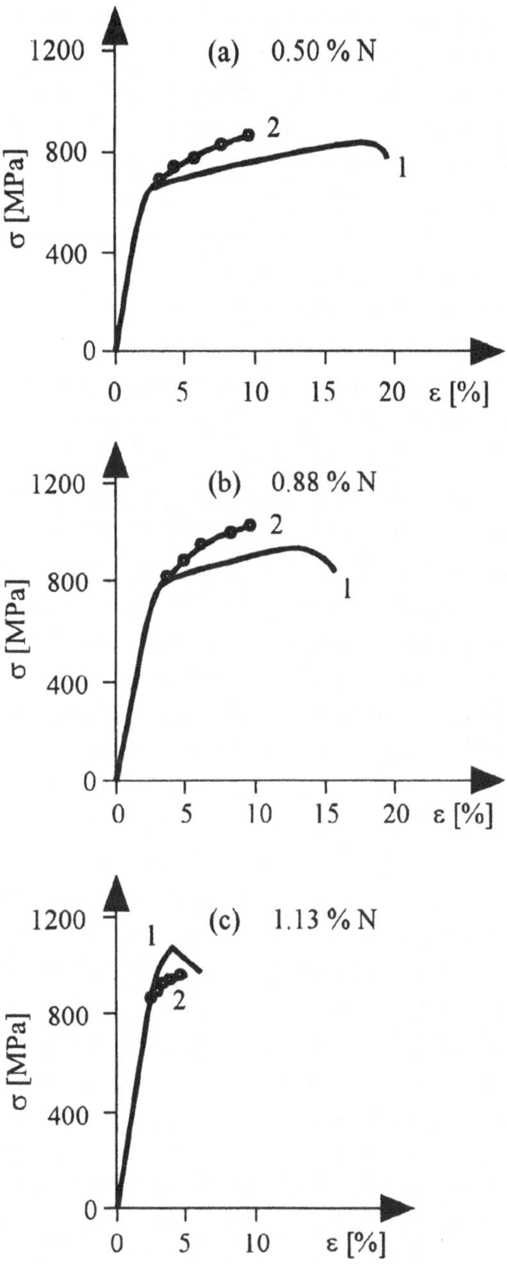

Figure 4. Static and cyclic stress - strain curves for the austenitic materials investigated

ably lower for high nitrogen steel modifications - 5 erg/cm^2 for Cr18Mn12N steel with 0.50% nitrogen versus 16 erg/cm^2 for Cr18Ni9Ti steel. For both cases split dislocations are typical and they move as a complex of two partial dislocations linked by stacking fault. Because of the higher value of stacking fault energy of Cr18Ni9Ti steel, the fault width is smaller and the applied stress can squeeze this complex back in such way that partial dislocations can merge into copmlete ones. This process allows cross slip and this way dislocation walls are formed. The dislocation splitting for the high nitrogen steels is much greater and even high values of the applied stress cannot squeeze these complexes. Also, during the dislocations movement, nitrogen atoms segregate on the stacking faults and additionaly stabilize these faults. Generally speaking the process is similar to dislocations movement in ordered solid solution.

2.4. PLASTIC ZONE AHEAD OF THE CRACK TIP INVESTIGATIONS

The plastic zone size and shape in front of the crack tip for the three modifications of this steel under cyclic loading conditions have been experimentally determined, [13]. The holographic interferometry method has been used. This method has been successfuly applied for deformation distribution establishing in the crack tip vicinity for some materials, [14-15]. Schematic presentation of the experimental set-up used is given in Fig. 6. Fatigue crack was developed with 1 Hz test frequency, stress ratio R=0 and stress amplitude σ_a = 24 kN and after that the specimen was mounted to the loading rig (item 5, Fig. 6), loaded to the maximum load of the cycle and after that unloaded to zero. The holographic pictures were taken using camera (item 11, Fig.6).

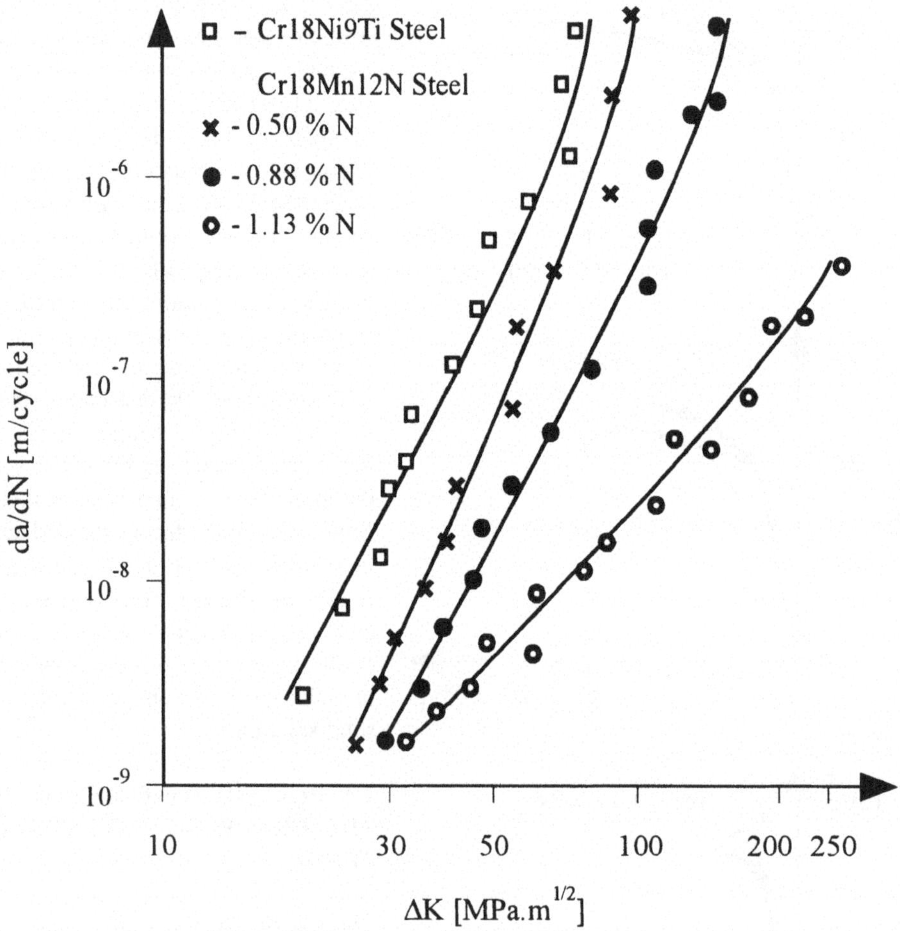

Figure 5. da/dN - ΔK curves for Cr18Ni10Ti steel and austenitic modifications of Cr18Mn12N steel

The holographic pictures obtained for the three modifications investigated are presented in Fig. 7. The strips in the photograph for Cr18NMn12 stainless steel with 0.50 % nitrogen content (type B) are the finest ones (Fig. 7, a) and at the same time their number is the greatest one compared to these for the other modifications. That means this modification has the largest plastic zone size compared to the others. The modification with 1.13% nitrogen content (type D) has the smallest one, (Fig. 7, c). The plastic zone shape does not differ significantly for the modifications investigated. This shape is similar to that one, accepted in the theoretical approach where the classical Von Mises yield criterion is used, in plane stress condition. The strip number from initial strip to the point of measurement N versus distance in the loading direction y curves for the three modifica-

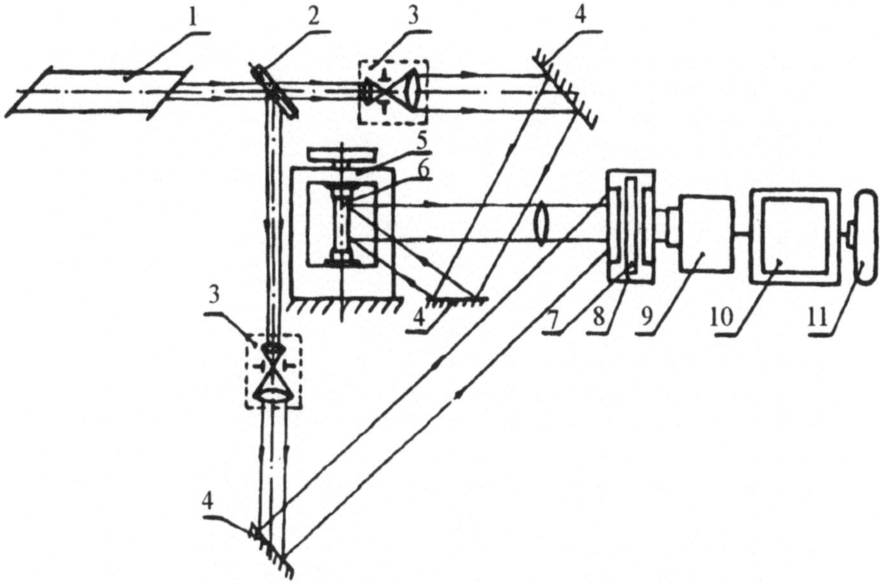

Figure 6. Schematic presentation of the experimental set-up
1 - laser; 2 - semi-transmission mirror; 3 - collimator, 4 - mirror;
5 - loading rig; 6 - specimen; 7 - holographic plate; 8 - immersion cell;
9 - TV camera; 10 - monitor; 11 - camera

tions are presented in Fig. 8. Measurements for distances of 0.5 mm (Fig.8, curves 1, 3 and 5) and 10.5 mm (Fig.8, curves 2, 4 and 6) from the crack tip along the x axis have been done. The curves in this Figure give a quantitative idea for the nitrogen influence on the fatigue plastic deformation zone size. It is evident that the increase of the nitrogen content in Cr18NMn12 stainless steel decreases the fatigue plastic deformation zone size, due to decrease of the cyclic ductility of this material.

2.5. FRACTOGRAPHIC AND MICROSTRUCTURAL INVESTIGATIONS

Fractographic and microstructural investigations have been carried out for the three austenitic modifications in order the influence of the nitrogen content on the mechanisms of fracture to be estimated, [9, 16]. Both Metallographic and Scanning Electron Microscopes have been used. Fracture surfaces of specimens cycled both under high- and low-cycle fatigue loading conditions were observed.

Increasing the nitrogen content leads to coarser relief of the fracture surface during high-cycle fatigue, Fig.9,10. The steel containing 1.13% nitrogen (type D) possesses not typical fatigue fracture surface with number of brittle tear-off ligaments and secondary cracks,Fig.10. Fatigue striations can be observed for the material with 0.50% nitrogen content (type B) and martensitic transition in the austenitic grains due to the intensive deformation processes also takes place, Fig 11.

Scanning Electron Microscope (SEM) observations also show that for Cr18NMn12 steel with 0.50% nitrogen content under high-cycle fatigue loading, the process of frac-

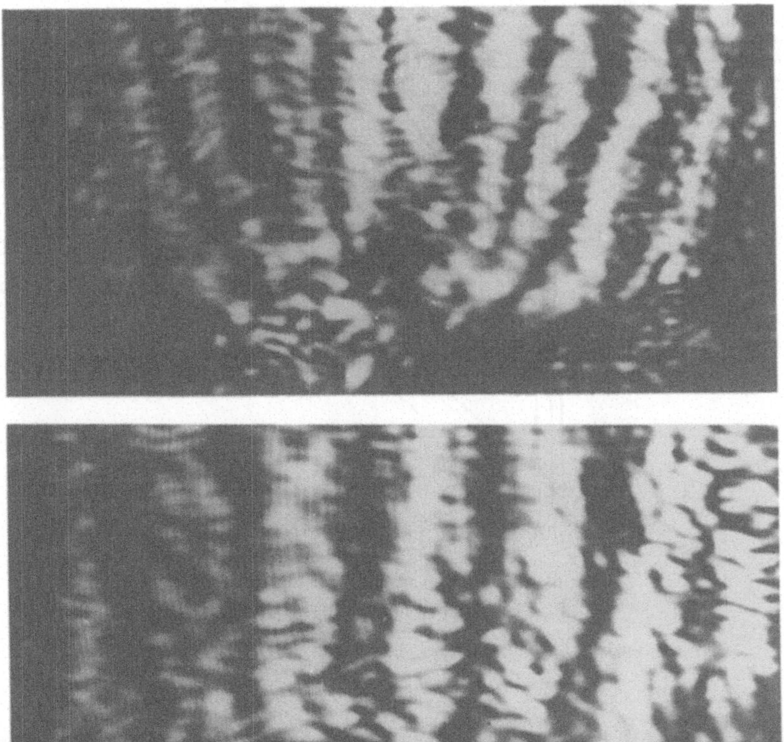

a

b

c

Figure 7. Holographic pictures for the austenitic materials investigated

a/ Cr18NMn12 with 0.50 % N, x5

b/ Cr18NMn12 with 0.88 % N, x5

c/ Cr18NMn12 with 1.13 % N, x5

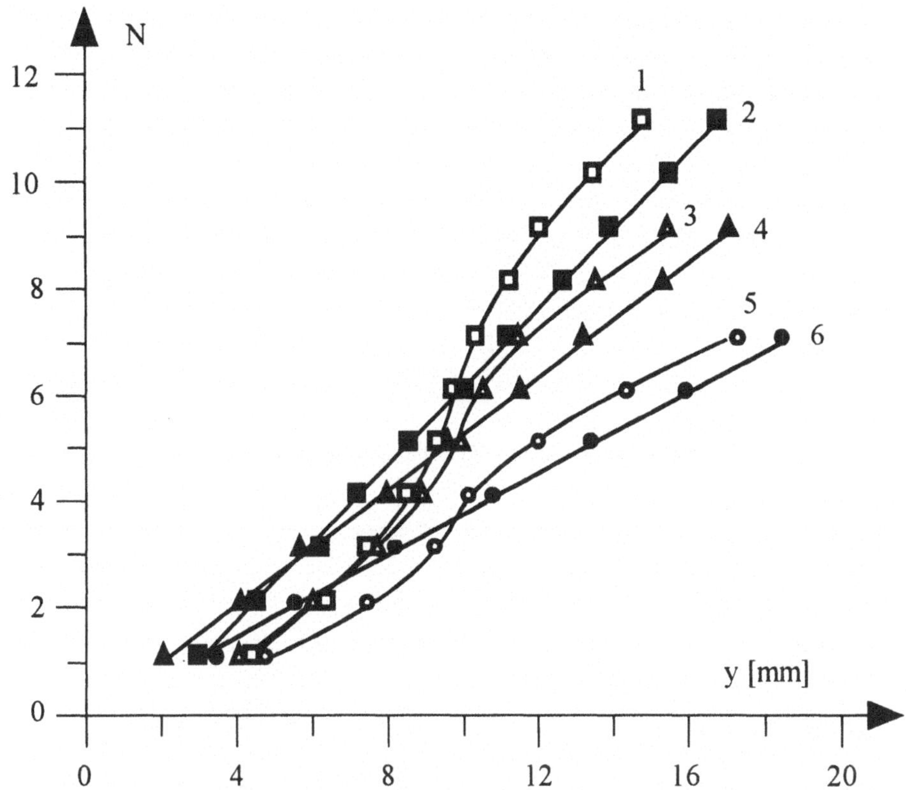

Figure 8. Strip number to the point of measurement N versus distance in the loading direction y curves for the austenitic materials investigated

1 - Cr18NMn12 with 0.50 % N, x = 0.5 mm

2 - Cr18NMn12 with 0.50 % N, x = 10.5 mm

3 - Cr18NMn12 with 0.88 % N, x = 0.5 mm

4 - Cr18NMn12 with 0.88 % N, x = 10.5 mm

5 - Cr18NMn12 with 1.13 % N, x = 0.5 mm

6 - Cr18NMn12 with 1.13 % N, x = 10.5 mm

ture is accompanied by fatigue striations formation and micropores coalescence, Fig. 12. Persistent slip bands can be observed at higher magnifications, Fig. 13. Increasing the nitrogen content to 0.88% leads to absence of micropores coalescence. Brittle tear-off ligaments situated at 45^0 from the loading direction exist, Fig.14. Fatigue striations absence for the type D material can be established, due to the low plasticity of this steel. The brittle type of fracture accompanied with secondary cracks formation is prevailing in this case, Fig. 15. Generally speaking nitrogen content increasing from 0.50% to 1.13% under high-cycle fatigue loading leads from relatively ductile to definitely brittle way of fracture. Enlarging of the non-metallic inclusions in the materials with higher nitrogen contents is the reason for increasing their influence on the local mechanisms of fracture.

The process of fracture for Cr18NMn12 steel with 0.50% nitrogen content under low-cycle fatigue loading is accompanied by micropores initiation, growth and coalescence,

Figure 9. Fracture surface for Cr18NMn12 specimen with 0.50% N

Figure 10. Fracture surface for Cr18NMn12 specimen with 1.13% N

Figure 11. Near to the crack path microstructure for Cr18NMn12 specimen with 0.50% N

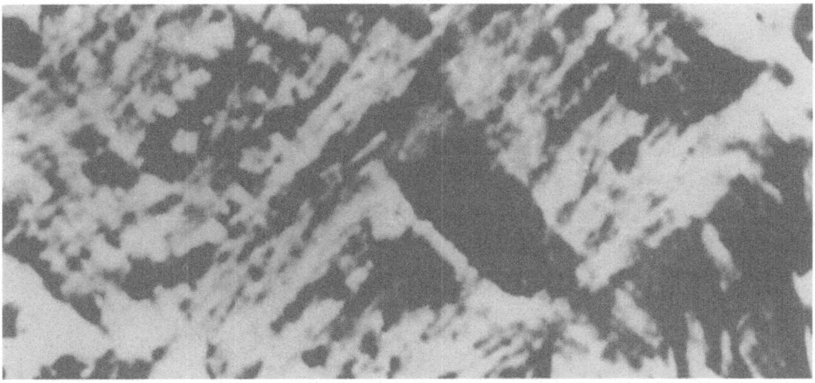

Figure 12. SEM photograph for Cr18NMn12 steel with 0.50% nitrogen content under high-cycle fatigue loading, x1360

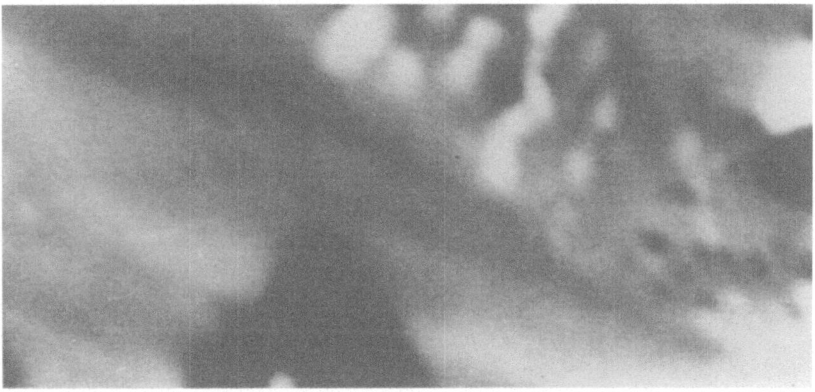

Figure 13. SEM photograph for Cr18NMn12 steel with 0.50% nitrogen content under high-cycle fatigue loading, x5450

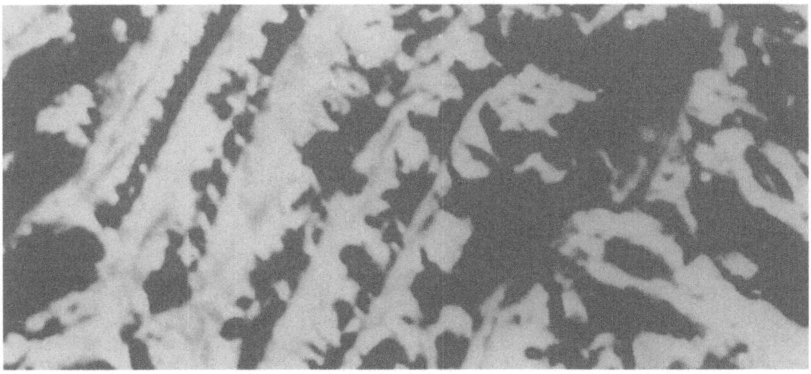

Figure 14. SEM photograph for Cr18NMn12 steel with 0.88% nitrogen content under high-cycle fatigue loading, x1360

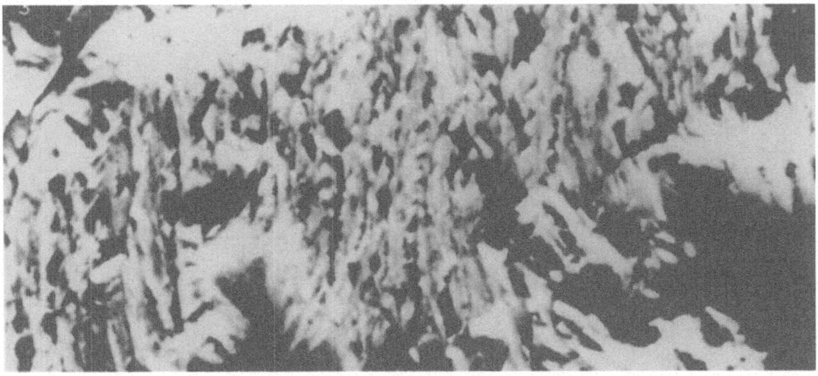

Figure 15. SEM photograph for Cr18NMn12 steel with 1.13% nitrogen content under high-cycle fatigue loading, x680

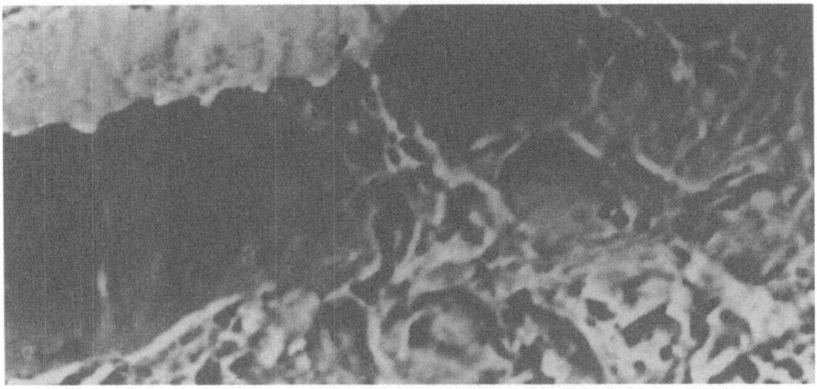

Figure 16. SEM photograph for Cr18NMn12 steel with 0.50% nitrogen content under low-cycle fatigue loading, x775

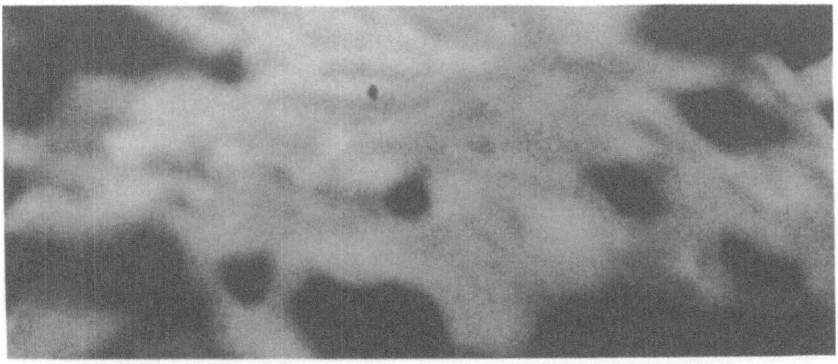

Figure 17. SEM photograph for Cr18NMn12 steel with 0.50% nitrogen content under low-cycle fatigue loading, x5450

Fig. 16. Well distinguished fatigue striations can rarely be observed. Here secondary cracks presence is established in spite of the relatively ductile type of fracture, Fig 17. The type C steel demonstrates coarser relief and mixed type of fracture with increased number of secondary cracks and facet tearings. The material with 1.13% nitrogen demonstrates brittle type of fracture accompanied with relatively large secondary cracks, Fig. 18. Generally speaking for the low-cycle fatigue case the same tendency of transition from relatively ductile to brittle way of fracture is observed but the secondary cracks formation process plays important role and this is one of the main mechanisms of fracture for this case.

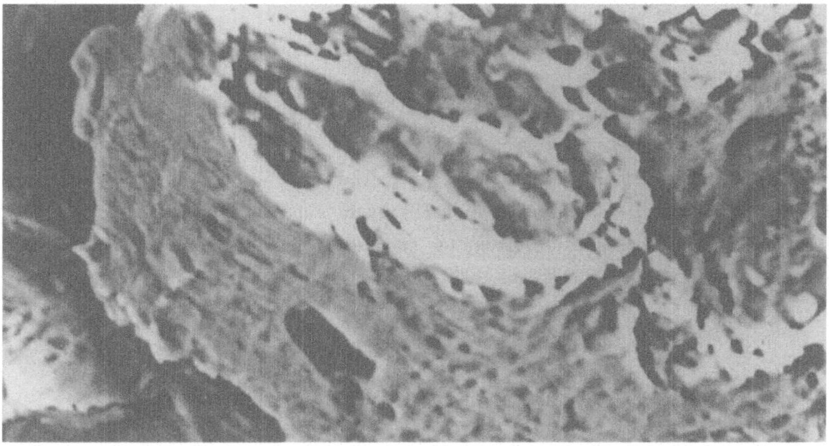

Figure 18. SEM photograph for Cr18NMn12 steel with 1.13% nitrogen content under low-cycle fatigue loading, x625

The investigations carried out for Cr18NMn12 stainless high nitrogen steel show that the nitrogen content variation in relatively wide interval gives the possibility to optimize the steel behaviour for specific fatigue loading conditions and the manufacturer has to account for this in order best results to be achieved.

3. References

1. Rashev, Ts. (1995) *High Nitrogen Steels - Metallurgy under Pressure*, Bulgarian Academy of Sciences Publishing House, Sofia
2. Davidov, Y., Vodenicharov, St., Penchev, H. and Minchev, K. (1987) Some Nitrogen Steels Behaviour under cyclic loadings, *Proc. of Higher Military School Scientific Session*, V.Tarnovo, Bulgaria, Vol.7, pp.66-72
3. Dixon, W. and Massey, F., Jr. (1957) *Introduction to Statistical Analysis*, McGraw-Hill, New York
4. Davidov, Y., Kortensky, G. (1995) Some mechanical properties establishing for nitrogen stainless steel, *Journal of Materials Science and Technology*, Vol. 4, No 1, pp. 28-34
5. Crubisis, V. and Sousino, C.M. (1983) Low-Cycle Fatigue at Room Temperature, in W. Dahl (editor), *Verhalten von Stahl bei Schwingender Beanspruchung*, Verlag Stahleisen M.B.H., Duesseldorf, pp. 529-550
6. Coffin, L.F. (1954) *Trans. ASME*, **76**, pp. 931-964
7. Manson S.S. (1954) *NASA Technical note* No 2933, pp. 38-57

8. Polak, J., Nemec, J. and Klesnil, M. (1984) Elastic-Plastic Fatigue Crack Growth, *Proc.of ICF-6*, New Delhi, India, Vol.3, pp.1831-1837
9. Davidov, Y., Kovacheva, R. and Vodenicharov, St. (1990) Fractographic and microsrtuctural investigations of crack propagation process for nitrogen stainless steels under cyclic loadings, *Scientific Report* No 51/1990, Inst. of Metal Science, Sofia, Bulgaria, pp. 12-19
10. Vodenicharov, St. (1992) Nitrogen contents effect on crack resistance of some steels, *Proc. Ninth European Conference on Fracture (ECF9)*, Varna, Bulgaria, Vol. I, pp. 73-77
11. Vodenicharov, St., Davidov, Y., Kalchevska, K. and Kolev, S. (1987) Nitrogen Influence on the Fatigue Crack Growth Rate in Austenitic Structures, *Proc. of Higher Military School Scientific Session*, V.Tarnovo, Bulgaria, Vol. 4, pp.38-44
12. Zlateva, G. (1996) Nitrogen influence on the type of fracture during cycling loading, Unpublished paper, Institute of Metal Science, Sofia, Bulgaria
13. Davidov, Y. (1996) Plastic zone size and shape in front of fatigue crack tip for nitrogen stainless steel, In press in Journal *"Theoretical and Applied Mechanics"*
14. Duddelar, T., O'Reagan, R. (1971) *Exp. Mech.*, No **11**, pp. 49-56
15. Marshall, S., Rixon, R. (1987) *Opt. and Lasers in Engng.*, No 7, pp.175-182
16. Davidov, Y. (1991) Mechanisms of Fracture for nitrogen stainless steels under cyclic loading, *Scientific Report* No 64/1991, Inst. of Metal Science, Sofia, Bulgaria, pp. 86-95

ENERGY-BASED APPROACH TO DAMAGE CUMULATION IN RANDOM FATIGUE

T. ŁAGODA, E. MACHA
Technical University of Opole
ul.Mikołajczyka 5, 45-233 Opole, Poland

Abstract

Strain energy density rate, i.e.power density of stress has been proposed as a new parameter applied for description of fatigue under random loadings. The authors show how to change a scale of standard stress and strain characteristics of cyclic fatigue of the materials with use of the power parameter. Fatigue life calculated with the stress parameter and power was compared with the data for 10HNAP obtained under uniaxial tension with non-Gaussian probability distribution and wide-band frequency spectrum. From the tests it appears that the power parameter seems to be efficient.

1. Introduction

The well-known energetic models of uniaxial and multiaxial fatigue are based on various forms of strain energy density per a cycle. For a large number of cycles elastic strain energy is applied and when a number of cycles is low, we use plastic strain energy. The total elastic and plastic strain energy density is applied for both high and low-cycle fatigue [2]. Under multiaxial loadings the models including not all the strain energy density, but only its part connected with the critical fracture plane are also used [1, 5].

2. Theoretical considerations

It is difficult to formulate energy based fatigue models for random loadings. It results from the difficulties occurring while counting of plastic strain energy density based on the areas of closed stress-strain hysteresis loops under random loadings. We can, however, use another fatigue parameter, namely a rate of strain energy density, i.e. power density of internal forces or stresses, p(t), where t - time.

In the case of adiabatic processes the power density is equal to work increment, δW, done by the internal force in elementary volume of the material at time dt

R.A. Smith (ed.), Reliability Assessment of Cyclically Loaded Engineering Structures, 435–442.

$$p(t) = \frac{\delta W}{dt} = \sigma(t)\dot{\varepsilon}(t) \tag{1}$$

where: σ(t) - stress, $\dot{\varepsilon}(t)$ - rate of the strain.
For elastic strains with sinusoidal histories caused by uniaxial stress we have

$$\sigma(t) = \sigma_a \sin\omega t; \quad \varepsilon(t) = \varepsilon_a \sin\omega t; \quad \dot{\varepsilon}(t) = \varepsilon_a \omega \cos\omega t \tag{2}$$

$$p(t) = 0.5\sigma_a \varepsilon_a \omega \sin 2\omega t = p_a \sin 2\omega t \tag{3}$$

where: $p_a = 0.5\sigma_a\varepsilon_a\omega = 0.5\varepsilon_a^2 E\omega = 0.5\sigma_a^2\omega / E$ (E -Young modulus).
Two cycles of p(t) history while one cycle of σ(t) can be observed. Double increase of frequency can be also observed in the case of random histories.
Strain energy density per cycle

$$\Delta W = \int_{cycle} p(t)dt = \int_{cycle} \sigma(t)\dot{\varepsilon}(t)dt = \int_{cycle} \sigma d\varepsilon \tag{4}$$

and its maximum value W_{max} can be calculated by integration of p(t) at time T/4 (T-period of stress cycle), i.e.:

$$W_{max} = W^{e+} = \int_0^{T/4} p(t)dt = 0.5\sigma_a \varepsilon_a \tag{5}$$

Multiplying W^{e+} by frequency ω, we obtain amplitude, p_a, expressed by (3).
In the case of elastic-plastic strains the power density, $p(t)$, can be understood as a sum of elastic, $p^e(t)$, and plastic, $p^p(t)$, components

$$p(t) = p^e(t) + p^p(t) \tag{6}$$

After integration of $p^e(t)$ according to (5) we have

$$W^{e+} = 0.5\sigma_a \varepsilon_a^e \tag{7}$$

The integral of plastic power density $p^p(t)$ while one full cycle of stress should be equal to the area of closed stress-strain hysteresis loop, i.e. plastic strain energy density, ΔW^p.Assuming that the cyclic stress-strain curve is of Ramberg-Osgood type, the energy density, ΔW^p, per cycle has been determined in an analytical way [2, 3] and we can write

$$\Delta W^p = \int_{cycle} p^p(t)dt = \frac{1-n'}{1+n'}\Delta\sigma\Delta\varepsilon^p = 4\frac{1-n'}{1+n'}\sigma_a\varepsilon_a^p \tag{8}$$

where n' is the cyclic strain hardening exponent.
In the elastic-plastic range, the maximum strain energy density takes the form

$$W_{max} = W^{e+} + 0.5\Delta W^p = \frac{\sigma_a}{2}\left[\varepsilon_a^e + 4\frac{1-n'}{1+n'}\varepsilon_a^p\right] \tag{9}$$

Thus, the power density amplitude is

$$p_a = p_a^e + p_a^p = W_{max}\omega = \frac{\sigma_a\varepsilon_a^e\omega}{2} + 2\frac{1-n'}{1+n'}\sigma_a\varepsilon_a^p\omega \tag{10}$$

Assuming p(t) as a fatigue damage parameter we can rescale standard characteristics of cyclic fatigue $(\sigma_a - N_f)$ and $(\varepsilon_a - N_f)$ and obtain the new ones, $(p_a - N_{fp})$ for ω = const, for instance $\omega = 1s^{-1}$. In the case of high-cycle fatigue, where the characteristic $(\sigma_a - N_f)$ is used, the axis σ_a should be replaced by $p_a = \sigma_a^2\omega/(2E)$, and the axis N_f – by $N_{fp} = 2N_f$.
In the case of low- and high cycle fatigue where the characteristics $(\varepsilon_a - N_f)$ is applied, a similar rescaling can be done.
From Manson-Coffin-Basquin equation and (10) we obtain

$$\varepsilon_a = \varepsilon_a^e + \varepsilon_a^p = \frac{\sigma'_f}{E}(2N_f)^b + \varepsilon'_f(2N_f)^c \tag{11}$$

$$p_a = p_a^e + p_a^p = \frac{\sigma_a\omega}{2}\left[\frac{\sigma'_f}{E}N_{fp}^b + 4\frac{1-n'}{1+n'}\varepsilon'_f N_{fp}^c\right] \tag{12}$$

where σ'_f, b - fatigue strength coefficient and exponent, ε_f', c - fatigue ductility coefficient and exponent. Assuming that

$$\sigma_a = \sigma'_f(2N_f)^b = \sigma'_f N_{fp}^b \tag{13}$$

equation (12) becomes a form of a new fatigue characteristic $(p_a - N_{fp})$

$$p_a = \frac{(\sigma'_f)^2\omega}{2E}N_{fp}^{2b} + 2\frac{1-n'}{1+n'}\omega\varepsilon'_f\sigma'_f N_{fp}^{b+c} \tag{14}$$

For random fatigue we must calculate a strain rate, $\dot{\varepsilon}(t)$, from the strain history; next is should be multiplied by random stress, σ(t), in order to obtain power density history.

Power density p(t) is also a random process now and its sample records can be schematized with the known algorithms of cycle counting and damages can be cumulated according to commonly used hypotheses.

3. Experiments

The proposed energetic approach for fatigue damage summation was verified for a long life time while tests of specimens made of 10HNAP steel under uniaxial cyclic and random loading with wide-band frequency spectrum, non-Gaussian probability distribution and zero expected value [4]. The following constants of the $\sigma_a - N_f$ curves were determined under uniaxial cyclic tension - compression

$$\lg N_f = A - m \lg \sigma_a = 29.69 - 9.82 \lg \sigma_a \tag{15}$$

σ_{af} = 252.33 ± $\delta\sigma_{af}$, $\delta\sigma_{af} = 18.75\text{MPa}$, $N_o = 1.281 \times 10^6$ cycles; A ∈ (22.39; 36.98); m ∈ (6.89; 12.75), σ_{af} - fatigue limit.

Let us substitute the histogram of amplitudes under random loadings by the weighed average stress amplitude

$$\sigma_{aw} = \left(\frac{1}{N_b} \sum_{i=1}^{k} n_i \sigma_{ai}^m \right)^{1/m} \quad \text{for} \quad \sigma_{ai} \geq 0.5\sigma_{af} \quad \text{when} \quad \max_i(\sigma_{ai}) \geq \sigma_{af} - \delta\sigma_{af} \tag{16}$$

where N_b - block length in observation time T_o, n_i - number of cycles at the stress amplitude σ_{ai}.

Fig.1 shows the weighed average stress amplitudes and corresponding numbers of destructive cycles. It has been shown that weighed average stress amplitudes are included in the scatter band with the factor of 3 of the Wohler curve.

After rescaling characteristics (15) by means of the power parameter (ω=1/s) we obtain the following equation of regression

$$\lg 2N_f = A - m \lg p_a = 31.49 - 4.91 \lg p_a \tag{17}$$

Fig.2 shows two curves of cyclic fatigue for 10HNAP steel expressed by stresses and power.

4. Comparison of the calculated fatigue lives with those from tests

The cycles were counted from stres σ(t) and power density p(t) histories with the rain-flow algorithm; damages were cumulated with Palmgren-Miner hypothesis, including amplitudes of 50% above the fatigue limit for stress histories and 25% for power histories. In Fig.3 calculated and experimental fatigue lives are compared, taking into account stresses and power. It has been shown that fatigue life calculated with the power density parameter is included in the scatter band with the factor of 3 of the observed fatigue lives. From the tests it appears that the power density parameter seems to be more efficient than the conventional stress parameter.

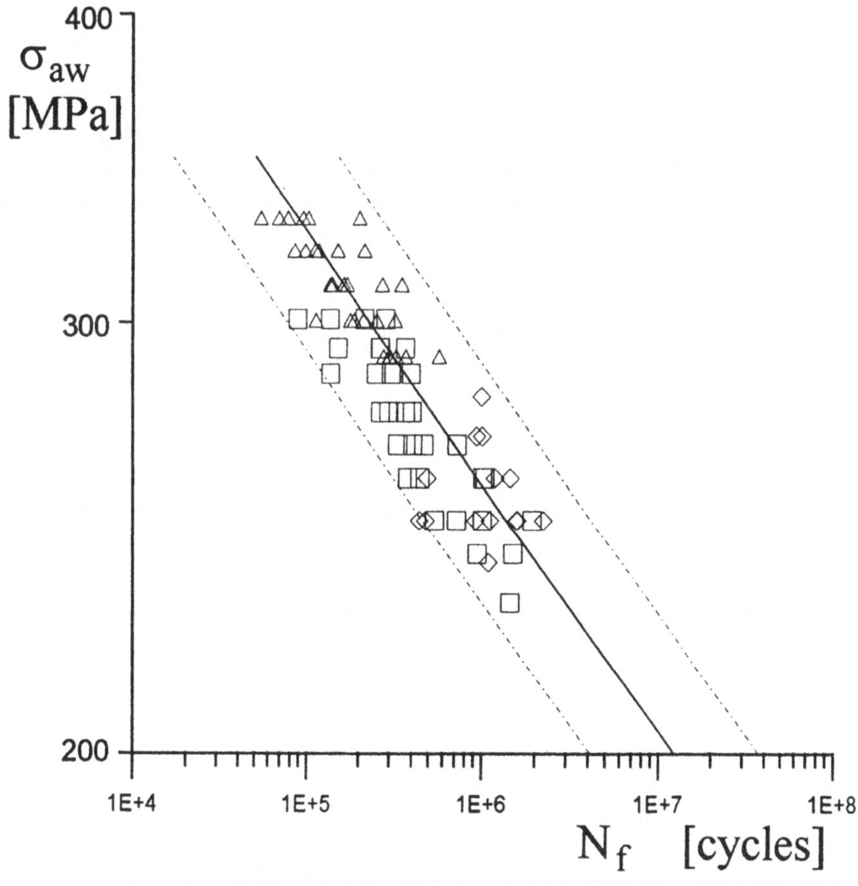

Fig.1 Weighed average stress amplitudes and corresponding numbers of cycles
Δ - uniaxial cyclic fatigue; ◊ - uniaxial cyclic fatigue at the level of fatigue limit; □ - uniaxial random fatigue;

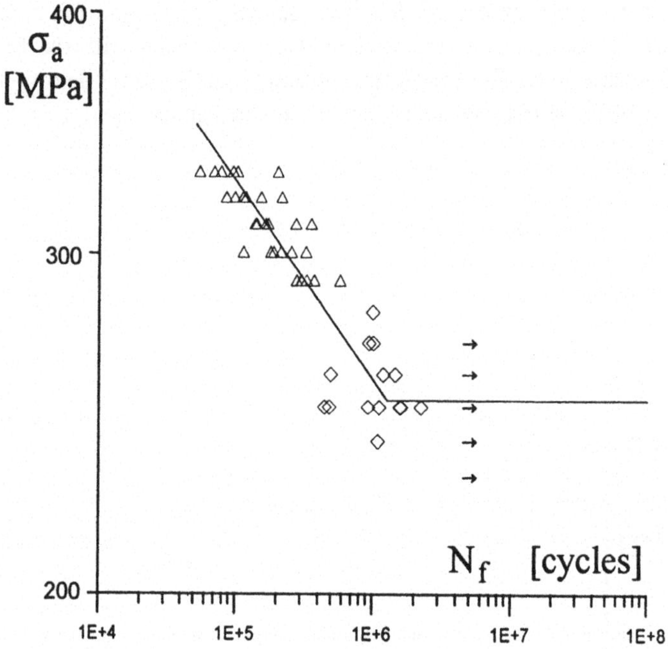

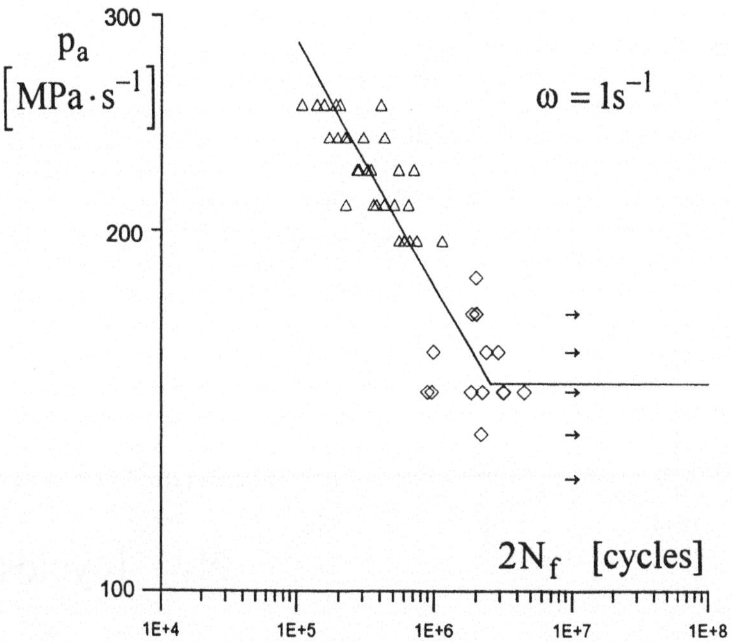

Fig.2 Curves of cyclic fatigue for 10HNAP steel expressed by stresses and power

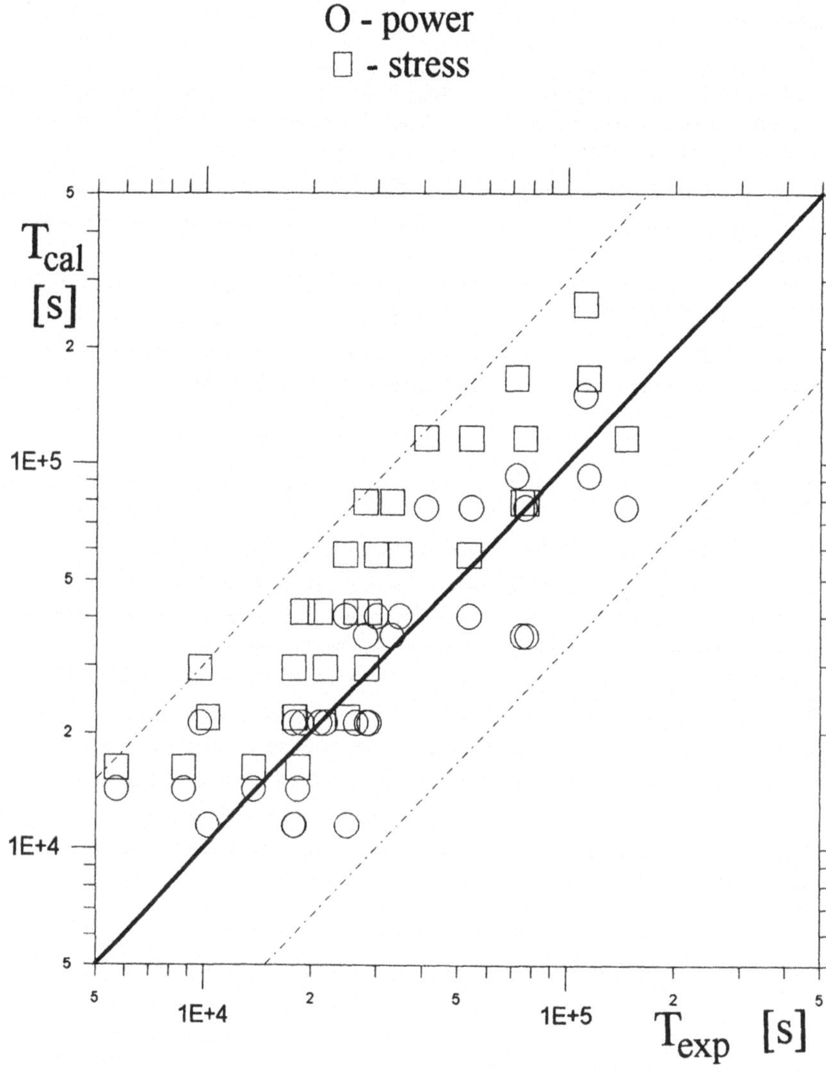

Fig.3 Comparison of calculated and experimental fatigue lives

5. Conclusions

1. It is difficult to count plastic strain energy basing on the closed histeresis loops under random loadings. The difficulties can be eliminated if a new fatigue parameter, i.e.strain energy density rate - power density of stresses - is used.
2. Application of the new parameter helps to determine a new fatigue characteristic of the material by rescaling of the widely used standard stress or strain characteristics for cyclic fatigue.
3. The experiments prove that long-life time calculated by means of the power density parameter for 10HNAP steel under uniaxial random tension-compression with non-Gaussian probability distribution and a wide-band frequency spectrum is included in the scatter band with the factor of 3, determined during cyclic tests.

6. References

1. Glinka G., Shen G., Plumtree A., (1995) A multiaxial fatigue strain energy density parameter related to the critical fracture plane, *Fatigue Fract.Engng.Mater.Struct.*, vol.18, No.1, pp.37-46
2. Gołoś K,. Ellyin F., (1988) A total strain energy density theory for cumulative fatigue damage, *Trans.ASME, Journal of Pressure Vessel Technology*, vol.110, No.1, pp.36-41
3. Halford G.R., (1966) The energy required for fatigue, Journal of Materials, vol.1, No.1, pp.3-18
4. Lachowicz C., Łagoda T., Macha E.,.(1996) Comparison of analytical and algorithmical methods for lifetime estimation in 10HNAP steel under random loadings, in: Fatigue '96, G.Lütjering and H.Nowack (Eds.), Proceedings of the sixth International Fatigue Congres, Berlin-Germany, vol. I, pp.595-600
5. Macha E., (1984) Fatigue failure criteria for materials under random triaxial state of stress, in S.R.Valuri et al (eds.), *Advances in Fracture Research, Proc.of the 6th Int.Conf. on Fracture (ICF6)*, New Delhi 1984,, Pergamon Press, Vol.3, pp.1895-1902.

The paper was partially financed within the grant KBN 7 T07B 013 10

APPLICATION OF A PROBABILISTIC APPROACH OF DURABILITY ANALYSIS TO GUST LOADED STRUCTURES AND SOME POSSIBLE EXTENSIONS

ANDREA PIERACCI
Assistant Professor at the University of Pisa
Department of Aerospace Engineering
Via Diotisalvi 2, 56126, Pisa, Italy

1. Abstract

The probabilistic approach to durability analysis of aircraft structures proposed by the U.S. Air Force for military aircrafts has been considered for application to gust loaded structures. An experimental program with simple specimens has been executed and the method applied with positive results. An extension of the model based upon a confidence level approach for the assessment of the quality of the parameters estimation has been proposed. This extension has the intent to allow the user to plan the number of experimental tests to be carried out to achieve a certain confidence level of the probability of crack exceedance, which is a key parameter of the U.S. Air Force method. This same approach might be followed to compare different sets of specimens using the Initial Fatigue Quality concept on a confidence level basis thus taking also the estimation error into account. Some example problems are discussed to show the application of the proposed approach.

2. Nomenclature

$a_{(0)}$ = equivalent initial flaw size, i.e. crack length at time zero
a_0 = reference crack length at time t_0
b = parameter of the crack growth law
$F_{a(0)}$ = cumulative probability function of EIFS
$\hat{F}_{a(0)}$ = estimated cumulative probability function of EIFS
F_T = cumulative probability function of TTCI
LS = Least Square
ML = Maximum Likelihood
N = number of elements in a data sample
Q = parameter of the crack growth law
t = service time
T = time to crack initiation

R.A. Smith (ed.), Reliability Assessment of Cyclically Loaded Engineering Structures, 443–466.

α = shape parameter of the Weibull probability distribution function

$\hat{\alpha}$ = estimated shape parameter of the Weibull probability distribution function

β = scale parameter of the Weibull probability distribution function

$\hat{\beta}$ = estimated scale parameter of the Weibull probability distribution function

ε = lower bound of the TTCI probability distribution function

$\hat{\phi}$ = estimated scale parameter of the EIFS cumulative probability distribution function

K = parameter related to the uniform distributed data set in the Monte Carlo simulation by using the Least Square method

K' = parameter related to the uniform distributed data set in the Monte Carlo simulation by using the Maximum Likelihood method

λ = parameter related to the uniform distributed data set in the Monte Carlo simulation by using the Least Square method

λ' = parameter related to the uniform distributed data set in the Monte Carlo simulation by using the Maximum Likelihood method

x_u = upper bound of the EIFS probability distribution function

3. Introduction

A probabilistic approach to durability analysis of aerospace structure has been proposed by the U.S. Air Force, [1-9]. Following this approach the distribution of the damage in a structure is evaluated by considering the population of cracks which nucleate from stress concentration zones in the analyzed component, for instance fastener holes, cut outs, etc. The crack length is modeled as a random variable. The user can monitor the evolution of the damage in the structure, through the estimation of the random variable defined above and therefore estimate quantitatively how the capability of the stucture of carrying load is degraded; through this knowledge one can decide the location in time for inspection and upgrading intervals. Using the U.S. Air Force approach one obtains the service time corresponding to a fixed probability of crack exceedance for a choosen crack length. Consequently it is possible to estimate the number of cracks in the structure with dimension greater than a reference one and a corresponding γ-level confidence interval. This approach has been proposed and extensively tested by the U.S. Air Force for application to military aircrafts, i.e. aircrafts mainly subjected to manouvre loading. Aim of this work is to apply the U.S. Air Force methodology to commercial aircrafts, i.e. to aircrafts structures primarly subjected to gust type loading histories. An experimental program has been carried out and applied to the test data obtained. A further study of the model based on the parameter estimation procedure, [10], has allowed to suggest some further extensions of the method in the field of inspection intervals definition, [11], as well as in the field of statistical test planning and of fatigue quality assessment and comparison, as will be shown in the sequel.

4. U.S. Air Force Initial Fatigue Quality Method

According to the U.S. Air Force method the dimension of the crack is modeled as a random variable. Each flaw, due to the limited range of length considered in comparison to the distance between the stress concentration zones where the cracks originate, is independent of the others originating from nearby locations; i.e. it is assumed that the situation studied is such that no interaction takes place. Once the probability distribution function of having a flaw with a dimension greater than a fixed value has been estimated, an assessment of the damage of the structure can be obtained. In fact, at a given time the probability of crack exceedance is known and, under the assumption of binomial distribution, an estimation is easily calculated of the number of cracks longer than a reference value. The inspection interval should be planned in such a way that, when the number of cracks with dimension greater than a reference value exceeds a fixed value, the aircraft must be grounded and the structure upgraded. In the case of riveted components, the rivets must be removed, the holes reamed and larger diameter rivets installed, thus restoring the fatigue quality level of the structure. The approach proposed by the U.S. Air Force makes use of two statistical distributions, see [8], namely: i) the distribution of the Time To Crack Initiation (TTCI) and ii) the distribution of the Equivalent Initial Flaw Size (EIFS), see Figure 1. The TTCI distribution represents the time necessary to reach a fixed length for the largest crack in the stress concentration zone. Once the TTCI distribution has been evaluated, one can obtain the EIFS distribution by back extrapolating the crack length to time zero, as will be explained in the sequel. The TTCI distribution is described by means of a three parameter Weibull distribution as follows:

$$F_T(t) \equiv P[T \leq t] = 1 - \exp\left[-\left(\frac{t-\varepsilon}{\beta}\right)^{\alpha}\right] \tag{1}$$

where $F_T(t)$ gives the probability that the crack has reached the reference crack length at a service time T smaller than t, whereas its complementary value, i.e.

$$1 - F_T(t) = P[T > t] = \exp\left[-\left(\frac{t-\varepsilon}{\beta}\right)^{\alpha}\right] \tag{2}$$

obviously represents the probability that the crack reaches the reference crack length at a time T greater than t.

The EIFS distribution is obtained from the TTCI distribution by using a back extrapolation method, see [8]. The following expression holds:

$$F_T(t) = 1 - F_{a(0)}(x) \tag{3}$$

where:

$$F_{a(0)}(x) \equiv P[a \leq x] \tag{4}$$

and:

$$\frac{da}{dt} = Qa^b \tag{5}$$

where a is the crack length and Q and b are parameters which depend on the component material and geometry and on the load history. In the present analysis it is assumed that the crack growth law can be exactly obtained so that Q and b can be considered constant, further it is assumed $b=1$ as in some cases examined in [8]. Considering Eqns. (1), (3) and (5), one obtains:

$$F_{a(0)} \equiv \exp\left\{-\left[\frac{\ln(x_u / x)}{\phi}\right]^{\alpha}\right\}; \qquad 0 \le x \le x_u$$

$$\equiv 1.0; \qquad x \ge x_u \tag{6}$$

where x_u is the EIFS upper bound limit, corresponding to the TTCI lower bound limit, and ϕ is related through Eqn.(5) to the parameters of the TTCI distribution function. Indeed the following relationship holds:

$$x_u = a_0 \exp(-Q\varepsilon) \tag{7}$$

$$\phi = Q\beta \tag{8}$$

The EIFS probability function is also called Weibull compatible distribution because of Eqn. (3), see [8].

Once the EIFS distribution function is known, one can obtain the probability of crack exceedance of a given crack length at any time using Eqns. (5) and (6), see [8]. Through the estimated probability of crack exceedance one can estimate the number of cracks with length greater than a fixed one assuming that the cracks are independent on each other due to their small dimension compared to their relative distance. Assuming that N is the number of stress concentration zones of the component studied, i.e. the locations where cracks can nucleate, and p the probability of exceeding a given crack length, the mean number of cracks having the dimension greater of that dimension at the service time t, $\overline{N}(t)$, and the corresponding standard deviation, $\sigma(t)$, are given by:

$$\overline{N}(t) = Np \tag{9}$$

$$\sigma(t) = Np\sqrt{1-p} \tag{10}$$

Consequentely the damage extent as a function of the service time is known. In case that the TTCI probability distribution function is known for stress concentration zones with different stress level, a global EIFS distribution can be obtained imposing the condition that all the TTCI distributions, once that are back extrapolated to time zero, give place to the same EIFS distribution, see [8] for further details.

5. Experimental Program

An experimental program has been carried out to evaluate the applicability of the model to commercial aircrafts. The material considered was 2024-T351. The specimens were simple dog-bone plain specimens with a central hole without any fitting to semplify the data acquisition. The load sequence used in the testing was the miniTWIST standard loading history, [12], which is a typical rapresentation of the vertical acceleration history at the wing root of a commercial aircraft. Three sets of specimens have been tested, named NTR80, NTR85 and NTR90, having respectively a mean flight stress of 80, 85 and 90 MPa. A first testing activity showed that it was difficult to build crack growth curves on the basis of fractographic data, due to problems in identifying striations and in univocally relating them to load cycles. Two different approaches were followed to solve this problem: first, marker blocks were introduced in the loading sequence, [13], so to have marker bands which could allow to identify correctly the relation between load cycles and striations on the fractographic surfaces. The approach followed was the one proposed in [14], in which the intent was to plan the introduction of a block loading sequence in the actual loading history. The single cycle of the marker block should not contribute to crack propagation, but should be such that the block of cycles as a whole will mark the crack surface. The application of this approach gave at first positive results, see Figure 2, but soon it became clear that the validation of the ininfluence of the marking loading sequence on the nucleation and propagation of the cracks would lead to an extensive and overcharging testing activity. A second approach was then applied; crack gages were bonded to the specimen surface near the crack. The crack gages were composed by copper wires annealed in an epossidic matrix; electric current was applied to the circuit. When the crack was advancing, the matrix, and thus the wires annealed in it, get broken and a potential drop was recorded at the end of the circuit. The potential drop, which was recorded during the test, was thus related through a calibration curve to the propagation of the crack. The crack gages allowed to record the crack propagation on the surface of the specimen which was then used to identify the striations on the crack surface due to the most stressing flight of the miniTWIST loading history, thus relating biunivocally crack length and applied loads (i.e. flight hours). With the aid of the crack gages it was possible to read the crack surfaces with a stereo optical microscope down to crack length smaller than 0.5 mm. The crack lengths vs. blocks of flights (one block = 4000 miniTWIST flights) for the three sets of specimens tested are reported at Tabs. 1-3. The crack lengths reported are the values interpoleted from the experimental ones as suggested in [8] for reducing all the experimental data to the same crack length.

6. Application of the U.S. Air Force Model to the Test Data

The U.S. Air Force model has been applied to the data obtained from the experimental testing in the crack length range of 0.5-1.5mm. Least Square parameter estimation has been used to evaluate the shape and scale parameters of the Weibull probability distribution function of the Initial Fatigue Quality for a given value of the upper bound of the EIFS distribution, x_u. Such value is usually given either by considerations on the technology applied and the workmanship and inspection techniques of the components tested, or it is obtained almost blindly by minimizing the root square error between the estimated cumulative probability distribution function and the experimental cumulative probability distribution of EIFS or TTCI. The crack growth law reported in Eqn. (5) with the assumption of having $b=1$ has been used to back extrapolate the nucleation data, i.e. the TTCI distribution, to service time zero, thus obtaining the experimental EIFS distribution. Using the Least Square method the parameters of the EIFS distribution based on all of the experimental data, i.e. of all the three sets togheter, have been obtained. The comparison between the estimated EIFS distribution and the experimental one is reported in Figure 3. The parameters estimated by the probabilistic model for the test data are reported in Tab. 4-6. The good correspondence between the experimental data and the model curve can not be used as a verification of the predictive quality of the model, but only to certify that the compatible Weibull probability distribution function seems to be a good choice for modelling the EIFS, as it was expected. Nothing more can be said to this regard, due to the fact that the experimental data reported are the ones used to identify the parameter of the EIFS distribution, so that they can not be used to evaluate the predictive capability of the model. A censored data analysis could have been used to help in evaluating the predictive capability of the model, but a more detailed analytical and numerical analysis was executed instead, as will be shown in the sequel.

7. Parameter Estimation of the TTCI and EIFS Probability Distribution Functions

The quality of the predictions which might be obtained applying the probabilistic model are strongly dependent on the capability of correctly evaluating the parameters of the EIFS and TTCI probability distribution functions. One first point to be investigated was to check if Least Square parameter estimation would actually give better results than Maximum Likelihood parameter estimation. Maximum Likelihood gives the lowest variance for unbiased estimation for large data sample, theorically for the number of data in the sample growing to infinity; no study was found which could help in assessing if LS was actually better than ML parameter estimation for limited data sample and which is the dependency of the quality of the estimation from the number of data considered. To achieve this Monte Carlo simulations were executed to compare the two estimation procedures; the results led to the conclusion that ML gives better results than LS for estimating the probability of crack exceedance obtained from a Weibull distribution, so that the use of ML estimation instead of LS was suggested in applying the U.S. Air Force probability method, [10]. The simulations gave results of general validity, because of the independency of the results obtained from the Weibull random number generating function used in the Monte Carlo calculations. This is a property of the LS and ML parameter estimation of the Weibull probability distribution function, which was analytically demonstrated, [15].

As shown in [10] and [15], the actual Weibull distributed data can be thought as if it has been obtained through a transformation from a set of uniform distributed data. Accordingly the LS estimated values of the parameters can be written in the following way:

$$\hat{\alpha} = K\alpha \tag{11}$$

$$\ln\hat{\beta} = \ln\beta + \frac{\lambda}{\alpha} \tag{12}$$

where the symbols with the carets represent the estimated values of the parameters, and the constants λ and K are only dependent on the transformation from the uniform distributed data set to the Weibull distributed data set.

Following [10], [15] and [16], for ML the estimated values of the parameters can be written as:

$$\hat{\alpha} = K'\alpha \tag{13}$$

$$\hat{\beta} = \beta\lambda'^{\frac{1}{K'\alpha}} \tag{14}$$

where the symbols have the same meaning as for the case of LS parameters estimation. The relations reported above between actual and extimated values of the parameters allow to obtain some results of general validity as will be shown in the sequel.

7.1 APPLICATIONS OF THE RELATIONS OBTAINED TO STATISTICAL TEST PLANNING

From the relations reported in the previous paragraph, it comes out that the relative error in probability of crack exceedance is independent of the Weibull random number generating function. In fact, indicating with P the probability of having a crack with a dimension greater than a given one at a given service time, i.e. $F_T(t)$, for LS estimation it is:

$$e(P) = \frac{P-\hat{P}}{P} = 1 - \frac{\exp\left\{-\left\{\left[-\ln(1-P)\right]\exp{-\lambda}\right\}\right\} - 1}{P} \tag{15}$$

while for ML estimation, it is:

$$e(P) = \frac{P-\hat{P}}{P} = 1 - \frac{\exp\left\{-\left\{\left[-\ln(1-P)^{K'}\right]/\lambda'\right\}\right\} - 1}{P}. \tag{16}$$

Monte Carlo simulations have been executed to evaluate the behaviour of the statistic $e(P)$ and to use this information to plan experimental testing according to the confidence required for the estimation. The total number of simulations was 64000 which was found to give stability in the results to be obtained. The results of the simulations are reported graphically in Figs. 4 through 9 in the following way: in the x-axis the values of the statistic $e(P)$ are reported, while on the y-axis the probability of having a relative error in probability lower than the value reported in the x-axis is reported. In Figures 4, 5 and 6 these results are reported for P= 0.05, 0.5 and 0.95 for LS parameter estimation, while in Figs. 7, 8 and 9, analogous results are reported for ML parameters estimation. The simulations have been done varying the data sample amplitude from 10 to 100 data increasing the amplitude of the sample of 10 data each time. The curves reported appear to move upward, so reducing the probability of being in error, increasing the amplitude of the data sample. The use of this curve is the following: if one wants to estimate the parameter of the TTCI distribution, and thus of the EIFS distribution, with a certain probability that it will give a relative error in probability of crack exceedance, $e(P)$, lower than a fixed value for a given P, through the estimation reported the user might evaluate which is the number of tests to be executed to achieve the required goal. For example, considering Figure 5 for LS estimation of a 0.5 probability of crack exceedance, with a set of 100 data we have 0.8 probability of doing an estimation with a relative error smaller than 10%, while with 10 data sample the probability of having an error lower than 10% is 0.4.

This kind of informations, easily obtainable also for other values of probability of crack exceedance, will allow an effective test planning which could account of the error in the results due to the parameter estimation.

7.2 APPLICATIONS OF THE RELATIONS OBTAINED TO FATIGUE QUALITY ASSESSMENT AND EVALUATION ON A CONFIDENCE LEVEL BASIS

The relations reported in Eqns. (11) through (14) could also be used to obtain confidence bounds of the values of the EIFS and TTCI probability distributions through Monte Carlo simulations.

Using the expressions of the relations between the estimated and the actual values of the parameters of the Weibull probability distribution function of Initial Fatigue Quality, it is easy to obtain the relationships between the estimated and actual value of the probability of crack exceedance for both ML and LS parameter estimation, as well as of the probability of having cracks with dimension smaller than a given value, which might also be a parameter of interest. For LS estimation, analogously to Eqns. (15) and (16), it comes out:

$$\hat{F}_{a(0)} = 1 - \exp - \left\{ \left[-\ln\left(F_{a(0)} \right) \right] \exp(-\lambda) \right\}^{K} . \tag{17}$$

In a similar way for ML estimation it is:

$$\hat{F}_{a(0)} = 1 - \exp\left\{ - \frac{\left[-\ln\left(F_{a(0)} \right) \right]^{K'}}{\lambda'} \right\} . \tag{18}$$

From the two relations reported above, it is clear that it is possible to evaluate the behaviour of the actual value of $F_{a(0)}$ through its estimated value and viceversa. Most important thing, these relations are not dependent on the actual values of the parameters of the Weibull (or compatible Weibull) probability distribution function. Monte Carlo simulations have been executed which allowed to obtain the relations between actual and estimated values of the cumulative probability distribution of the EIFS. Taking the complement to one, the probability of exceeding a certain crack length can also be obtained. The results of these simulations have been reported in a tabular way and used for estimating the most appropriate time interval for inspection and upgrading of an aircraft structure in [11]. These tabulated results can also be used to assess the fatigue quality of a structure and to compare on appropriate basis the fatigue quality of different structures. Assume that one wants to estimate the initial fatigue quality of a metallic aicraft structures. To quantitatively assess this, one would like to know which is the service time at which one has a certain probability of having cracks with length greater (or smaller) than a reference value, so to know the behaviour of the structure during service. To do this one has to estimate the parameters of the Initial Fatigue Quality, i.e. of the EIFS and TTCI probability distribution functions, and then apply Eqn. (5). The fact is that if one does so, the smaller the data sample amplitude, the greater the probability of having an incorrect estimate as might be seen considering Figs. 4 through 9 again. To overcame this difficulty the tables reported in [11] might be used. These tables report the actual value of the probability of having a crack with dimension smaller than a given one as a function of the estimated value of the probability for a given confidence level as a result of the Monte Carlo simulations. The use of these tables will be explained in the sequel through an example calculation. The knowledge of the confidence of the estimation obtained as a function of the data sample amplitude, will allow to have a more valuable comparison of the Initial Fatigue Quality.

8. Example Calculations

Consider the data obtained by the experimental program of the present investigation, i.e. the data reported at the tables 1-3. Consider first the NTR85 data set. For that set, see Tab. 5, it is: $\alpha = 2.638$, $Q\beta = 4.812$, $x_u = 0.5$, $a_0 = 0.5$. Assume to be interested in the percentile corresponding to a 0.9 probability of having cracks smaller than 0.5 mm, i.e. $F_{a(0)}(x) = P = 0.9$. Applying Eqn. (5), it comes out x=0.064 mm, where x is the equivalent initial flaw size for that probability of crack exceedance. If one considers all three data sets togheter minimizing the root square error, it is, Tab. 4-6, $\alpha = 3.557$ and$Q\beta = 7.616$ and it follows x=0.008 mm. The time to reach a crack length of 0.5 mm. for that percentile is then t =3.858 flight blocks for the 85 Mpa data set, while it is t =7.622 flight blocks for the whole data set. This first result brings into light the difference quality of the inference that might be executed varying the data sample amplitude on which the estimation is based.

Consider now the effects of the estimation errors; to evaluate a 0.9 probability of having cracks smaller than 0.5 mm with a 0.95 confidence level, the estimated value of that probability has to be 0.966, see Tab.7. Then, using Eqn. (5), it is x=0.13 and then t=2.529 flight blocks. Similarly, for the global data sample, it is $\hat{F}_{a(0)}$=0.944. From this last result it comes out: x=0.016 and t =6.402 flight blocks. These results show the different conclusions which are obtained using a confidence level approach.

Let us consider now some data sets taken from [17] (the British Units system will be followed for this example to be consistent with the cited reference); the first one, named AFXLR4, consist of the following TTCI data sample relative to a crack length of 0.035 inches, i.e. 4511, 6875, 7474, 9905, 10262, 12818, 14066, 16676, 18653, and 23536 flight hours and it is α=1.575, β =29872, $Q\beta$ =5.443, x_u = 0.025 inches [8]. This data set is relative to the application of the F-16 load spectrum to 7475-T7351 specimens with a 15% load transfer and mean stress level of 32 Ksi. Considering again a probability of 0.9, it comes out x=0.00678 inches, which requires a service time interval of 10966 flight hours to grow up to an hypothetical economic limit length of 0.05 inches. If one considers the lower 95% confidence bound, the necessary estimated value of the probability of crack exceedance is 0.966 which gives a crack length of 0.0131 inches and a service time of 7351 flight hours. For the second set, named AFXLR3 differring from the AFXLR4 data set for the fitting diameter, the TTCI data sample for a_0 = 0.035 inches is: 4824, 6555, 7486, 8108, 8823, 8923, 8923, 9921, 10804, 13786, 17200, and 17794 flight hours. Assuming as for the previuos data set, that it should be x_u = 0.025 inches, it comes out to be α=2.222, β =10201 and $Q\beta$ =2.532. Considering as before, P=0.9, it comes out to be x=0.00996 inches, which leads to a service time interval of 6523 flight hours. Considering the 0.95% confidence level, the estimated value of P has to be 0.963, which leads to x=0.0139 inches and to a service time of 5156 flight hours. Consider the global data set composed by the three sets of specimens, AFXLR4, AFXHLR4, AFXMLR4, having the same kind of fitting and percentage of load transfer but a different value of the mean stress level, it follows [8] : α=1.805, $Q\beta$ =2.155, x_u = 0.025 inches , so that for P=0.9 it is x=0.0134 inches; the service time interval apt to reach a crack length of 0.05 inches is then 7571 flight hours. Considering the 95% confidence value for a data set of 30 data, see Tab.7, the value of probability of crack exceedance to be estimated is equal to 0.945, so that it is x=0.0161 inches, which gives a value of the service time interval of 6516 flight hours.

These examples, apart from the effect of the data sample amplitude, show clearly the difference that occour if one considers a confidence level approach instead of the usual U.S. Air Force method procedure to evaluate the degradation of a structure during service on the basis of the EIFS probability distribution function. Because difference of the inference quality that can be obtained with the two different approaches might be considerable, the use of the confidence level approach for estimating fatigue quality and for assessing the degradation of a structure is recommended.

9. Conclusions

In the present work the probabilistic approach to durability analysis of metallic aircraft structures proposed by the U.S. Air Force for application to military aircrafts has been applied to gust loaded structures. An experimental program with simple specimens has been executed and the probabilistic approach applied to the test data. Further studies have been executed which allow to use some relations obtained in a related work by the author to build curves, herein reported, to be used in statistical planning of experimental tests. A confidence interval approach for assessing and compare the Initial Fatigue Quality is also suggested; such an approach could take the effect of the estimation errors into account. Examples calculations are reported to show the application of the proposed extension of the model.

10. Acknowledgements

The investigation reported has been supported by the National Research Councill of Italy under the framework of the *Progetto Finalizzato Trasporti 2.*

The author would like to thank his friend and collegue Giovanni Mengali for the valuable discussions and the continuous support during this and other research activities.

11. References

1. Manning, S.D. and Yang, J.N. (1984*) Durability Methods Development - Phase I Summary*, AFWAL-TR-83-3027, Vol. I, Wright-Patterson Air Force Base, Ohio.
2. Manning, S.D. and Yang, J.N. (1984) *Durability Analysis: State of the Art Assesment*, AFWAL-TR-83-3027, Vol. II, Wright-Patterson Air Force Base, Ohio.
3. Manning, S.D. and Yang, J.N. (1984*) Structural Durability Survey: State of the Art Assesment*, AFWAL-TR-83-3027, Vol. III, Wright-Patterson Air Force Base, Ohio.
4. Shinozuka, M. (1984) *Initial Quality Rapresentation*, AFWAL-TR-83-3027, Vol. IV, Wright-Patterson Air Force Base, Ohio.
5. Manning, S.D. and Yang, J.N. (1984) *Durability Analysis Methodology Development*, AFWAL-TR-83-3027, Vol. V, Wright-Patterson Air Force Base, Ohio.
6. Manning, S.D. and Yang, J.N. (1984) *Durability Analysis Methodology Development - Phase II Documentation*, AFWAL-TR-83-3027, Vol. VII, Wright-Patterson Air Force Base, Ohio.
7. Rudd, J.L., Manning, S.D., Yang, J.N. and Yee, B.G.W. (1982) Probabilistic Fracture Mechanics Analysis Methods for Structural Durability, *Proc. of the AGARD Meeting on Behaviour of Short Cracks in Airframe Components*, Toronto, Canada.
8. Manning, S.D. and Yang, J.N. (1984) *USAF Durability Design Handbook: Guidelines for the Analysis of Durable Aircraft Structures*, AFWAL-TR-83-3027, Wright-Patterson Air Force Base, Ohio.
9. Manning, S.D. and Yang, J.N. (1989) *USAF Durability Design Handbook: Guidelines for the Analysis of Durable Aircraft Structures*, AFWAL-TR-88-3119, Wright-Patterson Air Force Base, Ohio.

10. Pieracci, A. (1995) Parameter Estimation for Weibull Probability Distribution Function of Initial Fatigue Quality, *AIAA Journal*, **33**, 1574-1581.
11. Pieracci, A. (1996) A Confidence Interval Approach for the Determination of Inspection Intervals Based Upon Initial Fatigue Quality , to be published by the *AIAA Journal*.
12. Lowak, H., De Jonge, J.B., Franz, J., Scutz, D. (1979) MiniTWIST, a shortened version of TWIST, *Report NLR-MP 79018*, National Aerospace Laboratory, Amsterdam.
13. Salvetti, A., Pieracci, A., (1993) Approccio Statistico al Progetto a Durabilita' di Strutture Aeronautiche in Materiale Metallico, *Proceedings of the Ist National Conference of the National Research Council of Italy on the Progetto Finalizzato Trasporti 2*, **2**,1383-1398, Rome.
14. Palmberg, B., (1984) On the Use of Marker Loads for Fatigue Crack Growth Measurements'Fatigue Crack Topography, *AGARD CP 376*, Printing Services Limited, Loughton, Essex, England.
15. Pieracci, A., (1996) Some Useful Formulas for Monte Carlo Simulations Related to the Weibull and to the Exponential Probability Distribution Functions, submitted as a technical note to *Fatigue & Fracture of Engineering Materials & Structures - The International Journal.*
16. Thoman, R.T., Bain, L. J. and Antle, C.E. (1970) Maximum Likelihood Estimation, Exact Confidence Intervals for Reliability, and Tolerance Limits in the Weibull Distribution, *Technometrics*, **12**, 363-371.
17. Manning, S.D. and Yang, J.N. (1982) *Durability Methods Development - Volume VIII - Test and Fractographic Data*, AFFDL-TR-79-3123, Vol. VIII, Wright-Patterson Air Force Base, Ohio.

TABLE 1. TTCI in terms of blocks of 4000 miniTWIST flights with 90 Mpa mean flight stress

Specimen	a_0=0.50	a_0=0.75	a_0=1.00	a_0=1.25	a_0=1.50
MTG01	4.27900	4.94863	5.35574	5.62741	5.86924
MTG02	6.76736	7.55076	8.09621	8.56710	8.91968
MTG03	8.68363	8.56101	9.44982	9.62960	10.03884
MTG04	4.40813	4.50490	5.02930	5.82206	6.51837
MTG05	11.28446	11.89084	12.60330	13.10052	13.51449
MTG06	10.86254	11.21765	11.57811	11.94393	12.31512
MTG07	10.36754	10.65639	11.42927	11.85158	12.42024
MTG08	5.33190	5.85968	6.48667	7.00692	7.46317
MTG09	12.10553	12.41453	12.71822	13.01639	13.30904
MTG10	6.35359	7.05358	7.67637	8.08295	8.48835
MTG11	7.23432	7.52278	8.13254	8.62558	9.00441
AVERAGE	7.97073	8.38007	8.95959	9.38855	9.80554
Q			0.531		

TABLE 2. TTCI in terms of blocks of 4000 miniTWIST flights with 85 MPa mean flight stress

Specimen	a_0=0.50	a_0=0.75	a_0=1.00	a_0=1.25	a_0=1.50
MTG12	17.11864	18.58517	19.34306	20.62284	20.99646
MTG13	17.78670	18.65190	19.44115	19.68650	19.98964
MTG14	16.68597	18.09530	18.87727	19.88681	20.46416
MTG15	13.73783	15.12424	16.10791	16.87090	17.48321
MTG16	13.87880	16.22149	17.93523	19.50535	20.36742
MTG17	13.10291	14.16798	14.95786	15.46831	15.72858
MTG18	21.40420	22.37034	22.64584	23.22305	23.63475
MTG19	19.22305	20.27690	21.00128	21.68650	22.15716
MTG20	22.28771	22.85088	23.36108	23.75804	24.35421
MTG21	16.57977	17.04420	17.48728	18.12369	18.82064
AVERAGE	17.18056	18.33884	19.11579	19.88320	20.39962
Q			0.340		

TABLE 3. TTCI in terms of blocks of 4000 miniTWIST flights with 80 MPa mean flight stress

Specimen	a_0=0.50	a_0=0.75	a_0=1.00	a_0=1.25	a_0=1.50
MTG22	21.64404	22.13259	22.56943	23.01807	23.40190
MTG23	19.14598	19.49928	19.95683	20.43819	20.88884
MTG24	12.49647	12.77023	12.97437	13.21867	13.48265
MTG25	30.11700	30.70594	31.24160	31.64383	31.92467
MTG26	12.78860	13.90148	14.69412	15.31662	15.75269
MTG27	19.04680	19.87450	20.38058	21.11496	21.60917
MTG28	25.66944	26.61146	27.08444	27.79483	28.36435
MTG29	26.30709	27.20835	27.80800	28.19916	28.49310
MTG30	13.28654	14.36662	14.99871	15.55789	16.08952
MTG31	24.62572	25.52081	25.96382	26.51065	26.91117
AVERAGE	20.51277	21.25912	21.76719	22.28129	22.69181
Q			0.505		

TABLE 4. Root Square Error (RSE) between theoretical and experimental EIFS cumulative probability distribution function

a_0	0.5	0.75	1.0	1.25	1.5
x_u	0.5	0.75	1.0	1.25	1.5
RSE	0.079288	0.082836	0.085477	0.087425	0.087894

TABLE 5. EIFS parameters for single data set

a_0=0.5 SET	x_u=0.5 Q_i	ε_i	α_i	β_i	$Q_i\beta_i$
NTR90	0.531	0.0	2.638	9.055	4.812
NTR85	0.340	0.0	5.304	18.582	6.319
NTR80	0.505	0.0	2.975	23.156	11.716

TABLE 6. EIFS parameters for the global data set

a_0=0.5	x_u=0.5	$\alpha = 3.557$	$Q\beta = 7.616$

TABLE 7. Lower Confidence Bound of the EIFS distribution with 0.95 confidence level for Least Square parameter estimation, [11]

number of data	10	12	30	31
$\hat{F}_{a(0)}$	0.966	0.963	0.945	0.944
$F_{a(0)}$	0.9	0.9	0.9	0.9

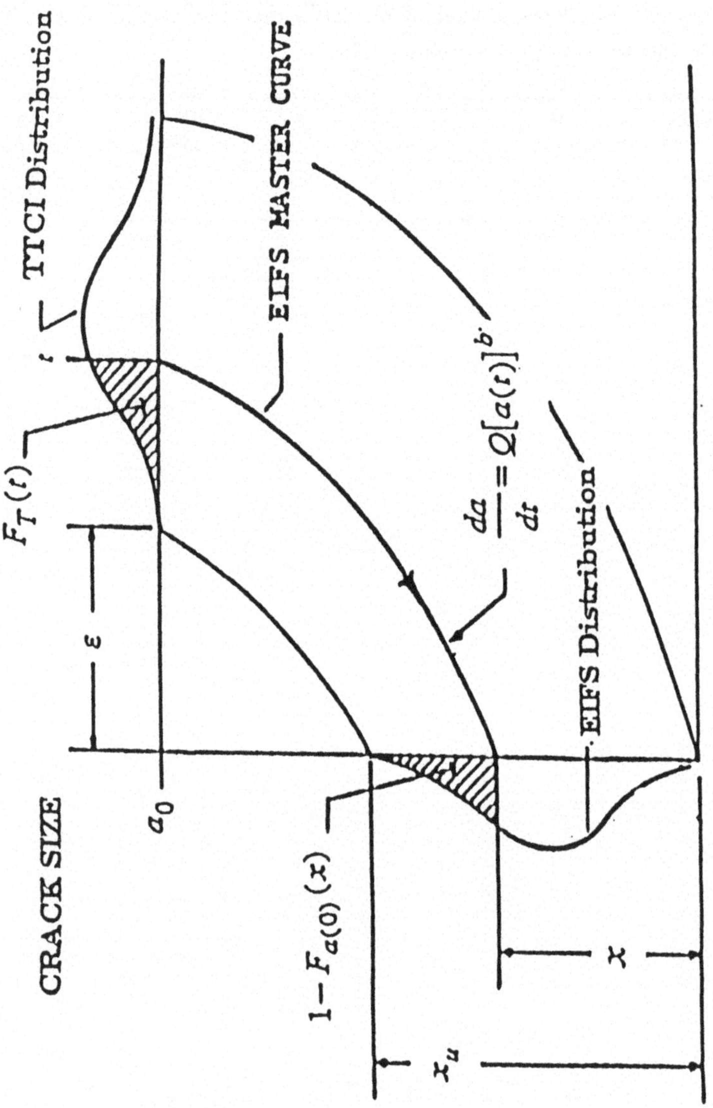

Figure 1. Initial Fatigue Quality,[8]

Fig. 2 Specimen fractographic surface

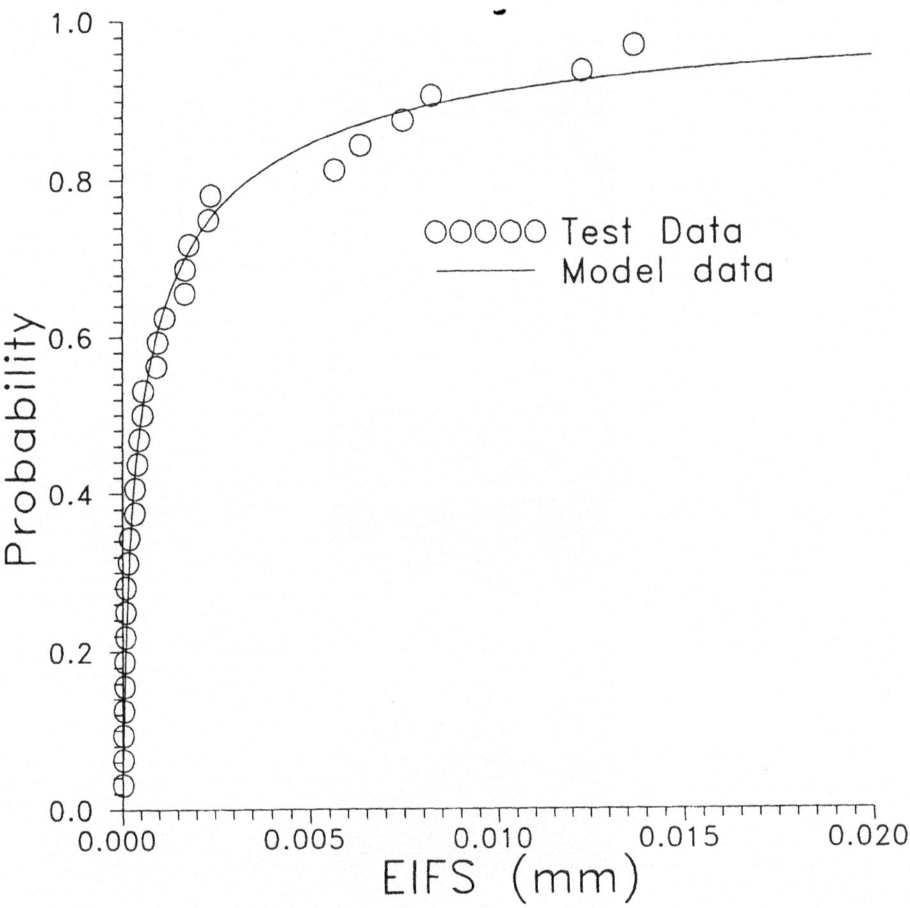

Figure3. Experimental and calculated EIFS cumulative probability distribution function

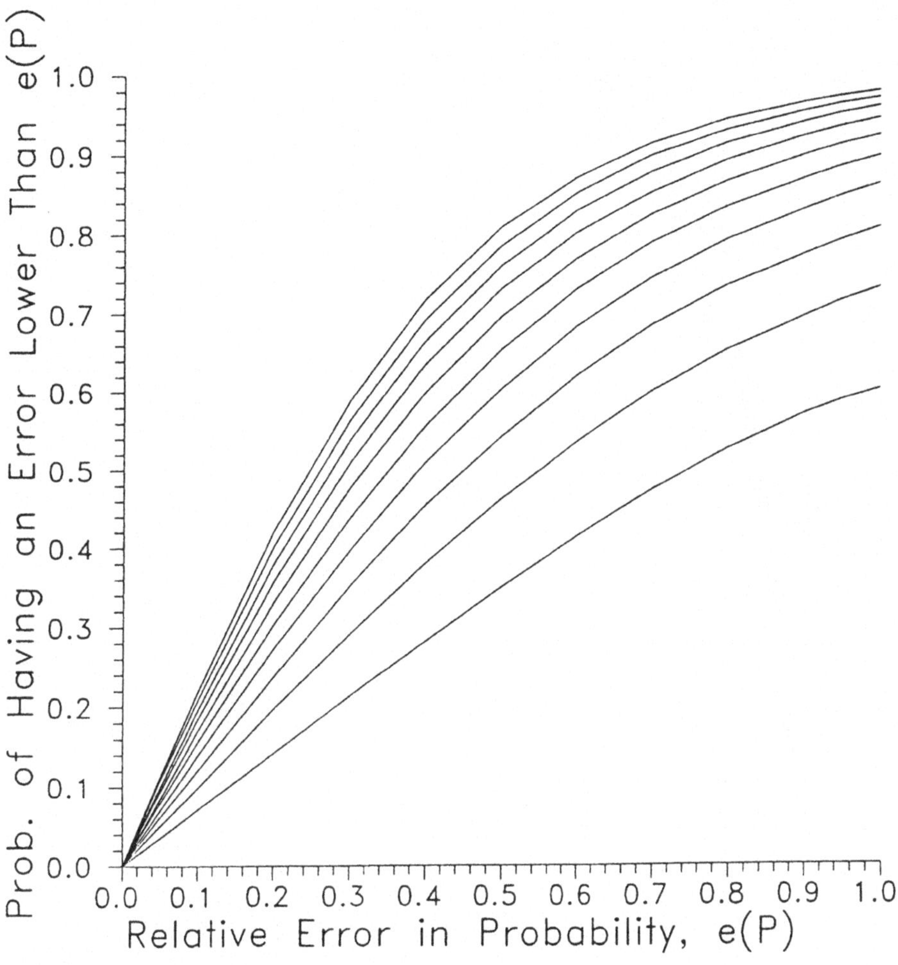

Figure 4. Relative error in probability for P=0.05 for LS parameter estimation. The data sample amplitude of the curves increases of 10 data moving upward. The lower curve is relative to a data sample amplitude of 10 data, the upper one of 100 data

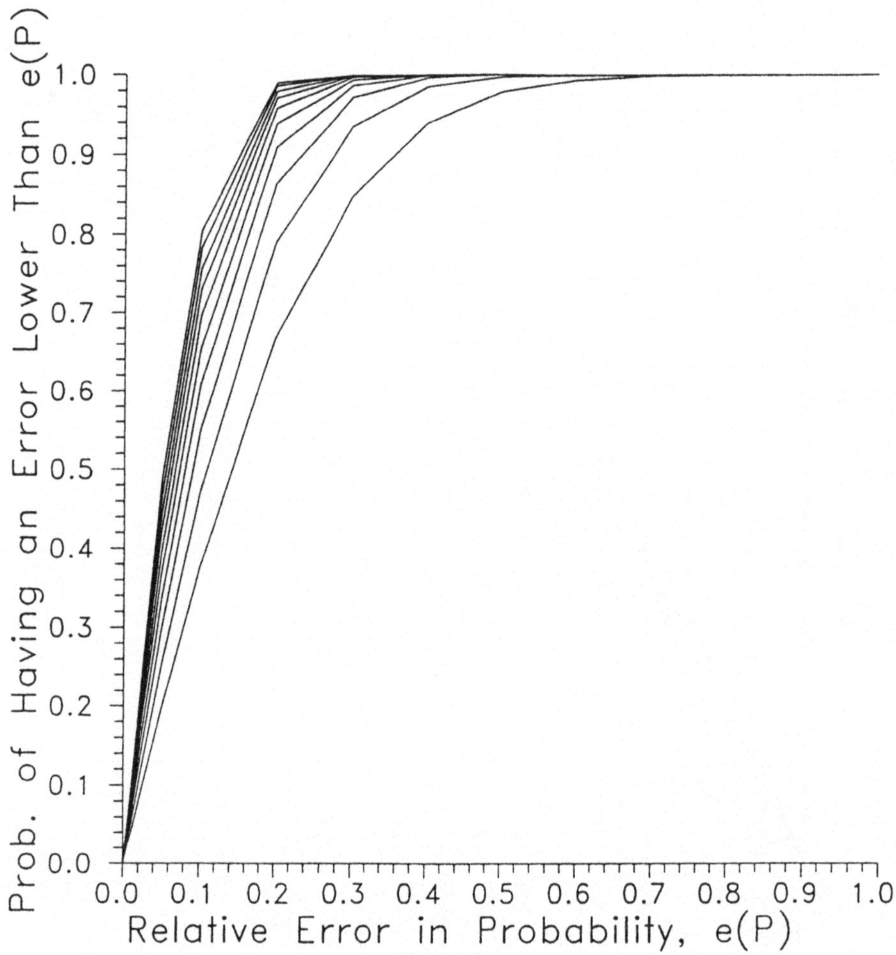

Figure 5. Relative error in probability for P=0.5 for LS parameter estimation. The data sample amplitude of the curves increases of 10 data moving upward. The lower curve is relative to a data sample amplitude of 10 data, the upper one of 100 data

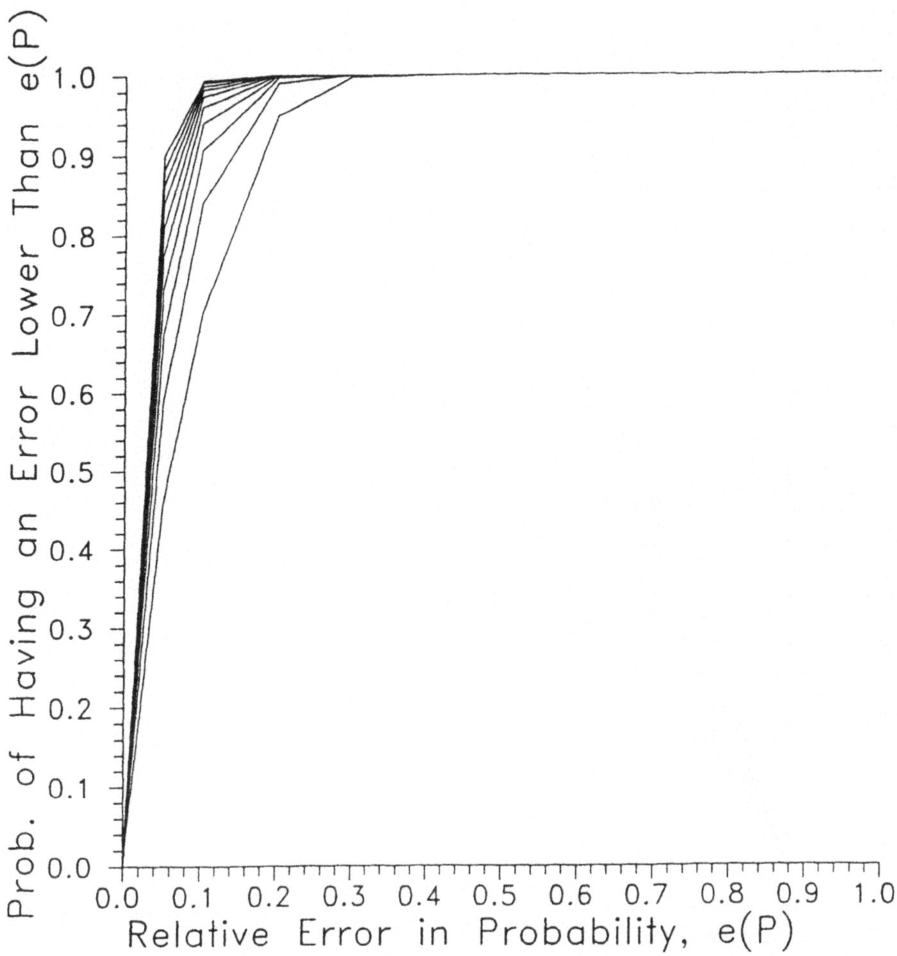

Figure 6. Relative error in probability for P=0.95 for LS parameter estimation. The data sample amplitude of the curves increases of 10 data moving upward. The lower curve is relative to a data sample amplitude of 10 data, the upper one of 100 data

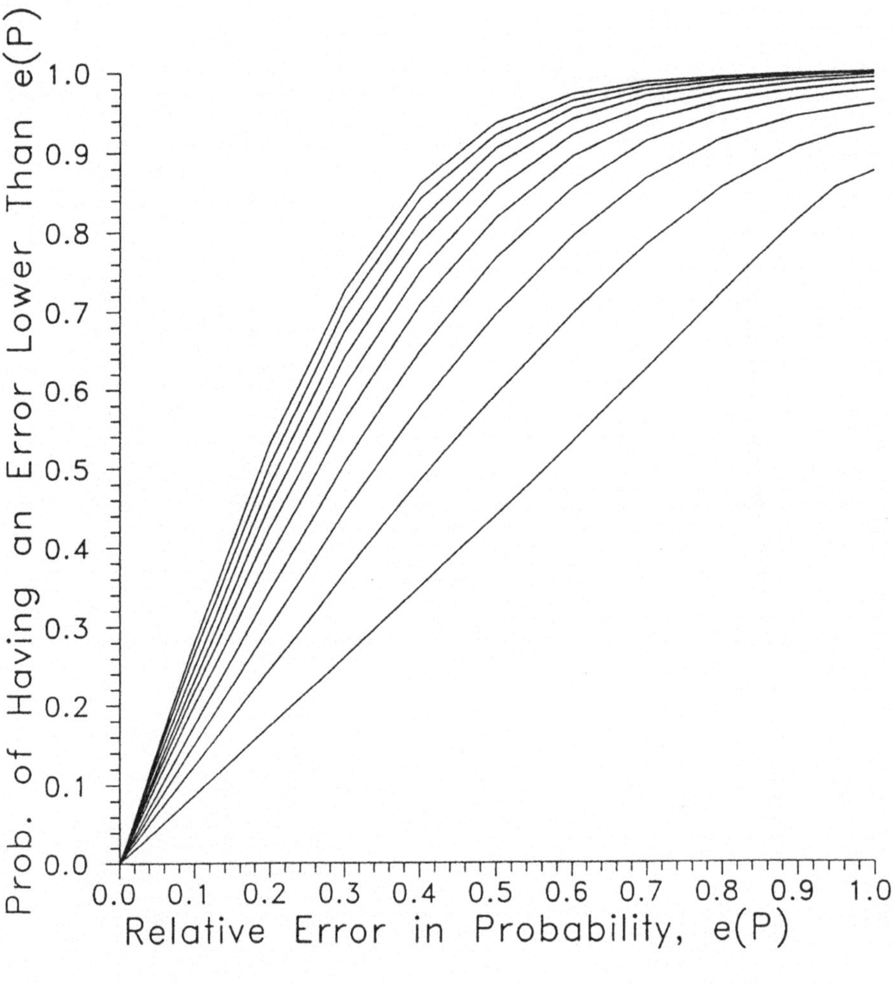

Figure 7. Relative error in probability for P=0.05 for ML parameter estimation. The data sample amplitude of the curves increases of 10 data moving upward. The lower curve is relative to a data sample amplitude of 10 data, the upper one of 100 data

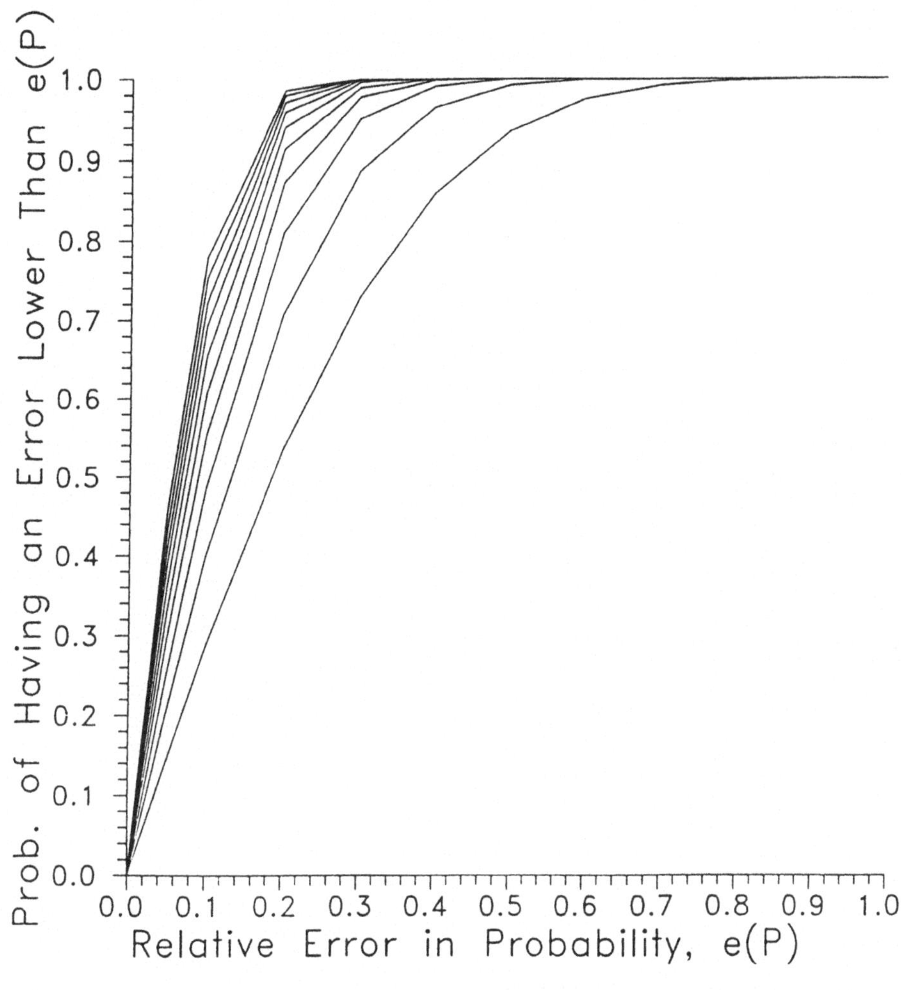

Figure 8. Relative error in probability for P=0.5 for ML parameter estimation. The data sample amplitude of the curves increases of 10 data moving upward. The lower curve is relative to a data sample amplitude of 10 data, the upper one of 100 data

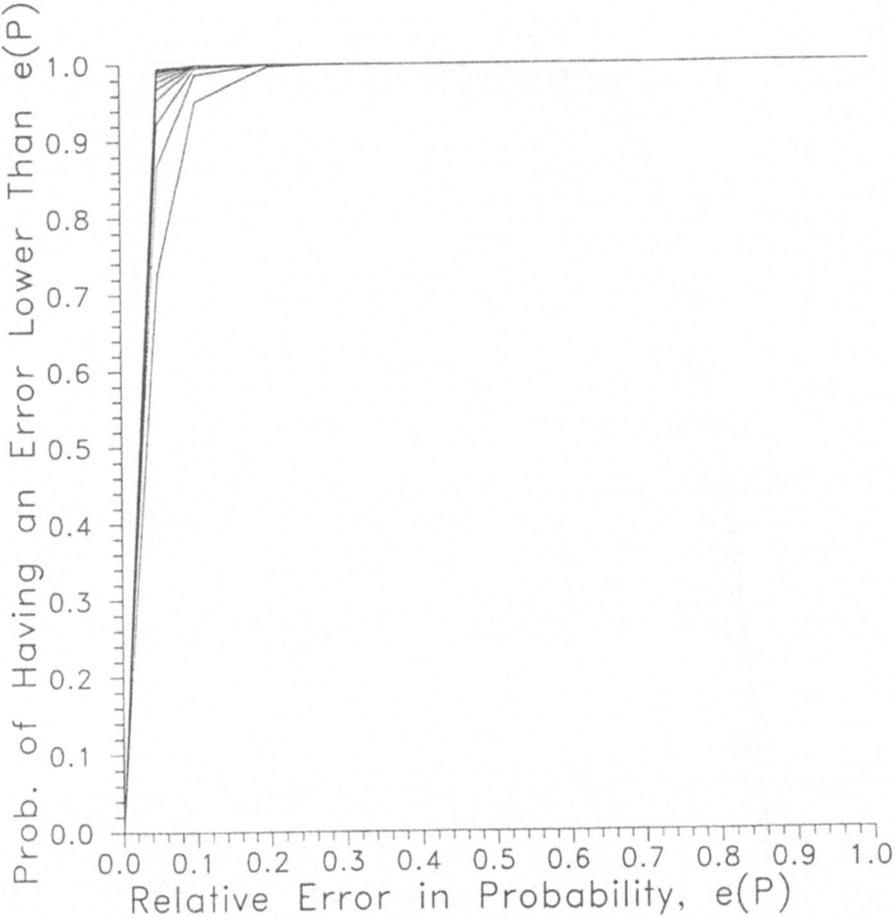

Figure 9. Relative error in probability for P=0.95 for ML parameter estimation. The data sample amplitude of the curves increases of 10 data moving upward. The lower curve is relative to a data sample amplitude of 10 data, the upper one of 100 data

The manufacturer's authorised representative in the EU is Springer Nature Customer Service Centre GmbH, Europaplatz 3, 69115 Heidelberg, Germany. If you have any concerns regarding our products, please contact ProductSafety@springernature.com

Printed and bound by CPI Group (UK) Ltd, Croydon, CR0 4YY
28/07/2026
02179215-0005